Praise for *Ein*

"An illuminating delight . . . This is a w̶
portrait with a human and immensely cha
A wonderfully rounded portrait of the e_ ̶ ̶ ̶ ̶ ̶ ̶ ̶ ̶Einstein personality."

—Janet Maslin, *The New York Times*

"Once again Walter Isaacson has produced a most valuable biography of a great man about whom much has already been written. It helps that he has had access to important new material. He met the challenge of dealing with his subject as a human being and describing profound ideas in physics. His biography is a pleasure to read and makes the great physicist come alive."

—Murray Gell-Mann, winner of the 1969 Nobel Prize in Physics and author of *The Quark and the Jaguar*

"Brilliant . . . An illuminating biography of Einstein."

—*Vanity Fair*

"This book does an amazing job getting the science right and the man revealed."

—Sylvester James Gates Jr., the John S. Toll Professor of Physics at the University of Maryland

"A triumph . . . Isaacson understands Einstein and explains his discoveries while sharing riveting personal detail."

—*People* (4 stars)

"Isaacson has given us a life, not just a mind, perhaps the greatest in the twentieth century, but also a personality, as imperfect and fallible as all the rest of us. This unique combination of sheer brilliance and human uncertainty makes this one of the great biographies of our time."

—Joseph J. Ellis, author of *Founding Brothers: The Revolutionary Generation*

"A narrative masterpiece . . . This is a great read by a great writer about a great man—a biographical perfect storm."

—Michael Shermer, *The New York Sun*

"Isaacson has a lovely sense of the poetry of physics. . . . Utterly absorbing."

—Susan Larson, *The Times-Picayune* (New Orleans)

"This book will be widely and deservedly admired. It is excellently readable and combines the personal and the scientific aspects of Einstein's life in a graceful way."
—Gerald Holton, the Mallinckrodt Research Professor of Physics at Harvard University and author of *Einstein, History, and Other Passions*

"An excellent book . . . Isaacson's biography is well researched and contains a surprising amount of new information about its enigmatic subject. . . . Einstein emerges as a flesh-and-blood figure—a human with good qualities and flaws. Even Einstein scholars will likely find here facts they hadn't known. . . . A major and authoritative work on one of the most interesting figures in the history of science."
—Amir D. Aczel, *The Boston Globe*

"Isaacson has admirably succeeded in weaving together the complex threads of Einstein's personal and scientific life to paint a superb portrait."
—Arthur I. Miller, author of *Einstein, Picasso: Space, Time, and the Beauty That Havoc Causes*

"Delightful . . . The most comprehensive English-language biography of Einstein for a general readership . . . Isaacson weaves it all into a seamless narrative."
—Sharon Begley, *Newsweek*

"Isaacson has written a crisp, engaging, and refreshing biography, one that beautifully masters the historical literature and offers many new insights into Einstein's work and life."
—Diana Kormos Buchwald, the general editor of The Collected Papers of Albert Einstein and professor of history at Caltech

"Expansive in scope and exhaustively researched. . . . Isaacson skillfully sheds new light on Einstein's personality. . . . Superb."
—Bob Van Brocklin, *The Sunday Oregonian*

"With unmatched narrative skill, Isaacson has managed the extraordinary feat of preserving Einstein's monumental stature while at the same time bringing him to such vivid life that we come to feel as if he could be walking in our midst. This is a terrific work."
—Doris Kearns Goodwin, author of *Team of Rivals* and *No Ordinary Time*, winner of the Pulitzer Prize for history

"Thoroughly researched and well written, *Einstein* does an excellent job of summarizing the concepts behind Einstein's theories. . . . Isaacson also does an excellent job illuminating Einstein's personality."
—Dennis O'Brien, *The Baltimore Sun*

"Isaacson has done a remarkable job conveying a sense of Einstein the man and also the fine details of Einstein's science. This is not only a compelling biography, one in which the next page always beckons, but an example of science writing at its best."
—Lawrence M. Krauss, the Ambrose Swasey Professor of Physics at Case Western Reserve and author of *Hiding in the Mirror*

"Isaacson brings the genius to life through letters, anecdotes, quotes and humor. . . . Isaacson has managed to make a science book read like a thriller."
—L. A. Lorek, *San Antonio Express-News*

"Isaacson's treatment of Einstein's scientific work is excellent: accurate, complete, and at just the right level of detail for the general reader. Taking advantage of the wealth of recently uncovered historical material, he has produced the most readable biography of Einstein yet."
—A. Douglas Stone, professor of physics at Yale

"Stimulating and provocative."
—Thomas L. Friedman, *The New York Times*

"Isaacson has triumphed . . . producing a thorough exploration of his subject's life, a skillful piece of scientific literature and a thumping good read. . . . It's one of the greatest stories of modern science and to his credit . . . Isaacson has done a first-rate job in telling it. This is, quite simply, a riveting read."
—Robin McKie, *The Guardian* (UK)

"Exemplary science writing . . . Isaacson exudes both a crisp precision and profundity that belie the difficulty of the physics Einstein created. He magisterially guides us through the man's expansive body of work that prefigured most modern physics. . . . Isaacson's tremendous scholarship in uncovering more of the less frequently discussed aspects of Einstein's character will stand as a benchmark for works to come."
—Joshua Roebke, *Seed*

"An accessible, fascinating account of one of the twentieth century's greatest figures . . . Like its subject, Walter Isaacson's ambitious biography of Albert Einstein radiates intelligence, wit and eloquence."
—Kathleen Krog, *The Miami Herald*

"A painstaking and reliable biography. You won't go wrong in reading and learning from it."
—Michael Dirda, *The Washington Post Book World*

"I found it hard to put down."
—Daniel Sutherland, *Chicago Tribune*

"A biography of Albert Einstein may seem daunting to many readers. Walter Isaacson gives you one that isn't. . . . Isaacson is a fluid writer whose narrative talents give Einstein an aura missing from many previous accounts of his life."
—Steve Weinberg, *The Houston Chronicle*

"Dramatic and revelatory."
—Bryan Appleyard, *Sunday Times* (London)

"Fascinating . . . a delicious read."
—Ian Stewart, *Winnipeg Free Press*

"Narrative nonfiction at its best . . . What the book also does is move the author up from the ranks of skilled narrator of history—one who seeks the story behind historical facts—and into the top tier of the craft to join the likes of David McCullough and Doris Kearns Goodwin."
—James Srodes, *The Washington Times*

"A fine, affectionate and determinedly lucid account of both Einstein's life and thought."
—Duane Davis, *Rocky Mountain News*

"A new biography offering hearty helpings alike of energy, mass, and light . . . To Isaacson's credit, *Einstein: His Life and Universe* conveys the dizzying concepts of physics in a way most lay readers can grasp."
—Erik Spanberg, *Christian Science Monitor*

"A triumphant biography . . . another coup for Isaacson."
—John Mark Eberhart, *The Kansas City Star*

ALSO BY WALTER ISAACSON

A Benjamin Franklin Reader

Benjamin Franklin: An American Life

Kissinger: A Biography

The Wise Men: Six Friends and the World They Made
(with Evan Thomas)

Pro and Con

EINSTEIN

HIS LIFE
AND UNIVERSE

WALTER
ISAACSON

SIMON & SCHUSTER PAPERBACKS
New York   London   Toronto   Sydney

To my father,
the nicest, smartest, and most moral man I know

SIMON & SCHUSTER PAPERBACKS
Rockefeller Center
1230 Avenue of the Americas
New York, NY 10020

First Simon & Schuster paperback edition May 2008

SIMON & SCHUSTER PAPERBACKS and colophon are registered trademarks
of Simon & Schuster, Inc.

For information about special discounts for bulk purchases,
please contact Simon & Schuster Special Sales at
1-800-456-6798 or business@simonandschuster.com.

Frontispiece: Ullstein Bilderdienst/The Granger Collection, New York

Illustration credits are on page 679.

Manufactured in the United States of America

10   9   8   7   6   5   4   3   2   1

Library of Congress Cataloging-in-Publication Data
Isaacson, Walter.
  Einstein : his life and universe / Walter Isaacson.
    p. cm.
  Includes bibliographical references and index.
    1. Einstein, Albert, 1879–1955.   2. Physicists—Biography.   3. Einstein, Albert,
1879–1955—Friends and associates.   4. Relativity (Physics).   5. Unified field
theories.   I. Title.

QC16.E5I76 2007
530.092—dc22
[B]   2006051264

ISBN-13: 978-0-7432-6473-0
ISBN-10:   0-7432-6473-8
ISBN-13: 978-0-7432-6474-7 (pbk)
ISBN-10:   0-7432-6474-6 (pbk)

In Santa Barbara, 1933

Life is like riding a bicycle.
To keep your balance you must keep moving.

—ALBERT EINSTEIN, IN A LETTER TO HIS SON EDUARD, FEBRUARY 5, 1930[1]

CONTENTS

ACKNOWLEDGMENTS

Diana Kormos Buchwald, the general editor of Einstein's papers, read this book meticulously and made copious comments and corrections through many drafts. In addition, she helped me get early and complete access to the wealth of new Einstein papers that became available in 2006, and guided me through them. She was also a gracious host and facilitator during my trips to the Einstein Papers Project at Caltech. She has a passion for her work and a delightful sense of humor, which would have pleased her subject.

Two of her associates were also very helpful in guiding me through the newly available papers as well as untapped riches in the older archival material. Tilman Sauer, who likewise checked and annotated this book, in particular vetted the sections on Einstein's quest for the equations of general relativity and his pursuit of a unified field theory. Ze'ev Rosenkranz, the historical editor of the papers, provided insights on Einstein's attitudes toward Germany and his Jewish heritage. He was formerly curator of the Einstein archives at Hebrew University in Jerusalem.

Barbara Wolff, who is now at those archives at Hebrew University, did a careful fact-checking of every page of the manuscript, making fastidious corrections large and small. She warned that she has a reputation as a nitpicker, but I am very grateful for each and every nit she found. I also appreciate the encouragement given by Roni Grosz, the curator there.

Brian Greene, the Columbia University physicist and author of *The*

Fabric of the Cosmos, was an indispensable friend and editor. He talked me through numerous revisions, honed the wording of the science passages, and read the final manuscript. He is a master of both science and language. In addition to his work on string theory, he and his wife, Tracy Day, are organizing an annual science festival in New York City, which will help spread the enthusiasm for physics so evident in his work and books.

Lawrence Krauss, professor of physics at Case Western Reserve and author of *Hiding in the Mirror,* also read my manuscript, vetted the sections on special relativity, general relativity, and cosmology, and offered many good suggestions and corrections. He, too, has an infectious enthusiasm for physics.

Krauss helped me enlist a protégé of his at Case, Craig J. Copi, who teaches relativity there. I hired him to do a thorough checking of the science and math, and I am grateful for his diligent edits.

Douglas Stone, professor of physics at Yale, also vetted the science in the book. A condensed matter theorist, he is writing what will be an important book on Einstein's contributions to quantum mechanics. In addition to checking my science sections, he helped me write the chapters on the 1905 light quanta paper, quantum theory, Bose-Einstein statistics, and kinetic theory.

Murray Gell-Mann, winner of the 1969 Nobel Prize in physics, was a delightful and passionate guide from the beginning to the end of this project. He helped me revise early drafts, edited and corrected the chapters on relativity and quantum mechanics, and helped draft sections that explained Einstein's objections to quantum uncertainty. With his combination of erudition and humor, and his feel for the personalities involved, he made the process a great joy.

Arthur I. Miller, emeritus professor of history and philosophy of science at University College, London, is the author of *Einstein, Picasso* and of *Empire of the Stars.* He read and reread the versions of my scientific chapters and helped with numerous revisions, especially on special relativity (about which he wrote a pioneering book), general relativity, and quantum theory.

Sylvester James Gates Jr., a physics professor at the University of Maryland, agreed to read my manuscript when he came out to Aspen for a conference on Einstein. He did a comprehensive edit filled with smart comments and rephrasing of certain scientific passages.

John D. Norton, a professor at the University of Pittsburgh, has specialized in tracing Einstein's thought process as he developed both special and then general relativity. He read these sections of my book, made edits, and offered useful comments. I am also grateful for guidance from two of his fellow scholars specializing in Einstein's development of his theories: Jürgen Renn of the Max Planck Institute in Berlin and Michel Janssen of the University of Minnesota.

George Stranahan, a founder of the Aspen Center for Physics, also agreed to read and review the manuscript. He was particularly helpful in editing the sections on the light quanta paper, Brownian motion, and the history and science of special relativity.

Robert Rynasiewicz, a philosopher of science at Johns Hopkins, read many of the science chapters and made useful suggestions about the quest for general relativity.

N. David Mermin, professor of theoretical physics at Cornell and author of *It's About Time: Understanding Einstein's Relativity,* edited and made corrections to the final version of the introductory chapter and chapters 5 and 6 on Einstein's 1905 papers.

Gerald Holton, professor of physics at Harvard, has been one of the pioneers in the study of Einstein, and he is still a guiding light. I am deeply flattered that he was willing to read my book, make comments, and offer generous encouragement. His Harvard colleague Dudley Herschbach, who has done so much for science education, also was supportive. Both Holton and Herschbach made useful comments on my draft and spent an afternoon with me in Holton's office going over suggestions and refining my descriptions of the historical players.

Ashton Carter, professor of science and international affairs at Harvard, kindly read and checked an early draft. Columbia University's Fritz Stern, author of *Einstein's German World,* provided encouragement and advice at the outset. Robert Schulmann, one of the original editors at the Einstein Papers Project, did likewise. And Jeremy Bernstein, who has written many fine books on Einstein, warned me how difficult the science would be. He was right, and I am grateful for that as well.

In addition, I asked two teachers of high school physics to give the book a careful reading to make sure the science was correct, and also comprehensible to those whose last physics course was in high school. Nancy Stravinsky Isaacson taught physics in New Orleans until, alas, Hurricane Katrina gave her more free time. David Derbes teaches physics at the University of Chicago Lab School. Their comments were very incisive and also aimed at the lay reader.

There is a corollary of the uncertainty principle that says that no matter how often a book is observed, some mistakes will remain. Those are my fault.

It also helped to have some nonscientific readers, who made very useful suggestions from a lay perspective on parts or all of the manuscript. These included William Mayer, Orville Wright, Daniel Okrent, Steve Weisman, and Strobe Talbott.

For twenty-five years, Alice Mayhew at Simon & Schuster has been my editor and Amanda Urban at ICM my agent. I can imagine no better partners, and they were again enthusiastic and helpful in their comments on the book. I also appreciate the help of Carolyn Reidy, David Rosenthal, Roger Labrie, Victoria Meyer, Elizabeth Hayes, Serena Jones, Mara Lurie, Judith Hoover, Jackie Seow, and Dana Sloan at Simon & Schuster. For their countless acts of support over the years, I am also grateful to Elliot Ravetz and Patricia Zindulka.

Natasha Hoffmeyer and James Hoppes translated for me Einstein's German correspondence and writing, especially the new material that had not yet been translated, and I appreciate their diligence. Jay Colton, who was photo editor for *Time*'s Person of the Century issue, also did a creative job tracking down pictures for this book.

I had two and a half other readers who were the most valuable of all. The first was my father, Irwin Isaacson, an engineer who instilled in me a love of science and is the smartest teacher I've ever had. I am grateful to him for the universe that he and my late mother created for me, and to my brilliant and wise stepmother, Julanne.

The other truly valuable reader was my wife, Cathy, who read every page with her usual wisdom, common sense, and curiosity. And the valuable half-a-reader was my daughter, Betsy, who as usual read selected portions of my book. The surety with which she made her pronouncements made up for the randomness of her reading. I love them both dearly.

MAIN CHARACTERS

MICHELE ANGELO BESSO (1873–1955). Einstein's closest friend. An engaging but unfocused engineer, he met Einstein in Zurich, then followed him to work at the Bern patent office. Served as a sounding board for the 1905 special relativity paper. Married Anna Winteler, sister of Einstein's first girlfriend.

NIELS BOHR (1885–1962). Danish pioneer of quantum theory. At Solvay conferences and subsequent intellectual trysts, he parried Einstein's enthusiastic challenges to his Copenhagen interpretation of quantum mechanics.

MAX BORN (1882–1970). German physicist and mathematician. Engaged in a brilliant, intimate correspondence with Einstein for forty years. Tried to convince Einstein to be comfortable with quantum mechanics; his wife, Hedwig, challenged Einstein on personal issues.

HELEN DUKAS (1896–1982). Einstein's loyal secretary, Cerberus-like guard, and housemate from 1928 until his death, and after that protector of his legacy and papers.

ARTHUR STANLEY EDDINGTON (1882–1944). British astrophysicist and champion of relativity whose 1919 eclipse observations dramatically confirmed Einstein's prediction of how much gravity bends light.

PAUL EHRENFEST (1880–1933). Austrian-born physicist, intense and insecure, who bonded with Einstein on a visit to Prague in 1912 and became a professor in Leiden, where he frequently hosted Einstein.

EDUARD EINSTEIN (1910–1965). Second son of Mileva Marić and Einstein. Smart and artistic, he obsessed about Freud and hoped to be a psychiatrist, but he succumbed to his own schizophrenic demons in his twenties and was institutionalized in Switzerland for much of the rest of his life.

ELSA EINSTEIN (1876–1936). Einstein's first cousin, second wife. Mother of Margot and Ilse Einstein from her first marriage to textile merchant Max Löwenthal. She and her daughters reverted to her maiden name, Einstein, after her 1908 divorce. Married Einstein in 1919. Smarter than she pretended to be, she knew how to handle him.

Hans Albert Einstein (1904–1973). First son of Mileva Marić and Einstein, a difficult role that he handled with grace. Studied engineering at Zurich Polytechnic. Married Frieda Knecht (1895–1958) in 1927. They had two sons, Bernard (1930–) and Klaus (1932–1938), and an adopted daughter, Evelyn (1941–). Moved to the United States in 1938 and eventually became a professor of hydraulic engineering at Berkeley. After Frieda's death, married Elizabeth Roboz (1904–1995) in 1959. Bernard has five children, the only known great-grandchildren of Albert Einstein.

Hermann Einstein (1847–1902). Einstein's father, from a Jewish family from rural Swabia. With his brother Jakob, he ran electrical companies in Munich and then Italy, but not very successfully.

Ilse Einstein (1897–1934). Daughter of Elsa Einstein from her first marriage. Dallied with adventurous physician Georg Nicolai and in 1924 married literary journalist Rudolph Kayser, who later wrote a book on Einstein using the pseudonym Anton Reiser.

Lieserl Einstein (1902–?). Premarital daughter of Einstein and Mileva Marić. Einstein probably never saw her. Likely left in her Serbian mother's hometown of Novi Sad for adoption and may have died of scarlet fever in late 1903.

Margot Einstein (1899–1986). Daughter of Elsa Einstein from her first marriage. A shy sculptor. Married Russian Dimitri Marianoff in 1930; no children. He later wrote a book on Einstein. She divorced him in 1937, moved in with Einstein at Princeton, and remained at 112 Mercer Street until her death.

Maria "Maja" Einstein (1881–1951). Einstein's only sibling, and among his closest confidantes. Married Paul Winteler, had no children, and in 1938 moved without him from Italy to Princeton to live with her brother.

Pauline Koch Einstein (1858–1920). Einstein's strong-willed and practical mother. Daughter of a prosperous Jewish grain dealer from Württemberg. Married Hermann Einstein in 1876.

Abraham Flexner (1866–1959). American education reformer. Founded the Institute for Advanced Study in Princeton and recruited Einstein there.

Philipp Frank (1884–1966). Austrian physicist. Succeeded his friend Einstein at German University of Prague and later wrote a book about him.

Marcel Grossmann (1878–1936). Diligent classmate at Zurich Polytechnic who took math notes for Einstein and then helped him get a job in the patent office. As professor of descriptive geometry at the Polytechnic, guided Einstein to the math he needed for general relativity.

Fritz Haber (1868–1934). German chemist and gas warfare pioneer who helped recruit Einstein to Berlin and mediated between him and Marić. A Jew who converted to Christianity in an attempt to be a good German, he preached to Einstein the virtues of assimilation, until the Nazis came to power.

CONRAD HABICHT (1876–1958). Mathematician and amateur inventor, member of the "Olympia Academy" discussion trio in Bern, and recipient of two famous 1905 letters from Einstein heralding forthcoming papers.

WERNER HEISENBERG (1901–1976). German physicist. A pioneer of quantum mechanics, he formulated the uncertainty principle that Einstein spent years resisting.

DAVID HILBERT (1862–1943). German mathematician who in 1915 raced Einstein to discover the mathematical equations for general relativity.

BANESH HOFFMANN (1906–1986). Mathematician and physicist who collaborated with Einstein in Princeton and later wrote a book about him.

PHILIPP LENARD (1862–1947). Hungarian-German physicist whose experimental observations on the photoelectric effect were explained by Einstein in his 1905 light quanta paper. Became an anti-Semite, Nazi, and Einstein hater.

HENDRIK ANTOON LORENTZ (1853–1928). Genial and wise Dutch physicist whose theories paved the way for special relativity. Became a father figure to Einstein.

MILEVA MARIĆ (1875–1948). Serbian physics student at Zurich Polytechnic who became Einstein's first wife. Mother of Hans Albert, Eduard, and Lieserl. Passionate and driven, but also brooding and increasingly gloomy, she triumphed over many, but not all, of the obstacles that then faced an aspiring female physicist. Separated from Einstein in 1914, divorced in 1919.

ROBERT ANDREWS MILLIKAN (1868–1953). American experimental physicist who confirmed Einstein's law of the photoelectric effect and recruited him to be a visiting scholar at Caltech.

HERMANN MINKOWSKI (1864–1909). Taught Einstein math at the Zurich Polytechnic, referred to him as a "lazy dog," and devised a mathematical formulation of special relativity in terms of four-dimensional spacetime.

GEORG FRIEDRICH NICOLAI, born Lewinstein (1874–1964). Physician, pacifist, charismatic adventurer, and seducer. A friend and doctor of Elsa Einstein and probable lover of her daughter Ilse, he wrote a pacifist tract with Einstein in 1915.

ABRAHAM PAIS (1918–2000). Dutch-born theoretical physicist who became a colleague of Einstein in Princeton and wrote a scientific biography of him.

MAX PLANCK (1858–1947). Prussian theoretical physicist who was an early patron of Einstein and helped recruit him to Berlin. His conservative instincts, both in life and in physics, made him a contrast to Einstein, but they remained warm and loyal colleagues until the Nazis took power.

ERWIN SCHRÖDINGER (1887–1961). Austrian theoretical physicist who was a pioneer of quantum mechanics but joined Einstein in expressing discomfort with the uncertainties and probabilities at its core.

MAURICE SOLOVINE (1875–1958). Romanian philosophy student in Bern who founded the "Olympia Academy" with Einstein and Habicht. Became Einstein's French publisher and lifelong correspondent.

LEÓ SZILÁRD (1898–1964). Hungarian-born physicist, charming and eccentric, who met Einstein in Berlin and patented a refrigerator with him. Conceived the nuclear chain reaction and cowrote the 1939 letter Einstein sent to President Franklin Roosevelt urging attention to the possibility of an atomic bomb.

CHAIM WEIZMANN (1874–1952). Russian-born chemist who emigrated to England and became president of the World Zionist Organization. In 1921, he brought Einstein to America for the first time, using him as the draw for a fundraising tour. Was first president of Israel, a post offered upon his death to Einstein.

THE WINTELER FAMILY. Einstein boarded with them while he was a student in Aarau, Switzerland. Jost Winteler was his history and Greek teacher; his wife, Rosa, became a surrogate mother. Of their seven children, Marie became Einstein's first girlfriend; Anna married Einstein's best friend, Michele Besso; and Paul married Einstein's sister, Maja.

HEINRICH ZANGGER (1874–1957). Professor of physiology at the University of Zurich. Befriended Einstein and Marić and helped mediate their disputes and divorce.

CHAPTER ONE

THE LIGHT-BEAM RIDER

"I promise you four papers," the young patent examiner wrote his friend. The letter would turn out to bear some of the most significant tidings in the history of science, but its momentous nature was masked by an impish tone that was typical of its author. He had, after all, just addressed his friend as "you frozen whale" and apologized for writing a letter that was "inconsequential babble." Only when he got around to describing the papers, which he had produced during his spare time, did he give some indication that he sensed their significance.[1]

"The first deals with radiation and the energy properties of light and is very revolutionary," he explained. Yes, it was indeed revolutionary. It argued that light could be regarded not just as a wave but also as a stream of tiny particles called quanta. The implications that would eventually arise from this theory—a cosmos without strict causality or certainty—would spook him for the rest of his life.

"The second paper is a determination of the true sizes of atoms." Even though the very existence of atoms was still in dispute, this was the most straightforward of the papers, which is why he chose it as the safest bet for his latest attempt at a doctoral thesis. He was in the process of revolutionizing physics, but he had been repeatedly thwarted in his efforts to win an academic job or even get a doctoral degree, which he hoped might get him promoted from a third- to a second-class examiner at the patent office.

The third paper explained the jittery motion of microscopic particles in liquid by using a statistical analysis of random collisions. In the process, it established that atoms and molecules actually exist.

"The fourth paper is only a rough draft at this point, and is an electrodynamics of moving bodies which employs a modification of the theory of space and time." Well, that was certainly more than inconsequential babble. Based purely on thought experiments—performed in his head rather than in a lab—he had decided to discard Newton's concepts of absolute space and time. It would become known as the Special Theory of Relativity.

What he did not tell his friend, because it had not yet occurred to him, was that he would produce a fifth paper that year, a short addendum to the fourth, which posited a relationship between energy and mass. Out of it would arise the best-known equation in all of physics: $E=mc^2$.

Looking back at a century that will be remembered for its willingness to break classical bonds, and looking ahead to an era that seeks to nurture the creativity needed for scientific innovation, one person stands out as a paramount icon of our age: the kindly refugee from oppression whose wild halo of hair, twinkling eyes, engaging humanity, and extraordinary brilliance made his face a symbol and his name a synonym for genius. Albert Einstein was a locksmith blessed with imagination and guided by a faith in the harmony of nature's handiwork. His fascinating story, a testament to the connection between creativity and freedom, reflects the triumphs and tumults of the modern era.

Now that his archives have been completely opened, it is possible to explore how the private side of Einstein—his nonconformist personality, his instincts as a rebel, his curiosity, his passions and detachments—intertwined with his political side and his scientific side. Knowing about the man helps us understand the wellsprings of his science, and vice versa. Character and imagination and creative genius were all related, as if part of some unified field.

Despite his reputation for being aloof, he was in fact passionate in both his personal and scientific pursuits. At college he fell madly in love with the only woman in his physics class, a dark and intense Serbian named Mileva Marić. They had an illegitimate daughter, then married and had two sons. She served as a sounding board for his scientific ideas and helped to check the math in his papers, but eventually their relationship disintegrated. Einstein offered her a deal. He would win the Nobel Prize someday, he said; if she gave him a divorce, he would give her the prize money. She thought for a week and accepted. Because his theories were so radical, it was seventeen years after his miraculous outpouring from the patent office before he was awarded the prize and she collected.

Einstein's life and work reflected the disruption of societal certainties and moral absolutes in the modernist atmosphere of the early twentieth century. Imaginative nonconformity was in the air: Picasso, Joyce, Freud, Stravinsky, Schoenberg, and others were breaking conventional bonds. Charging this atmosphere was a conception of the universe in which space and time and the properties of particles seemed based on the vagaries of observations.

Einstein, however, was not truly a relativist, even though that is how he was interpreted by many, including some whose disdain was tinged by anti-Semitism. Beneath all of his theories, including relativity, was a quest for invariants, certainties, and absolutes. There was a harmonious reality underlying the laws of the universe, Einstein felt, and the goal of science was to discover it.

His quest began in 1895, when as a 16-year-old he imagined what it would be like to ride alongside a light beam. A decade later came his miracle year, described in the letter above, which laid the foundations for the two great advances of twentieth-century physics: relativity and quantum theory.

A decade after that, in 1915, he wrested from nature his crowning glory, one of the most beautiful theories in all of science, the general theory of relativity. As with the special theory, his thinking had evolved through thought experiments. Imagine being in an enclosed elevator accelerating up through space, he conjectured in one of them. The effects you'd feel would be indistinguishable from the experience of gravity.

Gravity, he figured, was a warping of space and time, and he came up with the equations that describe how the dynamics of this curvature result from the interplay between matter, motion, and energy. It can be described by using another thought experiment. Picture what it would be like to roll a bowling ball onto the two-dimensional surface of a trampoline. Then roll some billiard balls. They move toward the bowling ball not because it exerts some mysterious attraction but because of the way it curves the trampoline fabric. Now imagine this happening in the four-dimensional fabric of space and time. Okay, it's not easy, but that's why we're no Einstein and he was.

The exact midpoint of his career came a decade after that, in 1925, and it was a turning point. The quantum revolution he had helped to launch was being transformed into a new mechanics that was based on uncertainties and probabilities. He made his last great contributions to quantum mechanics that year but, simultaneously, began to resist it. He would spend the next three decades, ending with some equations scribbled while on his deathbed in 1955, stubbornly criticizing what he regarded as the incompleteness of quantum mechanics while attempting to subsume it into a unified field theory.

Both during his thirty years as a revolutionary and his subsequent thirty years as a resister, Einstein remained consistent in his willingness to be a serenely amused loner who was comfortable not conforming. Independent in his thinking, he was driven by an imagination that broke from the confines of conventional wisdom. He was that odd breed, a reverential rebel, and he was guided by a faith, which he wore lightly and with a twinkle in his eye, in a God who would not play dice by allowing things to happen by chance.

Einstein's nonconformist streak was evident in his personality and politics as well. Although he subscribed to socialist ideals, he was too much of an individualist to be comfortable with excessive state control or centralized authority. His impudent instincts, which served him so well as a young scientist, made him allergic to nationalism, militarism, and anything that smacked of a herd mentality. And until Hitler caused him to revise his geopolitical equations, he was an instinctive pacifist who celebrated resistance to war.

His tale encompasses the vast sweep of modern science, from the infinitesimal to the infinite, from the emission of photons to the expansion of the cosmos. A century after his great triumphs, we are still living in Einstein's universe, one defined on the macro scale by his theory of relativity and on the micro scale by a quantum mechanics that has proven durable even as it remains disconcerting.

His fingerprints are all over today's technologies. Photoelectric cells and lasers, nuclear power and fiber optics, space travel, and even semiconductors all trace back to his theories. He signed the letter to Franklin Roosevelt warning that it may be possible to build an atom bomb, and the letters of his famed equation relating energy to mass hover in our minds when we picture the resulting mushroom cloud.

Einstein's launch into fame, which occurred when measurements made during a 1919 eclipse confirmed his prediction of how much gravity bends light, coincided with, and contributed to, the birth of a new celebrity age. He became a scientific supernova and humanist icon, one of the most famous faces on the planet. The public earnestly puzzled over his theories, elevated him into a cult of genius, and canonized him as a secular saint.

If he did not have that electrified halo of hair and those piercing eyes, would he still have become science's preeminent poster boy? Suppose, as a thought experiment, that he had looked like a Max Planck or a Niels Bohr. Would he have remained in their reputational orbit, that of a mere scientific genius? Or would he still have made the leap into the pantheon inhabited by Aristotle, Galileo, and Newton?[2]

The latter, I believe, is the case. His work had a very personal character, a stamp that made it recognizably his, the way a Picasso is recognizably a Picasso. He made imaginative leaps and discerned great principles through thought experiments rather than by methodical inductions based on experimental data. The theories that resulted were at times astonishing, mysterious, and counterintuitive, yet they contained notions that could capture the popular imagination: the relativity of space and time, $E=mc^2$, the bending of light beams, and the warping of space.

Adding to his aura was his simple humanity. His inner security was tempered by the humility that comes from being awed by nature. He could be detached and aloof from those close to him, but toward mankind in general he exuded a true kindness and gentle compassion.

Yet for all of his popular appeal and surface accessibility, Einstein also came to symbolize the perception that modern physics was something that ordinary laymen could not comprehend, "the province of priest-like experts," in the words of Harvard professor Dudley Herschbach.[3] It was not always thus. Galileo and Newton were both great geniuses, but their mechanical cause-and-effect explanation of the world was something that most thoughtful folks could grasp. In the eighteenth century of Benjamin Franklin and the nineteenth century of Thomas Edison, an educated person could feel some familiarity with science and even dabble in it as an amateur.

A popular feel for scientific endeavors should, if possible, be restored given the needs of the twenty-first century. This does not mean that every literature major should take a watered-down physics course or that a corporate lawyer should stay abreast of quantum mechanics. Rather, it means that an appreciation for the methods of science is a useful asset for a responsible citizenry. What science teaches us, very significantly, is the correlation between factual evidence and general theories, something well illustrated in Einstein's life.

In addition, an appreciation for the glories of science is a joyful trait for a good society. It helps us remain in touch with that childlike capacity for wonder, about such ordinary things as falling apples and elevators, that characterizes Einstein and other great theoretical physicists.[4]

That is why studying Einstein can be worthwhile. Science is inspiring and noble, and its pursuit an enchanting mission, as the sagas of its heroes remind us. Near the end of his life, Einstein was asked by the New York State Education Department what schools should emphasize. "In teaching history," he replied, "there should be extensive discussion of personalities who benefited mankind through independence of character and judgment."[5] Einstein fits into that category.

At a time when there is a new emphasis, in the face of global competition, on science and math education, we should also note the other part of Einstein's answer. "Critical comments by students should be taken in a friendly spirit," he said. "Accumulation of material should not stifle the student's independence." A society's competitive advantage will come not from how well its schools teach the multiplication and periodic tables, but from how well they stimulate imagination and creativity.

Therein lies the key, I think, to Einstein's brilliance and the lessons of his life. As a young student he never did well with rote learning. And later, as a theorist, his success came not from the brute strength of his mental processing power but from his imagination and creativity. He could construct complex equations, but more important, he knew that math is the language nature uses to describe her wonders. So he could visualize how equations were reflected in realities—how the electromagnetic field equations discovered by James Clerk Maxwell, for example, would manifest themselves to a boy riding alongside a light beam. As he once declared, "Imagination is more important than knowledge." [6]

That approach required him to embrace nonconformity. "Long live impudence!" he exulted to the lover who would later become his wife. "It is my guardian angel in this world." Many years later, when others thought that his reluctance to embrace quantum mechanics showed that he had lost his edge, he lamented, "To punish me for my contempt for authority, fate made me an authority myself." [7]

His success came from questioning conventional wisdom, challenging authority, and marveling at mysteries that struck others as mundane. This led him to embrace a morality and politics based on respect for free minds, free spirits, and free individuals. Tyranny repulsed him, and he saw tolerance not simply as a sweet virtue but as a necessary condition for a creative society. "It is important to foster individuality," he said, "for only the individual can produce the new ideas." [8]

This outlook made Einstein a rebel with a reverence for the harmony of nature, one who had just the right blend of imagination and wisdom to transform our understanding of the universe. These traits are just as vital for this new century of globalization, in which our success will depend on our creativity, as they were for the beginning of the twentieth century, when Einstein helped usher in the modern age.

CHAPTER TWO

CHILDHOOD

1879–1896

Maja, age 3, and Albert Einstein, 5

The Swabian

He was slow in learning how to talk. "My parents were so worried," he later recalled, "that they consulted a doctor." Even after he had begun using words, sometime after the age of 2, he developed a quirk that prompted the family maid to dub him "der Depperte," the dopey one, and others in his family to label him as "almost backwards." Whenever he had something to say, he would try it out on himself, whispering it softly until it sounded good enough to pronounce aloud. "Every sentence he uttered," his worshipful younger sister recalled, "no matter how routine, he repeated to himself softly, moving his lips." It was all very worrying, she said. "He had such difficulty with language that those around him feared he would never learn."[1]

His slow development was combined with a cheeky rebelliousness toward authority, which led one schoolmaster to send him packing and another to amuse history by declaring that he would never amount to much. These traits made Albert Einstein the patron saint of distracted school kids everywhere.[2] But they also helped to make him, or so he later surmised, the most creative scientific genius of modern times.

His cocky contempt for authority led him to question received wisdom in ways that well-trained acolytes in the academy never contemplated. And as for his slow verbal development, he came to believe that it allowed him to observe with wonder the everyday phenomena that others took for granted. "When I ask myself how it happened that I in particular discovered the relativity theory, it seemed to lie in the following circumstance," Einstein once explained. "The ordinary adult never bothers his head about the problems of space and time. These are things he has thought of as a child. But I developed so slowly that I began to wonder about space and time only when I was already grown up. Consequently, I probed more deeply into the problem than an ordinary child would have."[3]

Einstein's developmental problems have probably been exaggerated, perhaps even by himself, for we have some letters from his adoring grandparents saying that he was just as clever and endearing as every grandchild is. But throughout his life, Einstein had a mild form of echolalia, causing him to repeat phrases to himself, two or three times, especially if they amused him. And he generally preferred to think in pictures, most notably in famous thought experiments, such as imagining watching lightning strikes from a moving train or experiencing gravity while inside a falling elevator. "I very rarely think in words at all," he later told a psychologist. "A thought comes, and I may try to express it in words afterwards."[4]

Einstein was descended, on both parents' sides, from Jewish tradesmen and peddlers who had, for at least two centuries, made modest livings in the rural villages of Swabia in southwestern Germany. With each generation they had become, or at least so they thought, increasingly assimilated into the German culture that they loved. Although Jewish by cultural designation and kindred instinct, they displayed scant interest in the religion or its rituals.

Einstein regularly dismissed the role that his heritage played in shaping who he became. "Exploration of my ancestors," he told a friend late in life, "leads nowhere."[5] That's not fully true. He was blessed by being born into an independent-minded and intelligent family line that valued education, and his life was certainly affected, in ways both beautiful and tragic, by membership in a religious heritage that had a distinctive intellectual tradition and a history of being both outsiders and wanderers. Of course, the fact that he happened to be Jewish in Germany in the early twentieth century made him more of an outsider, and more of a wanderer, than he would have preferred— but that, too, became integral to who he was and the role he would play in world history.

Einstein's father, Hermann, was born in 1847 in the Swabian village of Buchau, whose thriving Jewish community was just beginning to enjoy the right to practice any vocation. Hermann showed "a marked inclination for mathematics,"[6] and his family was able to send him seventy-five miles north to Stuttgart for high school. But they could not afford to send him to a university, most of which were closed to Jews in any event, so he returned home to Buchau to go into trade.

A few years later, as part of the general migration of rural German Jews into industrial centers during the late nineteenth century, Hermann and his parents moved thirty-five miles away to the more prosperous town of Ulm, which prophetically boasted as its motto "Ulmenses sunt mathematici," the people of Ulm are mathematicians.[7]

There he became a partner in a cousin's featherbed company. He was "exceedingly friendly, mild and wise," his son would recall.[8] With a gentleness that blurred into docility, Hermann was to prove inept as a businessman and forever impractical in financial matters. But his docility did make him well suited to be a genial family man and good husband to a strong-willed woman. At age 29, he married Pauline Koch, eleven years his junior.

Pauline's father, Julius Koch, had built a considerable fortune as a grain dealer and purveyor to the royal Württemberg court. Pauline inherited his practicality, but she leavened his dour disposition with a teasing wit edged with sarcasm and a laugh that could be both infectious and wounding (traits she would pass on to her son). From all accounts, the match between Hermann and Pauline was a happy one, with her strong personality meshing "in complete harmony" with her husband's passivity.[9]

Their first child was born at 11:30 a.m. on Friday, March 14, 1879, in Ulm, which had recently joined, along with the rest of Swabia, the new German Reich. Initially, Pauline and Hermann had planned to name the boy Abraham, after his paternal grandfather. But they came to feel, he later said, that the name sounded "too Jewish."[10] So they kept the initial A and named him Albert Einstein.

Munich

In 1880, just a year after Albert's birth, Hermann's featherbed business foundered and he was persuaded to move to Munich by his brother Jakob, who had opened a gas and electrical supply company there. Jakob, the youngest of five siblings, had been able to get a higher education, unlike Hermann, and he had qualified as an engineer. As they competed for contracts to provide generators and electrical lighting to municipalities in southern Germany, Jakob was in charge of the technical side while Hermann provided a modicum of salesmanship skills plus, perhaps more important, loans from his wife's side of the family.[11]

Pauline and Hermann had a second and final child, a daughter, in November 1881, who was named Maria but throughout her life used instead the diminutive Maja. When Albert was shown his new sister for the first time, he was led to believe that she was like a wonderful toy that he would enjoy. His response was to look at her and exclaim, "Yes, but where are the wheels?"[12] It may not have been the most perceptive of questions, but it did show that during his third year his language challenges did not prevent him from making some memorable comments. Despite a few childhood squabbles, Maja was to become her brother's most intimate soul mate.

The Einsteins settled into a comfortable home with mature trees and an elegant garden in a Munich suburb for what was to be, at least through most of Albert's childhood, a respectable bourgeois existence. Munich had been architecturally burnished by mad King Ludwig II (1845–1886) and boasted a profusion of churches, art galleries, and concert halls that favored the works of resident Richard Wagner. In 1882, just after the Einsteins arrived, the city had about 300,000 residents, 85 percent of them Catholics and 2 percent of them Jewish, and it was the host of the first German electricity exhibition, at which electric lights were introduced to the city streets.

Einstein's back garden was often bustling with cousins and children. But he shied from their boisterous games and instead "occupied himself with quieter things." One governess nicknamed him "Father Bore." He was generally a loner, a tendency he claimed to cherish throughout his life, although his was a special sort of detachment that was interwoven with a relish for camaraderie and intellectual companionship. "From the very beginning he was inclined to separate himself from children his own age and to engage in daydreaming and meditative musing," according to Philipp Frank, a longtime scientific colleague.[13]

He liked to work on puzzles, erect complex structures with his toy building set, play with a steam engine that his uncle gave him, and build houses of cards. According to Maja, Einstein was able to construct card structures as high as fourteen stories. Even discounting the recollections of a star-struck younger sister, there was probably a lot of truth to her claim that "persistence and tenacity were obviously already part of his character."

He was also, at least as a young child, prone to temper tantrums. "At such moments his face would turn completely yellow, the tip of his nose snow-white, and he was no longer in control of himself," Maja remembers. Once, at age 5, he grabbed a chair and threw it at a tutor, who fled and never returned. Maja's head became the target of various hard objects. "It takes a sound skull," she later joked, "to be the sister of an intellectual." Unlike his persistence and tenacity, he eventually outgrew his temper.[14]

To use the language of psychologists, the young Einstein's ability to systemize (identify the laws that govern a system) was far greater than his ability to empathize (sense and care about what other humans are feeling), which have led some to ask if he might have exhibited mild symptoms of some developmental disorder.[15] However, it is important to note that, despite his aloof and occasionally rebellious manner, he did have the ability to make close friends and to empathize both with colleagues and humanity in general.

The great awakenings that happen in childhood are usually lost to memory. But for Einstein, an experience occurred when he was 4 or 5 that would alter his life and be etched forever in his mind—and in the history of science.

He was sick in bed one day, and his father brought him a compass. He later recalled being so excited as he examined its mysterious powers that he trembled and grew cold. The fact that the magnetic needle behaved as if influenced by some hidden force field, rather than through the more familiar mechanical method involving touch or contact, produced a sense of wonder that motivated him throughout his life. "I can still remember—or at least I believe I can remember—that this experience made a deep and lasting impression on me," he wrote on one of the many occasions he recounted the incident. "Something deeply hidden had to be behind things."[16]

"It's an iconic story," Dennis Overbye noted in *Einstein in Love*, "the young boy trembling to the invisible order behind chaotic reality." It has been told in the movie *IQ*, in which Einstein, played by Walter Matthau, wears the compass around his neck, and it is the focus of a children's book, *Rescuing Albert's Compass*, by Shulamith Oppenheim, whose father-in-law heard the tale from Einstein in 1911.[17]

After being mesmerized by the compass needle's fealty to an unseen field, Einstein would develop a lifelong devotion to field theories as a way to describe nature. Field theories use mathematical quantities, such as numbers or vectors or tensors, to describe how the conditions at any point in space will affect matter or another field. For example, in a gravitational or an electromagnetic field there are forces that could act on a particle at any point, and the equations of a field theory describe how these change as one moves through the region. The first paragraph of his great 1905 paper on special relativity begins with a consideration of the effects of electrical and magnetic fields; his theory of general relativity is based on equations that describe a gravitational field; and at the very end of his life he was doggedly scribbling further field equations in the hope that they would form the basis for a theory of everything. As the science historian Gerald Holton has noted, Einstein regarded "the classical concept of the field the greatest contribution to the scientific spirit." [18]

His mother, an accomplished pianist, also gave him a gift at around the same time, one that likewise would last throughout his life. She arranged for him to take violin lessons. At first he chafed at the mechanical discipline of the instruction. But after being exposed to Mozart's sonatas, music became both magical and emotional to him. "I believe that love is a better teacher than a sense of duty," he said, "at least for me." [19]

Soon he was playing Mozart duets, with his mother accompanying him on the piano. "Mozart's music is so pure and beautiful that I see it as a reflection of the inner beauty of the universe itself," he later told a friend. "Of course," he added in a remark that reflected his view of math and physics as well as of Mozart, "like all great beauty, his music was pure simplicity." [20]

Music was no mere diversion. On the contrary, it helped him think. "Whenever he felt that he had come to the end of the road or faced a difficult challenge in his work," said his son Hans Albert, "he would take refuge in music and that would solve all his difficulties." The violin thus proved useful during the years he lived alone in Berlin, wrestling with general relativity. "He would often play his violin in his kitchen late at night, improvising melodies while he pondered complicated problems," a friend recalled. "Then, suddenly, in the middle of playing, he would announce excitedly, 'I've got it!' As if by inspiration, the answer to the problem would have come to him in the midst of music." [21]

His appreciation for music, and especially for Mozart, may have reflected his feel for the harmony of the universe. As Alexander Moszkowski, who wrote a biography of Einstein in 1920 based on conversations with him, noted, "Music, Nature, and God became intermingled in him in a complex of feeling, a moral unity, the trace of which never vanished." [22]

Throughout his life, Albert Einstein would retain the intuition and the awe of a child. He never lost his sense of wonder at the magic of nature's phenomena—magnetic fields, gravity, inertia, acceleration, light beams—which grown-ups find so commonplace. He retained the ability to hold two thoughts in his mind simultaneously, to be puzzled when they conflicted, and to marvel when he could smell an underlying unity. "People like you and me never grow old," he wrote a friend later in life. "We never cease to stand like curious children before the great mystery into which we were born."[23]

School

In his later years, Einstein would tell an old joke about an agnostic uncle, who was the only member of his family who went to synagogue. When asked why he did so, the uncle would respond, "Ah, but you never know." Einstein's parents, on the other hand, were "entirely irreligious" and felt no compulsion to hedge their bets. They did not keep kosher or attend synagogue, and his father referred to Jewish rituals as "ancient superstitions."[24]

Consequently, when Albert turned 6 and had to go to school, his parents did not care that there was no Jewish one near their home. Instead he went to the large Catholic school in their neighborhood, the Petersschule. As the only Jew among the seventy students in his class, Einstein took the standard course in Catholic religion and ended up enjoying it immensely. Indeed, he did so well in his Catholic studies that he helped his classmates with theirs.[25]

One day his teacher brought a large nail to the class. "The nails with which Jesus was nailed to the cross looked like this," he said.[26] Nevertheless, Einstein later said that he felt no discrimination from the teachers. "The teachers were liberal and made no distinction based on denominations," he wrote. His fellow students, however, were a different matter. "Among the children at the elementary school, anti-Semitism was prevalent," he recalled.

Being taunted on his walks to and from school based on "racial characteristics about which the children were strangely aware" helped reinforce the sense of being an outsider, which would stay with him his entire life. "Physical attacks and insults on the way home from school were frequent, but for the most part not too vicious. Nevertheless, they were sufficient to consolidate, even in a child, a lively sense of being an outsider."[27]

When he turned 9, Einstein moved up to a high school near the center of Munich, the Luitpold Gymnasium, which was known as an enlightened institution that emphasized math and science as well as Latin and Greek. In addition, the school supplied a teacher to provide religious instruction for him and other Jews.

Despite his parents' secularism, or perhaps because of it, Einstein rather suddenly developed a passionate zeal for Judaism. "He was so fervent in his feelings that, on his own, he observed Jewish religious strictures in every detail," his sister recalled. He ate no pork, kept kosher dietary laws, and obeyed the strictures of the Sabbath, all rather difficult to do when the rest of his family had a lack of interest bordering on disdain for such displays. He even composed his own hymns for the glorification of God, which he sang to himself as he walked home from school.[28]

One widely held belief about Einstein is that he failed math as a student, an assertion that is made, often accompanied by the phrase "as everyone knows," by scores of books and thousands of websites designed to reassure underachieving students. It even made it into the famous "Ripley's Believe It or Not!" newspaper column.

Alas, Einstein's childhood offers history many savory ironies, but this is not one of them. In 1935, a rabbi in Princeton showed him a clipping of the Ripley's column with the headline "Greatest Living Mathematician Failed in Mathematics." Einstein laughed. "I never failed in mathematics," he replied, correctly. "Before I was fifteen I had mastered differential and integral calculus."[29]

In fact, he was a wonderful student, at least intellectually. In primary school, he was at the top of his class. "Yesterday Albert got his grades," his mother reported to an aunt when he was 7. "Once again he was ranked first." At the gymnasium, he disliked the mechanical learning of languages such as Latin and Greek, a problem exacerbated by what he later said was his "bad memory for words and texts." But even in these courses, Einstein consistently got top grades. Years later, when Einstein celebrated his fiftieth birthday and there were stories about how poorly the great genius had fared at the gymnasium, the school's current principal made a point of publishing a letter revealing how good his grades actually were.[30]

As for math, far from being a failure, he was "far above the school requirements." By age 12, his sister recalled, "he already had a predilection for solving complicated problems in applied arithmetic," and he decided to see if he could jump ahead by learning geometry and algebra on his own. His parents bought him the textbooks in advance so that he could master them over summer vacation. Not only did he learn the proofs in the books, he tackled the new theories by trying to prove them on his own. "Play and playmates were forgotten," she noted. "For days on end he sat alone, immersed in the search for a solution, not giving up before he had found it."[31]

His uncle Jakob Einstein, the engineer, introduced him to the joys of algebra. "It's a merry science," he explained. "When the animal that we are hunting cannot be caught, we call it *X* temporarily and continue to hunt until it is bagged." He went on to give the boy even more difficult challenges, Maja recalled, "with good-natured doubts about his ability to solve them." When Einstein triumphed, as he invariably did, he "was overcome with great happiness and was already then aware of the direction in which his talents were leading him."

Among the concepts that Uncle Jakob threw at him was the Pythagorean theorem (the square of the lengths of the legs of a right triangle add up to the square of the length of the hypotenuse). "After much effort I succeeded in 'proving' this theorem on the basis of the similarity of triangles," Einstein recalled. Once again he was thinking in pictures. "It seemed to me 'evident' that the relations of the sides of the right-angled triangles would have to be completely determined by one of the acute angles."[32]

Maja, with the pride of a younger sister, called Einstein's Pythagorean proof "an entirely original new one." Although perhaps new to him, it is hard to imagine that Einstein's approach, which was surely similar to the standard ones based on the proportionality of the sides of similar triangles, was completely original. Nevertheless, it did show Einstein's youthful appreciation that elegant theorems can be derived from simple axioms—and the fact that he was in little danger of failing math. "As a boy of 12, I was thrilled to see that it was possible to find out truth by reasoning alone, without the help of any outside experience," he told a reporter from a high school newspaper in Princeton years later. "I became more and more convinced that nature could be understood as a relatively simple mathematical structure."[33]

Einstein's greatest intellectual stimulation came from a poor medical student who used to dine with his family once a week. It was an old Jewish custom to take in a needy religious scholar to share the Sabbath meal; the Einsteins modified the tradition by hosting instead a medical student on Thursdays. His name was Max Talmud (later changed to Talmey, when he immigrated to the United States), and he began his weekly visits when he was 21 and Einstein was 10. "He was a pretty, dark-haired boy," remembered Talmud. "In all those years, I never saw him reading any light literature. Nor did I ever see him in the company of schoolmates or other boys his age."[34]

Talmud brought him science books, including a popular illustrated series called *People's Books on Natural Science*, "a work which I read with breathless attention," said Einstein. The twenty-one little volumes were written by Aaron Bernstein, who stressed the interrelations between biology and physics, and he reported in great detail the scientific experiments being done at the time, especially in Germany.[35]

In the opening section of the first volume, Bernstein dealt with the speed of light, a topic that obviously fascinated him. Indeed, he returned to it repeatedly in his subsequent volumes, including eleven essays on the topic in volume 8. Judging from the thought experiments that Einstein later used in creating his theory of relativity, Bernstein's books appear to have been influential.

For example, Bernstein asked readers to imagine being on a speeding train. If a bullet is shot through the window, it would seem that it was shot at an angle, because the train would have moved between the time the bullet entered one window and exited the window on the other side. Likewise, because of the speed of the earth through space, the same must be true of light going through a telescope. What was amazing, said Bernstein, was that experiments showed the same effect no matter how fast the source of the light was moving. In a sentence that, because of its relation to what Einstein would later famously conclude, seems to have made an impression, Bernstein declared, "Since each kind of light proves to be of exactly the same speed, the law of the speed of light can well be called the most general of all of nature's laws."

In another volume, Bernstein took his young readers on an imaginary trip through space. The mode of transport was the wave of an electric signal. His books celebrated the joyful wonders of scientific investigation and included such exuberant passages as this one written about the successful prediction of the location of the new planet Uranus: "Praised be this science! Praised be the men who do it! And praised be the human mind, which sees more sharply than does the human eye." [36]

Bernstein was, as Einstein would later be, eager to tie together all of nature's forces. For example, after discussing how all electromagnetic phenomena, such as light, could be considered waves, he speculated that the same may be true for gravity. A unity and simplicity, Bernstein wrote, lay beneath all the concepts applied by our perceptions. Truth in science consisted in discovering theories that described this underlying reality. Einstein later recalled the revelation, and the realist attitude, that this instilled in him as a young boy: "Out yonder there was this huge world, which exists independently of us human beings and which stands before us like a great, eternal riddle." [37]

Years later, when they met in New York during Einstein's first visit there, Talmud asked what he thought, in retrospect, of Bernstein's work. "A very good book," he said. "It has exerted a great influence on my whole development." [38]

Talmud also helped Einstein continue to explore the wonders of mathematics by giving him a textbook on geometry two years before he was scheduled to learn that subject in school. Later, Einstein would refer to it as "the sacred little geometry book" and speak of it with awe: "Here were assertions, as for example the intersection of the three altitudes of a triangle in one point, which—though by no means evident—could nevertheless be proved with such certainty that any doubt appeared to be out of the question. This lucidity and certainty made an indescribable impression upon me." Years later, in a lecture at Oxford, Einstein noted, "If Euclid failed to kindle your youthful enthusiasm, then you were not born to be a scientific thinker." [39]

When Talmud arrived each Thursday, Einstein delighted in showing him the problems he had solved that week. Initially, Talmud was able to help him, but he was soon surpassed by his pupil. "After a short time, a few months, he had worked through the whole book," Talmud recalled. "He thereupon devoted himself to higher mathematics . . . Soon the flight of his mathematical genius was so high that I could no longer follow." [40]

So the awed medical student moved on to introducing Einstein to philosophy. "I recommended Kant to him," he recalled. "At that time he was still a child, only thirteen years old, yet Kant's works, incomprehensible to ordinary mortals, seemed to be clear to him." Kant became, for a while, Einstein's favorite philosopher, and his *Critique of Pure Reason* eventually led him to delve also into David Hume, Ernst Mach, and the issue of what can be known about reality.

Einstein's exposure to science produced a sudden reaction against religion at age 12, just as he would have been readying for a bar mitzvah. Bernstein, in his popular science volumes, had reconciled science with religious inclination. As he put it, "The religious inclination lies in the dim consciousness that dwells in humans that all nature, including the humans in it, is in no way an accidental game, but a work of lawfulness, that there is a fundamental cause of all existence."

Einstein would later come close to these sentiments. But at the time, his leap away from faith was a radical one. "Through the reading of popular scientific books, I soon reached the conviction that much in the stories of the Bible could not be true. The consequence was a positively fanatic orgy of freethinking coupled with the impression that youth is intentionally being deceived by the state through lies; it was a crushing impression." [41]

As a result, Einstein avoided religious rituals for the rest of his life. "There arose in Einstein an aversion to the orthodox practice of the Jewish or any traditional religion, as well as to attendance at religious services, and this he has never lost," his friend Philipp Frank later noted. He did, however, retain from his childhood religious phase a profound reverence for the harmony and beauty of what he called the mind of God as it was expressed in the creation of the universe and its laws. [42]

Einstein's rebellion against religious dogma had a profound effect on his general outlook toward received wisdom. It inculcated an allergic reaction against all forms of dogma and authority, which was to affect both his politics and his science. "Suspicion against every kind of authority grew out of this experience, an attitude which has never again left me," he later said. Indeed, it was this comfort with being a nonconformist that would define both his science and his social thinking for the rest of his life.

He would later be able to pull off this contrariness with a grace that was generally endearing, once he was accepted as a genius. But it did not play so well when he was merely a sassy student at a Munich gymnasium. "He was very uncomfortable in school," according to his sister. He found the style of teaching—rote drills, impatience with questioning—to be repugnant. "The military tone of the school, the systematic training in the worship of authority that was supposed to accustom pupils at an early age to military discipline, was particularly unpleasant."[43]

Even in Munich, where the Bavarian spirit engendered a less regimented approach to life, this Prussian glorification of the military had taken hold, and many of the children loved to play at being soldiers. When troops would come by, accompanied by fifes and drums, kids would pour into the streets to join the parade and march in lockstep. But not Einstein. Watching such a display once, he began to cry. "When I grow up, I don't want to be one of those poor people," he told his parents. As Einstein later explained, "When a person can take pleasure in marching in step to a piece of music it is enough to make me despise him. He has been given his big brain only by mistake."[44]

The opposition he felt to all types of regimentation made his education at the Munich gymnasium increasingly irksome and contentious. The mechanical learning there, he complained, "seemed very much akin to the methods of the Prussian army, where a mechanical discipline was achieved by repeated execution of meaningless orders." In later years, he would liken his teachers to members of the military. "The teachers at the elementary school seemed to me like drill sergeants," he said, "and the teachers at the gymnasium like lieutenants."

He once asked C. P. Snow, the British writer and scientist, whether he was familiar with the German word *Zwang*. Snow allowed that he was; it meant constraint, compulsion, obligation, coercion. Why? In his Munich school, Einstein answered, he had made his first strike against *Zwang,* and it had helped define him ever since.[45]

Skepticism and a resistance to received wisdom became a hallmark of his life. As he proclaimed in a letter to a fatherly friend in 1901, "A foolish faith in authority is the worst enemy of truth."[46]

Throughout the six decades of his scientific career, whether leading the quantum revolution or later resisting it, this attitude helped shape Einstein's work. "His early suspicion of authority, which never wholly left him, was to prove of decisive importance," said Banesh Hoffmann, who was a collaborator of Einstein's in his later years. "Without it he would not have been able to develop the powerful independence of mind that gave him the courage to challenge established scientific beliefs and thereby revolutionize physics."[47]

This contempt for authority did not endear him to the German "lieutenants" who taught him at his school. As a result, one of his teachers proclaimed that his insolence made him unwelcome in class. When Einstein insisted that he had committed no offense, the teacher replied, "Yes, that is true, but you sit there in the back row and smile, and your mere presence here spoils the respect of the class for me."[48]

Einstein's discomfort spiraled toward depression, perhaps even close to a nervous breakdown, when his father's business suffered a sudden reversal of fortune. The collapse was a precipitous one. During most of Einstein's school years, the Einstein brothers' company had been a success. In 1885, it had two hundred employees and provided the first electrical lights for Munich's Oktoberfest. Over the next few years, it won the contract to wire the community of Schwabing, a Munich suburb of ten thousand people, using gas motors to drive twin dynamos that the Einsteins had designed. Jakob Einstein received six patents for improvements in arc lamps, automatic circuit breakers, and electric meters. The company was poised to rival Siemens and other power companies then flourishing. To raise capital, the brothers mortgaged their homes, borrowed more than 60,000 marks at 10 percent interest, and went deeply in debt.[49]

But in 1894, when Einstein was 15, the company went bust after it lost competitions to light the central part of Munich and other locations. His parents and sister, along with Uncle Jakob, moved to northern Italy—first Milan and then the nearby town of Pavia—where the company's Italian partners thought there would be more fertile territory for a smaller firm. Their elegant home was torn down by a developer to build an apartment block. Einstein was left behind in Munich, at the house of a distant relative, to finish his final three years of school.

It is not quite clear whether Einstein, in that sad autumn of 1894, was actually forced to leave the Luitpold Gymnasium or was merely politely encouraged to leave. Years later, he recalled that the teacher who had declared that his "presence spoils the respect of the class for me" had gone on to "express the wish that I leave the school." An early book by a member of his family said that it was his own decision. "Albert increasingly resolved not to remain in Munich, and he worked out a plan."

That plan involved getting a letter from the family doctor, Max Talmud's older brother, who certified that he was suffering from nervous exhaustion. He used this to justify leaving the school at Christmas vacation in 1894 and not returning. Instead, he took a train across the Alps to Italy and informed his "alarmed" parents that he was never going back to Germany. Instead, he promised, he would study on his own and attempt to gain admission to a technical college in Zurich the following autumn.

There was perhaps one other factor in his decision to leave Germany. Had he remained there until he was 17, just over a year away, he would have been required to join the army, a prospect that his sister said "he contemplated with dread." So, in addition to announcing that he would not go back to Munich, he would soon ask for his father's help in renouncing his German citizenship. [50]

Aarau

Einstein spent the spring and summer of 1895 living with his parents in their Pavia apartment and helping at the family firm. In the process, he was able to get a good feel for the workings of magnets, coils, and generated electricity. Einstein's work impressed his family. On one occasion, Uncle Jakob was having problems with some calculations for a new machine, so Einstein went to work on it. "After my assistant engineer and I had been racking our brain for days, that young sprig had got the whole thing in just fifteen minutes," Jakob reported to a friend. "You will hear of him yet."[51]

With his love of the sublime solitude found in the mountains, Einstein hiked for days in the Alps and Apennines, including an excursion from Pavia to Genoa to see his mother's brother Julius Koch. Wherever he traveled in northern Italy, he was delighted by the non-Germanic grace and "delicacy" of the people. Their "naturalness" was a contrast to the "spiritually broken and mechanically obedient automatons" of Germany, his sister recalled.

Einstein had promised his family that he would study on his own to get into the local technical college, the Zurich Polytechnic.* So he bought all three volumes of Jules Violle's advanced physics text and copiously noted his ideas in the margins. His work habits showed his ability to concentrate, his sister recalled. "Even in a large, quite noisy group, he could withdraw to the sofa, take pen and paper in hand, set the inkstand precariously on the armrest, and lose himself so completely in a problem that the conversation of many voices stimulated rather than disturbed him."[52]

That summer, at age 16, he wrote his first essay on theoretical physics, which he titled "On the Investigation of the State of the Ether in a Magnetic Field." The topic was important, for the notion of the ether would play a critical role in Einstein's career. At the time, scientists conceived of light simply as a wave, and so they assumed that the universe must contain an all-pervasive yet unseen substance that was doing the rippling and thus propagating the waves, just as water was the medium rippling up and down and thus propagating the waves in an ocean. They dubbed this the ether, and Einstein (at least for the time being) went along with the assumption. As he put it in his essay,

---

* The official name of the institution was the Eidgenössische Polytechnische Schule. In 1911, it gained the right to grant doctoral degrees and changed its name to the Eidgenössische Technische Hochschule, or the Swiss Federal Institute of Technology, referred to as the ETH. Einstein, then and later, usually called it the Züricher Polytechnikum, or the Zurich Polytechnic.

"An electric current sets the surrounding ether in a kind of momentary motion."

The fourteen-paragraph handwritten paper echoed Violle's text-book as well as some of the reports in the popular science magazines about Heinrich Hertz's recent discoveries about electromagnetic waves. In it, Einstein made suggestions for experiments that could explain "the magnetic field formed around an electric current." This would be interesting, he argued, "because the exploration of the elastic state of the ether in this case would permit us a look into the enigmatic nature of electric current."

The high school dropout freely admitted that he was merely making a few suggestions without knowing where they might lead. "As I was completely lacking in materials that would have enabled me to delve into the subject more deeply than by merely meditating about it, I beg you not to interpret this circumstance as a mark of superficiality," he wrote.[53]

He sent the paper to his uncle Caesar Koch, a merchant in Belgium, who was one of his favorite relatives and occasionally a financial patron. "It is rather naïve and imperfect, as might be expected from such a young fellow like myself," Einstein confessed with a pretense of humility. He added that his goal was to enroll the following fall at the Zurich Polytechnic, but he was concerned that he was younger than the age requirement. "I should be at least two years older."[54]

To help him get around the age requirement, a family friend wrote to the director of the Polytechnic, asking for an exception. The tone of the letter can be gleaned from the director's response, which expressed skepticism about admitting this "so-called 'child prodigy.' " Nevertheless, Einstein was granted permission to take the entrance exam, and he boarded the train for Zurich in October 1895 "with a sense of well-founded diffidence."

Not surprisingly, he easily passed the section of the exam in math and science. But he failed to pass the general section, which included sections on literature, French, zoology, botany, and politics. The Polytechnic's head physics professor, Heinrich Weber, suggested that Einstein stay in Zurich and audit his classes. Instead, Einstein decided, on the advice of the college's director, to spend a year preparing at the cantonal school in the village of Aarau, twenty-five miles to the west.[55]

It was a perfect school for Einstein. The teaching was based on the philosophy of a Swiss educational reformer of the early nineteenth century, Johann Heinrich Pestalozzi, who believed in encouraging students to visualize images. He also thought it important to nurture the "inner dignity" and individuality of each child. Students should be allowed to reach their own conclusions, Pestalozzi preached, by using a series of steps that began with hands-on observations and then proceeded to intuitions, conceptual thinking, and visual imagery.[56] It was even possible to learn—and truly understand—the laws of math and physics that way. Rote drills, memorization, and force-fed facts were avoided.

Einstein loved Aarau. "Pupils were treated individually," his sister recalled, "more emphasis was placed on independent thought than on punditry, and young people saw the teacher not as a figure of authority, but, alongside the student, a man of distinct personality." It was the opposite of the German education that Einstein had hated. "When compared to six years' schooling at a German authoritarian gymnasium," Einstein later said, "it made me clearly realize how much superior an education based on free action and personal responsibility is to one relying on outward authority."[57]

The visual understanding of concepts, as stressed by Pestalozzi and his followers in Aarau, became a significant aspect of Einstein's genius. "Visual understanding is the essential and only true means of teaching how to judge things correctly," Pestalozzi wrote, and "the learning of numbers and language must be definitely subordinated."[58]

Not surprisingly, it was at this school that Einstein first engaged in the visualized thought experiment that would help make him the greatest scientific genius of his time: he tried to picture what it would be like to ride alongside a light beam. "In Aarau I made my first rather childish experiments in thinking that had a direct bearing on the Special Theory," he later told a friend. "If a person could run after a light wave with the same speed as light, you would have a wave arrangement which could be completely independent of time. Of course, such a thing is impossible."[59]

This type of visualized thought experiments—*Gedankenexperi-*

ment—became a hallmark of Einstein's career. Over the years, he would picture in his mind such things as lightning strikes and moving trains, accelerating elevators and falling painters, two-dimensional blind beetles crawling on curved branches, as well as a variety of contraptions designed to pinpoint, at least in theory, the location and velocity of speeding electrons.

While a student in Aarau, Einstein boarded with a wonderful family, the Wintelers, whose members would long remain entwined in his life. There was Jost Winteler, who taught history and Greek at the school; his wife, Rosa, soon known to Einstein as Mamerl, or Mama; and their seven children. Their daughter Marie would become Einstein's first girlfriend. Another daughter, Anna, would marry Einstein's best friend, Michele Besso. And their son Paul would marry Einstein's beloved sister, Maja.

"Papa" Jost Winteler was a liberal who shared Einstein's allergy to German militarism and to nationalism in general. His edgy honesty and political idealism helped to shape Einstein's social philosophy. Like his mentor, Einstein would become a supporter of world federalism, internationalism, pacifism, and democratic socialism, with a strong devotion to individual liberty and freedom of expression.

More important, in the warm embrace of the Winteler family, Einstein became more secure and personable. Even though he still fancied himself a loner, the Wintelers helped him flower emotionally and open himself to intimacy. "He had a great sense of humor and at times could laugh heartily," recalled daughter Anna. In the evenings he would sometimes study, "but more often he would sit with the family around the table." [60]

Einstein had developed into a head-turning teenager who possessed, in the words of one woman who knew him, "masculine good looks of the type that played havoc at the turn of the century." He had wavy dark hair, expressive eyes, a high forehead, and jaunty demeanor. "The lower half of his face might have belonged to a sensualist who found plenty of reasons to love life."

One of his schoolmates, Hans Byland, later wrote a striking description of "the impudent Swabian" who made such a lasting impression. "Sure of himself, his gray felt hat pushed back on his thick, black hair, he strode energetically up and down in the rapid, I might say crazy, tempo of a restless spirit which carries a whole world in itself. Nothing escaped the sharp gaze of the large bright brown eyes. Whoever approached him was captivated by his superior personality. A mocking curl of his fleshy mouth with its protruding lower lip did not encourage Philistines to fraternize with him."

Most notably, Byland added, young Einstein had a sassy, sometimes intimidating wit. "He confronted the world spirit as a laughing philosopher, and his witty sarcasm mercilessly castigated all vanity and artificiality." [61]

Einstein fell in love with Marie Winteler at the end of 1895, just a few months after he moved in with her parents. She had just completed teacher training college and was living at home while waiting to take a job in a nearby village. She was just turning 18, he was still 16. The romance thrilled both families. Albert and Marie sent New Year's greetings to his mother; she replied warmly, "Your little letter, dear Miss Marie, brought me immense joy." [62]

The following April, when he was back home in Pavia for spring break, Einstein wrote Marie his first known love letter:

> Beloved sweetheart!
>   Many, many thanks sweetheart for your charming little letter, which made me endlessly happy. It was so wonderful to be able to press to one's heart such a bit of paper which two so dear little eyes have lovingly beheld and on which the dainty little hands have charmingly glided back and forth. I was now made to realize, my little angel, the meaning of homesickness and pining. But love brings much happiness—much more so than pining brings pain . . .
>   My mother has also taken you to her heart, even though she does not know you; I only let her read two of your charming little letters. And she always laughs at me because I am no longer attracted to the girls who were supposed to have enchanted me so much in the past. You mean more to my soul than the whole world did before.

To which his mother penned a postscript: "Without having read this letter, I send you cordial greetings!" [63]

Although he enjoyed the school in Aarau, Einstein turned out to be an uneven student. His admission report noted that he needed to do remedial work in chemistry and had "great gaps" in his knowledge of

French. By midyear, he still was required to "continue with private lessons in French & chemistry," and "the protest in French remains in effect." His father was sanguine when Jost Winteler sent him the midyear report. "Not all its parts fulfill my wishes and expectations," he wrote, "but with Albert I got used to finding mediocre grades along with very good ones, and I am therefore not disconsolate about them."[64]

Music continued to be a passion. There were nine violinists in his class, and their teacher noted that they suffered from "some stiffness in bowing technique here and there." But Einstein was singled out for praise: "One student, by the name of Einstein, even sparkled by rendering an adagio from a Beethoven sonata with deep understanding." At a concert in the local church, Einstein was chosen to play first violin in a piece by Bach. His "enchanting tone and incomparable rhythm" awed the second violinist, who asked, "Do you count the beats?" Einstein replied, "Heavens no, it's in my blood."

His classmate Byland recalled Einstein playing a Mozart sonata with such passion—"What fire there was in his playing!"—that it seemed like hearing the composer for the first time. Listening to him, Byland realized that Einstein's wisecracking, sarcastic exterior was a shell around a softer inner soul. "He was one of those split personalities who know how to protect, with a prickly exterior, the delicate realm of their intense personal life."[65]

Einstein's contempt for Germany's authoritarian schools and militarist atmosphere made him want to renounce his citizenship in that country. This was reinforced by Jost Winteler, who disdained all forms of nationalism and instilled in Einstein the belief that people should consider themselves citizens of the world. So he asked his father to help him drop his German citizenship. The release came through in January 1896, and for the time being he was stateless.[66]

He also that year became a person without a religious affiliation. In the application to renounce his German citizenship, his father had written, presumably at Albert's request, "no religious denomination." It was a statement Albert would also make when applying for Zurich residency a few years later, and on various occasions over the ensuing two decades.

His rebellion from his childhood fling with ardent Judaism, coupled with his feelings of detachment from Munich's Jews, had alienated him from his heritage. "The religion of the fathers, as I encountered it in Munich during religious instruction and in the synagogue, repelled rather than attracted me," he later explained to a Jewish historian. "The Jewish bourgeois circles that I came to know in my younger years, with their affluence and lack of a sense of community, offered me nothing that seemed to be of value."[67]

Later in life, beginning with his exposure to virulent anti-Semitism in the 1920s, Einstein would begin to reconnect with his Jewish identity. "There is nothing in me that can be described as a 'Jewish faith,' " he said, "however I am happy to be a member of the Jewish people." Later he would make the same point in more colorful ways. "The Jew who abandons his faith," he once said, "is in a similar position to a snail that abandons his shell. He is still a snail."[68]

His renunciation of Judaism in 1896 should, therefore, be seen not as a clean break but as part of a lifelong evolution of his feelings about his cultural identity. "At that time I would not even have understood what leaving Judaism could possibly mean," he wrote a friend the year before he died. "But I was fully aware of my Jewish origin, even though the full significance of belonging to Jewry was not realized by me until later."[69]

Einstein ended his year at the Aarau school in a manner that would have seemed impressive for anyone except one of history's great geniuses, scoring the second highest grades in his class. (Alas, the name of the boy who bested Einstein is lost to history.) On a 1 to 6 scale, with 6 being the highest, he scored a 5 or 6 in all of his science and math courses as well as in history and Italian. His lowest grade was a 3, in French.

That qualified him to take a series of exams, written and oral, that would permit him, if he passed, to enter the Zurich Polytechnic. On his German exam, he did a perfunctory outline of a Goethe play and scored a 5. In math, he made a careless mistake, calling a number "imaginary" when he meant "irrational," but still got a top grade. In physics, he arrived late and left early, completing the two-hour test in an hour and fifteen minutes; he got the top grade. Altogether, he ended up with a 5.5, the best grade among the nine students taking the exams.

The one section on which he did poorly was French. But his three-paragraph essay was, to those of us today, the most interesting part of all of his exams. The topic was "Mes Projets d'avenir," my plans for the future. Although the French was not memorable, the personal insights were:

> If I am lucky and pass my exams, I will enroll in the Zurich Polytechnic. I will stay there four years to study mathematics and physics. I suppose I will become a teacher in these fields of science, opting for the theoretical part of these sciences.
>
> Here are the reasons that have led me to this plan. They are, most of all, my personal talent for abstract and mathematical thinking . . . My desires have also led me to the same decision. That is quite natural; everybody desires to do that for which he has a talent. Besides, I am attracted by the independence offered by the profession of science.[70]

In the summer of 1896, the Einstein brothers' electrical business again failed, this time because they bungled getting the necessary water rights to build a hydroelectric system in Pavia. The partnership was dissolved in a friendly fashion, and Jakob joined a large firm as an engineer. But Hermann, whose optimism and pride tended to overwhelm any prudence, insisted on opening yet another new dynamo business, this time in Milan. Albert was so dubious of his father's prospects that he went to his relatives and suggested that they not finance him again, but they did.[71]

Hermann hoped that Albert would someday join him in the business, but engineering held little appeal for him. "I was originally supposed to become an engineer," he later wrote a friend, "but the thought of having to expend my creative energy on things that make practical everyday life even more refined, with a bleak capital gain as the goal, was unbearable to me. Thinking for its own sake, like music!"[72] And thus he headed off to the Zurich Polytechnic.

CHAPTER THREE

THE ZURICH POLYTECHNIC

1896–1900

The Impudent Scholar

The Zurich Polytechnic, with 841 students, was mainly a teachers' and technical college when 17-year-old Albert Einstein enrolled in October 1896. It was less prestigious than the neighboring University of Zurich and the universities in Geneva and Basel, all of which could grant doctoral degrees (a status that the Polytechnic, officially named the Eidgenössische Polytechnische Schule, would attain in 1911 when it became the Eidgenössische Technische Hochschule, or ETH). Nevertheless, the Polytechnic had a solid reputation in engineering and science. The head of the physics department, Heinrich Weber, had recently procured a grand new building, funded by the electronics magnate (and Einstein Brothers competitor) Werner von Siemens. It housed showcase labs famed for their precision measurements.

Einstein was one of eleven freshmen enrolled in the section that provided training "for specialized teachers in mathematics and physics." He lived in student lodgings on a monthly stipend of 100 Swiss francs from his Koch family relatives. Each month he put aside 20 of those francs toward the fee he would eventually have to pay to become a Swiss citizen.[1]

Theoretical physics was just coming into its own as an academic discipline in the 1890s, with professorships in the field sprouting up across Europe. Its pioneer practitioners—such as Max Planck in Berlin,

Hendrik Lorentz in Holland, and Ludwig Boltzmann in Vienna—combined physics with math to suggest paths where experimentalists had yet to tread. Because of this, math was supposed to be a major part of Einstein's required studies at the Polytechnic.

Einstein, however, had a better intuition for physics than for math, and he did not yet appreciate how integrally the two subjects would be related in the pursuit of new theories. During his four years at the Polytechnic, he got marks of 5 or 6 (on a 6-point scale) in all of his theoretical physics courses, but got only 4s in most of his math courses, especially those in geometry. "It was not clear to me as a student," he admitted, "that a more profound knowledge of the basic principles of physics was tied up with the most intricate mathematical methods."[2]

That realization would sink in a decade later, when he was wrestling with the geometry of his theory of gravity and found himself forced to rely on the help of a math professor who had once called him a lazy dog. "I have become imbued with great respect for mathematics," he wrote to a colleague in 1912, "the subtler part of which I had in my simple-mindedness regarded as pure luxury until now." Near the end of his life, he expressed a similar lament in a conversation with a younger friend. "At a very early age, I made an assumption that a successful physicist only needs to know elementary mathematics," he said. "At a later time, with great regret, I realized that the assumption of mine was completely wrong."[3]

His primary physics professor was Heinrich Weber, the one who a year earlier had been so impressed with Einstein that, even after he had failed his entrance exam to the Polytechnic, he urged him to stay in Zurich and audit his lectures. During Einstein's first two years at the Polytechnic, their mutual admiration endured. Weber's lectures were among the few that impressed him. "Weber lectured on heat with great mastery," he wrote during their second year. "One lecture after another of his pleases me." He worked in Weber's laboratory "with fervor and passion," took fifteen courses (five lab and ten classroom) with him, and scored well in them all.[4]

Einstein, however, gradually became disenchanted with Weber. He felt that the professor focused too much on the historical foundations of physics, and he did not deal much with contemporary frontiers.

"Anything that came after Helmholtz was simply ignored," one contemporary of Einstein complained. "At the close of our studies, we knew all the past of physics but nothing of the present and future."

Notably absent from Weber's lectures was any exploration of the great breakthroughs of James Clerk Maxwell, who, beginning in 1855, developed profound theories and elegant mathematical equations that described how electromagnetic waves such as light propagated. "We waited in vain for a presentation of Maxwell's theory," wrote another fellow student. "Einstein above all was disappointed."[5]

Given his brash attitude, Einstein didn't hide his feelings. And given his dignified sense of himself, Weber bristled at Einstein's ill-concealed disdain. By the end of their four years together they were antagonists.

Weber's irritation was yet another example of how Einstein's scientific as well as personal life was affected by the traits deeply bred into his Swabian soul: his casual willingness to question authority, his sassy attitude in the face of regimentation, and his lack of reverence for received wisdom. He tended to address Weber, for example, in a rather informal manner, calling him "Herr Weber" instead of "Herr Professor."

When his frustration finally overwhelmed his admiration, Professor Weber's pronouncement on Einstein echoed that of the irritated teacher at the Munich gymnasium a few years earlier. "You're a very clever boy, Einstein," Weber told him. "An extremely clever boy. But you have one great fault: you'll never let yourself be told anything."

There was some truth to that assessment. But Einstein was to show that, in the jangled world of physics at the turn of the century, this insouciant ability to tune out the conventional wisdom was not the worst fault to have.[6]

Einstein's impertinence also got him into trouble with the Polytechnic's other physics professor, Jean Pernet, who was in charge of experimental and lab exercises. In his course Physical Experiments for Beginners, Pernet gave Einstein a 1, the lowest possible grade, thus earning himself the historic distinction of having flunked Einstein in a physics course. Partly it was because Einstein seldom showed up for the course. At Pernet's written request, in March 1899 Einstein was given an official "director's reprimand due to lack of diligence in physics practicum."[7]

Why are you specializing in physics, Pernet asked Einstein one day, instead of a field like medicine or even law? "Because," Einstein replied, "I have even less talent for those subjects. Why shouldn't I at least try my luck with physics?"[8]

On those occasions when Einstein did deign to show up in Pernet's lab, his independent streak sometimes got him in trouble, such as the day he was given an instruction sheet for a particular experiment. "With his usual independence," his friend and early biographer Carl Seelig reports, "Einstein naturally flung the paper in the waste paper basket." He proceeded to pursue the experiment in his own way. "What do you make of Einstein?" Pernet asked an assistant. "He always does something different from what I have ordered."

"He does indeed, Herr Professor," the assistant replied, "but his solutions are right and the methods he uses are of great interest."[9]

Eventually, these methods caught up with him. In July 1899, he caused an explosion in Pernet's lab that "severely damaged" his right hand and required him to go to the clinic for stitches. The injury made it difficult for him to write for at least two weeks, and it forced him to give up playing the violin for even longer. "My fiddle had to be laid aside," he wrote to a woman he had performed with in Aarau. "I'm sure it wonders why it is never taken out of the black case. It probably thinks it has gotten a stepfather."[10] He soon resumed playing the violin, but the accident seemed to make him even more wedded to the role of theorist rather than experimentalist.

Despite the fact that he focused more on physics than on math, the professor who would eventually have the most positive impact on him was the math professor Hermann Minkowski, a square-jawed, handsome Russian-born Jew in his early thirties. Einstein appreciated the way Minkowski tied math to physics, but he avoided the more challenging of his courses, which is why Minkowski labeled him a lazy dog: "He never bothered about mathematics at all."[11]

Einstein preferred to study, based on his own interests and passions, with one or two friends.[12] Even though he was still priding himself on being "a vagabond and a loner," he began to hang around the coffee-houses and attend musical soirees with a congenial crowd of bohemian soul mates and fellow students. Despite his reputation for detachment, he forged lasting intellectual friendships in Zurich that became important bonds in his life.

Among these was Marcel Grossmann, a middle-class Jewish math wizard whose father owned a factory near Zurich. Grossmann took copious notes that he shared with Einstein, who was less diligent about attending lectures. "His notes could have been printed and published," Einstein later marveled to Grossmann's wife. "When it came time to prepare for my exams, he would always lend me those notebooks, and they were my savior. What I would have done without these books I would rather not speculate on."

Together Einstein and Grossmann smoked pipes and drank iced coffee while discussing philosophy at the Café Metropole on the banks of the Limmat River. "This Einstein will one day be a great man," Grossmann predicted to his parents. He would later help make that prediction true by getting Einstein his first job, at the Swiss Patent Office, and then aiding him with the math he needed to turn the special theory of relativity into a general theory.[13]

Because many of the Polytechnic lectures seemed out of date, Einstein and his friends read the most recent theorists on their own. "I played hooky a lot and studied the masters of theoretical physics with a holy zeal at home," he recalled. Among those were Gustav Kirchhoff on radiation, Hermann von Helmholtz on thermodynamics, Heinrich Hertz on electromagnetism, and Boltzmann on statistical mechanics.

He was also influenced by reading a lesser-known theorist, August Föppl, who in 1894 had written a popular text titled *Introduction to Maxwell's Theory of Electricity*. As science historian Gerald Holton has pointed out, Föppl's book is filled with concepts that would soon echo in Einstein's work. It has a section on "The Electrodynamics of Moving Conductors" that begins by calling into question the concept of "absolute motion." The only way to define motion, Föppl notes, is relative to another body. From there he goes on to consider a question concerning the induction of an electric current by a magnetic field: "if it is all the same whether a magnet moves in the vicinity of a resting electric circuit or whether it is the latter that moves while the magnet is at rest." Einstein would begin his 1905 special relativity paper by raising this same issue.[14]

Einstein also read, in his spare time, Henri Poincaré, the great French polymath who would come tantalizingly close to discovering the core concepts of special relativity. Near the end of Einstein's first year at the Polytechnic, in the spring of 1897, there was a mathematics conference in Zurich where the great Poincaré was due to speak. At the last minute he was unable to appear, but a paper of his was read there that contained what would become a famous proclamation. "Absolute space, absolute time, even Euclidean geometry, are not conditions to be imposed on mechanics," he wrote.[15]

The Human Side

One evening when Einstein was at home with his landlady, he heard someone playing a Mozart piano sonata. When he asked who it was, his landlady told him that it was an old woman who lived in the attic next door and taught piano. Grabbing his violin, he dashed out without putting on a collar or a tie. "You can't go like that, Herr Einstein," the landlady cried. But he ignored her and rushed into the neighboring house. The piano teacher looked up, shocked. "Go on playing," Einstein pleaded. A few moments later, the air was filled with the sounds of a violin accompanying the Mozart sonata. Later, the teacher asked who the intruding accompanist was. "Merely a harmless student," her neighbor reassured her.[16]

Music continued to beguile Einstein. It was not so much an escape as it was a connection: to the harmony underlying the universe, to the creative genius of the great composers, and to other people who felt comfortable bonding with more than just words. He was awed, both in music and in physics, by the beauty of harmonies.

Suzanne Markwalder was a young girl in Zurich whose mother hosted musical evenings featuring mostly Mozart. She played piano, while Einstein played violin. "He was very patient with my shortcomings," she recalled. "At the worst he used to say, 'There you are, stuck like the donkey on the mountain,' and he would point with his bow to the place where I had to come in."

What Einstein appreciated in Mozart and Bach was the clear architectural structure that made their music seem "deterministic" and, like his own favorite scientific theories, plucked from the universe rather than composed. "Beethoven created his music," Einstein once said, but "Mozart's music is so pure it seems to have been ever-present in the universe." He contrasted Beethoven with Bach: "I feel uncomfortable listening to Beethoven. I think he is too personal, almost naked. Give me Bach, rather, and then more Bach."

He also admired Schubert for his "superlative ability to express emotion." But in a questionnaire he once filled out, he was critical about other composers in ways that reflect some of his scientific sentiments: Handel had "a certain shallowness"; Mendelssohn displayed "considerable talent but an indefinable lack of depth that often leads to banality"; Wagner had a "lack of architectural structure I see as decadence"; and Strauss was "gifted but without inner truth." [17]

Einstein also took up sailing, a more solitary pursuit, in the glorious Alpine lakes around Zurich. "I still remember how when the breeze dropped and the sails drooped like withered leaves, he would take out his small notebook and he would start scribbling," recalled Suzanne Markwalder. "But as soon as there was a breath of wind he was immediately ready to start sailing again." [18]

The political sentiments he had felt as a boy—a contempt for arbitrary authority, an aversion to militarism and nationalism, a respect for individuality, a disdain for bourgeois consumption or ostentatious wealth, and a desire for social equality—had been encouraged by his landlord and surrogate father in Aarau, Jost Winteler. Now, in Zurich, he met a friend of Winteler's who became a similar political mentor: Gustav Maier, a Jewish banker who had helped arrange Einstein's first visit to the Polytechnic. With support from Winteler, Maier founded the Swiss branch of the Society for Ethical Culture, and Einstein was a frequent guest at their informal gatherings in Maier's home.

Einstein also came to know and like Friedrich Adler, the son of Austria's Social Democratic leader, who was studying in Zurich. Einstein later called him the "purest and most fervent idealist" he had ever met. Adler tried to get Einstein to join the Social Democrats. But it was not Einstein's style to spend time at meetings of organized institutions.[19]

His distracted demeanor, casual grooming, frayed clothing, and forgetfulness, which were later to make him appear to be the iconic absentminded professor, were already evident in his student days. He was known to leave behind clothes, and sometimes even his suitcase, when he traveled, and his inability to remember his keys became a running joke with his landlady. He once visited the home of family friends and, he recalled, "I left forgetting my suitcase. My host said to my parents, 'That man will never amount to anything because he can't remember anything.' "[20]

This carefree life as a student was clouded by the continued financial failings of his father, who, against Einstein's advice, kept trying to set up his own businesses rather than go to work for a salary at a stable company, as Uncle Jakob had finally done. "If I had my way, papa would have looked for salaried employment two years ago," he wrote his sister during a particularly gloomy moment in 1898 when his father's business seemed doomed to fail again.

The letter was unusually despairing, probably more than his parents' financial situation actually warranted:

> What depresses me most is the misfortune of my poor parents who have not had a happy moment for so many years. What further hurts me deeply is that as an adult man, I have to look on without being able to do anything. I am nothing but a burden to my family . . . It would be better off if I were not alive at all. Only the thought that I have always done what lay in my modest powers, and that I do not permit myself a single pleasure or distraction save for what my studies offer me, sustains me and sometimes protects me from despair.[21]

Perhaps this was all merely an attack of teenage angst. In any event, his father seemed to get through the crisis with his usual optimism. By the following February, he had won contracts for providing street lights to two small villages near Milan. "I am happy at the thought that the worst worries are over for our parents," Einstein wrote Maja. "If everyone lived such a way, namely like me, the writing of novels would never have been invented."[22]

Einstein's new bohemian life and old self-absorbed nature made it unlikely that he would continue his relationship with Marie Winteler, the sweet and somewhat flighty daughter of the family he had boarded with in Aarau. At first, he still sent her, via the mail, baskets of his laundry, which she would wash and then return. Sometimes there was not even a note attached, but she would cheerfully try to please him. In one letter she wrote of "crossing the woods in the pouring rain" to the post office to send back his clean clothes. "In vain did I strain my eyes for a little note, but the mere sight of your dear handwriting in the address was enough to make me happy."

When Einstein sent word that he planned to visit her, Marie was giddy. "I really thank you, Albert, for wanting to come to Aarau, and I don't have to tell you that I will be counting the minutes until that time," she wrote. "I could never describe, because there are no words for it, how blissful I feel ever since the dear soul of yours has come to live and weave in my soul. I love you for all eternity, sweetheart."

But he wanted to break off the relationship. In one of his first letters after arriving at the Zurich Polytechnic, he suggested that they refrain from writing each other. "My love, I do not quite understand a passage in your letter," she replied. "You write that you do not want to correspond with me any longer, but why not, sweetheart? . . . You must be quite annoyed with me if you can write so rudely." Then she tried to laugh off the problem: "But wait, you'll get some proper scolding when I get home."[23]

Einstein's next letter was even less friendly, and he complained about a teapot she had given him. "The matter of my sending you the stupid little teapot does not have to please you at all as long as you are going to brew some good tea in it," she replied. "Stop making that angry face which looked at me from all the sides and corners of the writing paper." There was a little boy in the school where she taught named Albert, she said, who looked like him. "I love him ever so much," she said. "Something comes over me when he looks at me and I always believe that you are looking at your little sweetheart."[24]

But then the letters from Einstein stopped, despite Marie's pleas. She even wrote his mother for advice. "The rascal has become frightfully lazy," Pauline Einstein replied. "I have been waiting in vain for news for these last three days; I will have to give him a thorough talking-to once he's here."[25]

Finally, Einstein declared the relationship over in a letter to Marie's mother, saying that he would not come to Aarau during his academic break that spring. "It would be more than unworthy of me to buy a few days of bliss at the cost of new pain, of which I have already caused too much to the dear child through my fault," he wrote.

He went on to give a remarkably introspective—and memorable—assessment of how he had begun to avoid the pain of emotional commitments and the distractions of what he called the "merely personal" by retreating into science:

> It fills me with a peculiar kind of satisfaction that now I myself have to taste some of the pain that I brought upon the dear girl through my thoughtlessness and ignorance of her delicate nature. Strenuous intellectual work and looking at God's nature are the reconciling, fortifying yet relentlessly strict angels that shall lead me through all of life's troubles. If only I were able to give some of this to the good child. And yet, what a peculiar way this is to weather the storms of life—in many a lucid moment I appear to myself as an ostrich who buries his head in the desert sand so as not to perceive the danger.[26]

Einstein's coolness toward Marie Winteler can seem, from our vantage, cruel. Yet relationships, especially those of teenagers, are hard to judge from afar. They were very different from each other, particularly intellectually. Marie's letters, especially when she was feeling insecure, often descended into babble. "I'm writing a lot of rubbish, isn't that so, and in the end you'll not even read it to the finish (but I don't believe that)," she wrote in one. In another, she said, "I do not think about myself, sweetheart, that's quite true, but the only reason for this is that I do not think at all, except when it comes to some tremendously stupid calculation that requires, for a change, that I know more than my pupils."[27]

Whoever was to blame, if either, it was not surprising that they ended up on different paths. After her relationship with Einstein ended, Marie lapsed into a nervous depression, often missing days of teaching, and a few years later married the manager of a watch factory. Einstein, on the other hand, rebounded from the relationship by falling into the arms of someone who was just about as different from Marie as could be imagined.

Mileva Marić

Mileva Marić was the first and favorite child of an ambitious Serbian peasant who had joined the army, married into modest wealth, and then dedicated himself to making sure that his brilliant daughter was able to prevail in the male world of math and physics. She spent most of her childhood in Novi Sad, a Serbian city then held by Hungary,[28] and attended a variety of ever more demanding schools, at each of which she was at the top of her class, culminating when her father convinced the all-male Classical Gymnasium in Zagreb to let her enroll. After graduating there with the top grades in physics and math, she made her way to Zurich, where she became, just before she turned 21, the only woman in Einstein's section of the Polytechnic.

More than three years older than Einstein, afflicted with a congenital hip dislocation that caused her to limp, and prone to bouts of tuberculosis and despondency, Mileva Marić was known for neither her looks nor her personality. "Very smart and serious, small, delicate, brunette, ugly," is how one of her female friends in Zurich described her.

But she had qualities that Einstein, at least during his romantic scholar years, found attractive: a passion for math and science, a brooding depth, and a beguiling soul. Her deep-set eyes had a haunting intensity, her face an enticing touch of melancholy.[29] She would become, over time, Einstein's muse, partner, lover, wife, bête noire, and antagonist, and she would create an emotional field more powerful than that of anyone else in his life. It would alternately attract and repulse him with a force so strong that a mere scientist like himself would never be able to fathom it.

They met when they both entered the Polytechnic in October 1896, but their relationship took a while to develop. There is no sign, from their letters or recollections, that they were anything more than classmates that first academic year. They did, however, decide to go hiking together in the summer of 1897. That fall, "frightened by the new feelings she was experiencing" because of Einstein, Marić decided to leave the Polytechnic temporarily and instead audit classes at Heidelberg University.[30]

Her first surviving letter to Einstein, written a few weeks after she moved to Heidelberg, shows glimmers of a romantic attraction but also highlights her self-confident nonchalance. She addresses Einstein with the formal *Sie* in German, rather than the more intimate *du*. Unlike Marie Winteler, she teasingly makes the point that she has not been obsessing about him, even though he had written an unusually long letter to her. "It's now been quite a while since I received your letter," she said, "and I would have replied immediately and thanked you for the sacrifice of writing four long pages, would have also told of the joy you provided me through our trip together, but you said I should write to you someday when I happened to be bored. And I am very obedient, and I waited and waited for boredom to set in; but so far my waiting has been in vain."

Distinguishing Marić even more from Marie Winteler was the intellectual intensity of her letters. In this first one, she enthused over the lectures she had been attending of Philipp Lenard, then an assistant professor at Heidelberg, on kinetic theory, which explains the properties of gases as being due to the actions of millions of individual molecules. "Oh, it was really neat at the lecture of Professor Lenard yesterday," she wrote. "He is talking now about the kinetic theory of heat and gases. So, it turns out that the molecules of oxygen move with a velocity of over 400 meters per second, then the good professor calculated and calculated . . . and it finally turned out even though molecules do move with this velocity, they travel a distance of only 1/100 of a hairbreadth."

Kinetic theory had not yet been fully accepted by the scientific establishment (nor, for that matter, had even the existence of atoms and molecules), and Marić's letter indicated that she did not have a deep understanding of the subject. In addition, there was a sad irony: Lenard would be one of Einstein's early inspirations but later one of his most hateful anti-Semitic tormentors.

Marić also commented on ideas Einstein had shared in his earlier letter about the difficulty mortals have in comprehending the infinite. "I do not believe that the structure of the human brain is to be blamed

for the fact that man cannot grasp infinity," she wrote. "Man is very capable of imagining infinite happiness, and he should be able to grasp
the infinity of space—I think that should be much easier." There is a
slight echo of Einstein's escape from the "merely personal" into the
safety of scientific thinking: finding it easier to imagine infinite space
than infinite happiness.

Yet Marić was also, it is clear from her letter, thinking of Einstein in
a more personal way. She had even talked to her adoring and protective
father about him. "Papa gave me some tobacco to take with me and I
was supposed to hand it to you personally," she said. "He wanted so
much to whet your appetite for our little land of outlaws. I told him all
about you—you must absolutely come back with me someday. The two
of you would really have a lot to talk about!" The tobacco, unlike Marie
Winteler's teapot, was a present Einstein would likely have wanted, but
Marić teased that she wasn't sending it. "You would have to pay duty on
it, and then you would curse me."[31]

That conflicting admixture of playfulness and seriousness, of insouciance and intensity, of intimacy and detachment—so peculiar yet also
so evident in Einstein as well—must have appealed to him. He urged
her to return to Zurich. By February 1898, she had made up her mind
to do so, and he was thrilled. "I'm sure you won't regret your decision,"
he wrote. "You should come back as soon as possible."

He gave her a thumbnail of how each of the professors was performing (admitting that he found the one teaching geometry to be "a
little impenetrable"), and he promised to help her catch up with the aid
of the lecture notes he and Marcel Grossmann had kept. The one
problem was that she would probably not be able to get her "old pleasant room" at the nearby pension back. "Serves you right, you little runaway!"[32]

By April she was back, in a boarding house a few blocks from his,
and now they were a couple. They shared books, intellectual enthusiasms, intimacies, and access to each other's apartments. One day, when
he again forgot his key and found himself locked out of his own place,
he went to hers and borrowed her copy of a physics text. "Don't be
angry with me," he said in the little note he left her. Later that year, a

similar note left for her added, "If you don't mind, I'd like to come over this evening to read with you."[33]

Friends were surprised that a sensuous and handsome man such as Einstein, who could have almost any woman fall for him, would find himself with a short and plain Serbian who had a limp and exuded an air of melancholy. "I would never be brave enough to marry a woman unless she were absolutely healthy," a fellow student said to him. Einstein replied, "But she has such a lovely voice."[34]

Einstein's mother, who had adored Marie Winteler, was similarly dubious about the dark intellectual who had replaced her. "Your photograph had quite an effect on my old lady," Einstein wrote from Milan, where he was visiting his parents during spring break of 1899. "While she studied it carefully, I said with the deepest sympathy: 'Yes, yes, she certainly is a clever one.' I've already had to endure much teasing about this."[35]

It is easy to see why Einstein felt such an affinity for Marić. They were kindred spirits who perceived themselves as aloof scholars and outsiders. Slightly rebellious toward bourgeois expectations, they were both intellectuals who sought as a lover someone who would also be a partner, colleague, and collaborator. "We understand each other's dark souls so well, and also drinking coffee and eating sausages, etcetera," Einstein wrote her.

He had a way of making the *etcetera* sound roguish. He closed another letter: "Best wishes etc., especially the latter." After being apart for a few weeks, he listed the things he liked to do with her: "Soon I'll be with my sweetheart again and can kiss her, hug her, make coffee with her, scold her, study with her, laugh with her, walk with her, chat with her, and ad infinitum!" They took pride in sharing a quirkiness. "I'm the same old rogue as I've always been," he wrote, "full of whims and mischief, and as moody as ever!"[36]

Above all, Einstein loved Marić for her mind. "How proud I will be to have a little Ph.D. for a sweetheart," he wrote to her at one point. Science and romance seemed to be interwoven. While on vacation with his family in 1899, Einstein lamented in a letter to Marić, "When I read Helmholtz for the first time I could not—and still cannot—believe that I was doing so without you sitting next to me. I enjoy working together and I find it soothing and also less boring."

Indeed, most of their letters mixed romantic effusions with scientific enthusiasms, often with an emphasis on the latter. In one letter, for example, he foreshadowed not only the title but also some of the concepts of his great paper on special relativity. "I am more and more convinced that the electrodynamics of moving bodies as it is presented today does not correspond to reality and that it will be possible to present it in a simpler way," he wrote. "The introduction of the term 'ether' into theories of electricity has led to the conception of a medium whose motion can be described without, I believe, being able to ascribe physical meaning to it."[37]

Even though this mix of intellectual and emotional companionship appealed to him, every now and then he recalled the enticement of the simpler desire represented by Marie Winteler. And with the tactlessness that masqueraded for him as honesty (or perhaps because of his puckish desire to torment), he let Marić know it. After his 1899 summer vacation, he decided to take his sister to enroll in school in Aarau, where Marie lived. He wrote Marić to assure her that he would not spend much time with his former girlfriend, but the pledge was written in a way that was, perhaps intentionally, more unsettling than reassuring. "I won't be going to Aarau as often now that the daughter I was so madly in love with four years ago is coming back home," he said. "For the most part I feel quite secure in my high fortress of calm. But I know that if I saw her a few more times, I would certainly go mad. Of that I am certain, and I fear it like fire."

But the letter goes on, happily for Marić, with a description of what they would do when they met back in Zurich, a passage in which Einstein showed once again why their relationship was so special. "The first thing we'll do is climb the Ütliberg," he said, referring to a high point just out of town. There they would be able to "take pleasure in unpacking our memories" of the things they had done together on other hiking trips. "I can already imagine the fun we will have," he wrote. Finally, with a flourish only they could have fully appreciated, he concluded, "And then we'll start in on Helmholtz's electromagnetic theory of light."[38]

In the ensuing months, their letters became even more intimate and passionate. He began calling her Doxerl (Dollie), as well as "my wild little rascal" and "my street urchin"; she called him Johannzel (Johnnie) and "my wicked little sweetheart." By the start of 1900, they were using the familiar *du* with one another, a process that began with a little note from her that reads, in full:

> My little Johnnie,
>   Because I like you so much, and because you're so far away that I can't give you a little kiss, I'm writing this letter to ask if you like me as much as I do you? Answer me immediately.
>
> A thousand kisses from your
Dollie[39]

Graduation, August 1900

Academically, things were also going well for Einstein. In his intermediate exams in October 1898, he had finished first in his class, with an average of 5.7 out of a possible 6. Finishing second, with a 5.6, was his friend and math note-taker Marcel Grossmann.[40]

To graduate, Einstein had to do a research thesis. He initially proposed to Professor Weber that he do an experiment to measure how fast the earth was moving through the ether, the supposed substance that allowed light waves to propagate through space. The accepted wisdom, which he would famously destroy with his special theory of relativity, was that if the earth were moving through this ether toward or away from the source of a light beam, we'd be able to detect a difference in the observed speed of the light.

During his visit to Aarau at the end of his summer vacation of 1899, he worked on this issue with the rector of his old school there. "I had a good idea for investigating the way in which a body's relative motion with respect to the ether affects the velocity of the propagation of light," he wrote Marić. His idea involved building an apparatus that would use angled mirrors "so that light from a single source would be reflected in two different directions," sending one part of the beam in the direction of the earth's movement and the other part of the beam perpendicular to it. In a lecture on how he discovered relativity, Einstein recalled that his idea was to split a light beam, reflect it in different directions, and see if there was "a difference in energy depending on whether or not the direction was along the earth's motion through the ether." This could be done, he posited, by "using two thermoelectric piles to examine the difference of the heat generated in them."[41]

Weber rejected the proposal. What Einstein did not fully realize was that similar experiments had already been done by many others, including the Americans Albert Michelson and Edward Morley, and none had been able to detect any evidence of the perplexing ether—or that the speed of light varied depending on the motion of the observer or the light source. After discussing the topic with Weber, Einstein read a paper delivered the previous year by Wilhelm Wien, which briefly described thirteen experiments that had been conducted to detect the ether, including the Michelson-Morley one.

Einstein sent Professor Wien his own speculative paper on that topic and asked him to write him back. "He'll write me via the Polytechnic," Einstein predicted to Marić. "If you see a letter there for me, you may go ahead and open it." There is no evidence that Wien ever wrote back.[42]

Einstein's next research proposal involved exploring the link between the ability of different materials to conduct heat and to conduct electricity, something that was suggested by the electron theory. Weber apparently did not like that idea either, so Einstein was reduced, along with Marić, to doing a study purely on heat conduction, which was one of Weber's specialties.

Einstein later dismissed their graduation research papers as being of "no interest to me." Weber gave Einstein and Marić the two lowest essay grades in the class, a 4.5 and a 4.0, respectively; Grossmann, by comparison, got a 5.5. Adding annoyance to that injury, Weber said that Einstein had not written his on the proper regulation paper, and he forced him to copy the entire essay over again.[43]

Despite the low mark on his essay, Einstein was able to eke by with a 4.9 average in his final set of grades, placing him fourth in his class of five. Although history refutes the delicious myth that he flunked math in high school, at least it does offer as a consolation the amusement that he graduated college near the bottom of his class.

At least he graduated. His 4.9 average was just enough to let him get his diploma, which he did officially in July 1900. Mileva Marić, however, managed only a 4.0, by far the lowest in the class, and was not allowed to graduate. She determined that she would try again the following year.[44]

Not surprisingly, Einstein's years at the Polytechnic were marked by his pride at casting himself as a nonconformist. "His spirit of independence asserted itself one day in class when the professor mentioned a mild disciplinary measure just taken by the school's authorities," a classmate recalled. Einstein protested. The fundamental requirement of education, he felt, was the "need for intellectual freedom."[45]

Throughout his life, Einstein would speak lovingly of the Zurich Polytechnic, but he also would note that he did not like the discipline that was inherent in the system of examinations. "The hitch in this was, of course, that one had to cram all this stuff into one's mind for the examinations, whether one liked it or not," he said. "This coercion had such a deterring effect that, after I had passed the final examination, I found the consideration of any scientific problems distasteful to me for an entire year."[46]

In reality, that was neither possible nor true. He was cured within weeks, and he ended up taking with him some science books, including texts by Gustav Kirchhoff and Ludwig Boltzmann, when he joined his mother and sister later that July for their summer holiday in the Swiss Alps. "I've been studying a great deal," he wrote Marić, "mainly Kirchhoff's notorious investigations of the motion of the rigid body." He admitted that his resentment over the exams had already worn off. "My nerves have calmed down enough so that I'm able to work happily again," he said. "How are yours?"[47]

CHAPTER FOUR

THE LOVERS

1900–1904

With Mileva and Hans Albert Einstein, 1904

Summer Vacation, 1900

Newly graduated, carrying his Kirchhoff and other physics books, Einstein arrived at the end of July 1900 for his family's summer vacation in Melchtal, a village nestled in the Swiss Alps between Lake Lucerne and the border with northern Italy. In tow was his "dreadful aunt," Julia Koch. They were met at the train station by his mother and sister, who smothered him with kisses, and then all piled into a carriage for the ride up the mountain.

As they neared the hotel, Einstein and his sister got off to walk. Maja confided that she had not dared to discuss with their mother his relationship with Mileva Marić, known in the family as "the Dollie affair" after his nickname for her, and she asked him to "go easy on Mama." It was not in Einstein's nature, however, "to keep my big mouth shut," as he later put it in his letter to Marić about the scene, nor was it in his nature to protect Marić's feelings by sparing her all the dramatic details about what ensued.[1]

He went to his mother's room and, after hearing about his exams, she asked him, "So, what will become of your Dollie now?"

"My wife," Einstein answered, trying to affect the same nonchalance that his mother had used in her question.

His mother, Einstein recalled, "threw herself on the bed, buried her head in the pillow, and wept like a child." She was finally able to regain her composure and proceeded to go on the attack. "You are ruining your future and destroying your opportunities," she said. "No decent family will have her. If she gets pregnant you'll really be in a mess."

At that point, it was Einstein's turn to lose his composure. "I vehemently denied we had been living in sin," he reported to Marić, "and scolded her roundly."

Just as he was about to storm out, a friend of his mother's came in, "a small, vivacious lady, an old hen of the most pleasant variety." They promptly segued into the requisite small talk: about the weather, the new guests at the spa, the ill-mannered children. Then they went off to eat and play music.

Such periods of storm and calm alternated throughout the vacation. Every now and then, just when Einstein thought that the crisis had receded, his mother would revisit the topic. "Like you, she's a book, but you ought to have a wife," she scolded at one point. Another time she brought up the fact that Marić was 24 and he was then only 21. "By the time you're 30, she'll be an old witch."

Einstein's father, still working back in Milan, weighed in with "a moralistic letter." The thrust of his parents' views—at least when applied to the situation of Mileva Marić rather than Marie Winteler—was that a wife was "a luxury" affordable only when a man was making a comfortable living. "I have a low opinion of that view of a relationship between a man and wife," he told Marić, "because it makes the wife and the prostitute distinguishable only insofar as the former is able to secure a lifelong contract."[2]

Over the ensuing months, there would be times when it seemed as if his parents had decided to accept their relationship. "Mama is slowly resigning herself," Einstein wrote Marić in August. Likewise in September: "They seem to have reconciled themselves to the inevitable. I think they will both come to like you very much once they get to know you." And once again in October: "My parents have retreated, grudgingly and with hesitation, from the battle of Dollie—now that they have seen that they'll lose it."[3]

But repeatedly, after each period of acceptance, their resistance would flare up anew, randomly leaping into a higher state of frenzy. "Mama often cries bitterly and I don't have a single moment of peace," he wrote at the end of August. "My parents weep for me almost as if I had died. Again and again they complain that I have brought misfortune upon myself by my devotion to you. They think you are not healthy."[4]

His parents' dismay had little to do with the fact that Marić was not Jewish, for neither was Marie Winteler, nor that she was Serbian, although that certainly didn't help her cause. Primarily, it seems, they considered her an unsuitable wife for many of the reasons that some of Einstein's friends did: she was older, somewhat sickly, had a limp, was plain looking, and was an intense but not a star intellectual.

All of this emotional pressure stoked Einstein's rebellious instincts and his passion for his "wild street urchin," as he called her. "Only now do I see how madly in love with you I am!" The relationship, as expressed in their letters, remained equal parts intellectual and emotional, but the emotional part was now filled with a fire unexpected from a self-proclaimed loner. "I just realized that I haven't been able to kiss you for an entire month, and I long for you so terribly much," he wrote at one point.

During a quick trip to Zurich in August to check on his job prospects, he found himself walking around in a daze. "Without you, I lack self-confidence, pleasure in my work, pleasure in life—in short, without you my life is not life." He even tried his hand at a poem for her, which began: "Oh my! That Johnnie boy! / So crazy with desire / While thinking of his Dollie / His pillow catches fire."[5]

Their passion, however, was an elevated one, at least in their minds. With the lonely elitism of young German coffeehouse denizens who have read the philosophy of Schopenhauer once too often, they unabashedly articulated the mystical distinction between their own rarefied spirits and the baser instincts and urges of the masses. "In the case of my parents, as with most people, the senses exercise a direct control over the emotions," he wrote her amid the family wars of August. "With us, thanks to the fortunate circumstances in which we live, the enjoyment of life is vastly broadened."

To his credit, Einstein reminded Marić (and himself) that "we mustn't forget that many existences like my parents' make our existence possible." The simple and honest instincts of people like his parents had ensured the progress of civilization. "Thus I am trying to protect my parents without compromising anything that is important to me—and that means you, sweetheart!"

In his attempt to please his mother, Einstein became a charming son at their grand hotel in Melchtal. He found the endless meals excessive and the "overdressed" patrons to be "indolent and pampered," but he dutifully played his violin for his mother's friends, made polite conversation, and feigned a cheerful mood. It worked. "My popularity among the guests here and my music successes act as a balm on my mother's heart."[6]

As for his father, Einstein decided that the best way to assuage him, as well as to draw off some of the emotional charge generated by his relationship with Marić, was to visit him back in Milan, tour some of his new power plants, and learn about the family firm "so I can take Papa's place in an emergency." Hermann Einstein seemed so pleased that he promised to take his son to Venice after the inspection tour. "I'm leaving for Italy on Saturday to partake of the 'holy sacraments' administered by my father, but the valiant Swabian* is not afraid."

Einstein's visit with his father went well, for the most part. A distant yet dutiful son, he had fretted mightily about each family financial crisis, perhaps even more than his father did. But business was good for the moment, and that lifted Hermann Einstein's spirits. "My father is a completely different man now that he has no more financial worries," Einstein wrote Marić. Only once did the "Dollie affair" intrude

---

* The phrase "valiant Swabian," used often by Einstein to refer to himself, comes from the poem "Swabian Tale" by Ludwig Uhland.

enough to make him consider cutting short his visit, but this threat so alarmed his father that Einstein stuck to the original plans. He seemed flattered that his father appreciated both his company and his willingness to pay attention to the family business.[7]

Even though Einstein occasionally denigrated the idea of being an engineer, it was possible that he could have followed that course at the end of the summer of 1900—especially if, on their trip to Venice, his father had asked him to, or if fate intervened so that he was needed to take his father's place. He was, after all, a low-ranked graduate of a teaching college without a teaching job, without any research accomplishments, and certainly without academic patrons.

Had he made such a choice in 1900, Einstein would have likely become a good enough engineer, but probably not a great one. Over the ensuing years he would dabble with inventions as a hobby and come up with some good concepts for devices ranging from noiseless refrigerators to a machine that measured very low voltage electricity. But none resulted in a significant engineering breakthrough or marketplace success. Though he would have been a more brilliant engineer than his father or uncle, it is not clear that he would have been any more financially successful.

Among the many surprising things about the life of Albert Einstein was the trouble he had getting an academic job. Indeed, it would be an astonishing nine years after his graduation from the Zurich Polytechnic in 1900—and four years after the miracle year in which he not only upended physics but also finally got a doctoral dissertation accepted—before he would be offered a job as a junior professor.

The delay was not due to a lack of desire on his part. In the middle of August 1900, between his family vacation in Melchtal and his visit to his father in Milan, Einstein stopped back in Zurich to see about getting a post as an assistant to a professor at the Polytechnic. It was typical that each graduate would find, if he wanted, some such role, and Einstein was confident it would happen. In the meantime, he rejected a friend's offer to help him get a job at an insurance company, dismissing it as "an eight hour day of mindless drudgery." As he told Marić, "One must avoid stultifying affairs."[8]

The problem was that the two physics professors at the Polytechnic were acutely aware of his impudence but not of his genius. Getting a job with Professor Pernet, who had reprimanded him, was not even a consideration. As for Professor Weber, he had developed such an allergy to Einstein that, when no other graduates of the physics and math department were available to become his assistant, he instead hired two students from the engineering division.

That left math professor Adolf Hurwitz. When one of Hurwitz's assistants got a job teaching at a high school, Einstein exulted to Marić: "This means I will become Hurwitz's servant, God willing." Unfortunately, he had skipped most of Hurwitz's classes, a slight that apparently had not been forgotten.[9]

By late September, Einstein was still staying with his parents in Milan and had not received an offer. "I plan on going to Zurich on October 1 to talk with Hurwitz personally about the position," he said. "It's certainly better than writing."

While there, he also planned to look for possible tutoring jobs that could tide them over while Marić prepared to retake her final exams. "No matter what happens, we'll have the most wonderful life in the world. Pleasant work and being together—and what's more, we now answer to no one, can stand on our own two feet, and enjoy our youth to the utmost. Who could have it any better? When we have scraped together enough money, we can buy bicycles and take a bike tour every couple of weeks."[10]

Einstein ended up deciding to write Hurwitz instead of visiting him, which was probably a mistake. His two letters do not stand as models for future generations seeking to learn how to write a job application. He readily conceded that he did not show up at Hurwitz's calculus classes and was more interested in physics than math. "Since lack of time prevented me from taking part in the mathematics seminar," he rather lamely said, "there is nothing in my favor except the fact that I attended most of the lectures offered." Rather presumptuously, he said he was eager for an answer because "the granting of citizenship in Zurich, for which I have applied, has been made conditional upon my proving that I have a permanent job."[11]

Einstein's impatience was matched by his confidence. "Hurwitz still hasn't written me more," he said only three days after sending his letter, "but I have hardly any doubt that I will get the position." He did not. Indeed, he managed to become the only person graduating in his section of the Polytechnic who was not offered a job. "I was suddenly abandoned by everyone," he later recalled.[12]

By the end of October 1900 he and Marić were both back in Zurich, where he spent most of his days hanging out at her apartment, reading and writing. On his citizenship application that month, he wrote "none" on the question asking his religion, and for his occupation he wrote, "I am giving private lessons in mathematics until I get a permanent position."

Throughout that fall, he was able to find only eight sporadic tutoring jobs, and his relatives had ended their financial support. But Einstein put up an optimistic front. "We support ourselves by private lessons, if we can ever pick up some, which is still very doubtful," he wrote a friend of Marić's. "Isn't this a journeyman's or even a gypsy's life? But I believe that we will remain cheerful in it as ever."[13] What kept him happy, in addition to Marić's presence, were the theoretical papers he was writing on his own.

Einstein's First Published Paper

The first of these papers was on a topic familiar to most school kids: the capillary effect that, among other things, causes water to cling to the side of a straw and curve upward. Although he later called this essay "worthless," it is interesting from a biographical perspective. Not only is it Einstein's first published paper, but it shows him heartily embracing an important premise—one not yet fully accepted—that would be at the core of much of his work over the next five years: that molecules (and their constituent atoms) actually exist, and that many natural phenomena can be explained by analyzing how these particles interact with one another.

During his vacation in the summer of 1900, Einstein had been reading the work of Ludwig Boltzmann, who had developed a theory of gases based on the behavior of countless molecules bouncing around. "The Boltzmann is absolutely magnificent," he enthused to Marić in September. "I am firmly convinced of the correctness of the principles of his theory, i.e., I am convinced that in the case of gases we are really dealing with discrete particles of definite finite size which move according to certain conditions."[14]

To understand capillarity, however, required looking at the forces acting between molecules in a liquid, not a gas. Such molecules attract one another, which accounts for the surface tension of a liquid, or the fact that drops hold together, as well as for the capillary effect. Einstein's idea was that these forces might be analogous to Newton's gravitational forces, in which two objects are attracted to each other in proportion to their mass and in inverse proportion to their distance from one another.

Einstein looked at whether the capillary effect showed such a relationship to the atomic weight of various liquid substances. He was encouraged, so he decided to see if he could find some experimental data to test the theory further. "The results on capillarity I recently obtained in Zurich seem to be entirely new despite their simplicity," he wrote Marić. "When we're back in Zurich we'll try to get some empirical data on this subject . . . If this yields a law of nature, we'll send the results to the *Annalen*."[15]

He did end up sending the paper in December 1900 to the *Annalen der Physik*, Europe's leading physics journal, which published it the following March. Written without the elegance or verve of his later papers, it conveyed what is at best a tenuous conclusion. "I started from the simple idea of attractive forces among the molecules, and I tested the consequences experimentally," he wrote. "I took gravitational forces as an analogy." At the end of the paper, he declares limply, "The question of whether and how our forces are related to gravitational forces must therefore be left completely open for the time being."[16]

The paper elicited no comments and contributed nothing to the history of physics. Its basic conjecture was wrong, as the distance dependence is not the same for differing pairs of molecules.[17] But it did get him published for the first time. That meant that he now had a printed article to attach to the job-seeking letters with which he was beginning to spam professors all over Europe.

In his letter to Marić, Einstein had used the term "we" when discussing plans to publish the paper. In two letters written the month after it appeared, Einstein referred to "our theory of molecular forces" and "our investigation." Thus was launched a historical debate over how much credit Marić deserves for helping Einstein devise his theories.

In this case, she mainly seemed to be involved in looking up some data for him to use. His letters conveyed his latest thoughts on molecular forces, but hers contained no substantive science. And in a letter to her best friend, Marić sounded as if she had settled into the role of supportive lover rather than scientific partner. "Albert has written a paper in physics that will probably be published very soon in the *Annalen der Physik*," she wrote. "You can imagine how very proud I am of my darling. This is not just an everyday paper, but a very significant one. It deals with the theory of liquids."[18]

Jobless Anguish

It had been almost four years since Einstein had renounced his German citizenship, and ever since then he had been stateless. Each month, he put aside some money toward the fee he would need to pay to become a Swiss citizen, a status he deeply desired. One reason was that he admired the Swiss system, its democracy, and its gentle respect for individuals and their privacy. "I like the Swiss because, by and large, they are more humane than the other people among whom I have lived," he later said.[19] There were also practical reasons; in order to work as a civil servant or a teacher in a state school, he would have to be a Swiss citizen.

The Zurich authorities examined him rather thoroughly, and they even sent to Milan for a report on his parents. By February 1901, they were satisfied, and he was made a citizen. He would retain that designation his entire life, even as he accepted citizenships in Germany (again), Austria, and the United States. Indeed, he was so eager to be a Swiss citizen that he put aside his antimilitary sentiments and presented himself, as required, for military service. He was rejected for having sweaty feet ("hyperidrosis ped"), flat feet ("pes planus"), and varicose veins ("varicosis"). The Swiss Army was, apparently, quite discriminating, and so his military service book was stamped "unfit."[20]

A few weeks after he got his citizenship, however, his parents insisted that he come back to Milan and live with them. They had decreed, at the end of 1900, that he could not stay in Zurich past Easter unless he got a job there. When Easter came, he was still unemployed.

Marić, not unreasonably, assumed that his summons to Milan was due to his parents' antipathy toward her. "What utterly depressed me was the fact that our separation had to come about in such an unnatural way, on account of slanders and intrigues," she wrote her friend. With an absentmindedness he was later to make iconic, Einstein left behind in Zurich his nightshirt, toothbrush, comb, hairbrush (back then he used one), and other toiletries. "Send everything along to my sister," he instructed Marić, "so she can bring them home with her." Four days later, he added, "Hold on to my umbrella for the time being. We'll figure out something to do with it later."[21]

Both in Zurich and then in Milan, Einstein churned out job-seeking letters, ever more pleading, to professors around Europe. They were accompanied by his paper on the capillary effect, which proved not particularly impressive; he rarely even received the courtesy of a response. "I will soon have graced every physicist from the North Sea to the southern tip of Italy with my offer," he wrote Marić.[22]

By April 1901, Einstein was reduced to buying a pile of postcards with postage-paid reply attachments in the forlorn hope that he would, at least, get an answer. In the two cases where these postcard pleas have survived, they have become, rather amusingly, prized collectors' items. One of them, to a Dutch professor, is now on display in the Leiden Museum for the History of Science. In both cases, the return-reply attachment was not used; Einstein did not even get the courtesy of a rejection. "I leave no stone unturned and do not give up my sense of humor," he wrote his friend Marcel Grossmann. "God created the donkey and gave him a thick skin."[23]

Among the great scientists Einstein wrote was Wilhelm Ostwald, professor of chemistry in Leipzig, whose contributions to the theory of dilution were to earn him a Nobel Prize. "Your work on general chemistry inspired me to write the enclosed article," Einstein said. Then flattery turned to plaintiveness as he asked "whether you might have use for a mathematical physicist." Einstein concluded by pleading: "I am without money, and only a position of this kind would enable me to continue my studies." He got no answer. Einstein wrote again two weeks later using the pretext "I am not sure whether I included my address" in the earlier letter. "Your judgment of my paper matters very much to me." There was still no answer.[24]

Einstein's father, with whom he was living in Milan, quietly shared his son's anguish and tried, in a painfully sweet manner, to help. When no answer came after the second letter to Ostwald, Hermann Einstein took it upon himself, without his son's knowledge, to make an unusual and awkward effort, suffused with heart-wrenching emotion, to prevail upon Ostwald himself:

> Please forgive a father who is so bold as to turn to you, esteemed Herr Professor, in the interest of his son. Albert is 22 years old, he studied at the Zurich Polytechnic for four years, and he passed his exam with flying colors last summer. Since then he has been trying unsuccessfully to get a position as a teaching assistant, which would enable him to continue his education in physics. All those in a position to judge praise his talents; I can assure you that he is extraordinarily studious and diligent and clings with great love to his science. He therefore feels profoundly unhappy about his current lack of a job, and he becomes more and more convinced that he has gone off the tracks with his career. In addition, he is oppressed by the thought that he is a burden on us, people of modest means. Since it is you whom my son seems to admire and esteem more than any other scholar in physics, it is you to whom I have taken the liberty of turning with the humble request to read his paper and to write to him, if possible, a few words of encouragement, so that he might recover his joy in living and working. If, in addition, you could secure him an assistant's position, my gratitude would know no bounds. I beg you to forgive me for my impudence in writing you, and my son does not know anything about my unusual step.[25]

Ostwald still did not answer. However, in one of history's nice ironies, he would become, nine years later, the first person to nominate Einstein for the Nobel Prize.

Einstein was convinced that his nemesis at the Zurich Polytechnic, physics professor Heinrich Weber, was behind the difficulties. Having hired two engineers rather than Einstein as his own assistant, he was apparently now giving him unfavorable references. After applying for a job with Göttingen professor Eduard Riecke, Einstein despaired to Marić: "I have more or less given up the position as lost. I cannot believe that Weber would let such a good opportunity pass without doing some mischief." Marić advised him to write Weber, confronting him directly, and Einstein reported back that he had. "He should at least know that he cannot do these things behind my back. I wrote to him that I know that my appointment now depends on his report alone."

It didn't work. Einstein again got turned down. "Riecke's rejection hasn't surprised me," he wrote Marić. "I'm completely convinced that Weber is to blame." He became so discouraged that, at least for the moment, he felt it futile to continue his search. "Under these circumstances it no longer makes sense to write further to professors, since, should things get far enough along, it is certain they would all enquire with Weber, and he would again give a poor reference." To Grossmann he lamented, "I could have found a job long ago had it not been for Weber's underhandedness."[26]

To what extent did anti-Semitism play a role? Einstein came to believe that it was a factor, which led him to seek work in Italy, where he felt it was not so pronounced. "One of the main obstacles in getting a position is absent here, namely anti-Semitism, which in German-speaking countries is as unpleasant as it is a hindrance," he wrote Marić. She, in turn, lamented to her friend about her lover's difficulties. "You know my sweetheart has a sharp tongue and moreover he is a Jew."[27]

In his effort to find work in Italy, Einstein enlisted one of the friends he had made while studying in Zurich, an engineer named Michele Angelo Besso. Like Einstein, Besso was from a middle-class Jewish family that had wandered around Europe and eventually settled in Italy. He was six years older than Einstein, and by the time they met he had already graduated from the Polytechnic and was working for an engineering firm. He and Einstein forged a close friendship that would last for the rest of their lives (they died within weeks of each other in 1955).

Over the years, Besso and Einstein would share both the most intimate personal confidences and the loftiest scientific notions. As Einstein wrote in one of the 229 extant letters they exchanged, "Nobody else is so close to me, nobody knows me so well, nobody is so kindly disposed to me as you are."[28]

Besso had a delightful intellect, but he lacked focus, drive, and diligence. Like Einstein, he had once been asked to leave high school because of his insubordinate attitude (he sent a petition complaining about a math teacher). Einstein called Besso "an awful weakling . . . who cannot rouse himself to any action in life or scientific creation, but who has an extraordinarily fine mind whose working, though disorderly, I watch with great delight."

Einstein had introduced Besso to Anna Winteler of Aarau, Marie's sister, whom he ended up marrying. By 1901 he had moved to Trieste with her. When Einstein caught up with him, he found Besso as smart, as funny, and as maddeningly unfocused as ever. He had recently been asked by his boss to inspect a power station, and he decided to leave the night before to make sure that he arrived on time. But he missed his train, then failed to get there the next day, and finally arrived on the third day—"but to his horror realizes that he has forgotten what he's supposed to do." So he sent a postcard back to the office asking them to resend his instructions. It was the boss's assessment that Besso was "completely useless and almost unbalanced."

Einstein's assessment of Besso was more loving. "Michele is an awful schlemiel," he reported to Marić, using the Yiddish word for a hapless bumbler. One evening, Besso and Einstein spent almost four hours talking about science, including the properties of the mysterious ether and "the definition of absolute rest." These ideas would burst into bloom four years later, in the relativity theory that he would devise with Besso as his sounding board. "He's interested in our research," Einstein wrote Marić, "though he often misses the big picture by worrying about petty considerations."

Besso had some connections that could, Einstein hoped, be useful. His uncle was a mathematics professor at the polytechnic in Milan, and Einstein's plan was to have Besso provide an introduction: "I'll grab him by the collar and drag him to his uncle, where I'll do the talking myself." Besso was able to persuade his uncle to write letters on Einstein's behalf, but nothing came of the effort. Instead, Einstein spent most of 1901 juggling temporary teaching assignments and some tutoring.[29]

It was Einstein's other close friend from Zurich, his classmate and math note-taker Marcel Grossmann, who ended up finally getting

Einstein a job, though not one that would have been expected. Just when Einstein was beginning to despair, Grossmann wrote that there was likely to be an opening for an examiner at the Swiss Patent Office, located in Bern. Grossmann's father knew the director and was willing to recommend Einstein.

"I was deeply moved by your devotion and compassion, which did not let you forget your luckless friend," Einstein replied. "I would be delighted to get such a nice job and that I would spare no effort to live up to your recommendation." To Marić he exulted: "Just think what a wonderful job this would be for me! I'll be mad with joy if something should come of that."

It would take months, he knew, before the patent-office job would materialize, assuming that it ever did. So he accepted a temporary post at a technical school in Winterthur for two months, filling in for a teacher on military leave. The hours would be long and, worse yet, he would have to teach descriptive geometry, neither then nor later his strongest field. "But the valiant Swabian is not afraid," he proclaimed, repeating one of his favorite poetic phrases.[30]

In the meantime, he and Marić would have the chance to take a romantic vacation together, one that would have fateful consequences.

Lake Como, May 1901

"You absolutely must come see me in Como, you little witch," Einstein wrote Marić at the end of April 1901. "You'll see for yourself how bright and cheerful I've become and how all my brow-knitting is gone."

The family disputes and frustrating job search had caused him to be snappish, but he promised that was now over. "It was only out of nervousness that I was mean to you," he apologized. To make it up to her, he proposed that they should have a romantic and sensuous tryst in one of the world's most romantic and sensuous places: Lake Como, the grandest of the jewel-like Alpine finger lakes high on the border of Italy and Switzerland, where in early May the lush foliage bursts forth under majestic snow-capped peaks.

"Bring my blue dressing-gown so we can wrap ourselves up in it," he said. "I promise you an outing the likes of which you've never seen."[31]

Marić quickly accepted, but then changed her mind; she had received a letter from her family in Novi Sad "that robs me of all desire, not only for having fun, but for life itself." He should make the trip on his own, she sulked. "It seems I can have nothing without being punished." But the next day she changed her mind again. "I wrote you a little card yesterday while in the worst of moods because of a letter I received. But when I read your letter today I became a bit more cheerful, since I see how much you love me, so I think we'll take that trip after all."[32]

And thus it was that early on the morning of Sunday, May 5, 1901, Albert Einstein was waiting for Mileva Marić at the train station in the village of Como, Italy, "with open arms and a pounding heart." They spent the day there, admiring its gothic cathedral and walled old town, then took one of the stately white steamers that hop from village to village along the banks of the lake.

They stopped to visit Villa Carlotta, the most luscious of all the famous mansions that dot the shore, with its frescoed ceilings, a version of Antonio Canova's erotic sculpture *Cupid and Psyche,* and five hundred species of plants. Marić later wrote a friend how much she admired "the splendid garden, which I preserved in my heart, the more so because we were not allowed to swipe a single flower."

After spending the night in an inn, they decided to hike through the mountain pass to Switzerland, but found it still covered with up to twenty feet of snow. So they hired a small sleigh, "the kind they use that has just enough room for two people in love with each other, and a coachman stands on a little plank in the rear and prattles all the time and calls you 'signora,' " Marić wrote. "Could you think of anything more beautiful?"

The snow was falling merrily, as far as the eye could see, "so that this cold, white infinity gave me the shivers and I held my sweetheart firmly in my arms under the coats and shawls covering us." On the way down, they stomped and kicked at the snow to produce little avalanches, "so as to properly scare the world below."[33]

A few days later, Einstein recalled "how beautiful it was the last time you let me press your dear little person against me in that most natural way."[34] And in that most natural way, Mileva Marić became pregnant with Albert Einstein's child.

After returning to Winterthur, where he was a substitute teacher, Einstein wrote Marić a letter that made reference to her pregnancy. Oddly—or perhaps not oddly at all—he began by delving into matters scientific rather than personal. "I just read a wonderful paper by Lenard on the generation of cathode rays by ultraviolet light," he started. "Under the influence of this beautiful piece I am filled with such happiness and joy that I must share some of it with you." Einstein would soon revolutionize science by building on Lenard's paper to produce a theory of light quanta that explained this photoelectric effect. Even so, it is rather surprising, or at least amusing, that when he rhapsodized about sharing "happiness and joy" with his newly pregnant lover, he was referring to a paper on beams of electrons.

Only after this scientific exultation came a brief reference to their expected child, whom Einstein referred to as a boy: "How are you darling? How's the boy?" He went on to display an odd notion of what parenting would be like: "Can you imagine how pleasant it will be when we're able to work again, completely undisturbed, and with no one around to tell us what to do!"

Most of all, he tried to be reassuring. He would find a job, he pledged, even if it meant going into the insurance business. They would create a comfortable home together. "Be happy and don't fret, darling. I won't leave you and will bring everything to a happy conclusion. You just have to be patient! You will see that my arms are not so bad to rest in, even if things are beginning a little awkwardly." [35]

Marić was preparing to retake her graduation exams, and she was hoping to go on to get a doctorate and become a physicist. Both she and her parents had invested enormous amounts, emotionally and financially, in that goal over the years. She could have, if she had wished, terminated her pregnancy. Zurich was then a center of a burgeoning birth control industry, which included a mail-order abortion drug firm based there.

Instead, she decided that she wanted to have Einstein's child—even though he was not yet ready or willing to marry her. Having a child out of wedlock was rebellious, given their upbringings, but not uncommon. The official statistics for Zurich in 1901 show that 12 percent of births were illegitimate. Residents who were Austro-Hungarian, moreover, were much more likely to get pregnant while unmarried. In southern Hungary, 33 percent of births were illegitimate. Serbs had the highest rate of illegitimate births, Jews by far the lowest.[36]

The decision caused Einstein to focus on the future. "I will look for a position *immediately,* no matter how humble it is," he told her. "My scientific goals and my personal vanity will not prevent me from accepting even the most subordinate position." He decided to call Besso's father as well as the director of the local insurance company, and he promised to marry her as soon as he settled into a job. "Then no one can cast a stone on your dear little head."

The pregnancy could also resolve, or so he hoped, the issues they faced with their families. "When your parents and mine are presented with a fait accompli, they'll just have to reconcile themselves to it as best they can."[37]

Marić, bedridden in Zurich with pregnancy sickness, was thrilled. "So, sweetheart, you want to look for a job immediately? And have me move in with you!" It was a vague proposal, but she immediately pronounced herself "happy" to agree. "Of course it mustn't involve accepting a really bad position, darling," she added. "That would make me feel terrible." At her sister's suggestion she tried to convince Einstein to visit her parents in Serbia for the summer vacation. "It would make me so happy," she begged. "And when my parents see the two of us physically in front of them, all their doubts will evaporate."[38]

But Einstein, to her dismay, decided to spend the summer vacation again with his mother and sister in the Alps. As a result, he was not there to help and encourage her at the end of July 1901 when she retook her exams. Perhaps as a consequence of her pregnancy and personal situation, Mileva ended up failing for the second time, once again getting a 4.0 out of 6 and once again being the only one in her group not to pass.

Thus it was that Mileva Marić found herself resigned to giving up her dream of being a scientific scholar. She visited her home in Serbia—alone—and told her parents about her academic failure and her pregnancy. Before leaving, she asked Einstein to send her father a letter describing their plans and, presumably, pledging to marry her.

"Will you send me the letter so I can see what you've written?" she asked. "By and by I'll give him the necessary information, the unpleasant news as well."[39]

Disputes with Drude and Others

Einstein's impudence and contempt for convention, traits that were abetted by Marić, were evident in his science as well as in his personal life in 1901. That year, the unemployed enthusiast engaged in a series of tangles with academic authorities.

The squabbles show that Einstein had no qualms about challenging those in power. In fact, it seemed to infuse him with glee. As he proclaimed to Jost Winteler in the midst of his disputes that year, "Blind respect for authority is the greatest enemy of truth." It would prove a worthy credo, one suitable for being carved on his coat of arms if he had ever wanted such a thing.

His struggles that year also reveal something more subtle about Einstein's scientific thinking: he had an urge—indeed, a compulsion—to unify concepts from different branches of physics. "It is a glorious feeling to discover the unity of a set of phenomena that seem at first to be completely separate," he wrote to his friend Grossmann as he embarked that spring on an attempt to tie his work on capillarity to Boltzmann's theory of gases. That sentence, more than any other, sums up the faith that underlay Einstein's scientific mission, from his first paper until his last scribbled field equations, guiding him with the same sure sense that was displayed by the needle of his childhood compass.[40]

Among the potentially unifying concepts that were mesmerizing Einstein, and much of the physics world, were those that sprang from kinetic theory, which had been developed in the late nineteenth century by applying the principles of mechanics to phenomena such as heat transfer and the behavior of gases. This involved regarding a gas, for example, as a collection of a huge number of tiny particles—in this case, molecules made up of one or more atoms—that careen around freely and occasionally collide with one another.

Kinetic theory spurred the growth of statistical mechanics, which describes the behavior of a large number of particles using statistical calculations. It was, of course, impossible to trace each molecule and each collision in a gas, but knowing the statistical behavior gave a workable theory of how billions of molecules behaved under varying conditions.

Scientists proceeded to apply these concepts not only to the behavior of gases, but also to phenomena that occurred in liquids and solids, including electrical conductivity and radiation. "The opportunity arose to apply the methods of the kinetic theory of gases to completely different branches of physics," Einstein's close friend Paul Ehrenfest, himself an expert in the field, later wrote. "Above all, the theory was applied to the motion of electrons in metals, to the Brownian motion of microscopically small particles in suspensions, and to the theory of blackbody radiation."[41]

Although many scientists were using atomism to explore their own specialties, for Einstein it was a way to make connections, and develop unifying theories, between a variety of disciplines. In April 1901, for example, he adapted the molecular theories he had used to explain the capillary effect in liquids and applied them to the diffusion of gas molecules. "I've got an extremely lucky idea, which will make it possible to apply our theory of molecular forces to gases as well," he wrote Marić. To Grossmann he noted, "I am now convinced that my theory of atomic attractive forces can also be extended to gases."[42]

Next he became interested in the conduction of heat and electricity, which led him to study Paul Drude's electron theory of metals. As the Einstein scholar Jürgen Renn notes, "Drude's electron theory and Boltzmann's kinetic theory of gas do not just happen to be two arbitrary subjects of interest to Einstein, but rather they share an important common property with several other of his early research topics: they are two examples of the application of atomistic ideas to physical and chemical problems."[43]

Drude's electron theory posited that there are particles in metal that move freely, as molecules of gas do, and thereby conduct both heat and electricity. When Einstein looked into it, he was pleased with it in parts. "I have a study in my hands by Paul Drude on the electron theory, which is written to my heart's desire, even though it contains some very sloppy things," he told Marić. A month later, with his usual lack of deference to authority, he declared, "Perhaps I'll write to Drude privately to point out his mistakes."

And so he did. In a letter to Drude in June, Einstein pointed out what he thought were two mistakes. "He will hardly have anything sensible to refute me with," Einstein gloated to Marić, "because my objections are very straightforward." Perhaps under the charming illusion that showing an eminent scientist his purported lapses is a good method for getting a job, Einstein included a request for one in his letter.[44]

Surprisingly, Drude replied. Not surprisingly, he dismissed Einstein's objections. Einstein was outraged. "It is such manifest proof of the wretchedness of its author that no further comment by me is necessary," Einstein said when forwarding Drude's reply to Marić. "From now on I'll no longer turn to such people, and will instead attack them mercilessly in the journals, as they deserve. It is no wonder that little by little one becomes a misanthrope."

Einstein also vented his frustration to Jost Winteler, his father figure from Aarau, in a letter that included his declaration about a blind respect for authority being the greatest enemy of truth. "He responds by pointing out that another 'infallible' colleague of his shares his opinion. I'll soon make it hot for the man with a masterly publication."[45]

The published papers of Einstein do not identify this "infallible" colleague cited by Drude, but some sleuthing by Renn has turned up a letter from Marić that declares it to be Ludwig Boltzmann.[46] That explains why Einstein proceeded to immerse himself in Boltzmann's writings. "I have been engrossed in Boltzmann's works on the kinetic theory of gases," he wrote Grossmann in September, "and these last few days I wrote a short paper myself that provides the missing keystone in the chain of proofs that he started."[47]

Boltzmann, then at the University of Leipzig, was Europe's master of statistical physics. He had helped to develop the kinetic theory and defend the faith that atoms and molecules actually exist. In doing so, he found it necessary to reconceive the great Second Law of Thermodynamics. This law has many equivalent formulations. It says that heat flows naturally from hot to cold, but not the reverse. Another way to describe the Second Law is in terms of entropy, the degree of disorder and randomness in a system. Any spontaneous process tends to increase the entropy of a system. For example, perfume molecules drift out of an open bottle and into a room but don't, at least in our common experience, spontaneously gather themselves together and all drift back into the bottle.

The problem for Boltzmann was that mechanical processes, such as molecules bumping around, could each be reversed, according to Newton. So a spontaneous decrease in entropy would, at least in theory, be possible. The absurdity of positing that diffused perfume molecules could gather back into a bottle, or that heat could flow from a cold body to a hot one spontaneously, was flung against Boltzmann by opponents, such as Wilhelm Ostwald, who did not believe in the reality of atoms and molecules. "The proposition that all natural phenomena can ultimately be reduced to mechanical ones cannot even be taken as a useful working hypothesis: it is simply a mistake," Ostwald declared. "The irreversibility of natural phenomena proves the existence of processes that cannot be described by mechanical equations."

Boltzmann responded by revising the Second Law so that it was not absolute but merely a statistical near-certainty. It was theoretically possible that millions of perfume molecules could randomly bounce around in a way that they all put themselves back into a bottle at a certain moment, but that was exceedingly unlikely, perhaps trillions of times less likely than that a new deck of cards shuffled a hundred times would end up back in its pristine rank-and-suit precise order.[48]

When Einstein rather immodestly declared in September 1901 that he was filling in a "keystone" that was missing in Boltzmann's chain of proofs, he said he planned to publish it soon. But first, he sent a paper to the *Annalen der Physik* that involved an electrical method for investigating molecular forces, which used calculations derived from experiments others had done using salt solutions and an electrode.[49]

Then he published his critique of Boltzmann's theories. He noted that they worked well in explaining heat transfer in gases but had not yet been properly generalized for other realms. "Great as the achievements of the kinetic theory of heat have been in the domain of gas theory," he wrote, "the science of mechanics has not yet been able to produce an adequate foundation for the general theory of heat." His aim was "to close this gap."[50]

This was all quite presumptuous for an undistinguished Polytechnic student who had not been able to get either a doctorate or a job. Einstein himself later admitted that these papers added little to the body of physics wisdom. But they do indicate what was at the heart of his 1901 challenges to Drude and Boltzmann. Their theories, he felt, did not live up to the maxim he had proclaimed to Grossmann earlier that year about how glorious it was to discover an underlying unity in a set of phenomena that seem completely separate.

In the meantime, in November 1901, Einstein had submitted an attempt at a doctoral dissertation to Professor Alfred Kleiner at the University of Zurich. The dissertation has not survived, but Marić told a friend that "it deals with research into the molecular forces in gases using various known phenomena." Einstein was confident. "He won't dare reject my dissertation," he said of Kleiner, "otherwise the short-sighted man is of little use to me."[51]

By December Kleiner had not even responded, and Einstein started worrying that perhaps the professor's "fragile dignity" might make him uncomfortable accepting a dissertation that denigrated the work of such masters as Drude and Boltzmann. "If he dares to reject my dissertation, then I'll publish his rejection along with my paper and make a fool of him," Einstein said. "But if he accepts it, then we'll see what good old Herr Drude has to say."

Eager for a resolution, he decided to go see Kleiner personally. Rather surprisingly, the meeting went well. Kleiner admitted he had not yet read the dissertation, and Einstein told him to take his time. They then proceeded to discuss various ideas that Einstein was developing, some of which would eventually bear fruit in his relativity theory. Kleiner promised Einstein that he could count on him for a recommendation the next time a teaching job came up. "He's not quite as stupid as I'd thought," was Einstein's verdict. "Moreover, he's a good fellow."[52]

Kleiner may have been a good fellow, but he did not like Einstein's dissertation when he finally got around to reading it. In particular, he was unhappy about Einstein's attack on the scientific establishment. So he rejected it; more precisely, he told Einstein to withdraw it voluntarily, which permitted him to get back his 230 franc fee. According to a book written by Einstein's stepson-in-law, Kleiner's action was "out of consideration to his colleague Ludwig Boltzmann, whose train of reasoning Einstein had sharply criticized." Einstein, lacking such sensitivity, was persuaded by a friend to send the attack directly to Boltzmann.[53]

Lieserl

Marcel Grossmann had mentioned to Einstein that there was likely to be a job at the patent office for him, but it had not yet materialized. So five months later, he gently reminded Grossmann that he still needed help. Noticing in the newspaper that Grossmann had won a job teaching at a Swiss high school, Einstein expressed his "great joy" and then plaintively added, "I, too, applied for that position, but I did it only so that I wouldn't have to tell myself that I was too faint-hearted to apply."[54]

In the fall of 1901, Einstein took an even humbler job as a tutor at a little private academy in Schaffhausen, a village on the Rhine twenty miles north of Zurich. The work consisted solely of tutoring a rich English schoolboy who was there. To be taught by Einstein would someday seem a bargain at any price. But at the time, the proprietor of the school, Jacob Nüesch, was getting the bargain. He was charging the child's family 4,000 francs a year, while paying Einstein only 150 francs a month, plus providing room and board.

Einstein continued to promise Marić that she would "get a good husband as soon as this becomes feasible," but he was now despairing about the patent job. "The position in Bern has not yet been advertised so that I am really giving up hope for it."[55]

Marić was eager to be with him, but her pregnancy made it impossible for them to be together in public. So she spent most of November at a small hotel in a neighboring village. Their relationship was becoming strained. Despite her pleas, Einstein came only infrequently to visit her, often claiming that he did not have the spare money. "You'll surely surprise me, right?" she begged after getting yet another note canceling a visit. Her pleadings and anger alternated, often in the same letter:

If you only knew how terribly homesick I am, you would surely come. Are you really out of money? That's nice! The man earns 150 francs, has room and board provided, and at the end of the month doesn't have a cent to his name! . . . Don't use that as an excuse for Sunday, please. If you don't get any money by then, I will send you some . . . If you only knew how much I want to see you again! I think about you all day long, and even more at night.[56]

Einstein's impatience with authority soon pitted him against the proprietor of the academy. He tried to cajole his tutee to move to Bern with him and pay him directly, but the boy's mother balked. Then Einstein asked Nüesch to give him his meal money in cash so that he would not have to eat with his family. "You know what our conditions are," Nüesch replied. "There is no reason to deviate from them."

A surly Einstein threatened to find new arrangements, and Nüesch backed down in a rage. In a line that could be considered yet another maxim for his life, Einstein recounted the scene to Marić and exulted, "Long live impudence! It is my guardian angel in this world."

That night, as he sat down for his last meal at the Nüesch household, he found a letter for him next to his soup plate. It was from his real-life guardian angel, Marcel Grossmann. The position at the patent office, Grossmann wrote, was about to be advertised, and Einstein was sure to get it. Their lives were soon to be "brilliantly changed for the better," an excited Einstein wrote Marić. "I'm dizzy with joy when I think about it," he said. "I'm even happier for you than for myself. Together we'd surely be the happiest people on the earth."

That still left the issue of what to do about their baby, who was due to be born in less than two months, by early February 1902. "The only problem that would remain to be solved would be how to keep our Lieserl with us," Einstein (who had begun referring to their unborn child as a girl) wrote to Marić, who had returned home to have the baby at her parents' house in Novi Sad. "I wouldn't want to have to give her up." It was a noble intention on his part, yet he knew that it would be difficult for him to show up for work in Bern with an illegitimate child. "Ask your Papa; he's an experienced man, and knows the world better than your overworked, impractical Johnnie." For good measure, he declared that the baby, when born, "shouldn't be stuffed with cow milk, because it might make her stupid." Marić's milk would be more nourishing, he said.[57]

Although he was willing to consult Marić's family, Einstein had no intention of letting his own family know that his mother's worst fears about his relationship—a pregnancy and possible marriage—were materializing. His sister seemed to realize that he and Marić were secretly planning to be married, and she told this to members of the Winteler family in Aarau. But none of them showed any sign of suspecting that a child was involved. Einstein's mother learned about the purported engagement from Mrs. Winteler. "We are resolutely against Albert's relationship with Fraulein Marić, and we don't ever wish to have anything to do with her," Pauline Einstein lamented.[58]

Einstein's mother even took the extraordinary step of writing a nasty letter, signed also by her husband, to Marić's parents. "This lady," Marić lamented to a friend about Einstein's mother, "seems to have set as her life's goal to embitter as much as possible not only my life but also that of her son. I could not have thought it possible that there could exist such heartless and outright wicked people! They felt no compunctions about writing a letter to my parents in which they reviled me in a manner that was a disgrace."[59]

The official advertisement announcing the patent office opportunity finally appeared in December 1901. The director, Friedrich Haller, apparently tailored the specifications so that Einstein would get the job. Candidates did not need a doctorate, but they must have mechanical training and also know physics. "Haller put this in for my sake," Einstein told Marić.

Haller wrote Einstein a friendly letter making it clear that he was the prime candidate, and Grossmann called to congratulate him. "There's no doubt anymore," Einstein exulted to Marić. "Soon you'll be my happy little wife, just watch. Now our troubles are over. Only now that this terrible weight is off my shoulders do I realize how much I love you . . . Soon I'll be able to take my Dollie in my arms and call her my own in front of the whole world."[60]

He made her promise, however, that marriage would not turn them into a comfortable bourgeois couple: "We'll diligently work on science together so we don't become old philistines, right?" Even his sister, he felt, was becoming "so crass" in her approach to creature comforts. "You'd better not get that way," he told Marić. "It would be terrible. You must always be my witch and street urchin. Everyone but you seems foreign to me, as if they were separated from me by an invisible wall."

In anticipation of getting the patent-office job, Einstein abandoned the student he had been tutoring in Schaffhausen and moved to Bern in late January 1902. He would be forever grateful to Grossmann, whose aid would continue in different ways over the next few years. "Grossmann is doing his dissertation on a subject that is related to non-Euclidean geometry," Einstein noted to Marić. "I don't know exactly what it is."[61]

A few days after Einstein arrived in Bern, Mileva Marić, staying at her parents' home in Novi Sad, gave birth to their baby, a girl whom they called Lieserl. Because the childbirth was so difficult, Marić was unable to write to him. Her father sent Einstein the news.

"Is she healthy, and does she cry properly?" Einstein wrote Marić. "What are her eyes like? Which one of us does she more resemble? Who is giving her milk? Is she hungry? She must be completely bald. I love her so much and don't even know her yet!" Yet his love for their new baby seemed to exist mainly in the abstract, for it was not quite enough to induce him to make the train trip to Novi Sad.[62]

Einstein did not tell his mother, sister, or any of his friends about the birth of Lieserl. In fact, there is no indication that he *ever* told them about her. Never once did he publicly speak of her or acknowledge that she even existed. No mention of her survives in any correspondence, except for a few letters between Einstein and Marić, and these were suppressed and hidden until 1986, when scholars and the editors of his papers were completely surprised to learn of Lieserl's existence.*

But in his letter to Marić right after Lieserl's birth, the baby

---

* The letters were discovered by John Stachel of the Einstein Papers Project among a cache of four hundred family letters that were stored in a California safe deposit box by the second wife of Einstein's son Hans Albert Einstein, whose first wife had brought them to California after she went to Zurich to clean out Mileva Marić's apartment following her death in 1948.

brought out Einstein's wry side. "She's certainly able to cry already, but won't know how to laugh until much later," he said. "Therein lies a profound truth."

Fatherhood also focused him on the need to make some money while he waited to get the patent-office job. So the next day an ad appeared in the newspaper: "Private lessons in Mathematics and Physics . . . given most thoroughly by Albert Einstein, holder of the federal Polytechnic teacher's diploma . . . Trial lessons free."

Lieserl's birth even caused Einstein to display a domestic, nesting instinct not previously apparent. He found a large room in Bern and drew for Marić a sketch of it, complete with diagrams showing the bed, six chairs, three cabinets, himself ("Johnnie"), and a couch marked "look at that!"[63] However, Marić was not going to be moving into it with him. They were not married, and an aspiring Swiss civil servant could not be seen cohabitating in such a way. Instead, after a few months, Marić moved back to Zurich to wait for him to get a job and, as promised, marry her. She did not bring Lieserl with her.

Einstein and his daughter apparently never laid eyes on each other. She would merit, as we shall see, just one brief mention in their surviving correspondence less than two years later, in September 1903, and then not be referred to again. In the meantime, she was left back in Novi Sad with her mother's relatives or friends so that Einstein could maintain both his unencumbered lifestyle and the bourgeois respectability he needed to become a Swiss official.

There is a cryptic hint that the person who took custody of Lieserl may have been Marić's close friend, Helene Kaufler Savić, whom she had met in 1899 when they lived in the same rooming house in Zurich. Savić was from a Viennese Jewish family and had married an engineer from Serbia in 1900. During her pregnancy, Marić had written her a letter pouring out all of her woes, but she tore it up before mailing it. She was glad she had done so, she explained to Einstein two months before Lieserl's birth, because "I don't think we should say anything about Lieserl yet." Marić added that Einstein should write Savić a few words now and then. "We must now treat her very nicely. She'll have to help us in something important, after all."[64]

The Patent Office

As he was waiting to be offered the job at the patent office, Einstein ran into an acquaintance who was working there. The job was boring, the person complained, and he noted that the position Einstein was waiting to get was "the lowest rank," so at least he didn't have to worry that anyone else would apply for it. Einstein was unfazed. "Certain people find everything boring," Einstein told Marić. As for the disdain about being on the lowest rung, Einstein told her that they should feel just the opposite: "We couldn't care less about being on top!"[65]

The job finally came through on June 16, 1902, when a session of the Swiss Council officially elected him "provisionally as a Technical Expert Class 3 of the Federal Office for Intellectual Property with an annual salary of 3,500 francs," which was actually more than what a junior professor would make.[66]

His office in Bern's new Postal and Telegraph Building was near the world-famous clock tower over the old city gate (see p. 107). As he turned left out of his apartment on his way to work, Einstein walked past it every day. The clock was originally built shortly after the city was founded in 1191, and an astronomical contraption featuring the positions of the planets was added in 1530. Every hour, the clock would put on its show: out would come a dancing jester ringing bells, then a parade of bears, a crowing rooster, and an armored knight, followed by Father Time with his scepter and hourglass.

The clock was the official timekeeper for the nearby train station, the one from which all of the other clocks that lined the platform were synchronized. The moving trains arriving from other cities, where the local time was not always standardized, would reset their own clocks by looking up at the Bern clock tower as they sped into town.[67]

So it was that Albert Einstein would end up spending the most creative seven years of his life—even after he had written the papers that reoriented physics—arriving at work at 8 a.m., six days a week, and examining patent applications. "I am frightfully busy," he wrote a friend a few months later. "Every day I spend eight hours at the office and at least one hour of private lessons, and then, in addition, I do some scientific work." Yet it would be wrong to think that poring over applications for patents was drudgery. "I enjoy my work at the office very much, because it is uncommonly diversified."[68]

He soon learned that he could work on the patent applications so quickly that it left time for him to sneak in his own scientific thinking during the day. "I was able to do a full day's work in only two or three hours," he recalled. "The remaining part of the day, I would work out my own ideas." His boss, Friedrich Haller, was a man of good-natured, growling skepticism and genial humor who graciously ignored the sheets of paper that cluttered Einstein's desk and vanished into his drawer when people came to see him. "Whenever anybody would come by, I would cram my notes into my desk drawer and pretend to work on my office work."[69]

Indeed, we should not feel sorry for Einstein that he found himself exiled from the cloisters of academe. He came to believe that it was a benefit to his science, rather than a burden, to work instead in "that worldly cloister where I hatched my most beautiful ideas."[70]

Every day, he would do thought experiments based on theoretical premises, sniffing out the underlying realities. Focusing on real-life questions, he later said, "stimulated me to see the physical ramifications of theoretical concepts."[71] Among the ideas that he had to consider for patents were dozens of new methods for synchronizing clocks and coordinating time through signals sent at the speed of light.[72]

In addition, his boss Haller had a credo that was as useful for a creative and rebellious theorist as it was for a patent examiner: "You have to remain critically vigilant." Question every premise, challenge conventional wisdom, and never accept the truth of something merely because everyone else views it as obvious. Resist being credulous. "When you pick up an application," Haller instructed, "think that everything the inventor says is wrong."[73]

Einstein had grown up in a family that created patents and tried to apply them in business, and he found the process to be fulfilling. It reinforced one of his ingenious talents: the ability to conduct thought experiments in which he could visualize how a theory would play out in practice. It also helped him peel off the irrelevant facts that surrounded a problem.[74]

Had he been consigned instead to the job of an assistant to a professor, he might have felt compelled to churn out safe publications and be overly cautious in challenging accepted notions. As he later noted, originality and creativity were not prime assets for climbing academic ladders, especially in the German-speaking world, and he would have felt pressure to conform to the prejudices or prevailing wisdom of his patrons. "An academic career in which a person is forced to produce scientific writings in great amounts creates a danger of intellectual superficiality," he said.[75]

As a result, the happenstance that landed him on a stool at the Swiss Patent Office, rather than as an acolyte in academia, likely reinforced some of the traits destined to make him successful: a merry skepticism about what appeared on the pages in front of him and an independence of judgment that allowed him to challenge basic assumptions. There were no pressures or incentives among the patent examiners to behave otherwise.

The Olympia Academy

Maurice Solovine, a Romanian studying philosophy at the University of Bern, bought a newspaper while on a stroll one day during Easter vacation of 1902 and noticed Einstein's advertisement offering tutorials in physics ("trial lessons free"). A dapper dilettante with close-cropped hair and a raffish goatee, Solovine was four years older than Einstein, but he had yet to decide whether he wanted to be a philosopher, a physicist, or something else. So he went to the address, rang the bell, and a moment later a loud voice thundered "In here!" Einstein made an immediate impression. "I was struck by the extraordinary brilliance of his large eyes," Solovine recalled.[76]

Their first discussion lasted almost two hours, after which Einstein followed Solovine into the street, where they talked for a half-hour more. They agreed to meet the next day. At the third session, Einstein announced that conversing freely was more fun than tutoring for pay. "You don't have to be tutored in physics," he said. "Just come see me when you want and I will be glad to talk with you." They decided to read the great thinkers together and then discuss their ideas.

Their sessions were joined by Conrad Habicht, a banker's son and former student of mathematics at the Zurich Polytechnic. Poking a little fun at pompous scholarly societies, they dubbed themselves the Olympia Academy. Einstein, even though he was the youngest, was designated the president, and Solovine prepared a certificate with a drawing of an Einstein bust in profile beneath a string of sausages. "A man perfectly and clearly erudite, imbued with exquisite, subtle and elegant knowledge, steeped in the revolutionary science of the cosmos," the dedication declared.[77]

Generally their dinners were frugal repasts of sausage, Gruyère cheese, fruit, and tea. But for Einstein's birthday, Solovine and Habicht decided to surprise him by putting three plates of caviar on the table. Einstein was engrossed in analyzing Galileo's principle of inertia, and as he talked he took mouthful after mouthful of his caviar without seeming to notice. Habicht and Solovine exchanged furtive glances. "Do you realize what you've been eating?" Solovine finally asked.

"For goodness' sake," Einstein exclaimed. "So that was the famous caviar!" He paused for a moment, then added, "Well, if you offer gourmet food to peasants like me, you know they won't appreciate it."

After their discussions, which could last all night, Einstein would sometimes play the violin and, in the summertime, they occasionally climbed a mountain on the outskirts of Bern to watch the sunrise. "The sight of the twinkling stars made a strong impression on us and led to discussions of astronomy," Solovine recalled. "We would marvel at the sun as it came slowly toward the horizon and finally appeared in all of its splendor to bathe the Alps in a mystic rose." Then they would wait for the mountain café to open so they could drink dark coffee before hiking down to start work.

Solovine once skipped a session scheduled for his apartment because he was enticed instead to a concert by a Czech quartet. As a peace offering he left behind, as his note written in Latin proclaimed, "hard boiled eggs and a salutation." Einstein and Habicht, knowing how much Solovine hated tobacco, took revenge by smoking pipes and cigars in Solovine's room and piling his furniture and dishes on the bed. "Thick smoke and a salutation," they wrote in Latin. Solovine says he was "almost overwhelmed" by the fumes when he returned. "I thought I would suffocate. I opened the window wide and began to remove from the bed the mound of things that reached almost to the ceiling." [78]

Solovine and Habicht would become Einstein's lifelong friends, and he would later reminisce with them about "our cheerful 'Academy,' which was less childish than those respectable ones which I later got to know at close quarters." In response to a joint postcard sent from Paris by his two colleagues on his seventy-fourth birthday, he paid tribute to it: "Your members created you to make fun of your long-established sister Academies. How well their mockery hit the mark I have learned to appreciate fully through long years of careful observation." [79]

The Academy's reading list included some classics with themes that Einstein could appreciate, such as Sophocles' searing play about the defiance of authority, *Antigone,* and Cervantes' epic about stubbornly tilting at windmills, *Don Quixote.* But mostly the three academicians read books that explored the intersection of science and philosophy: David Hume's *A Treatise of Human Nature,* Ernst Mach's *Analysis of the Sensations* and *Mechanics and Its Development,* Baruch Spinoza's *Ethics,* and Henri Poincaré's *Science and Hypothesis.* [80] It was from reading these authors that the young patent examiner began to develop his own philosophy of science.

The most influential of these, Einstein later said, was the Scottish empiricist David Hume (1711–1776). In the tradition of Locke and Berkeley, Hume was skeptical about any knowledge other than what could be directly perceived by the senses. Even the apparent laws of causality were suspect to him, mere habits of the mind; a ball hitting another may behave the way that Newton's laws predict time after time after time, yet that was not, strictly speaking, a reason to believe that it would happen that way the next time. "Hume saw clearly that certain concepts, for example that of causality, cannot be deduced from our perceptions of experience by logical methods," Einstein noted.

A version of this philosophy, sometimes called positivism, denied the validity of any concepts that went beyond descriptions of phenomena that we directly experience. It appealed to Einstein, at least initially. "The theory of relativity suggests itself in positivism," he said. "This line of thought had a great influence on my efforts, most specifically Mach and even more so Hume, whose *Treatise of Human Nature* I studied avidly and with admiration shortly before discovering the theory of relativity."[81]

Hume applied his skeptical rigor to the concept of time. It made no sense, he said, to speak of time as having an absolute existence that was independent of observable objects whose movements permitted us to define time. "From the succession of ideas and impressions we form the idea of time," Hume wrote. "It is not possible for time alone ever to make its appearance." This idea that there is no such thing as absolute time would later echo in Einstein's theory of relativity. Hume's specific thoughts about time, however, had less influence on Einstein than his more general insight that it is dangerous to talk about concepts that are not definable by perceptions and observations.[82]

Einstein's views on Hume were tempered by his appreciation for Immanuel Kant (1724–1804), the German metaphysician he had been introduced to, back when he was a schoolboy, by Max Talmud. "Kant took the stage with an idea that signified a step towards the solution of Hume's dilemma," Einstein said. Some truths fit into a category of "definitely assured knowledge" that was "grounded in reason itself."

In other words, Kant distinguished between two types of truths: (1) analytic propositions, which derive from logic and "reason itself" rather than from observing the world; for example, all bachelors are unmarried, two plus two equals four, and the angles of a triangle always add up to 180 degrees; and (2) synthetic propositions, which are based on experience and observations; for example, Munich is bigger than

Bern, all swans are white. Synthetic propositions could be revised by new empirical evidence, but not analytic ones. We may discover a black swan but not a married bachelor or (at least so Kant thought) a triangle with 181 degrees. As Einstein said of Kant's first category of truths: "This is held to be the case, for example, in the propositions of geometry and in the principle of causality. These and certain other types of knowledge . . . do not previously have to be gained from sense data, in other words they are a priori knowledge."

Einstein initially found it wondrous that certain truths could be discovered by reason alone. But he soon began to question Kant's rigid distinction between analytic and synthetic truths. "The objects with which geometry deals seemed to be of no different type than the objects of sensory perception," he recalled. And later he would reject outright this Kantian distinction. "I am convinced that this differentiation is erroneous," he wrote. A proposition that seems purely analytic—such as the angles of a triangle adding up to 180 degrees—could turn out to be false in a non-Euclidean geometry or in a curved space (such as would be the case in the general theory of relativity). As he later said of the concepts of geometry and causality, "Today everyone knows, of course, that the mentioned concepts contain nothing of the certainty, of the inherent necessity, which Kant had attributed to them."[83]

Hume's empiricism was carried a step further by Ernst Mach (1838–1916), the Austrian physicist and philosopher whose writings Einstein read at the urging of Michele Besso. He became one of the favorite authors of the Olympia Academy, and he helped to instill in Einstein the skepticism about received wisdom and accepted conventions that would become a hallmark of his creativity. Einstein would later proclaim, in words that could be used to describe himself as well, that Mach's genius was partly due to his "incorruptible skepticism and independence."[84]

The essence of Mach's philosophy was this, in Einstein's words: "Concepts have meaning only if we can point to objects to which they refer and to the rules by which they are assigned to these objects."[85] In other words, for a concept to make sense you need an operational definition of it, one that describes how you would observe the concept in operation. This would bear fruit for Einstein when, a few years later, he and Besso would talk about what observation would give meaning to the apparently simple concept that two events happened "simultaneously."

The most influential thing that Mach did for Einstein was to apply this approach to Newton's concepts of "absolute time" and "absolute space." It was impossible to define these concepts, Mach asserted, in terms of observations you could make. Therefore they were meaningless. Mach ridiculed Newton's "conceptual monstrosity of absolute space"; he called it "purely a thought-thing which cannot be pointed to in experience." [86]

The final intellectual hero of the Olympia Academy was Baruch Spinoza (1632–1677), the Jewish philosopher from Amsterdam. His influence was primarily religious: Einstein embraced his concept of an amorphous God reflected in the awe-inspiring beauty, rationality, and unity of nature's laws. But like Spinoza, Einstein did not believe in a personal God who rewarded and punished and intervened in our daily lives.

In addition, Einstein drew from Spinoza a faith in determinism: a sense that the laws of nature, once we could fathom them, decreed immutable causes and effects, and that God did not play dice by allowing any events to be random or undetermined. "All things are determined by the necessity of divine nature," Spinoza declared, and even when quantum mechanics seemed to show that was wrong, Einstein steadfastly believed it was right. [87]

Marrying Mileva

Hermann Einstein was not destined to see his son become anything more successful than a third-class patent examiner. In October 1902, when Hermann's health began to decline, Einstein traveled to Milan to be with him at the end. Their relationship had long been a mix of alienation and affection, and it concluded on that note as well. "When the end came," Einstein's assistant Helen Dukas later said, "Hermann asked all of them to leave the room, so he could die on his own."

Einstein felt, for the rest of his life, a sense of guilt about that moment, which encapsulated his inability to forge a true bond with his father. For the first time, he was thrown into a daze, "overwhelmed by a feeling of desolation." He later called his father's death the deepest shock he had ever experienced. The event did, however, solve one important issue. On his deathbed, Hermann Einstein gave his permission, finally, for his son to marry Mileva Marić.[88]

Einstein's Olympia Academy colleagues, Maurice Solovine and Conrad Habicht, convened in special session on January 6, 1903, to serve as witnesses at the tiny civil ceremony in the Bern registrar's office where Albert Einstein married Mileva Marić. No family members—not Einstein's mother or sister, nor Marić's parents—came to Bern. The tight group of intellectual comrades celebrated together at a restaurant that evening, and then Einstein and Marić went back to his apartment together. Not surprisingly, he had forgotten his key and had to wake his landlady.[89]

"Well, now I am a married man and I am living a very pleasant cozy life with my wife," he reported to Michele Besso two weeks later. "She takes excellent care of everything, cooks well, and is always cheerful." For her part, Marić* reported to her own best friend, "I am even closer to my sweetheart, if it is at all possible, than I was in our Zurich days." Occasionally she would attend sessions of the Olympia Academy, but mainly as an observer. "Mileva, intelligent and reserved, listened intently but never intervened in our discussions," Solovine recalled.

Nevertheless, clouds began to form. "My new duties are taking their toll," Marić said of her housekeeping chores and role as a mere onlooker when science was discussed. Einstein's friends felt that she was becoming even more gloomy. At times she seemed laconic, and distrustful as well. And Einstein, at least so he claimed in retrospect, had already become wary. He had felt an "inner resistance" to marrying Marić, he later claimed, but had overcome it out of a "sense of duty."

---

* Once married, she usually used the name Mileva Einstein-Marić. After they were divorced, she eventually resumed using Mileva Marić. To avoid confusion, I refer to her as Marić throughout.

Marić soon began to look for ways to restore the magic to their relationship. She hoped that they would escape the bourgeois drudgery
that seemed inherent in the household of a Swiss civil servant and, instead, find some opportunity to recapture their old bohemian academic
life. They decided—or at least so Marić hoped—that Einstein would
find a teaching job somewhere far away, perhaps near their forsaken
daughter. "We will try anywhere," she wrote to her friend in Serbia.
"Do you think, for example, that in Belgrade people of our kind could
find something?" Marić said they would do anything academic, even
teaching German in a high school. "You see, we still have that old en-
terprising spirit." [90]

As far as we know, Einstein never went to Serbia to seek a job or to
see his baby. A few months into their marriage, in August 1903, the secret cloud hovering over their lives suddenly cast a new pall. Marić received word that Lieserl, then 19 months old, had come down with
scarlet fever. She boarded a train for Novi Sad. When it stopped in
Salzburg, she bought a postcard of a local castle and jotted a note,
which she mailed from the stop in Budapest: "It is going quickly, but it
is hard. I don't feel at all well. What are you doing, little Jonzile, write
me soon, will you? Your poor Dollie." [91]

Apparently, the child was given up for adoption. The only clue we
have is a cryptic letter Einstein wrote Marić in September, after she
had been in Novi Sad for a month: "I am very sorry about what happened with Lieserl. Scarlet fever often leaves some lasting trace behind. If only everything passes well. How is Lieserl registered? We
must take great care, lest difficulties arise for the child in the future." [92]

Whatever the motivation Einstein may have had for asking the
question, neither Lieserl's registration documents nor any other paper
trace of her existence is known to have survived. Various researchers,
Serbian and American, including Robert Schulmann of the Einstein
Papers Project and Michele Zackheim, who wrote a book about
searching for Lieserl, have fruitlessly scoured churches, registries, syn-
agogues, and cemeteries.

All evidence about Einstein's daughter was carefully erased. Almost
every one of the letters between Einstein and Marić in the summer and

fall of 1902, many of which presumably dealt with Lieserl, were destroyed. Those between Marić and her friend Helene Savić during that period were intentionally burned by Savić's family. For the rest of their lives, even after they divorced, Einstein and his wife did all they could, with surprising success, to cover up not only the fate of their first child but her very existence.

One of the few facts that have escaped this black hole of history is that Lieserl was still alive in September 1903. Einstein's expression of worry, in his letter to Marić that month, about potential difficulties "for the child in the future," makes this clear. The letter also indicates that she had been given up for adoption by then, because in it Einstein spoke of the desirability of having a "replacement" child.

There are two plausible explanations about the fate of Lieserl. The first is that she survived her bout of scarlet fever and was raised by an adoptive family. On a couple of occasions later in his life, when women came forward claiming (falsely, it turned out) to be illegitimate children of his, Einstein did not dismiss the possibility out of hand, although given the number of affairs he had, this is no indication that he thought they might be Lieserl.

One possibility, favored by Schulmann, is that Marić's friend Helene Savić adopted Lieserl. She did in fact raise a daughter Zorka, who was blind from early childhood (perhaps a result of scarlet fever), was never married, and was shielded by her nephew from people who sought to interview her. Zorka died in the 1990s.

The nephew who protected Zorka, Milan Popović, rejects this possibility. In a book he wrote on the friendship and correspondence between Marić and his grandmother Helene Savić, *In Albert's Shadow*, Popović asserted, "A theory has been advanced that my grandmother adopted Lieserl, but an examination of my family's history renders this groundless." He did not, however, produce any documentary evidence, such as his aunt's birth certificate, to back up this contention. His mother burned most of Helene Savić's letters, including any that had dealt with Lieserl. Popović's own theory, based partly on the family stories recalled by a Serbian writer named Mira Alečković, is that Lieserl died of scarlet fever in September 1903, after Einstein's letter of that month. Michele Zackheim, in her book describing her hunt for Lieserl, comes to a similar conclusion.[93]

Whatever happened added to Marić's gloom. Shortly after Einstein died, a writer named Peter Michelmore, who knew nothing of Lieserl, published a book that was based in part on conversations with Einstein's son Hans Albert Einstein. Referring to the year right after their marriage, Michelmore noted, "Something had happened between the two, but Mileva would say only that it was 'intensely personal.' Whatever it was, she brooded about it, and Albert seemed to be in some ways responsible. Friends encouraged Mileva to talk about her problem and get it out in the open. She insisted that it was too personal and kept it a secret all her life—a vital detail in the story of Albert Einstein that still remains shrouded in mystery."[94]

The illness that Marić complained about in her postcard from Budapest was likely because she was pregnant again. When she found out that indeed she was, she worried that this would anger her husband. But Einstein expressed happiness on hearing the news that there would soon be a replacement for their daughter. "I'm not the least bit angry that poor Dollie is hatching a new chick," he wrote. "In fact, I'm happy about it and had already given some thought to whether I shouldn't see to it that you get a new Lieserl. After all, you shouldn't be denied that which is the right of all women."[95]

Hans Albert Einstein was born on May 14, 1904. The new child lifted Marić's spirits and restored some joy to her marriage, or so at least she told her friend Helene Savić: "Hop over to Bern so I can see you again and I can show you my dear little sweetheart, who is also named Albert. I cannot tell you how much joy he gives me when he laughs so cheerfully on waking up or when he kicks his legs while taking a bath."

Einstein was "behaving with fatherly dignity," Marić noted, and he spent time making little toys for his baby son, such as a cable car he constructed from matchboxes and string. "That was one of the nicest toys I had at the time and it worked," Hans Albert could still recall when he was an adult. "Out of little string and matchboxes and so on, he could make the most beautiful things."[96]

Milos Marić was so overjoyed with the birth of a grandson that he came to visit and offered a sizable dowry, reported in family lore (likely with some exaggeration) to be 100,000 Swiss francs. But Einstein declined it, saying he had not married his daughter for money, Milos Marić later recounted with tears in his eyes. In fact, Einstein was beginning to do well enough on his own. After more than a year at the patent office, he had been taken off probationary status.[97]

CHAPTER FIVE

THE MIRACLE YEAR:

Quanta and Molecules, 1905

At the Patent Office, 1905

Turn of the Century

"There is nothing new to be discovered in physics now," the revered Lord Kelvin reportedly told the British Association for the Advancement of Science in 1900. "All that remains is more and more precise measurement."[1] He was wrong.

The foundations of classical physics had been laid by Isaac Newton (1642–1727) in the late seventeenth century. Building on the discoveries of Galileo and others, he developed laws that described a very comprehensible mechanical universe: a falling apple and an orbiting moon were governed by the same rules of gravity, mass, force, and motion. Causes produced effects, forces acted upon objects, and in theory everything could be explained, determined, and predicted. As the mathematician and astronomer Laplace exulted about Newton's universe, "An intelligence knowing all the forces acting in nature at a given instant, as well as the momentary positions of all things in the universe, would be able to comprehend in one single formula the motions of the largest bodies as well as the lightest atoms in the world; to him nothing would be uncertain, the future as well as the past would be present to his eyes."[2]

Einstein admired this strict causality, calling it "the profoundest characteristic of Newton's teaching."[3] He wryly summarized the history of physics: "In the beginning (if there was such a thing) God created Newton's laws of motion together with the necessary masses and forces." What especially impressed Einstein were "the achievements of mechanics in areas that apparently had nothing to do with mechanics," such as the kinetic theory he had been exploring, which explained the behavior of gases as being caused by the actions of billions of molecules bumping around.[4]

In the mid-1800s, Newtonian mechanics was joined by another great advance. The English experimenter Michael Faraday (1791–1867), the self-taught son of a blacksmith, discovered the properties of electrical and magnetic fields. He showed that an electric current produced magnetism, and then he showed that a changing magnetic field could produce an electric current. When a magnet is moved near a wire loop, or vice versa, an electric current is produced.[5]

Faraday's work on electromagnetic induction permitted inventive entrepreneurs like Einstein's father and uncle to create new ways of combining spinning wire coils and moving magnets to build electricity generators. As a result, young Albert Einstein had a profound physical feel for Faraday's fields and not just a theoretical understanding of them.

The bushy-bearded Scottish physicist James Clerk Maxwell (1831–1879) subsequently devised wonderful equations that specified, among other things, how changing electric fields create magnetic fields and how changing magnetic fields create electrical ones. A changing electric field could, in fact, produce a changing magnetic field that could, in turn, produce a changing electric field, and so on. The result of this coupling was an electromagnetic wave.

Just as Newton had been born the year that Galileo died, so Einstein was born the year that Maxwell died, and he saw it as part of his mission to extend the work of the Scotsman. Here was a theorist who had shed prevailing biases, let mathematical melodies lead him into unknown territories, and found a harmony that was based on the beauty and simplicity of a field theory.

All of his life, Einstein was fascinated by field theories, and he described the development of the concept in a textbook he wrote with a colleague:

> A new concept appeared in physics, the most important invention since Newton's time: the field. It needed great scientific imagination to realize that it is not the charges nor the particles but the field in the space between the charges and the particles that is essential for the description of physical phenomena. The field concept proved successful when it led to the formulation of Maxwell's equations describing the structure of the electromagnetic field.[6]

At first, the electromagnetic field theory developed by Maxwell seemed compatible with the mechanics of Newton. For example, Maxwell believed that electromagnetic waves, which include visible light, could be explained by classical mechanics—if we assume that the universe is suffused with some unseen, gossamer "light-bearing ether" that serves as the physical substance that undulates and oscillates to propagate the electromagnetic waves, comparable to the role water plays for ocean waves and air plays for sound waves.

By the end of the nineteenth century, however, fissures had begun to develop in the foundations of classical physics. One problem was that scientists, as hard as they tried, could not find any evidence of our motion through this supposed light-propagating ether. The study of radiation—how light and other electromagnetic waves emanate from physical bodies—exposed another problem: strange things were happening at the borderline where Newtonian theories, which described the mechanics of discrete particles, interacted with field theory, which described all electromagnetic phenomena.

Up until then, Einstein had published five little-noted papers. They had earned him neither a doctorate nor a teaching job, even at a high school. Had he given up theoretical physics at that point, the scientific community would not have noticed, and he might have moved up the ladder to become the head of the Swiss Patent Office, a job in which he would likely have been very good indeed.

There was no sign that he was about to unleash an *annus mirabilis* the like of which science had not seen since 1666, when Isaac Newton, holed up at his mother's home in rural Woolsthorpe to escape the plague that was devastating Cambridge, developed calculus, an analysis of the light spectrum, and the laws of gravity.

But physics was poised to be upended again, and Einstein was poised to be the one to do it. He had the brashness needed to scrub away the layers of conventional wisdom that were obscuring the cracks in the foundation of physics, and his visual imagination allowed him to make conceptual leaps that eluded more traditional thinkers.

The breakthroughs that he wrought during a four-month frenzy from March to June 1905 were heralded in what would become one of the most famous personal letters in the history of science. Conrad Habicht, his fellow philosophical frolicker in the Olympia Academy, had just moved away from Bern, which, happily for historians, gave a reason for Einstein to write to him in late May.

Dear Habicht,
    Such a solemn air of silence has descended between us that I almost feel as if I am committing a sacrilege when I break it now with some inconsequential babble . . .
    So, what are you up to, you frozen whale, you smoked, dried, canned piece of soul . . . ? Why have you still not sent me your dissertation? Don't you know that I am one of the 1½ fellows who would read it with interest and pleasure, you wretched man? I promise you four papers in return. The first deals with radiation and the energy properties of light and is very revolutionary, as you will see if you send me your work first. The second paper is a determination of the true sizes of atoms . . . The third proves that bodies on the order of magnitude 1/1000 mm, suspended in liquids, must already perform an observable random motion that is produced by thermal motion. Such movement of suspended bodies has actually been observed by physiologists who call it Brownian molecular motion. The fourth paper is only a rough draft at this point, and is an electrodynamics of moving bodies which employs a modification of the theory of space and time.[7]

Light Quanta, March 1905

As Einstein noted to Habicht, it was the first of these 1905 papers, not the famous final one expounding a theory of relativity, that deserved the designation "revolutionary." Indeed, it may contain the most revolutionary development in the history of physics. Its suggestion that light comes not just in waves but in tiny packets—quanta of light that were later dubbed "photons"—spirits us into strange scientific mists that are far murkier, indeed more spooky, than even the weirdest aspects of the theory of relativity.

Einstein recognized this in the slightly odd title he gave to the paper, which he submitted on March 17, 1905, to the *Annalen der Physik:* "On a Heuristic Point of View Concerning the Production and Transformation of Light."[8] Heuristic? It means a hypothesis that serves as a guide and gives direction in solving a problem but is not considered proven. From this first sentence he ever published about quantum theory until his last such sentence, which came in a paper exactly fifty years later, just before he died, Einstein regarded the concept of the quanta and all of its unsettling implications as heuristic at best: provisional and incomplete and not fully compatible with his own intimations of underlying reality.

At the heart of Einstein's paper were questions that were bedeviling physics at the turn of the century, and in fact have done so from the time of the ancient Greeks until today: Is the universe made up of particles, such as atoms and electrons? Or is it an unbroken continuum, as a gravitational or electromagnetic field seems to be? And if both methods of describing things are valid at times, what happens when they intersect?

Since the 1860s, scientists had been exploring just such a point of intersection by analyzing what was called "blackbody radiation." As anyone who has played with a kiln or a gas burner knows, the glow from a material such as iron changes color as it heats up. First it appears to radiate mainly red light; as it gets hotter, it glows more orange, and then white and then blue. To study this radiation, Gustav Kirchhoff and others devised a closed metal container with a tiny hole to let a little light escape. Then they drew a graph of the intensity of each wavelength when the device reached equilibrium at a certain temperature. No matter what the material or shape of the container's walls, the results were the same; the shape of the graphs depended only on the temperature.

There was, alas, a problem. No one could fully account for the basis of the mathematical formula that would produce the hill-like shape of these graphs.

When Kirchhoff died, his professorship at the University of Berlin was given to Max Planck. Born in 1858 into an ancient German family of great scholars, theologians, and lawyers, Planck was many things that Einstein was not: with his pince-nez glasses and meticulous dress, he was very proudly German, somewhat shy, steely in his resolve, conservative by instinct, and formal in his manner. "It is difficult to imagine two men of more different attitudes," their mutual friend Max Born later said. "Einstein a citizen of the whole world, little attached to the people around him, independent of the emotional background of the society in which he lived—Planck deeply rooted in the traditions of his family and nation, an ardent patriot, proud of the greatness of German history and consciously Prussian in his attitude to the state."[9]

His conservatism made Planck skeptical about the atom, and of particle (rather than wave and continuous field) theories in general. As he wrote in 1882, "Despite the great success that the atomic theory has so far enjoyed, ultimately it will have to be abandoned in favor of the assumption of continuous matter." In one of our planet's little ironies, Planck and Einstein would share the fate of laying the groundwork for quantum mechanics, and then both would flinch when it became clear that it undermined the concepts of strict causality and certainty they both worshipped.[10]

In 1900, Planck came up with an equation, partly using what he called "a fortuitous guess," that described the curve of radiation wavelengths at each temperature. In doing so he accepted that Boltzmann's statistical methods, which he had resisted, were correct after all. But the equation had an odd feature: it required the use of a constant, which was an unexplained tiny quantity (approximately 6.62607 x

10^{-34} joule-seconds), that needed to be included for it to come out right. It was soon dubbed Planck's constant, *h*, and is now known as one of the fundamental constants of nature.

At first Planck had no idea what, if any, physical meaning this mathematical constant had. But then he came up with a theory that, he thought, applied not to the nature of light itself but to the action that occurred when the light was absorbed or emitted by a piece of matter. He posited that the surface of anything that was radiating heat and light—such as the walls in a blackbody device—contained "vibrating molecules" or "harmonic oscillators," like little vibrating springs.[11] These harmonic oscillators could absorb or emit energy only in the form of discrete packets or bundles. These packets or bundles of energy came only in fixed amounts, determined by Planck's constant, rather than being divisible or having a continuous range of values.

Planck considered his constant a mere calculational contrivance that explained the process of emitting or absorbing light but did not apply to the fundamental nature of light itself. Nevertheless, the declaration he made to the Berlin Physical Society in December 1900 was momentous: "We therefore regard—and this is the most essential point of the entire calculation—energy to be composed of a very definite number of equal finite packages."[12]

Einstein quickly realized that quantum theory could undermine classical physics. "All of this was quite clear to me shortly after the appearance of Planck's fundamental work," he wrote later. "All of my attempts to adapt the theoretical foundation of physics to this knowledge failed completely. It was as if the ground had been pulled out from under us, with no firm foundation to be seen anywhere."[13]

In addition to the problem of explaining what Planck's constant was really all about, there was another curiosity about radiation that needed to be explained. It was called the photoelectric effect, and it occurs when light shining on a metal surface causes electrons to be knocked loose and emitted. In the letter he wrote to Marić right after he learned of her pregnancy in May 1901, Einstein enthused over a "beautiful piece" by Philipp Lenard that explored this topic.

Lenard's experiments found something unexpected. When he increased the *frequency* of the light—moving from infrared heat and red light up in frequency to violet and ultraviolet—the emitted electrons sped out with much more energy. Then, he increased the *intensity* of the light by using a carbon arc light that could be made brighter by a factor of 1,000. The brighter, more intense light had a lot more energy, so it seemed logical that the electrons emitted would have more energy and speed away faster. But that did not occur. More intense light produced more electrons, but the energy of each remained the same. This was something that the wave theory of light did not explain.

Einstein had been pondering the work of Planck and Lenard for four years. In his final paper of 1904, "On the General Molecular Theory of Heat," he discussed how the average energy of a system of molecules fluctuates. He then applied this to a volume filled with radiation, and found that experimental results were comparable. His concluding phrase was, "I believe that this agreement must not be ascribed to chance."[14] As he wrote to his friend Conrad Habicht just after finishing that 1904 paper, "I have now found in a most simple way the relation between the size of elementary quanta of matter and the wavelengths of radiation." He was thus primed, so it seems, to form a theory that the radiation field was made up of quanta.[15]

In his 1905 light quanta paper, published a year later, he did just that. He took the mathematical quirk that Planck had discovered, interpreted it literally, related it to Lenard's photoelectric results, and analyzed light as if it *really was* made up of pointlike particles—light quanta, he called them—rather than being a continuous wave.

Einstein began his paper by describing the great distinction between theories based on particles (such as the kinetic theory of gases) and theories that involve continuous functions (such as the electromagnetic fields of the wave theory of light). "There exists a profound formal difference between the theories that physicists have formed about gases and other ponderable bodies, and Maxwell's theory of electromagnetic processes in so-called empty space," he noted. "While we consider the state of a body to be completely determined by the positions and velocities of a very large, yet finite, number of atoms and electrons, we make use of continuous spatial functions to describe the electromagnetic state of a given volume."[16]

Before he made his case for a particle theory of light, he emphasized that this would *not* make it necessary to scrap the wave theory, which would continue to be useful as well. "The wave theory of light, which operates with continuous spatial functions, has worked well in the representation of purely optical phenomena and will probably never be replaced by another theory."

His way of accommodating both a wave theory and a particle theory was to suggest, in a "heuristic" way, that our observation of waves involve statistical averages of the positions of what could be countless particles. "It should be kept in mind," he said, "that the optical observations refer to time averages rather than instantaneous values."

Then came what may be the most revolutionary sentence that Einstein ever wrote. It suggests that light is made up of discrete particles or packets of energy: "According to the assumption to be considered here, when a light ray is propagated from a point, the energy is not continuously distributed over an increasing space but consists of a finite number of energy quanta which are localized at points in space and which can be produced and absorbed only as complete units."

Einstein explored this hypothesis by determining whether a volume of blackbody radiation, which he was now assuming consisted of discrete quanta, might in fact behave like a volume of gas, which he knew consisted of discrete particles. First, he looked at the formulas that showed how the entropy of a gas changes when its volume changes. Then he compared this to how the entropy of blackbody radiation changes as its volume changes. He found that the entropy of the radiation "varies with volume according to the same law as the entropy of an ideal gas."

He did a calculation using Boltzmann's statistical formulas for entropy. The statistical mechanics that described a dilute gas of particles was mathematically the same as that for blackbody radiation. This led Einstein to declare that the radiation "behaves thermodynamically as if it consisted of mutually independent energy quanta." It also provided a way to calculate the energy of a "particle" of light at a particular frequency, which turned out to be in accord with what Planck had found.[17]

Einstein went on to show how the existence of these light quanta could explain what he graciously called Lenard's "pioneering work" on the photoelectric effect. If light came in discrete quanta, then the energy of each one was determined simply by the frequency of the light multiplied by Planck's constant. If we assume, Einstein suggested, "that a light quantum transfers its entire energy to a single electron," then it follows that light of a higher frequency would cause the electrons to emit with more energy. On the other hand, increasing the intensity of the light (but not the frequency) would simply mean that more electrons would be emitted, but the energy of each would be the same.

That was precisely what Lenard had found. With a trace of humility or tentativeness, along with a desire to show that his conclusions had been deduced theoretically rather than induced entirely from experimental data, Einstein declared of his paper's premise that light consists of tiny quanta: "As far as I can see, our conception does not conflict with the properties of the photoelectric effect observed by Mr. Lenard."

By blowing on Planck's embers, Einstein had turned them into a flame that would consume classical physics. What precisely did Einstein produce that made his 1905 paper a discontinuous—one is tempted to say quantum—leap beyond the work of Planck?

In effect, as Einstein noted in a paper the following year, his role was that he figured out the physical significance of what Planck had discovered.[18] For Planck, a reluctant revolutionary, the quantum was a mathematical contrivance that explained how energy was emitted and absorbed when it interacted with matter. But he did not see that it related to a physical reality that was inherent in the nature of light and the electromagnetic field itself. "One can interpret Planck's 1900 paper to mean only that the quantum hypothesis is used as a *mathematical* convenience introduced in order to calculate a statistical distribution, not as a new *physical* assumption," write science historians Gerald Holton and Steven Brush.[19]

Einstein, on the other hand, considered the light quantum to be a feature of reality: a perplexing, pesky, mysterious, and sometimes maddening quirk in the cosmos. For him, these quanta of energy (which in 1926 were named photons)[20] existed even when light was moving through a vacuum. "We wish to show that Mr. Planck's determination of the elementary quanta is to some extent independent of his theory of blackbody radiation," he wrote. In other words, Einstein argued that the particulate nature of light was a property of the light itself and not just some description of how the light interacts with matter.[21]

Even after Einstein published his paper, Planck did not accept his leap. Two years later, Planck warned the young patent clerk that he had gone too far, and that quanta described a process that occurred during emission or absorption, rather than some real property of radiation in a vacuum. "I do not seek the meaning of the 'quantum of action' (light quantum) in the vacuum but at the site of absorption and emission," he advised.[22]

Planck's resistance to believing that the light quanta had a physical reality persisted. Eight years after Einstein's paper was published, Planck proposed him for a coveted seat in the Prussian Academy of Sciences. The letter he and other supporters wrote was filled with praise, but Planck added: "That he might sometimes have overshot the target in his speculations, as for example in his light quantum hypothesis, should not be counted against him too much."[23]

Just before he died, Planck reflected on the fact that he had long recoiled from the implications of his discovery. "My futile attempts to fit the elementary quantum of action somehow into classical theory continued for a number of years and cost me a great deal of effort," he wrote. "Many of my colleagues saw in this something bordering on a tragedy."

Ironically, similar words would later be used to describe Einstein. He became increasingly "aloof and skeptical" about the quantum discoveries he pioneered, Born said of Einstein. "Many of us regard this as a tragedy."[24]

Einstein's theory produced a law of the photoelectric effect that was experimentally testable: the energy of emitted electrons would depend on the frequency of the light according to a simple mathematical formula involving Planck's constant. The formula was subsequently shown to be correct. The physicist who did the crucial experiment was Robert Millikan, who would later head the California Institute of Technology and try to recruit Einstein.

Yet even after he verified Einstein's photoelectric formulas, Millikan still rejected the theory. "Despite the apparently complete success of the Einstein equation," he declared, "the physical theory on which it was designed to be the symbolic expression is found so untenable that Einstein himself, I believe, no longer holds to it."[25]

Millikan was wrong to say that Einstein's formulation of the photo-electric effect had been abandoned. In fact, it was specifically for dis-covering the law of the photoelectric effect that Einstein would win his only Nobel Prize. With the advent of quantum mechanics in the 1920s, the reality of the photon became a fundamental part of physics.

However, on the larger point Millikan was right. Einstein would increasingly find the eerie implications of the quantum—and of the wave-particle duality of light—to be deeply unsettling. In a letter he wrote near the end of his life to his dear friend Michele Besso, after quantum mechanics had been accepted by almost every living physi-cist, Einstein would lament, "All these fifty years of pondering have not brought me any closer to answering the question, What are light quanta?"[26]

Doctoral Dissertation on the Size of Molecules, April 1905

Einstein had written a paper that would revolutionize science, but he had not yet been able to earn a doctorate. So he tried one more time to get a dissertation accepted.

He realized that he needed a safe topic, not a radical one like quanta or relativity, so he chose the second paper he was working on, titled "A New Determination of Molecular Dimensions," which he completed on April 30 and submitted to the University of Zurich in July.[27]

Perhaps out of caution and deference to the conservative approach of his adviser, Alfred Kleiner, he generally avoided the innovative sta-tistical physics featured in his previous papers (and in his Brownian motion paper completed eleven days later) and relied instead mainly on classical hydrodynamics.[28] Yet he was still able to explore how the behavior of countless tiny particles (atoms, molecules) are reflected in observable phenomena, and conversely how observable phenomena can tell us about the nature of those tiny unseen particles.

Almost a century earlier, the Italian scientist Amedeo Avogadro

(1776–1856) had developed the hypothesis—correct, as it turned out—that equal volumes of any gas, when measured at the same temperature and pressure, will have the same number of molecules. That led to a difficult quest: figuring out just how many this was.

The volume usually chosen is that occupied by a mole of the gas (its molecular weight in grams), which is 22.4 liters at standard temperature and pressure. The number of molecules under such conditions later became known as Avogadro's number. Determining it precisely was, and still is, rather difficult. A current estimate is approximately 6.02214×10^{23}. (This is a big number: that many unpopped popcorn kernels when spread across the United States would cover the country nine miles deep.) [29]

Most previous measurements of molecules had been done by studying gases. But as Einstein noted in the first sentence of his paper, "The physical phenomena observed in liquids have thus far not served for the determination of molecular sizes." In this dissertation (after a few math and data corrections were later made), Einstein was the first person able to get a respectable result using liquids.

His method involved making use of data about viscosity, which is how much resistance a liquid offers to an object that tries to move through it. Tar and molasses, for example, are highly viscous. If you dissolve sugar in water, the solution's viscosity increases as it gets more syrupy. Einstein envisioned the sugar molecules gradually diffusing their way through the smaller water molecules. He was able to come up with two equations, each containing the two unknown variables—the size of the sugar molecules and the number of them in the water—that he was trying to determine. He could then solve for these unknown variables. Doing so, he got a result for Avogadro's number that was 2.1×10^{23}.

That, unfortunately, was not very close. When he submitted his paper to the *Annalen der Physik* in August, right after it had been accepted by Zurich University, the editor Paul Drude (who was blissfully unaware of Einstein's earlier desire to ridicule him) held up its publication because he knew of some better data on the properties of sugar solutions. Using this new data, Einstein came up with a result that was closer to correct: 4.15×10^{23}.

A few years later, a French student tested the approach experimentally and discovered something amiss. So Einstein asked an assistant in Zurich to look at it all over again. He found a minor error, which when corrected produced a result of 6.56 x 10^{23}, which ended up being quite respectable.[30]

Einstein later said, perhaps half-jokingly, that when he submitted his thesis, Professor Kleiner rejected it for being too short, so he added one more sentence and it was promptly accepted. There is no documentary evidence for this.[31] Either way, his thesis actually became one of his most cited and practically useful papers, with applications in such diverse fields as cement mixing, dairy production, and aerosol products. And even though it did not help him get an academic job, it did make it possible for him to become known, finally, as Dr. Einstein.

Brownian Motion, May 1905

Eleven days after finishing his dissertation, Einstein produced another paper exploring evidence of things unseen. As he had been doing since 1901, he relied on statistical analysis of the random actions of invisible particles to show how they were reflected in the visible world.

In doing so, Einstein explained a phenomenon, known as Brownian motion, that had been puzzling scientists for almost eighty years: why small particles suspended in a liquid such as water are observed to jiggle around. And as a byproduct, he pretty much settled once and for all that atoms and molecules actually existed as physical objects.

Brownian motion was named after the Scottish botanist Robert Brown, who in 1828 had published detailed observations about how minuscule pollen particles suspended in water can be seen to wiggle and wander when examined under a strong microscope. The study was replicated with other particles, including filings from the Sphinx, and a variety of explanations was offered. Perhaps it had something to do with tiny water currents or the effect of light. But none of these theories proved plausible.

With the rise in the 1870s of the kinetic theory, which used the random motions of molecules to explain things like the behavior of gases, some tried to use it to explain Brownian motion. But because the suspended particles were 10,000 times larger than a water molecule, it seemed that a molecule would not have the power to budge the particle any more than a baseball could budge an object that was a half-mile in diameter.[32]

Einstein showed that even though one collision could not budge a particle, the effect of millions of random collisions per second could explain the jig observed by Brown. "In this paper," he announced in his first sentence, "it will be shown that, according to the molecular-kinetic theory of heat, bodies of a microscopically visible size suspended in liquids must, as a result of thermal molecular motions, perform motions of such magnitudes that they can be easily observed with a microscope."[33]

He went on to say something that seems, on the surface, somewhat puzzling: his paper was not an attempt to explain the observations of Brownian motion. Indeed, he acted as if he wasn't even sure that the motions he deduced from his theory were the same as those observed by Brown: "It is possible that the motions to be discussed here are identical with so-called Brownian molecular motion; however, the data available to me on the latter are so imprecise that I could not form a judgment on the question." Later, he distanced his work even further from intending to be an explanation of Brownian motion: "I discovered that, according to atomistic theory, there would have to be a movement of suspended microscopic particles open to observations, without knowing that observations concerning the Brownian motion were already long familiar."[34]

At first glance his demurral that he was dealing with Brownian motion seems odd, even disingenuous. After all, he had written Conrad Habicht a few months earlier, "Such movement of suspended bodies has actually been observed by physiologists who call it Brownian molecular motion." Yet Einstein's point was both true and significant: his paper did not start with the observed facts of Brownian motion and build toward an explanation of it. Rather, it was a continuation of his earlier statistical analysis of how the actions of molecules could be manifest in the visible world.

In other words, Einstein wanted to assert that he had produced a theory that was deduced from grand principles and postulates, not a theory that was constructed by examining physical data (just as he had made plain that his light quanta paper had not *started* with the photo-electric effect data gathered by Philipp Lenard). It was a distinction he would also make, as we shall soon see, when insisting that his theory of relativity did not derive merely from trying to explain experimental results about the speed of light and the ether.

Einstein realized that a bump from a single water molecule would not cause a suspended pollen particle to move enough to be visible. However, at any given moment, the particle was being hit from all sides by thousands of molecules. There would be some moments when a lot more bumps happened to hit one particular side of the particle. Then, in another moment, a different side might get the heaviest barrage.

The result would be random little lurches that would result in what is known as a random walk. The best way for us to envision this is to imagine a drunk who starts at a lamppost and lurches one step in a random direction every second. After two such lurches he may have gone back and forth to return to the lamp. Or he may be two steps away in the same direction. Or he may be one step west and one step northeast. A little mathematical plotting and charting reveals an interesting thing about such a random walk: statistically, the drunk's distance from the lamp will be proportional to the square root of the number of seconds that have elapsed.[35]

Einstein realized that it was neither possible nor necessary to measure each zig and zag of Brownian motion, nor to measure the particle's velocity at any moment. But it was rather easy to measure the total distances of randomly lurching particles as these distances grew over time.

Einstein wanted concrete predictions that could be tested, so he used both his theoretical knowledge and experimental data about viscosity and diffusion rates to come up with precise predictions showing the distance a particle should move depending on its size and the temperature of the liquid. For example, he predicted, in the case of a particle with a diameter of one thousandth of a millimeter in water at 17 degrees centigrade, "the mean displacement in one minute would be about 6 microns."

Here was something that could actually be tested, and with great consequence. "If the motion discussed here can be observed," he wrote, "then classical thermodynamics can no longer be viewed as strictly valid." Better at theorizing than at conducting experiments, Einstein ended his paper with a charming exhortation: "Let us hope that a researcher will soon succeed in solving the problem presented here, which is so important for the theory of heat."

Within months, a German experimenter named Henry Seidentopf, using a powerful microscope, confirmed Einstein's predictions. For all practical purposes, the physical reality of atoms and molecules was now conclusively proven. "At the time atoms and molecules were still far from being regarded as real," the theoretical physicist Max Born later recalled. "I think that these investigations of Einstein have done more than any other work to convince physicists of the reality of atoms and molecules."[36]

As lagniappe, Einstein's paper also provided yet another way to determine Avogadro's number. "It bristles with new ideas," Abraham Pais said of the paper. "The final conclusion, that Avogadro's number can essentially be determined from observations with an ordinary microscope, never fails to cause a moment of astonishment even if one has read the paper before and therefore knows the punch line."

A strength of Einstein's mind was that it could juggle a variety of ideas simultaneously. Even as he was pondering dancing particles in a liquid, he had been wrestling with a different theory that involved moving bodies and the speed of light. A day or so after sending in his Brownian motion paper, he was talking to his friend Michele Besso when a new brainstorm struck. It would produce, as he wrote Habicht in his famous letter of that month, "a modification of the theory of space and time."

CHAPTER SIX

SPECIAL RELATIVITY

1905

The Bern Clock Tower

The Background

Relativity is a simple concept. It asserts that the fundamental laws of physics are the same whatever your state of motion.

For the *special* case of observers moving at a *constant velocity,* this concept is pretty easy to accept. Imagine a man in an armchair at home and a woman in an airplane gliding very smoothly above. Each can pour a cup of coffee, bounce a ball, shine a flashlight, or heat a muffin in a microwave and have the same laws of physics apply.

In fact, there is no way to determine which of them is "in motion" and which is "at rest." The man in the armchair could consider himself at rest and the plane in motion. And the woman in the plane could consider herself at rest and the earth as gliding past. There is no experiment that can prove who is right.

Indeed, there is no absolute right. All that can be said is that each is moving relative to the other. And of course, both are moving very rapidly relative to other planets, stars, and galaxies.*

The special theory of relativity that Einstein developed in 1905 applies only to this special case (hence the name): a situation in which the observers are moving at a constant velocity relative to one another—uniformly in a straight line at a steady speed—referred to as an "inertial reference system."[1]

It's harder to make the more general case that a person who is accelerating or turning or rotating or slamming on the brakes or moving in an arbitrary manner is not in some form of absolute motion, because coffee sloshes and balls roll away in a different manner than for people on a smoothly gliding train, plane, or planet. It would take Einstein a decade more, as we shall see, to come up with what he called a *general* theory of relativity, which incorporated accelerated motion into a theory of gravity and attempted to apply the concept of relativity to it.[2]

The story of relativity best begins in 1632, when Galileo articulated the principle that the laws of motion and mechanics (the laws of electromagnetism had not yet been discovered) were the same in all constant-velocity reference frames. In his *Dialogue Concerning the Two Chief World Systems,* Galileo wanted to defend Copernicus's idea that the earth does not rest motionless at the center of the universe with everything else revolving around it. Skeptics contended that if the earth was moving, as Copernicus said, we'd feel it. Galileo refuted this with a brilliantly clear thought experiment about being inside the cabin of a smoothly sailing ship:

> Shut yourself up with some friend in the main cabin below decks on some large ship, and have with you there some flies, butterflies, and other small flying animals. Have a large bowl of water with some fish in

---

* A person "at rest" on the equator is actually spinning with the earth's rotation at 1,040 miles per hour and orbiting with the earth around the sun at 67,000 miles per hour. When I refer to these observers being at a constant velocity, I am ignoring the change in velocity that arises from being on a rotating and orbiting planet, which would not affect most common experiments. (See Miller 1999, 25.)

it; hang up a bottle that empties drop by drop into a wide vessel beneath it. With the ship standing still, observe carefully how the little animals fly with equal speed to all sides of the cabin. The fish swim indifferently in all directions; the drops fall into the vessel beneath; and, in throwing something to your friend, you need throw it no more strongly in one direction than another, the distances being equal; jumping with your feet together, you pass equal spaces in every direction. When you have observed all these things carefully, have the ship proceed with any speed you like, so long as the motion is uniform and not fluctuating this way and that. You will discover not the least change in all the effects named, nor could you tell from any of them whether the ship was moving or standing still.[3]

There is no better description of relativity, or at least of how that principle applies to systems that are moving at a constant velocity relative to each other.

Inside Galileo's ship, it is easy to have a conversation, because the air that carries the sound waves is moving smoothly along with the people in the chamber. Likewise, if one of Galileo's passengers dropped a pebble into a bowl of water, the ripples would emanate the same way they would if the bowl were resting on shore; that's because the water propagating the ripples is moving smoothly along with the bowl and everything else in the chamber.

Sound waves and water waves are easily explained by classical mechanics. They are simply a traveling disturbance in some medium. That is why sound cannot travel through a vacuum. But it can travel through such things as air or water or metal. For example, sound waves move through room temperature air, as a vibrating disturbance that compresses and rarefies the air, at about 770 miles per hour.

Deep inside Galileo's ship, sound and water waves behave as they do on land, because the air in the chamber and the water in the bowls are moving at the same velocity as the passengers. But now imagine that you go up on deck and look at the waves out in the ocean, or that you measure the speed of the sound waves from the horn of another boat. The speed at which these waves come toward you depends on your motion relative to the medium (the water or air) propagating them.

In other words, the speed at which an ocean wave reaches you will depend on how fast you are moving through the water toward or away from the source of the wave. The speed of a sound wave relative to you will likewise depend on your motion relative to the air that's propagating the sound wave.

Those relative speeds add up. Imagine that you are standing in the ocean as the waves come toward you at 10 miles per hour. If you jump on a Jet Ski and head directly into the waves at 40 miles per hour, you will see them moving toward you and zipping past you at a speed (relative to you) of 50 miles per hour. Likewise, imagine that sound waves are coming at you from a distant boat horn, rippling through still air at 770 miles per hour toward the shore. If you jump on your Jet Ski and head toward the horn at 40 miles per hour, the sound waves will be moving toward you and zipping past you at a speed (relative to you) of 810 miles per hour.

All of this led to a question that Einstein had been pondering since age 16, when he imagined riding alongside a light beam: Does light behave the same way?

Newton had conceived of light as primarily a stream of emitted particles. But by Einstein's day, most scientists accepted the rival theory, propounded by Newton's contemporary Christiaan Huygens, that light should be considered a wave.

A wide variety of experiments had confirmed the wave theory by the late nineteenth century. For example, Thomas Young did a famous experiment, now replicated by high school students, showing how light passing through two slits produces an interference pattern that resembles that of water waves going through two slits. In each case, the crests and troughs of the waves emanating from each slit reinforce each other in some places and cancel each other out in some places.

James Clerk Maxwell helped to enshrine this wave theory when he successfully conjectured a connection between light, electricity, and magnetism. He came up with equations that described the behavior of electric and magnetic fields, and when they were combined they predicted electromagnetic waves. Maxwell found that these electromagnetic waves had to travel at a certain speed: approximately 186,000 miles per second.* That was the speed that scientists had already measured for light, and it was obviously not a mere coincidence.[4]

It became clear that light was the visible manifestation of a whole spectrum of electromagnetic waves. This includes what we now call AM radio signals (with a wavelength of 300 yards), FM radio signals (3 yards), and microwaves (3 inches). As the wavelengths get shorter (and the frequency of the wave cycles thus increases), they produce the spectrum of visible light, ranging from red (25 millionths of an inch) to violet (14 millionths of an inch). Even shorter wavelengths produce ultraviolet rays, X-rays, and gamma rays. When we speak of "light" and the "speed of light," we mean all electromagnetic waves, not just the ones that are visible to our eyes.

That raised some big questions: What was the medium that was propagating these waves? And their speed of 186,000 miles per second was a speed *relative to what*?

The answer, it seemed, was that light waves are a disturbance of an unseen medium, which was called the ether, and that their speed is relative to this ether. In other words, the ether was for light waves something akin to what air was for sound waves. "It appeared beyond question that light must be interpreted as a vibratory process in an elastic, inert medium filling up universal space," Einstein later noted.[5]

This ether, unfortunately, needed to have many puzzling properties. Because light from distant stars is able to reach the earth, the ether had to pervade the entire known universe. It had to be so gossamer and, shall we say, so ethereal that it had no effect on planets and feathers floating through it. Yet it had to be stiff enough to allow a wave to vibrate through it at an enormous speed.

All of this led to the great ether hunt of the late nineteenth century. If light was indeed a wave rippling through the ether, then you should see the waves going by you at a faster speed if you were moving *through*

---

* More precisely, 186,282.4 miles per second or 299,792,458 meters per second, in a vacuum. Unless otherwise specified, the "speed of light" is for light in a vacuum and refers to all electromagnetic waves, visible or not. This is also, as Maxwell discovered, the speed of electricity through a wire.

the ether toward the light source. Scientists devised all sorts of ingenious devices and experiments to detect such differences.

They used a variety of suppositions of how the ether might behave. They looked for it as if it were motionless and the earth passed freely through it. They looked for it as if the earth dragged parts of it along in a blob, the way it does its own atmosphere. They even considered the unlikely possibility that the earth was the only thing at rest with respect to the ether, and that everything else in the cosmos was spinning around, including the other planets, the sun, the stars, and presumably poor Copernicus in his grave.

One experiment, which Einstein later called "of fundamental importance in the special theory of relativity,"[6] was by the French physicist Hippolyte Fizeau, who sought to measure the speed of light in a moving medium. He split a light beam with a half-silvered angled mirror that sent one part of the beam through water in the direction of the water's flow and the other part against the flow. The two parts of the beam were then reunited. If one route took longer, then the crests and troughs of its waves would be out of sync with the waves of the other beam. The experimenters could tell if this happened by looking at the interference pattern that resulted when the waves were rejoined.

A different and far more famous experiment was done in Cleveland in 1887 by Albert Michelson and Edward Morley. They built a contraption that similarly split a light beam and sent one part back and forth to a mirror at the end of an arm facing in the direction of the earth's movement and the other part back and forth along an arm at a 90-degree angle to it. Once again, the two parts of the beam were then rejoined and the interference pattern analyzed to see if the path that was going up against the supposed ether wind would take longer.

No matter who looked, or how they looked, or what suppositions they made about the behavior of the ether, no one was able to detect the elusive substance. No matter which way anything was moving, the speed of light was observed to be exactly the same.

So scientists, somewhat awkwardly, turned their attention to coming up with explanations about why the ether existed but was undetectable in any experiment. Most notably, in the early 1890s Hendrik Lorentz—the cosmopolitan and congenial Dutch father figure of theoretical physics—and, independently, the Irish physicist George Fitzgerald came up with the hypothesis that solid objects contracted slightly when they moved through the ether. The Lorentz-Fitzgerald contraction would shorten everything, including the measuring arms used by Michelson and Morley, and it would do so by just the exact amount to make the effect of the ether on light undetectable.

Einstein felt that the situation "was very depressing." Scientists found themselves unable to explain electromagnetism using the Newtonian "mechanical view of nature," he said, and this "led to a fundamental dualism which in the long run was insupportable."[7]

Einstein's Road to Relativity

"A new idea comes suddenly and in a rather intuitive way," Einstein once said. "But," he hastened to add, "intuition is nothing but the outcome of earlier intellectual experience."[8]

Einstein's discovery of special relativity involved an intuition based on a decade of intellectual as well as personal experiences.[9] The most important and obvious, I think, was his deep understanding and knowledge of theoretical physics. He was also helped by his ability to visualize thought experiments, which had been encouraged by his education in Aarau. Also, there was his grounding in philosophy: from Hume and Mach he had developed a skepticism about things that could not be observed. And this skepticism was enhanced by his innate rebellious tendency to question authority.

Also part of the mix—and probably reinforcing his ability to both visualize physical situations and to cut to the heart of concepts—was the technological backdrop of his life: helping his uncle Jakob to refine the moving coils and magnets in a generator; working in a patent office that was being flooded with applications for new methods of coordinating clocks; having a boss who encouraged him to apply his skepticism; living near the clock tower and train station and just above the telegraph office in Bern just as Europe was using electrical signals to synchronize clocks within time zones; and having as a sounding board his engineer friend Michele Besso, who worked with him at the patent office, examining electromechanical devices.[10]

The ranking of these influences is, of course, a subjective judgment. After all, even Einstein himself could not be sure how the process unfolded. "It is not easy to talk about how I arrived at the theory of relativity," he said. "There were so many hidden complexities to motivate my thought."[11]

One thing we can note with some confidence is Einstein's main starting point. He repeatedly said that his path toward the theory of relativity began with his thought experiment at age 16 about what it would be like to ride at the speed of light alongside a light beam. This produced a "paradox," he said, and it troubled him for the next ten years:

> If I pursue a beam of light with the velocity c (velocity of light in a vacuum), I should observe such a beam of light as an electromagnetic field at rest though spatially oscillating. There seems to be no such thing, however, neither on the basis of experience nor according to Maxwell's equations. From the very beginning it appeared to me intuitively clear that, judged from the standpoint of such an observer, everything would have to happen according to the same laws as for an observer who, relative to the earth, was at rest. For how should the first observer know or be able to determine that he is in a state of fast uniform motion? One sees in this paradox the germ of the special relativity theory is already contained.[12]

This thought experiment did not necessarily undermine the ether theory of light waves. An ether theorist could imagine a frozen light beam. But it violated Einstein's intuition that the laws of optics should obey the principle of relativity. In other words, Maxwell's equations, which specify the speed of light, should be the same for all observers in constant-velocity motion. The emphasis that Einstein placed on this memory indicates that the idea of a frozen light beam—or frozen electromagnetic waves—seemed instinctively wrong to him.[13]

In addition, the thought experiment suggests that he sensed a conflict between Newton's laws of mechanics and the constancy of the speed of light in Maxwell's equations. All of this instilled in him "a state of psychic tension" that he found deeply unnerving. "At the very beginning, when the special theory of relativity began to germinate in me, I was visited by all sorts of nervous conflicts," he later recalled. "When young, I used to go away for weeks in a state of confusion."[14]

There was also a more specific "asymmetry" that began to bother him. When a magnet moves relative to a wire loop, an electric current is produced. As Einstein knew from his experience with his family's generators, the amount of this electric current is exactly the same whether the magnet is moving while the coil seems to be sitting still, or the coil is moving while the magnet seems to be sitting still. He also had studied an 1894 book by August Föppl, *Introduction to Maxwell's Theory of Electricity*. It had a section specifically on "The Electrodynamics of Moving Conductors" that questioned whether, when induction occurs, there should be any distinction between whether the magnet or the conducting coil is said to be in motion.[15]

"But according to the Maxwell-Lorentz theory," Einstein recalled, "the theoretical interpretation of the phenomenon is very different for the two cases." In the first case, Faraday's law of induction said that the motion of the magnet through the ether created an electric field. In the second case, Lorentz's force law said a current was created by the motion of the conducting coil through the magnetic field. "The idea that these two cases should essentially be different was unbearable to me," Einstein said.[16]

Einstein had been wrestling for years with the concept of the ether, which theoretically determined the definition of "at rest" in these electrical induction theories. As a student at the Zurich Polytechnic in 1899, he had written to Mileva Marić that "the introduction of the term 'ether' into theories of electricity has led to the conception of a medium whose motion can be described without, I believe, being able to ascribe physical meaning to it."[17] Yet that very month he was on vacation in Aarau working with a teacher at his old school on ways to detect the ether. "I had a good idea for investigating the way in which a body's relative motion with respect to the ether affects the velocity of the propagation of light," he told Marić.

Professor Weber told Einstein that his approach was impractical. Probably at Weber's suggestion, Einstein then read a paper by Wilhelm Wien that described the null results of thirteen ether-detection experiments, including those by Michelson and Morley and by

Fizeau.[18] He also learned about the Michelson-Morley experiment by reading, sometime before 1905, Lorentz's 1895 book, *Attempt at a Theory of Electrical and Optical Phenomena in Moving Bodies*. In this book, Lorentz goes through various failed attempts to detect the ether as a prelude to developing his theory of contractions.[19]

"Induction and Deduction in Physics"

So what effect did the Michelson-Morley results—which showed no evidence of the ether and no difference in the observed speed of light no matter in what direction the observer was moving—have on Einstein as he was incubating his ideas on relativity? To hear him tell it, almost none at all. In fact, at times he would even recollect (incorrectly) that he had not even known of the experiment before 1905. Einstein's inconsistent statements over the next fifty years about the influence of Michelson-Morley are useful in that they remind us of the caution needed when writing history based on dimming recollections.[20]

Einstein's trail of contradictory statements begins with an address he gave in Kyoto, Japan, in 1922, when he noted that Michelson's failure to detect an ether was "the first path that led me to what we call the principle of special relativity." In a toast at a 1931 dinner in Pasadena honoring Michelson, Einstein was gracious to the eminent experimenter, yet subtly circumspect: "You uncovered an insidious defect in the ether theory of light, as it then existed, and stimulated the ideas of Lorentz and Fitzgerald, out of which the Special Theory of Relativity developed."[21]

Einstein described his thought process in a series of talks with the Gestalt psychology pioneer Max Wertheimer, who later called the Michelson-Morley results "crucial" to Einstein's thinking. But as Arthur I. Miller has shown, this assertion was probably motivated by Wertheimer's goal of using Einstein's tale as a way to illustrate the tenets of Gestalt psychology.[22]

Einstein further confused the issue in the last few years of his life by giving a series of statements on the subject to a physicist named Robert Shankland. At first he said he had read of Michelson-Morley only *after* 1905, then he said he had read about it in Lorentz's book *before* 1905, and finally he added, "I guess I just took it for granted that it was true."[23]

That final point is the most significant one because Einstein made it often. He simply took for granted, by the time he started working seriously on relativity, that there was no need to review all the ether-drift experiments because, based on his starting assumptions, all attempts to detect the ether were doomed to failure.[24] For him, the significance of these experimental results was to reinforce what he already believed: that Galileo's relativity principle applied to light waves.[25]

This may account for the scant attention he gave to the experiments in his 1905 paper. He never mentioned the Michelson-Morley experiment by name, even where it would have been relevant, nor the Fizeau experiment using moving water. Instead, right after discussing the relativity of the magnet-and-coil movements, he merely flicked in a phrase about "the unsuccessful attempts to detect a motion of the earth relative to the light medium."

Some scientific theories depend primarily on induction: analyzing a lot of experimental findings and then finding theories that explain the empirical patterns. Others depend more on deduction: starting with elegant principles and postulates that are embraced as holy and then deducing the consequences from them. All scientists blend both approaches to differing degrees. Einstein had a good feel for experimental findings, and he used this knowledge to find certain fixed points upon which he could construct a theory.[26] But his emphasis was primarily on the deductive approach.[27]

Remember how in his Brownian motion paper he so oddly, yet accurately, downplayed the role that experimental findings played in what was essentially a theoretical deduction? There was a similar situation with his relativity theory. What he implied about Brownian motion he said explicitly about relativity and Michelson-Morley: "I was pretty much convinced of the validity of the principle before I knew of this experiment and its results."

Indeed, all three of his epochal papers in 1905 begin by asserting his intention to pursue a deductive approach. He opens each one by pointing out some oddity caused by jostling theories, rather than some unexplained set of experimental data. He then postulates grand principles while minimizing the role played by data, be it on Brownian motion or blackbody radiation or the speed of light.[28]

In a 1919 essay called "Induction and Deduction in Physics," he described his preference for the latter approach:

> The simplest picture one can form about the creation of an empirical science is along the lines of an inductive method. Individual facts are selected and grouped together so that the laws that connect them become apparent . . . However, the big advances in scientific knowledge originated in this way only to a small degree . . . The truly great advances in our understanding of nature originated in a way almost diametrically opposed to induction. The intuitive grasp of the essentials of a large complex of facts leads the scientist to the postulation of a hypothetical basic law or laws. From these laws, he derives his conclusions.[29]

His appreciation for this approach would grow. "The deeper we penetrate and the more extensive our theories become," he would declare near the end of his life, "the less empirical knowledge is needed to determine those theories."[30]

By the beginning of 1905, Einstein had begun to emphasize deduction rather than induction in his attempt to explain electrodynamics. "By and by, I despaired of the possibility of discovering the true laws by means of constructive efforts based on experimentally known facts," he later said. "The longer and the more despairingly I tried, the more I came to the conviction that only the discovery of a universal formal principle could lead us to assured results."[31]

The Two Postulates

Now that Einstein had decided to pursue his theory from the top down, by deriving it from grand postulates, he had a choice to make: What postulates—what basic assumptions of general principle—would he start with?[32]

His first postulate was the principle of relativity, which asserted that all of the fundamental laws of physics, even Maxwell's equations governing electromagnetic waves, are the same for all observers moving at constant velocity relative to each other. Put more precisely, they are the same for all inertial reference systems, the same for someone at rest relative to the earth as for someone traveling at a uniform velocity on a train or spaceship. He had nurtured his faith in this postulate beginning with his thought experiment about riding alongside a light beam: "From the very beginning it appeared to me intuitively clear that, judged from the standpoint of such an observer, everything would have to happen according to the same laws as for an observer who, relative to the earth, was at rest."

For a companion postulate, involving the velocity of light, Einstein had at least two options:

1. He could go with an emission theory, in which light would shoot from its source like particles from a gun. There would be no need for an ether. The light particles could zoom through emptiness. Their speed would be relative to the source. If this source was racing toward you, its emissions would come at you faster than if it was racing away. (Imagine a pitcher who can throw a ball at 100 miles per hour. If he throws it at you from a car racing toward you it will come at you faster than if he throws it from a car racing away.) In other words, starlight would be emitted from a star at 186,000 miles per second; but if that star was heading toward earth at 10,000 miles per second, the speed of its light would be 196,000 miles per second relative to an observer on earth.

2. An alternative was to postulate that the speed of light was a constant 186,000 miles per second irrespective of the motion of the source that emitted it, which was more consistent with a wave theory. By analogy with sound waves, a fire truck siren does not throw its sound at you faster when it's rushing toward you than it does when it's standing still. In either case, the sound travels through the air at 770 miles per hour.*

---

* If the source of sound is rushing toward you, the waves will not get to you any faster. However, in what is known as the Doppler effect, the waves will be compressed and the interval between them will be smaller. The decreased wavelength means a higher frequency, which results in a higher-pitched sound (or a lower one, when the siren passes by and starts moving away). A similar effect happens with light. If the source is moving toward you, the wavelength decreases (and frequency increases) so it is shifted to the blue end of the spectrum. Light from a source moving away will be red-shifted.

For a while, Einstein explored the emission theory route. This approach was particularly appealing if you conceived of light as behaving like a stream of quanta. And as noted in the previous chapter, that concept of light quanta was precisely what Einstein had propounded in March 1905, just when he was wrestling with his relativity theory.[33]

But there were problems with this approach. It seemed to entail abandoning Maxwell's equations and the wave theory. If the velocity of a light wave depended on the velocity of the source that emitted it, then the light wave must somehow encode within it this information. But experiments and Maxwell's equations indicated that was not the case.[34]

Einstein tried to find ways to modify Maxwell's equations so that they would fit an emission theory, but the quest became frustrating. "This theory requires that everywhere and in each fixed direction light waves of a different velocity of propagation should be possible," he later recalled. "It may be impossible to set up a reasonable electromagnetic theory that accomplishes such a feat."[35]

In addition, scientists had not been able to find any evidence that the velocity of light depended on that of its source. Light coming from any star seemed to arrive at the same speed.[36]

The more Einstein thought about an emission theory, the more problems he encountered. As he explained to his friend Paul Ehrenfest, it was hard to figure out what would happen when light from a "moving" source was refracted or reflected by a screen at rest. Also, in an emission theory, light from an accelerating source might back up on itself.

So Einstein rejected the emission theory in favor of postulating that the speed of a light beam was constant no matter how fast its source was moving. "I came to the conviction that all light should be defined by frequency and intensity alone, completely independently of whether it comes from a moving or from a stationary light source," he told Ehrenfest.[37]

Now Einstein had two postulates: "the principle of relativity" and this new one, which he called "the light postulate." He defined it carefully: "Light always propagates in empty space with a definite velocity V that is independent of the state of motion of the emitting body."[38]

For example, when you measure the velocity of light coming from the headlight of a train, it will always be a constant 186,000 miles per second, even if the train is rushing toward you or backing away from you.

Unfortunately, this light postulate seemed to be incompatible with the principle of relativity. Why? Einstein later used the following thought experiment to explain his apparent dilemma.

Imagine that "a ray of light is sent along the embankment" of a railway track, he said. A man standing on the embankment would measure its speed as 186,000 miles per second as it zipped past him. But now imagine a woman who is riding in a very fast train carriage that is racing away from the light source at 2,000 miles per second. We would assume that she would observe the beam to be zipping past her at only 184,000 miles per second. "The velocity of propagation of a ray of light relative to the carriage thus comes out smaller," Einstein wrote.

"But this result comes into conflict with the principle of relativity," he added. "For, like every other general law of nature, the law of the transmission of light must, according to the principle of relativity, be the same when the railway carriage is the reference body as it is when the embankment is the reference body." In other words, Maxwell's equations, which determine the speed at which light propagates, should operate the same way in the moving carriage as on the embankment. There should be no experiment you can do, including measuring the speed of light, to distinguish which inertial frame of reference is "at rest" and which is moving at a constant velocity.[39]

This was an odd result. A woman racing along the tracks toward or away from the source of a light beam should see that beam zip by her with the exact same speed as an observer standing on the embankment would see that same beam zip by him. The woman's speed relative to the train would vary, depending on whether she was running toward it or away from it. But her speed relative to the light beam coming from the train's headlight would be invariant. All of this made the two postulates, Einstein thought, "seemingly incompatible." As he later explained in a lecture on how he came to his theory, "the constancy of the velocity of light is not consistent with the law of the addition of velocities. The result was that I had to spend almost one year in fruitless thoughts."[40]

By combining the light postulate with the principle of relativity, it meant that an observer would measure the speed of light as the same whether the source was moving toward or away from him, or whether he was moving toward or away from the source, or both, or neither. The speed of light would be the same whatever the motion of the observer and the source.

That is where matters stood in early May 1905. Einstein had embraced the relativity principle and elevated it to a postulate. Then, with a bit more trepidation, he had adopted as a postulate that the velocity of light was independent of the motion of its source. And he puzzled over the apparent dilemma that an observer racing up a track toward a light would see the beam coming at him with the same velocity as when he was racing away from the light—and with the same velocity as someone standing still on the embankment would observe the same beam.

"In view of this dilemma, there appears to be nothing else to do than to abandon either the principle of relativity or the simple law of the propagation of light," Einstein wrote.[41]

Then something delightful happened. Albert Einstein, while talking with a friend, took one of the most elegant imaginative leaps in the history of physics.

"The Step"

It was a beautiful day in Bern, Einstein later remembered, when he went to visit his best friend Michele Besso, the brilliant but unfocused engineer he had met while studying in Zurich and then recruited to join him at the Swiss Patent Office. Many days they would walk to work together, and on this occasion Einstein told Besso about the dilemma that was dogging him.

"I'm going to give it up," Einstein said at one point. But as they discussed it, Einstein recalled, "I suddenly understood the key to the problem." The next day, when he saw Besso, Einstein was in a state of great excitement. He skipped any greeting and immediately declared, "Thank you. I've completely solved the problem."[42]

Only five weeks elapsed between that eureka moment and the day that Einstein sent off his most famous paper, "On the Electrodynamics of Moving Bodies." It contained no citations of other literature, no mention of anyone else's work, and no acknowledgments except for the charming one in the last sentence: "Let me note that my friend and colleague M. Besso steadfastly stood by me in my work on the problem discussed here, and that I am indebted to him for several valuable suggestions."

So what was the insight that struck him while talking to Besso? "An analysis of the concept of time was my solution," Einstein said. "Time cannot be absolutely defined, and there is an inseparable relation between time and signal velocity."

More specifically, the key insight was that two events that appear to be simultaneous to one observer will not appear to be simultaneous to another observer who is moving rapidly. And there is no way to declare that one of the observers is really correct. In other words, there is no way to declare that the two events are truly simultaneous.

Einstein later explained this concept using a thought experiment involving moving trains. Suppose lightning bolts strike the train track's embankment at two distant places, A and B. If we declare that they struck simultaneously, what does that mean?

Einstein realized that we need an operational definition, one we can actually apply, and that would require taking into account the speed of light. His answer was that we would define the two strikes as simultaneous if we were standing exactly halfway between them and the light from each reached us at the exact same time.

But now let us imagine how the event looks to a train passenger who is moving rapidly along the track. In a 1916 book written to explain this to nonscientists, he used the following drawing, in which the long train is the line on the top:

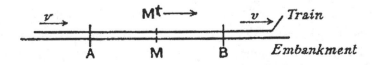

Suppose that at the exact instant (from the viewpoint of the person on the embankment) when lightning strikes at points A and B, there is a passenger at the midpoint of the train, M^t, just passing the observer who is at the midpoint alongside the tracks, M. If the train was motionless relative to the embankment, the passenger inside would see the lightning flashes simultaneously, just as the observer on the embankment would.

But if the train is moving to the right relative to the embankment, the observer inside will be rushing closer toward place B while the light signals are traveling. Thus he will be positioned slightly to the right by the time the light arrives; as a result, he will see the light from the strike at place B *before* he will see the light from the strike at place A. So he will assert that lightning hit at B before it did so at A, and the strikes were not simultaneous.

"We thus arrive at the important result: Events that are simultaneous with reference to the embankment are not simultaneous with respect to the train," said Einstein. The principle of relativity says that there is no way to decree that the embankment is "at rest" and the train "in motion." We can say only that they are in motion relative to each other. So there is no "real" or "right" answer. There is no way to say that any two events are "absolutely" or "really" simultaneous.[43]

This is a simple insight, but also a radical one. It means that *there is no absolute time.* Instead, all moving reference frames have their own relative time. Although Einstein refrained from saying that this leap was as truly "revolutionary" as the one he made about light quanta, it did in fact transform science. "This was a change in the very foundation of physics, an unexpected and very radical change that required all the courage of a young and revolutionary genius," noted Werner Heisenberg, who later contributed to a similar feat with his principle of quantum uncertainty.[44]

In his 1905 paper, Einstein used a vivid image, which we can imagine him conceiving as he watched the trains moving into the Bern station past the rows of clocks that were synchronized with the one atop the town's famed tower. "Our judgments in which time plays a part are always judgments of simultaneous events," he wrote. "If, for instance, I say, 'That train arrives here at 7 o'clock,' I mean something like this:

'The pointing of the small hand of my watch to 7 and the arrival of the train are simultaneous events.' " Once again, however, observers who are moving rapidly relative to one another will have a different view on whether two distant events are simultaneous.

The concept of absolute time—meaning a time that exists in "reality" and tick-tocks along independent of any observations of it—had been a mainstay of physics ever since Newton had made it a premise of his *Principia* 216 years earlier. The same was true for absolute space and distance. "Absolute, true, and mathematical time, of itself and from its own nature, flows equably without relation to anything external," he famously wrote in Book 1 of the *Principia*. "Absolute space, in its own nature, without relation to anything external, remains always similar and immovable."

But even Newton seemed discomforted by the fact that these concepts could not be directly observed. "Absolute time is not an object of perception," he admitted. He resorted to relying on the presence of God to get him out of the dilemma. "The Deity endures forever and is everywhere present, and by existing always and everywhere, He constitutes duration and space."[45]

Ernst Mach, whose books had influenced Einstein and his fellow members of the Olympia Academy, lambasted Newton's notion of absolute time as a "useless metaphysical concept" that "cannot be produced in experience." Newton, he charged, "acted contrary to his expressed intention only to investigate actual facts."[46]

Henri Poincaré also pointed out the weakness of Newton's concept of absolute time in his book *Science and Hypothesis,* another favorite of the Olympia Academy. "Not only do we have no direct intuition of the equality of two times, we do not even have one of the simultaneity of two events occurring in different places," he wrote.[47]

Both Mach and Poincaré were, it thus seems, useful in providing a foundation for Einstein's great breakthrough. But he owed even more, he later said, to the skepticism he learned from the Scottish philosopher David Hume regarding mental constructs that were divorced from purely factual observations.

Given the number of times in his papers that he uses thought experiments involving moving trains and distant clocks, it is also logical to surmise that he was helped in visualizing and articulating his thoughts by the trains that moved past Bern's clock tower and the rows of synchronized clocks on the station platform. Indeed, there is a tale that involves him discussing his new theory with friends by pointing to (or at least referring to) the synchronized clocks of Bern and the unsynchronized steeple clock visible in the neighboring village of Muni.[48]

Peter Galison provides a thought-provoking study of the technological ethos in his book *Einstein's Clocks, Poincaré's Maps*. Clock coordination was in the air at the time. Bern had inaugurated an urban time network of electrically synchronized clocks in 1890, and a decade later, by the time Einstein had arrived, finding ways to make them more accurate and coordinate them with clocks in other cities became a Swiss passion.

In addition, Einstein's chief duty at the patent office, in partnership with Besso, was evaluating electromechanical devices. This included a flood of applications for ways to synchronize clocks by using electric signals. From 1901 to 1904, Galison notes, there were twenty-eight such patents issued in Bern.

One of them, for example, was called "Installation with Central Clock for Indicating the Time Simultaneously in Several Places Separated from One Another." A similar application arrived on April 25, just three weeks before Einstein had his breakthrough conversation with Besso; it involved a clock with an electromagnetically controlled pendulum that could be coordinated with another such clock through an electric signal. What these applications had in common was that they used signals that traveled at the speed of light.[49]

We should be careful not to overemphasize the role played by the technological backdrop of the patent office. Although clocks are part of Einstein's description of his theory, his point is about the difficulties that observers *in relative motion* have in using light signals to synchronize them, something that was not an issue for the patent applicants.[50]

Nevertheless, it is interesting to note that almost the entire first two sections of his relativity paper deal directly and in vivid practical detail (in a manner so different from the writings of, say, Lorentz and Maxwell) with the two real-world technological phenomena he knew best. He writes about the generation of "electric currents of the same

magnitude" due to the "equality of relative motion" of coils and magnets, and the use of "a light signal" to make sure that "two clocks are
synchronous."

As Einstein himself stated, his time in the patent office "stimulated
me to see the physical ramifications of theoretical concepts."[51] And
Alexander Moszkowski, who compiled a book in 1921 based on conversations with Einstein, noted that Einstein believed there was "a definite connection between the knowledge acquired at the patent office
and the theoretical results."[52]

"On the Electrodynamics of Moving Bodies"

Now let's look at how Einstein articulated all of this in the famous
paper that the *Annalen der Physik* received on June 30, 1905. For all its
momentous import, it may be one of the most spunky and enjoyable
papers in all of science. Most of its insights are conveyed in words and
vivid thought experiments, rather than in complex equations. There is
some math involved, but it is mainly what a good high school senior
could comprehend. "The whole paper is a testament to the power of
simple language to convey deep and powerfully disturbing ideas," says
the science writer Dennis Overbye.[53]

The paper starts with the "asymmetry" that a magnet and wire loop
induce an electric current based only on their relative motion to one
another, but since the days of Faraday there had been two different theoretical explanations for the current produced depending on whether it
was the magnet or the loop that was in motion.[54] "The observable phenomenon here depends only on the relative motion of the conductor
and the magnet," Einstein writes, "whereas the customary view draws a
sharp distinction between the two cases in which either the one or the
other of these bodies is in motion."[55]

The distinction between the two cases was based on the belief,
which most scientists still held, that there was such a thing as a state of
"rest" with respect to the ether. But the magnet-and-coil example,
along with every observation made on light, "suggest that the phenomena of electrodynamics as well as of mechanics possess no properties
corresponding to the idea of absolute rest." This prompts Einstein to

raise "to the status of a postulate" the principle of relativity, which holds that the laws of mechanics and electrodynamics are the same in all reference systems moving at constant velocity relative to one another.

Einstein goes on to propound the other postulate upon which his theory was premised: the constancy of the speed of light "independent of the state of motion of the emitting body." Then, with the casual stroke of a pen, and the marvelously insouciant word "superfluous," the rebellious patent examiner dismissed two generations' worth of accrued scientific dogma: "The introduction of a 'light ether' will prove to be superfluous, inasmuch as the view to be developed here will not require a 'space at absolute rest.' "

Using these two postulates, Einstein explained the great conceptual step he had taken during his talk with Besso. "Two events which, viewed from a system of coordinates, are simultaneous, can no longer be looked upon as simultaneous events when envisaged from a system which is in motion relative to that system." In other words, there is no such thing as absolute simultaneity.

In phrases so simple as to be seductive, Einstein pointed out that time itself can be defined only by referring to simultaneous events, such as the small hand of a watch pointing to 7 as a train arrives. The obvious yet still astonishing conclusion: with no such thing as absolute simultaneity, there is no such thing as "real" or absolute time. As he later put it, "There is no audible tick-tock everywhere in the world that can be considered as time." [56]

Moreover, this realization also meant overturning the other assumption that Newton made at the beginning of his *Principia*. Einstein showed that if time is relative, so too are space and distance: "If the man in the carriage covers the distance w in a unit of time—*measured from the train*—then this distance—*as measured from the embankment*—is not necessarily also equal to w." [57]

Einstein explained this by asking us to picture a rod that has a certain length when it is measured while it is stationary relative to the observer. Now imagine that the rod is moving. How long is the rod?

One way to determine this is by moving alongside the rod, at the same speed, and superimposing a measuring stick on it. But how long would the rod be if measured by someone *not* in motion with it? In that case, a way to measure the moving rod would be to determine, based on synchronized stationary clocks, the precise location of each end of the rod at a specific moment, and then use a stationary ruler to measure the distance between these two points. Einstein shows that these methods will produce *different* results.

Why? Because the two stationary clocks have been synchronized by a stationary observer. But what happens if an observer who is moving as fast as the rod tries to synchronize those clocks? She would synchronize them differently, because she would have a different perception of simultaneity. As Einstein put it, "Observers moving with the moving rod would thus find that the two clocks were not synchronous, while observers in the stationary system would declare the clocks to be synchronous."

Another consequence of special relativity is that a person standing on the platform will observe that time goes more slowly on a train speeding past. Imagine that on the train there is a "clock" made up of a mirror on the floor and one on the ceiling and a beam of light that bounces up and down between them. From the perspective of a woman on the train, the light goes straight up and then straight down. But from the perspective of a man standing on the platform, it appears that the light is starting at the bottom but moving on a diagonal to get to the ceiling mirror, which has zipped ahead a tiny bit, then bouncing down on a diagonal back to the mirror on the floor, which has in turn zipped ahead a tiny bit. For both observers, the speed of the light is the same (that is Einstein's great given). The man on the track observes the distance the light has to travel as being longer than the woman on the train observes it to be. Thus, from the perspective of the man on the track, time is going by more slowly inside the speeding train.[58]

Another way to picture this is to use Galileo's ship. Imagine a light beam being shot down from the top of the mast to the deck. To an observer on the ship, the light beam will travel the exact length of the mast. To an observer on land, however, the light beam will travel a diagonal formed by the length of the mast plus the distance (it's a *fast* ship) that the ship has traveled forward during the time it took the light to get from the top to the bottom of the mast. To both observers,

the speed of light is the same. To the observer on land, it traveled farther before it reached the deck. In other words, the exact same event (a
light beam sent from the top of the mast hitting the deck) took longer
when viewed by a person on land than by a person on the ship.[59]

This phenomenon, called time dilation, leads to what is known as
the twin paradox. If a man stays on the platform while his twin sister
takes off in a spaceship that travels long distances at nearly the speed of
light, when she returns she would be younger than he is. But because
motion is relative, this seems to present a paradox. The sister on the
spaceship might think it's her brother on earth who is doing the fast
traveling, and when they are rejoined she would expect to observe that
it was *he* who did not age much.

Could they each come back younger than the other one? Of course
not. The phenomenon does not work in both directions. Because the
spaceship does not travel at a *constant velocity*, but instead must turn
around, it's the twin on the spaceship, not the one on earth, who would
age more slowly.

The phenomenon of time dilation has been experimentally confirmed, even by using test clocks on commercial planes. But in our normal life, it has no real impact, because our motion relative to any other
observer is never anything near the speed of light. In fact, if you spent
almost your entire life on an airplane, you would have aged merely
0.00005 seconds or so less than your twin on earth when you returned,
an effect that would likely be counteracted by a lifetime spent eating
airline food.[60]

Special relativity has many other curious manifestations. Think
again about that light clock on the train. What happens as the train approaches the speed of light relative to an observer on the platform? It
would take almost forever for a light beam in the train to bounce from
the floor to the moving ceiling and back to the moving floor. Thus time
on the train would almost stand still from the perspective of an observer on the platform.

As an object approaches the speed of light, its apparent mass also
increases. Newton's law that force equals mass times acceleration still
holds, but as the apparent mass increases, more and more force will
produce less and less acceleration. There is no way to apply enough

force to push even a pebble faster than the speed of light. That's the ultimate speed limit of the universe, and no particle or piece of information can go faster than that, according to Einstein's theory.

With all this talk of distance and duration being relative depending on the observer's motion, some may be tempted to ask: So which observer is "right"? Whose watch shows the "actual" time elapsed? Which length of the rod is "real"? Whose notion of simultaneity is "correct"?

According to the special theory of relativity, all inertial reference frames are equally valid. It is not a question of whether rods *actually* shrink or time *really* slows down; all we know is that observers in different states of motion will measure things differently. And now that we have dispensed with the ether as "superfluous," there is no designated "rest" frame of reference that has preference over any other.

One of Einstein's clearest explanations of what he had wrought was in a letter to his Olympia Academy colleague Solovine:

> The theory of relativity can be outlined in a few words. In contrast to the fact, known since ancient times, that movement is perceivable only as *relative* movement, physics was based on the notion of *absolute* movement. The study of light waves had assumed that one state of movement, that of the light-carrying ether, is distinct from all others. All movements of bodies were supposed to be relative to the light-carrying ether, which was the incarnation of absolute rest. But after efforts to discover the privileged state of movement of this hypothetical ether through experiments had failed, it seemed that the problem should be restated. That is what the theory of relativity did. It assumed that there are no privileged physical states of movement and asked what consequences could be drawn from this.

Einstein's insight, as he explained it to Solovine, was that we must discard concepts that "have no link with experience," such as "absolute simultaneity" and "absolute speed."[61]

It is very important to note, however, that the theory of relativity does not mean that "everything is relative." It does not mean that everything is subjective.

Instead, it means that measurements of time, including duration and simultaneity, can be relative, depending on the motion of the observer. So can the measurements of space, such as distance and length.

But there is a union of the two, which we call spacetime, and that remains invariant in all inertial frames. Likewise, there are things such as the speed of light that remain invariant.

In fact, Einstein briefly considered calling his creation Invariance Theory, but the name never took hold. Max Planck used the term *Relativtheorie* in 1906, and by 1907 Einstein, in an exchange with his friend Paul Ehrenfest, was calling it *Relativitätstheorie*.

One way to understand that Einstein was talking about invariance, rather than declaring everything to be relative, is to think about how far a light beam would travel in a given period of time. That distance would be the speed of light multiplied by the amount of time it traveled. If we were on a platform observing this happening on a train speeding by, the elapsed time would appear shorter (time seems to move more slowly on the moving train), and the distance would appear shorter (rulers seem to be contracted on the moving train). But there is a relationship between the two quantities—a relationship between the measurements of space and of time—that remains invariant, whatever your frame of reference.[62]

A more complex way to understand this is the method used by Hermann Minkowski, Einstein's former math teacher at the Zurich Polytechnic. Reflecting on Einstein's work, Minkowski uttered the expression of amazement that every beleaguered student wants to elicit someday from condescending professors. "It came as a tremendous surprise, for in his student days Einstein had been a lazy dog," Minkowski told physicist Max Born. "He never bothered about mathematics at all."[63]

Minkowski decided to give a formal mathematical structure to the theory. His approach was the same one suggested by the time traveler on the first page of H. G. Wells's great novel *The Time Machine*, published in 1895: "There are really four dimensions, three which we call the three planes of Space, and a fourth, Time." Minkowski turned all events into mathematical coordinates in four dimensions, with time as the fourth dimension. This permitted transformations to occur, but the mathematical relationships between the events remained invariant.

Minkowski dramatically announced his new mathematical approach in a lecture in 1908. "The views of space and time which I wish to lay before you have sprung from the soil of experimental physics, and therein lies their strength," he said. "They are radical. Henceforth space by itself, and time by itself, are doomed to fade away into mere shadows, and only a kind of union of the two will preserve an independent reality."[64]

Einstein, who was still not yet enamored of math, at one point described Minkowski's work as "superfluous learnedness" and joked, "Since the mathematicians have grabbed hold of the theory of relativity, I myself no longer understand it." But he in fact came to admire Minkowski's handiwork and wrote a section about it in his popular 1916 book on relativity.

What a wonderful collaboration it could have been! But at the end of 1908, Minkowski was taken to the hospital, fatally stricken with peritonitis. Legend has it that he declared, "What a pity that I have to die in the age of relativity's development."[65]

Once again, it's worth asking why Einstein discovered a new theory and his contemporaries did not. Both Lorentz and Poincaré had already come up with many of the components of Einstein's theory. Poincaré even questioned the absolute nature of time.

But neither Lorentz nor Poincaré made the full leap: that there is no need to posit an ether, that there is no absolute rest, that time is relative based on an observer's motion, and so is space. Both men, the physicist Kip Thorne says, "were groping toward the same revision of our notions of space and time as Einstein, but they were groping through a fog of misperceptions foisted on them by Newtonian physics."

Einstein, by contrast, was able to cast off Newtonian misconceptions. "His conviction that the universe loves simplification and beauty, and his willingness to be guided by this conviction, even if it meant destroying the foundations of Newtonian physics, led him, with a clarity of thought that others could not match, to his new description of space and time."[66]

Poincaré never made the connection between the relativity of simultaneity and the relativity of time, and he "drew back when on the brink" of understanding the full ramifications of his ideas about local time. Why did he hesitate? Despite his interesting insights, he was too

much of a traditionalist in physics to display the rebellious streak ingrained in the unknown patent examiner.[67] "When he came to the decisive step, his nerve failed him and he clung to old habits of thought
and familiar ideas of space and time," Banesh Hoffmann said of Poincaré. "If this seems surprising, it is because we underestimate the bold-
ness of Einstein in stating the principle of relativity as an axiom and, by
keeping faith with it, changing our notion of space and time."[68]

A clear explanation of Poincaré's limitations and Einstein's bold-
ness comes from one of Einstein's successors as a theoretical physicist
at the Institute for Advanced Study in Princeton, Freeman Dyson:

> The essential difference between Poincaré and Einstein was that Poin-
> caré was by temperament conservative and Einstein was by tempera-
> ment revolutionary. When Poincaré looked for a new theory of
> electromagnetism, he tried to preserve as much as he could of the old.
> He loved the ether and continued to believe in it, even when his own
> theory showed that it was unobservable. His version of relativity theory
> was a patchwork quilt. The new idea of local time, depending on the
> motion of the observer, was patched onto the old framework of absolute
> space and time defined by a rigid and immovable ether. Einstein, on the
> other hand, saw the old framework as cumbersome and unnecessary and
> was delighted to be rid of it. His version of the theory was simpler and
> more elegant. There was no absolute space and time and there was no
> ether. All the complicated explanations of electric and magnetic forces
> as elastic stresses in the ether could be swept into the dustbin of history,
> together with the famous old professors who still believed in them.[69]

As a result, Poincaré expressed a principle of relativity that contained certain similarities to Einstein's, but it had a fundamental difference. Poincaré retained the existence of the ether, and the speed of light
was, for him, constant only when measured by those at rest to this presumed ether's frame of reference.[70]

Even more surprising, and revealing, is the fact that Lorentz and
Poincaré never were able to make Einstein's leap even *after* they read
his paper. Lorentz still clung to the existence of the ether and its "at
rest" frame of reference. In a lecture in 1913, which he reprinted in his
1920 book *The Relativity Principle,* Lorentz said, "According to Einstein, it is meaningless to speak of motion relative to the ether. He likewise denies the existence of absolute simultaneity. As far as this lecturer

is concerned, he finds a certain satisfaction in the older interpretations, according to which the ether possesses at least some substantiality, space and time can be sharply separated, and simultaneity without further specification can be spoken of."[71]

For his part, Poincaré seems never to have fully understood Einstein's breakthrough. Even in 1909, he was still insisting that relativity theory required a third postulate, which was that "a body in motion suffers a deformation in the direction in which it was displaced." In fact, the contraction of rods is not, as Einstein showed, some separate hypothesis involving a real deformation, but rather the consequence of accepting Einstein's theory of relativity.

Until his death in 1912, Poincaré never fully gave up the concept of the ether or the notion of absolute rest. Instead, he spoke of the adoption of "the principle of relativity according to Lorentz." He never fully understood or accepted the basis of Einstein's theory. "Poincaré stood steadfast and held to his position that in the world of perceptions there was an absoluteness of simultaneity," notes the science historian Arthur I. Miller.[72]

His Partner

"How happy and proud I will be when the two of us together will have brought our work on the relative motion to a conclusion!" Einstein had written his lover Mileva Marić back in 1901.[73] Now it had been brought to that conclusion, and Einstein was so exhausted when he finished a draft in June that "his body buckled and he went to bed for two weeks," while Marić "checked the article again and again."[74]

Then they did something unusual: they celebrated together. As soon as he finished all four of the papers that he had promised in his memorable letter to Conrad Habicht, he sent his old colleague from the Olympia Academy another missive, this one a postcard signed by his wife as well. It read in full: "Both of us, alas, dead drunk under the table."[75]

All of which raises a question more subtle and contentious than that posed by the influences of Lorentz and Poincaré: What was Mileva Marić's role?

That August, they took a vacation together in Serbia to see her friends and family. While there, Marić was proud and also willing to accept part of the credit. "Not long ago we finished a very significant work that will make my husband world famous," she told her father, according to stories later recorded there. Their relationship seemed restored, for the time being, and Einstein happily praised his wife's help. "I need my wife," he told her friends in Serbia. "She solves all the mathematical problems for me."[76]

Some have contended that Marić was a full-fledged collaborator, and there was even a report, later discredited,[77] that an early draft version of his relativity paper had her name on it as well. At a 1990 conference in New Orleans, the American Association for the Advancement of Science held a panel on the issue at which Evan Walker, a physicist and cancer researcher from Maryland, debated John Stachel, the leader of the Einstein Papers Project. Walker presented the various letters referring to "our work," and Stachel replied that such phrases were clearly romantic politeness and that there was "no evidence at all that she contributed any ideas of her own."

The controversy, understandably, fascinated both scientists and the press. Columnist Ellen Goodman wrote a wry commentary in the *Boston Globe*, in which she judiciously laid out the evidence, and the *Economist* did a story headlined "The Relative Importance of Mrs. Einstein." Another conference followed in 1994 at the University of Novi Sad, where organizer Professor Rastko Maglić contended that it was time "to emphasize Mileva's merit in order to ensure a deserved place in the history of science for her." The public discussion culminated with a PBS documentary, *Einstein's Wife*, in 2003, that was generally balanced, although it gave unwarranted credence to the report that her name had been on the original manuscript.[78]

From all the evidence, Marić was a sounding board, though not as important in that role as Besso. She also helped check his math, although there is no evidence that she came up with any of the mathematical concepts. In addition, she encouraged him and (what at times was more difficult) put up with him.

For both the sake of colorful history and the emotional resonance it would have, it would be fun if we could go even further than this. But instead, we must follow the less exciting course of being confined to the evidence. None of their many letters, to each other or to friends, mentions a single instance of an idea or creative concept relating to relativity that came from Marić.

Nor did she ever—even to her family and close friends while in the throes of their bitter divorce—claim to have made any substantive contributions to Einstein's theories. Her son Hans Albert, who remained devoted to her and lived with her during the divorce, gave his own version that was reflected in a book by Peter Michelmore, and it seems to reflect what Marić told her son: "Mileva helped him solve certain mathematical problems, but no one could assist with the creative work, the flow of ideas."[79]

There is, in fact, no need to exaggerate Marić's contributions in order to admire, honor, and sympathize with her as a pioneer. To give her credit beyond what she ever claimed, says the science historian Gerald Holton, "only detracts both from her real and significant place in history and from the tragic unfulfillment of her early hopes and promise."

Einstein admired the pluck and courage of a feisty female physicist who had emerged from a land where women were generally not allowed to go into that field. Nowadays, when the same issues still reverberate across a century of time, the courage that Marić displayed by entering and competing in the male-dominated world of physics and math is what should earn her an admired spot in the annals of scientific history. This she deserves without inflating the importance of her collaboration on the special theory of relativity.[80]

The E=mc² *Coda, September 1905*

Einstein had raised the curtain on his miracle year in his letter to his Olympia Academy mate Conrad Habicht, and he celebrated its climax with his one-sentence drunken postcard to him. In September, he wrote yet another letter to Habicht, this one trying to entice him to come work at the patent office. Einstein's reputation as a lone wolf was somewhat artificial. "Perhaps it would be possible to smuggle you in among the patent slaves," he said. "You probably would find it relatively pleasant. Would you actually be ready and willing to come? Keep in mind that besides the eight hours of work, each day also has eight hours for fooling around, and then there's also Sunday. I would love to have you here."

As with his letter six months earlier, Einstein went on to reveal quite casually a momentous scientific breakthrough, one that would be expressed by the most famous equation in all of science:

> One more consequence of the electrodynamics paper has also crossed my mind. Namely, the relativity principle, together with Maxwell's equations, requires that mass be a direct measure of the energy contained in a body. Light carries mass with it. With the case of radium there should be a noticeable reduction of mass. The thought is amusing and seductive; but for all I know, the good Lord might be laughing at the whole matter and might have been leading me up the garden path.[81]

Einstein developed the idea with a beautiful simplicity. The paper that the *Annalen der Physik* received from him on September 27, 1905, "Does the Inertia of a Body Depend on Its Energy Content?," involved only three steps that filled merely three pages. Referring back to his special relativity paper, he declared, "The results of an electrodynamic investigation recently published by me in this journal lead to a very interesting conclusion, which will be derived here."[82]

Once again, he was deducing a theory from principles and postulates, not trying to explain the empirical data that experimental physicists studying cathode rays had begun to gather about the relation of mass to the velocity of particles. Coupling Maxwell's theory with the relativity theory, he began (not surprisingly) with a thought experiment. He calculated the properties of two light pulses emitted in opposite directions by a body at rest. He then calculated the properties of these light pulses when observed from a moving frame of reference. From this he came up with equations regarding the relationship between speed and mass.

The result was an elegant conclusion: mass and energy are different manifestations of the same thing. There is a fundamental interchangeability between the two. As he put it in his paper, "The mass of a body is a measure of its energy content."

The formula he used to describe this relationship was also strikingly simple: "If a body emits the energy L in the form of radiation, its mass decreases by L/V^2." Or, to express the same equation in a different manner: $L=mV^2$. Einstein used the letter L to represent energy until 1912, when he crossed it out in a manuscript and replaced it with the more common E. He also used V to represent the velocity of light, before changing to the more common c. So, using the letters that soon became standard, Einstein had come up with his memorable equation:

$$E=mc^2$$

Energy equals mass times the square of the speed of light. The speed of light, of course, is huge. Squared it is almost inconceivably bigger. That is why a tiny amount of matter, if converted completely into energy, has an enormous punch. A kilogram of mass would convert into approximately 25 billion kilowatt hours of electricity. More vividly: the energy in the mass of one raisin could supply most of New York City's energy needs for a day.[83]

As usual, Einstein ended by proposing experimental ways to confirm the theory he had just derived. "Perhaps it will prove possible," he wrote, "to test this theory using bodies whose energy content is variable to a high degree, e.g., salts of radium."

CHAPTER SEVEN

THE HAPPIEST THOUGHT

1906–1909

Recognition

Einstein's 1905 burst of creativity was astonishing. He had devised a revolutionary quantum theory of light, helped prove the existence of atoms, explained Brownian motion, upended the concept of space and time, and produced what would become science's best known equation. But not many people seemed to notice at first. According to his sister, Einstein had hoped that his flurry of essays in a preeminent journal would lift him from the obscurity of a third-class patent examiner and provide some academic recognition, perhaps even an academic job. "But he was bitterly disappointed," she noted. "Icy silence followed the publication."[1]

That was not exactly true. A small but respectable handful of physicists soon took note of Einstein's papers, and one of these turned out to be, as good fortune would have it, the most important possible admirer he could attract: Max Planck, Europe's revered monarch of theoretical physics, whose mysterious mathematical constant explaining blackbody radiation Einstein had transformed into a radical new reality of nature. As the editorial board member of *Annalen der Physik* responsible for theoretical submissions, Planck had vetted Einstein's papers, and the one on relativity had "immediately aroused my lively attention," he later recalled. As soon as it was published, Planck gave a lecture on relativity at the University of Berlin.[2]

Planck became the first physicist to build on Einstein's theory. In an article published in the spring of 1906, he argued that relativity conformed to the principle of least action, a foundation of physics that holds that light or any object moving between two points should follow the easiest path.[3]

Planck's paper not only contributed to the development of relativity theory; it also helped to legitimize it among other physicists. Whatever disappointment Maja Einstein had detected in her brother dissipated. "My papers are much appreciated and are giving rise to further investigations," he exulted to Solovine. "Professor Planck has recently written to me about that."[4]

The proud patent examiner was soon exchanging letters with the eminent professor. When another theorist challenged Planck's contention that relativity theory conformed to the principle of least action, Einstein took Planck's side and sent him a card saying so. Planck was pleased. "As long as the proponents of the principle of relativity constitute such a modest little band as is now the case," he replied to Einstein, "it is doubly important that they agree among themselves." He added that he hoped to visit Bern the following year and meet Einstein personally.[5]

Planck did not end up coming to Bern, but he did send his earnest assistant, Max Laue.* He and Einstein had already been corresponding about Einstein's light quanta paper, with Laue saying that he agreed with "your heuristic view that radiation can be absorbed and emitted only in specific finite quanta."

However, Laue insisted, just as Planck had, that Einstein was wrong to assume that these quanta were a characteristic of the radiation itself. Instead, Laue contended that the quanta were merely a description of the way that radiation was emitted or absorbed by a piece of matter. "This is not a characteristic of electromagnetic processes in a vacuum but rather of the emitting or absorbing matter," Laue wrote, "and hence radiation does not consist of light quanta as it says in section six of your first paper."[6] (In that section, Einstein had said that the

---

* Later, upon his father's death, he became Max von Laue.

radiation "behaves thermodynamically as if it consisted of mutually independent energy quanta.")

When Laue was preparing to visit in the summer of 1907, he was surprised to discover that Einstein was not at the University of Bern but was working at the patent office on the third floor of the Post and Telegraph Building. Meeting Einstein there did not lessen his wonder. "The young man who came to meet me made so unexpected an impression on me that I did not believe he could possibly be the father of the relativity theory," Laue said, "so I let him pass." After a while, Einstein came wandering through the reception area again, and Laue finally realized who he was.

They walked and talked for hours, with Einstein at one point offering a cigar that, Laue recalled, "was so unpleasant that I 'accidentally' dropped it into the river." Einstein's theories, on the other hand, made a pleasing impression. "During the first two hours of our conversation he overthrew the entire mechanics and electrodynamics," Laue noted. Indeed, he was so enthralled that over the next four years he would publish eight papers on Einstein's relativity theory and become a close friend.[7]

Some theorists found the amazing flurry of papers from the patent office to be uncomfortably abstract. Arnold Sommerfeld, later a friend, was among the first to suggest there was something Jewish about Einstein's theoretical approach, a theme later picked up by anti-Semites. It lacked due respect for the notion of order and absolutes, and it did not seem solidly grounded. "As remarkable as Einstein's papers are," he wrote Lorentz in 1907, "it still seems to me that something almost unhealthy lies in this unconstruable and impossible to visualize dogma. An Englishman would hardly have given us this theory. It might be here too, as in the case of Cohn, the abstract conceptual character of the Semite expresses itself."[8]

None of this interest made Einstein famous, nor did it get him any job offers. "I was surprised to read that you must sit in an office for eight hours a day," wrote yet another young physicist who was planning to visit. "History is full of bad jokes."[9] But because he had finally earned his doctorate, he had at least gotten promoted from a third-class to a second-class technical expert at the patent office, which came with a hefty 1,000-franc raise to an annual salary of 4,500 francs.[10]

His productivity was startling. In addition to working six days a week at the patent office, he continued his torrent of papers and reviews: six in 1906 and ten more in 1907. At least once a week he played in a string quartet. And he was a good father to the 3-year-old son he proudly labeled "impertinent." As Marić wrote to her friend Helene Savić, "My husband often spends his free time at home just playing with the boy."[11]

Beginning in the summer of 1907, Einstein also found time to dabble in what might have become, if the fates had been more impish, a new career path: as an inventor and salesman of electrical devices like his uncle and father. Working with Olympia Academy member Conrad Habicht and his brother Paul, Einstein developed a machine to amplify tiny electrical charges so they could be measured and studied. It had more academic than practical purpose; the idea was to create a lab device that would permit the study of small electrical fluctuations.

The concept was simple. When two strips of metal move close to each other, an electric charge on one will induce an opposite charge on the other. Einstein's idea was to use a series of strips that would induce the charge ten times and then transfer that to another disc. The process would be repeated until the original minuscule charge would be multiplied by a large number and thus be easily measurable. The trick was making the contraption actually work.[12]

Given his heritage, breeding, and years in the patent office, Einstein had the background to be an engineering genius. But as it turned out, he was better suited to theorizing. Fortunately, Paul Habicht was a good machinist, and by August 1907 he had a prototype of the *Maschinchen,* or little machine, ready to be unveiled. "I am astounded at the lightning speed with which you built the *Maschinchen,*" Einstein wrote. "I'll show up on Sunday." Unfortunately, it didn't work. "I am driven by *murderous* curiosity as to what you're up to," Einstein wrote a month later as they tried to fix things.

Throughout 1908, letters flew back and forth between Einstein and the Habichts, filled with complex diagrams and a torrent of ideas for how to make the device work. Einstein published a description in a journal, which produced, for a while, a potential sponsor. Paul Habicht was able to build a better version by October, but it had trouble keeping a charge. He brought the machine to Bern, where Einstein commandeered a lab in one of the schools and dragooned a local mechanic. By November the machine seemed to be working. It took another year or so to get a patent and begin to make some versions for sale. But even then, it never truly caught hold or found a market, and Einstein eventually lost interest.[13]

These practical exploits may have been fun, but Einstein's glorious isolation from the priesthood of academic physicists was starting to have more drawbacks than advantages. In a paper he wrote in the spring of 1907, he began by exuding a joyful self-assurance about having neither the library nor the inclination to know what other theorists had written on the topic. "Other authors might have already clarified part of what I am going to say," he wrote. "I felt I could dispense with doing a literature search (which would have been very troublesome for me), especially since there is good reason to hope that others will fill this gap." However, when he was commissioned to write a major yearbook piece on relativity later that year, there was slightly less cockiness in his warning to the editor that he might not be aware of all the literature. "Unfortunately I am not in a position to acquaint myself about everything that has been published on this subject," he wrote, "because the library is closed in my free time."[14]

That year he applied for a position at the University of Bern as a *privatdozent*, a starter rung on the academic ladder, which involved giving lectures and collecting a small fee from anyone who felt like showing up. To become a professor at most European universities, it helped to serve such an apprenticeship. With his application Einstein enclosed seventeen papers he had published, including the ones on relativity and light quanta. He was also expected to include an unpublished paper known as a *habilitation* thesis, but he decided not to bother writing one, as this requirement was sometimes waived for those who had "other outstanding achievements."

Only one professor on the faculty committee supported hiring him without requiring him to write a new thesis, "in view of the important scientific achievements of Herr Einstein." The others disagreed, and the requirement was not waived. Not surprisingly, Einstein considered the matter "amusing." He did not write the special *habilitation* or get the post.[15]

The Equivalence of Gravity and Acceleration

Einstein's road to the general theory of relativity began in November 1907, when he was struggling against a deadline to finish an article for a science yearbook explaining his special theory of relativity. Two limitations of that theory still bothered him: it applied only to uniform constant-velocity motion (things felt and behaved differently if your speed or direction was changing), and it did not incorporate Newton's theory of gravity.

"I was sitting in a chair in the patent office at Bern when all of a sudden a thought occurred to me," he recalled. "If a person falls freely, he will not feel his own weight." That realization, which "startled" him, launched him on an arduous eight-year effort to generalize his special theory of relativity and "impelled me toward a theory of gravitation."[16] Later, he would grandly call it "the happiest* thought in my life."[17]

The tale of the falling man has become an iconic one, and in some accounts it actually involves a painter who fell from the roof of an apartment building near the patent office.[18] In fact, probably like other great tales of gravitational discovery—Galileo dropping objects from the Tower of Pisa and the apple falling on Newton's head[19]—it was embellished in popular lore and was more of a thought experiment than a real occurrence. Despite Einstein's propensity to focus on science rather than the merely personal, even he was not likely to watch a real human plunging off a roof and think of gravitational theory, much less call it the happiest thought in his life.

Einstein refined his thought experiment so that the falling man was in an enclosed chamber, such as an elevator in free fall above the earth.

---

* The German phrase he used was "der glücklichste Gedanke," which has usually been translated as "happiest" thought, but perhaps in this context is more properly translated as "luckiest" or "most fortunate."

In this falling chamber (at least until it crashed), the man would feel weightless. Any objects he emptied from his pocket and let loose would float alongside him.

Looking at it another way, Einstein imagined a man in an enclosed chamber floating in deep space "far removed from stars and other appreciable masses." He would experience the same perceptions of weightlessness. "Gravitation naturally does not exist for this observer. He must fasten himself with strings to the floor, otherwise the slightest impact against the floor will cause him to rise slowly towards the ceiling."

Then Einstein imagined that a rope was hooked onto the roof of the chamber and pulled up with a constant force. "The chamber together with the observer then begin to move 'upwards' with a uniformly accelerated motion." The man inside will feel himself pressed to the floor. "He is then standing in the chest in exactly the same way as anyone stands in a room of a house on our earth." If he pulls something from his pocket and lets go, it will fall to the floor "with an accelerated relative motion" that is the same no matter the weight of the object—just as Galileo discovered to be the case for gravity. "The man in the chamber will thus come to the conclusion that he and the chest are in a gravitational field. Of course he will be puzzled for a moment as to why the chest does not fall in this gravitational field. Just then, however, he discovers the hook in the middle of the lid of the chest and the rope which is attached to it, and he consequently comes to the conclusion that the chamber is suspended at rest in the gravitational field."

"Ought we to smile at the man and say that he errs in his conclusion?" Einstein asked. Just as with special relativity, there was no right or wrong perception. "We must rather admit that his mode of grasping the situation violates neither reason nor known mechanical laws." [20]

A related way that Einstein addressed this same issue was typical of his ingenuity: he examined a phenomenon that was so very well-known that scientists rarely puzzled about it. Every object has a "gravitational mass," which determines its weight on the earth's surface or, more generally, the tug between it and any other object. It also has an "inertial mass," which determines how much force must be applied to it in order to make it accelerate. As Newton noted, the inertial mass of an object is always the same as its gravitational mass, even though they are defined differently. This was obviously more than a mere coincidence, but no one had fully explained why.

Uncomfortable with two explanations for what seemed to be one phenomenon, Einstein probed the equivalence of inertial mass and gravitational mass using his thought experiment. If we imagine that the enclosed elevator is being accelerated upward in a region of outer space where there is no gravity, then the downward force felt by the man inside (or the force that tugs downward on an object hanging from the ceiling by a string) is due to *inertial* mass. If we imagine that the enclosed elevator is at rest in a gravitational field, then the downward force felt by the man inside (or the force that tugs downward on an object hanging from the ceiling by a string) is due to *gravitational* mass. But inertial mass always equals gravitational mass. "From this correspondence," said Einstein, "it follows that it is impossible to discover by experiment whether a given system of coordinates is accelerated, or whether . . . the observed effects are due to a gravitational field."[21]

Einstein called this "the equivalence principle."[22] The local effects of gravity and of acceleration are equivalent. This became a foundation for his attempt to generalize his theory of relativity so that it was not restricted just to systems that moved with a uniform velocity. The basic insight that he would develop over the next eight years was that "the effects we ascribe to gravity and the effects we ascribe to acceleration are both produced by one and the same structure."[23]

Einstein's approach to general relativity again showed how his mind tended to work:

- He was disquieted when there were two seemingly unrelated theories for the same observable phenomenon. That had been the case with the moving coil or moving magnet producing the same observable electric current, which he resolved with the special theory of relativity. Now it was the case with the differing definitions of inertial mass and gravitational mass, which he began to resolve by building on the equivalence principle.
- He was likewise uncomfortable when a theory made distinctions

that could not be observed in nature. That had been the case with observers in uniform motion: there was no way of determining who was at rest and who was in motion. Now it was also, apparently, the case for observers in accelerated motion: there was no way of telling who was accelerating and who was in a gravitational field.

- He was eager to generalize theories rather than settling for having them restricted to a special case. There should not, he felt, be one set of principles for the special case of constant-velocity motion and a different set for all other types of motion. His life was a constant quest for unifying theories.

In November 1907, working against the deadline imposed by the *Yearbook of Radioactivity and Electronics*, Einstein tacked on a fifth section to his article on relativity that sketched out his new ideas. "So far we have applied the principle of relativity . . . only to nonaccelerated reference systems," he began. "Is it conceivable that the principle of relativity applies to systems that are accelerated relative to each other?"

Imagine two environments, he said, one being accelerated and the other resting in a gravitational field.[24] There is no physical experiment you can do that would tell these situations apart. "In the discussion that follows, we shall therefore assume the complete physical equivalence of a gravitational field and a corresponding acceleration of the reference system."

Using various mathematical calculations that can be made about an accelerated system, Einstein proceeded to show that, if his notions were correct, clocks would run more slowly in a more intense gravitational field. He also came up with many predictions that could be tested, including that light should be bent by gravity and that the wavelength of light emitted from a source with a large mass, such as the sun, should increase slightly in what has become known as the gravitational redshift. "On the basis of some ruminating, which, though daring, does have something going for it, I have arrived at the view that the gravitational difference might be the cause of the shift to the red end of the spectrum," he explained to a colleague. "A bending of light rays by gravity also follows from these arguments."[25]

It would take Einstein another eight years, until November 1915, to work out the fundamentals of this theory and find the math to express it. Then it would take another four years before the most vivid of his predictions, the extent to which gravity would bend light, was verified by dramatic observations. But at least Einstein now had a vision, one that started him on the road toward one of the most elegant and impressive achievements in the history of physics: the general theory of relativity.

Winning a Professorship

By the beginning of 1908, even as such academic stars as Max Planck and Wilhelm Wien were writing to ask for his insights, Einstein had tempered his aspirations to be a university professor. Instead, he had begun, believe it or not, to seek work as a high school teacher. "This craving," he told Marcel Grossmann, who had helped him get the patent-office job, "comes only from my ardent wish to be able to continue my private scientific work under easier conditions."

He was even eager to go back to the Technical School in Winterhur, where he had briefly been a substitute teacher. "How does one go about this?" he asked Grossmann. "Could I possibly call on somebody and talk him into the great worth of my admirable person as a teacher and a citizen? Wouldn't I make a bad impression on him (no Swiss-German dialect, my Semitic appearance, etc.)?" He had written papers that were transforming physics, but he did not know if that would help. "Would there be any point in my stressing my scientific papers on that occasion?"[26]

He also responded to an advertisement for a "teacher of mathematics and descriptive geometry" at a high school in Zurich, noting in his application "that I would be ready to teach physics as well." He ended up deciding to enclose all of the papers he had written thus far, including the special theory of relativity. There were twenty-one applicants. Einstein did not even make the list of three finalists.[27]

So Einstein finally overcame his pride and decided to write a thesis in order to become a *privatdozent* at Bern. As he explained to the patron there who had supported him, "The conversation I had with you in the city library, as well as the advice of several friends, has induced me to change my decision for the second time and to try my luck with a *habilitation* at the University of Bern after all."[28]

The paper he submitted, an extension of his revolutionary work on light quanta, was promptly accepted, and at the end of February 1908, he was made a *privatdozent.* He had finally scaled the walls, or at least the outer wall, of academe. But his post neither paid enough nor was important enough for him to give up his job at the patent office. His lectures at the University of Bern thus became simply one more thing for him to do.

His topic for the summer of 1908 was the theory of heat, held on Tuesday and Saturday at 7 a.m., and he initially attracted only three attendees: Michele Besso and two other colleagues who worked at the postal building. In the winter session he switched to the theory of radiation, and his three coworkers were joined by an actual student named Max Stern. By the summer of 1909, Stern was the only attendee, and Einstein canceled his lecturing. He had, in the meantime, begun to adopt his professorial look: both his hair and clothing became a victim of nature's tendency toward randomness.[29]

Alfred Kleiner, the University of Zurich physics professor who helped Einstein get his doctorate, had encouraged him to pursue the *privatdozent* position.[30] He also had waged a long effort, which succeeded in 1908, to convince the Zurich authorities to increase the university's stature by creating a new position in theoretical physics. It was not a full professorship; instead, it was an associate professorship under Kleiner.

It was the obvious post for Einstein, but there was one obstacle. Kleiner had another candidate in mind: his assistant Friedrich Adler, a pale and passionate political activist who had become friends with Einstein when they were both at the Polytechnic. Adler, whose father was the leader of the Social Democratic Party in Austria, was more disposed to political philosophy than theoretical physics. So he went to see Kleiner one morning in June 1908, and the two of them concluded that Adler was not right for the job and Einstein was.

In a letter to his father, Adler recounted the conversation and said that Einstein "had no understanding how to relate to people" and had been "treated by the professors at the Polytechnic with outright contempt." But Adler said he deserved the job because of his genius and was likely to get it. "They have a bad conscience over how they treated him earlier. The scandal is being felt not only here but in Germany that such a man would have to sit in the patent office."[31]

Adler made sure that the Zurich authorities, and for that matter everyone else, knew that he was officially stepping aside for his friend. "If it is possible to get a man like Einstein for our university, it would be absurd to appoint me," he wrote. That resolved the political issue for the councilor in charge of education, who was a partisan Social Democrat. "Ernst would have liked Adler, since he was a fellow party member," Einstein explained to Michele Besso. "But Adler's statements about himself and me made it impossible."[32]

So, at the end of June 1908, Kleiner traveled from Zurich to Bern to audit one of Einstein's *privatdozent* lectures and, as Einstein put it, "size up the beast." Alas, it was not a great show. "I really did not lecture divinely," Einstein lamented to a friend, "partly because I was not well prepared, partly because being investigated got on my nerves a bit." Kleiner sat listening with a wrinkled brow, and after the lecture he informed Einstein that his teaching style was not good enough to qualify him for the professorship. Einstein calmly claimed that he considered the job "quite unnecessary."[33]

Kleiner went back to Zurich and reported that Einstein "holds monologues" and was "a long way from being a teacher." That seemed to end his chances. As Adler informed his powerful father, "The situation has therefore changed, and the Einstein business is closed." Einstein pretended to be sanguine. "The business with the professorship fell through, but that's all right with me," he wrote a friend. "There are enough teachers even without me."[34]

In fact Einstein was upset, and he became even more so when he heard that Kleiner's criticism of his teaching skills was being widely circulated, even in Germany. So he wrote to Kleiner, angrily reproaching him "for spreading unfavorable rumors about me." He was already finding it difficult to get a proper academic job, and Kleiner's assessment would make it impossible.

There was some validity to Kleiner's criticism. Einstein was never an inspired teacher, and his lectures tended to be regarded as disorganized until his celebrity ensured that every stumble he made was transformed into a charming anecdote. Nevertheless, Kleiner relented. He said that he would be pleased to help him get the Zurich job if he could only show "some teaching ability."

Einstein replied by suggesting that he come to Zurich to give a full-fledged (and presumably well-prepared) lecture to the physics society there, which he did in February 1909. "I was lucky," Einstein reported soon after. "Contrary to my habit, I lectured well on that occasion."[35] When he went to call on Kleiner afterward, the professor intimated that a job offer would soon follow.

A few days after Einstein returned to Bern, Kleiner provided his official recommendation to the University of Zurich faculty. "Einstein ranks among the most important theoretical physicists and has been recognized as such since his work on the relativity principle," he wrote. As for Einstein's teaching skills, he said as politely as possible that they were ripe for improvement: "Dr. Einstein will prove his worth also as a teacher, because he is too intelligent and too conscientious not to be open to advice when necessary."[36]

One issue was Einstein's Jewishness. Some faculty members considered this a potential problem, but they were assured by Kleiner that Einstein did not exhibit the "unpleasant peculiarities" supposedly associated with Jews. Their conclusion is a revealing look at both the anti-Semitism of the time and the attempts to rise above it:

> The expressions of our colleague Kleiner, based on several years of personal contact, were all the more valuable for the committee as well as for the faculty as a whole since Herr Dr. Einstein is an Israelite and since precisely to the Israelites among scholars are inscribed (in numerous cases not entirely without cause) all kinds of unpleasant peculiarities of character, such as intrusiveness, impudence, and a shopkeeper's mentality in the perception of their academic position. It should be said, however, that also among the Israelites there exist men who do not exhibit a trace of these disagreeable qualities and that it is not proper, therefore, to disqualify a man only because he happens to be a Jew. Indeed, one occasionally finds people also among non-Jewish scholars who in regard to a commercial perception and utilization of their academic profession develop qualities that are usually considered as specifically Jewish. Therefore, neither the committee nor the faculty as a whole considered it compatible with its dignity to adopt anti-Semitism as a matter of policy.[37]

The secret faculty vote in late March 1909 was ten in favor and one abstention. Einstein was offered his first professorship, four years after he had revolutionized physics. Unfortunately, his proposed salary was less than what he was making at the patent office, so he declined. Finally, the Zurich authorities raised their offer, and Einstein accepted. "So, now I too am an official member of the guild of whores," he exulted to a colleague.[38]

One person who saw a newspaper notice about Einstein's appointment was a Basel housewife named Anna Meyer-Schmid. Ten years earlier, when she was an unmarried girl of 17, they had met during one of Einstein's vacations with his mother at the Hotel Paradies. Most of the guests had seemed to him "philistines," but he took a liking to Anna and even wrote a poem in her album: "What should I inscribe for you here? / I could think of many things / Including a kiss / On your tiny little mouth / If you're angry about it / Do not start to cry / The best punishment / Is to give me one too." He signed it, "Your rascally friend."[39]

In response to a congratulatory postcard from her, Einstein replied with a polite and mildly suggestive letter. "I probably cherish the memory of the lovely weeks that I was allowed to spend near you in the Paradies more than you do," he wrote. "So now I've become such a big schoolmaster that my name is even mentioned in the newspapers. But I have remained a simple fellow." He noted that he had married his college friend Marić, but he gave her his office address. "If you ever happen to be in Zurich and have time, look me up there; it would give me great pleasure."[40]

Whether or not Einstein intended his response to hover uncertainly between innocence and suggestiveness, Anna's eyes apparently snapped it into the latter position. She wrote a letter back, which Marić intercepted. Her jealousy aroused, Marić then wrote a letter to Anna's husband claiming (wishfully more than truthfully) that Einstein was outraged by Anna's "inappropriate letter" and brazen attempt to rekindle a relationship.

Einstein ended up having to calm matters with an apology to the husband. "I am very sorry if I have caused you distress by my careless behavior," he wrote. "I answered the congratulatory card your wife sent me on the occasion of my appointment too heartily and thereby re-awakened the old affection we had for each other. But this was not done with impure intentions. The behavior of your wife, for whom I have the greatest respect, was totally honorable. It was wrong of my wife—and excusable only on account of extreme jealousy—to behave—without my knowledge—the way she did."

Although the incident itself was of no consequence, it marked a turn in Einstein's relationship with Marić. In his eyes, her brooding jealousy was making her darker. Decades later, still rankling at Marić's behavior, he wrote to Anna's daughter asserting, with a brutal bluntness, that his wife's jealousy had been a pathological flaw typical of a woman of such "uncommon ugliness."[41]

Marić indeed had a jealous streak. She resented not only her husband's flirtations with other women but also the time he spent with male colleagues. Now that he had become a professor, she succumbed to a professional envy that was understandable given her own curtailed scientific career. "With that kind of fame, he does not have much time left for his wife," she told her friend Helene Savić. "You wrote that I must be jealous of science. But what can you do? One gets the pearl, the other the box."

In particular, Marić worried that her husband's fame would make him colder and more self-centered. "I am very happy for his success, because he really does deserve it," she wrote in another letter. "I only hope that fame does not exert a detrimental influence on his human side."[42]

In one sense, Marić's worries proved unwarranted. Even as his fame increased exponentially, Einstein would retain a personal simplicity, an unaffected style, and at least a veneer of genial humility. But viewed from a different reference frame, there were transformations to his human side. Sometime around 1909, he began drifting apart from his wife. His resistance to chains and bonds increasingly led him to escape into his work while taking a detached approach to the realm he dismissed as "the merely personal."

On one of his last days working at the patent office, he received a large envelope with an elegant sheet covered in what seemed to be Latin calligraphy. Because it seemed odd and impersonal, he threw it in the wastebasket. It was, in fact, an invitation to be one of those receiving an honorary doctorate at the July 1909 commemoration of the founding of Geneva's university, and authorities there finally got a friend of Einstein to persuade him to attend. Einstein brought only a straw hat and an informal suit, so he stood out rather strangely, both in the parade and at the opulent formal dinner that night. Amused by the whole situation, he turned to the patrician seated next to him and speculated about the austere Protestant Reformation leader who had founded the university: "Do you know what Calvin would have done had he been here?" The gentleman, befuddled, said no. Einstein replied, "He would have erected an enormous stake and had us all burnt for our sinful extravagance." As Einstein later recalled, "The man never addressed another word to me." [43]

Light Can Be Wave and Particle

Also at the end of the summer of 1909, Einstein was invited to address the annual *Naturforscher* conference, the preeminent meeting of German-speaking scientists, which was held that year in Salzburg. Organizers had put both relativity and the quantum nature of light on the agenda, and they expected him to speak on the former. Instead, Einstein decided that he preferred to emphasize what he considered the more pressing issue: how to interpret quantum theory and reconcile it with the wave theory of light that Maxwell had so elegantly formulated.

After his "happiest thought" at the end of 1907 about how the equivalence of gravity and acceleration might lead to a generalization of relativity theory, Einstein had put that subject aside to focus instead on what he called "the radiation problem" (i.e., quantum theory). The more he thought about his "heuristic" notion that light was made up of quanta, or indivisible packets, the more he worried that he and Planck had wrought a revolution that would destroy the classical foundations of physics, especially Maxwell's equations. "I have come to this pessimistic view mainly as a result of endless, vain efforts to interpret . . . Planck's constant in an intuitive way," he wrote a fellow physicist early in 1908. "I even seriously doubt that it will be possible to maintain the general validity of Maxwell's equations." [44] (As it turned out, his love of Maxwell's equations was well placed. They are among the few elements of theoretical physics to remain unchanged by both the relativity and quantum revolutions that Einstein helped launch.)

When Einstein, still not officially a professor, arrived at the Salzburg conference in September 1909, he finally met Max Planck and other giants that he had known only through letters. On the afternoon of the third day, he stepped in front of more than a hundred famed scientists and delivered a speech that Wolfgang Pauli, who was to become a pioneer of quantum mechanics, later pronounced "one of the landmarks in the development of theoretical physics."

Einstein began by explaining how the wave theory of light was no longer complete. Light (or any radiation) could also be regarded, he said, as a beam of particles or packets of energy, which he said was akin to what Newton had posited. "Light has certain basic properties that can be understood more readily from the standpoint of the Newtonian emission theory than from the standpoint of the wave theory," he declared. "I thus believe that the next phase of theoretical physics will bring us a theory of light that can be interpreted as a kind of fusion of the wave and of the emission theories of light."

Combining particle theory with wave theory, he warned, would bring "a profound change." This was not a good thing, he feared. It could undermine the certainties and determinism inherent in classical physics.

For a moment, Einstein mused that perhaps such a fate could be avoided by accepting Planck's more limited interpretation of quanta: that they were features only of how radiation was emitted and absorbed by a surface rather than a feature of the actual light wave as it propagated through space. "Would it not be possible," he asked, "to retain at least the equations for the propagation of radiation and conceive only the processes of emission and absorption differently?" But after comparing the behavior of light to the behavior of gas molecules, as he had done in his 1905 light quanta paper, Einstein concluded that, alas, this was not possible.

As a result, Einstein said, light must be regarded as behaving like both an undulating wave and a stream of particles. "These two structural properties simultaneously displayed by radiation," he declared at the end of his talk, "should not be considered as mutually incompatible."[45]

It was the first well-conceived promulgation of the wave-particle duality of light, and it had implications as profound as Einstein's earlier theoretical breakthroughs. "Is it possible to combine energy quanta and the wave principles of radiation?" he merrily wrote to a physicist friend. "Appearances are against it, but the Almighty—it seems—managed the trick."[46]

A vibrant discussion followed Einstein's speech, led by Planck himself. Still unwilling to embrace the physical reality underlying the mathematical constant that he had devised nine years earlier, or to accept the revolutionary ramifications envisioned by Einstein, Planck now played protector of the old order. He admitted that radiation involved discrete "quanta, which are to be conceived as atoms of action." But he insisted that these quanta existed *only* as part of the process of radiation being emitted or absorbed. "The question is where to look for these quanta," he said. "According to Mr. Einstein, it would be necessary to conceive that free radiation in a vacuum, and thus the light waves themselves consist of atomistic quanta, and hence force us to give up Maxwell's equations. This seems to me a step that is not yet necessary."[47]

Within two decades, Einstein would assume a similar role as protector of the old order. Indeed, he was already looking for ways out of the eerie dilemmas raised by quantum theory. "I am very hopeful that I will solve the radiation problem, and that I will do so without light quanta," he wrote a young physicist he was working with.[48]

It was all too mystifying, at least for the time being. So as he moved up the professorial ranks in the German-speaking universities of Europe, he turned his attention back to the topic that was uniquely his own, relativity, and for a while became a refugee from the wonderland of the quanta. As he lamented to a friend, "The more successes the quantum theory enjoys, the sillier it looks."[49]

CHAPTER EIGHT

THE WANDERING PROFESSOR

1909–1914

Zurich, 1909

As a self-assured 17-year-old, Einstein had enrolled at the Zurich Polytechnic and met Mileva Marić, the woman he would marry. Now, in October 1909, at age 30, he was returning to that city to take up his post as a junior professor at the nearby University of Zurich.

Their homecoming restored, at least temporarily, some of the romance to their relationship. Marić was thrilled to be back in their original nesting ground, and by the end of their first month there she became pregnant again.

The apartment they rented was in a building where, they happily discovered, Friedrich Adler and his wife lived, and the couples became even closer friends. "They run a bohemian household," Adler wrote his father approvingly. "The more I talk to Einstein, the more I realize that my favorable opinion of him was justified."

The two men discussed physics and philosophy most evenings, often retreating to the attic of the three-story building so they would not be disturbed by children or spouses. Adler introduced Einstein to the work of Pierre Duhem, whose 1906 book *La Théorie Physique* Adler had just published in German. Duhem offered a more holistic approach than Mach did to the relationship between theories and experimental evidence, one that seemed to influence Einstein as he staked out his own philosophy of science.[1]

158

Adler particularly respected Einstein's "most independent" mind. There was, he told his father, a nonconformist streak in Einstein that reflected an inner security but not an arrogance. "We find ourselves in agreement on questions that the majority of physicists would not even understand," Adler boasted.[2]

Einstein tried to persuade Adler to focus on science rather than be enticed into politics. "Be a little patient," he said. "You will certainly be my successor in Zurich one day." (Einstein was already assuming that he would move on to a more prestigious university.) But Adler ignored the advice and decided to become an editor at the Social Democratic Party newspaper. Loyalty to a party, Einstein felt, meant surrendering some independence of thought. Such conformity confounded him. "How an intelligent man can subscribe to a party I find a complete mystery," Einstein later lamented about Adler.[3]

Einstein was also reunited with his former classmate and note-taker Marcel Grossmann, who had helped him get his job at the patent office and was now a professor of math at their old Polytechnic. Einstein would often visit Grossmann after lunch for help with the complex geometry and calculus he needed to extend relativity into a more general field theory.

Einstein was even able to forge a friendship with the other distinguished math professor at the Polytechnic, Adolf Hurwitz, whose classes he had often skipped and who had spurned his plea for a job. Einstein became a regular at the Sunday music recitals at Hurwitz's home. When Hurwitz told him during a walk one day that his daughter had been given a math homework problem she did not understand, Einstein showed up that afternoon to help her solve it.[4]

As Kleiner predicted, Einstein's teaching talents improved. He was not a polished lecturer, but instead used informality to his advantage. "When he took his chair in shabby attire with trousers too short for him, we were skeptical," recalled Hans Tanner, who attended most of Einstein's Zurich lectures. Instead of prepared notes, Einstein used a card-sized strip of paper with scribbles. So the students got to watch him develop his thoughts as he spoke. "We obtained some insight into his working technique," said Tanner. "We certainly appreciated this more than any stylistically perfect lecture."

At each step of the way, Einstein would pause and ask the students if they were following him, and he even permitted interruptions. "This comradely contact between teacher and student was, at that time, a rare occurrence," according to Adolf Fisch, another who attended the lectures. Sometimes he would take a break and let the students gather around him for casual conversation. "With an impulsiveness and naturalness he would take students by the arm to discuss things," recalled Tanner.

During one lecture, Einstein found himself momentarily stumped about the steps needed to complete a calculation. "There must be some silly mathematical transformation that I can't find for a moment," he said. "Can one of you gentlemen see it?" Not surprisingly, none of them could. So Einstein continued: "Then leave a quarter of a page. We won't lose any time." Ten minutes later, Einstein interrupted himself in the middle of another point and exclaimed, "I've got it." As Tanner later marveled, "During the complicated development of his theme he had still found time to reflect upon the nature of that particular mathematical transformation."

At the end of many of his evening lectures, Einstein would ask, "Who's coming to the Café Terasse?" There, with an informal cadre on a terrace overlooking the Limmat River, they would talk until closing time.

On one occasion, Einstein asked if anyone wanted to come back to his apartment. "This morning I received some work from Planck in which there must be a mistake," he said. "We could read it together." Tanner and another student took him up on the offer and followed him home. There they all pored over Planck's paper. "See if you can spot the fault while I make some coffee," he said.

After a while, Tanner replied, "You must be mistaken, Herr Professor, there is no error in it."

"Yes, there is," Einstein said, pointing to some discrepancies in the data, "for otherwise that and that would become that and that." It was a vivid example of Einstein's great strength: he could look at a complex mathematical equation, which for others was merely an abstraction, and picture the physical reality that lay behind it.

Tanner was astounded. "Let's write to Professor Planck," he suggested, "and tell him of the mistake."

Einstein had by then become slightly more tactful, especially with those he placed on a pedestal, such as Planck and Lorentz. "We won't tell him he made a mistake," he said. "The result is correct, but the proof is faulty. We'll simply write and tell him how the real proof should run. The main thing is the content, not the mathematics."[5]

Despite his work on his machine to measure electrical charges, Einstein had become a confirmed theorist rather than experimental physicist. When he was asked during his second year as a professor to supervise laboratory work, he was dismayed. He hardly dared, he told Tanner, "pick up a piece of apparatus for fear it might blow up." To another eminent professor he confided, "My fears regarding the laboratory were rather well founded."[6]

As he was finishing his first academic year at Zurich, in July 1910, Marić gave birth, again with difficulty, to their second son, named Eduard and called Tete. She was ill for weeks afterward. Her doctor, contending that she was overworked, suggested that Einstein find a way to make more money and pay for a maid. Marić was annoyed and protective. "Isn't it clear to anyone that my husband works himself half dead?" she said. Instead, her mother came down from Novi Sad to help.[7]

Throughout his life, Einstein would sometimes appear aloof toward his two sons, especially Eduard, who suffered from increasingly severe mental illness as he grew older. But when they were young, he tended to be a good father. "When my mother was busy around the house, father would put aside his work and watch over us for hours, bouncing us on his knee," Hans Albert later recalled. "I remember he would tell us stories—and he often played the violin in an effort to keep us quiet."

One of his strengths as a thinker, if not as a parent, was that he had the ability, and the inclination, to tune out all distractions, a category that to him sometimes included his children and family. "Even the loudest baby-crying didn't seem to disturb Father," Hans Albert said. "He could go on with his work completely impervious to noise."

One day his student Tanner came for a visit and found Einstein in his study poring over a pile of papers. He was writing with his right hand and holding Eduard with his left. Hans Albert was playing with toy bricks and trying to get his attention. "Wait a minute, I've nearly finished," Einstein said, as he handed Eduard to Tanner and kept scribbling his equations. "It gave me," said Tanner, "a glimpse into his immense powers of concentration."[8]

Prague, 1911

Einstein had been in Zurich less than six months when he received, in March 1910, a solicitation to consider a more prestigious job: a full professorship at the German part of the University of Prague. Both the university and the academic position were a step up; however, moving from the familiar and friendly Zurich to the less congenial Prague would be disruptive for his family. For Einstein, the professional considerations outweighed the personal ones.

He was again going through difficult periods at home. "The bad mood that you noticed in me had nothing to do with you," he wrote to his mother, who was now living in Berlin. "To dwell on the things that depress or anger us does not help in overcoming them. One must knock them down alone."

His scientific work, on the other hand, was giving him great pleasure, and he expressed excitement about his possible new opportunity. "It is most probable that I will be offered the position of full professor at a large university with a significantly better salary than I now have."[9]

When word of Einstein's possible move spread in Zurich, fifteen of his students, led by Hans Tanner, signed a petition urging officials there "to do your utmost to keep this outstanding researcher and teacher at our university." They stressed the importance of having a professor in "this newly created discipline" of theoretical physics, and they extolled him personally in effusive terms. "Professor Einstein has an amazing talent for presenting the most difficult problems of theoretical physics so clearly and so comprehensibly that it is a great delight for us to follow his lectures, and he is so good at establishing a perfect rapport with his audience."[10]

The Zurich authorities were so eager to keep him that they raised his salary from its current 4,500 francs, which was the same as he made as a patent examiner, to 5,500 francs. Those attempting to lure him to Prague, on the other hand, were having a more difficult time.

The faculty department at Prague had settled on Einstein as its first choice and forwarded the recommendation to the education ministry in Vienna. (Prague was then part of the Austro-Hungarian Empire, and such an appointment had to be approved by Emperor Franz Joseph and his ministers.) The report was accompanied by the highest possible recommendation from the best possible authority, Max Planck. Einstein's theory of relativity "probably exceeds in audacity everything that has been achieved so far in speculative science," Planck proclaimed. "This principle has brought about a revolution in our physical picture of the world that can be compared only to that produced by Copernicus." In a comment that might later have seemed prescient to Einstein, Planck added, "Non-Euclidean geometry is child's play by comparison."[11]

Planck's imprimatur should have been enough. But it wasn't. The ministry decided that it preferred the second-place candidate, Gustav Jaumann, who had two advantages: he was Austrian, and he was not Jewish. "I did not get the call to Prague," Einstein lamented to a friend in August. "I was proposed by the faculty, but because of my Semitic origin the ministry did not approve."

Jaumann, however, soon discovered that he was the faculty's second choice, and he erupted. "If Einstein has been proposed as the first choice because of the belief that he has greater achievements to his credit," he declared, "then I will have nothing to do with a university that chases after modernity and does not appreciate merit." So by October 1910, Einstein could confidently declare that his own appointment was "almost certain."

There was one final hurdle, also dealing with religion. Being a Jew was a disadvantage; being a nonbeliever who claimed *no* religion was a disqualifier. The empire required that all of its servants, including professors, be a member of some religion. On his official forms, Einstein had written that he had none. "Einstein is as unpractical as a child in cases like this," Friedrich Adler's wife noted.

As it turned out, Einstein's desire for the job was greater than his ornery impracticality. He agreed to write "Mosaic" as his faith, and he also accepted Austro-Hungarian citizenship, with the proviso that he was allowed to remain a Swiss citizen as well. Along with the German citizenship that he had forsaken but that would soon be foisted back on him, that meant he had held, off and on, three citizenships by the age of 32. In January 1911, he was officially appointed to the post, with a pay twice what he had been making before his recent raise. He agreed to move to Prague that March.[12]

Einstein had two scientific heroes he had never met—Ernst Mach and Hendrik Lorentz—and he was able to visit them both before his move to Prague. When he went to Vienna for his formal presentation to the ministers there, he called on Mach, who lived in a suburb of that city. The aging physicist and preacher of empiricism, who so deeply influenced the Olympia Academy and instilled in Einstein a skepticism about unobservable concepts such as absolute time, had a gnarly beard and gnarlier personality. "Please speak loudly to me," he barked when Einstein entered his room. "In addition to my other unpleasant characteristics I am also almost stone deaf."

Einstein wanted to convince Mach of the reality of atoms, which the old man had long rejected as being imaginary constructs of the human mind. "Let us suppose that by assuming the existence of atoms in a gas we were able to predict an observable property of this gas that could not be predicted on the basis of non-atomistic theory," Einstein asked. "Would you then accept such a hypothesis?"

"If with the help of the atomic hypothesis one could actually establish a connection between several observable properties which without it would remain isolated, then I should say that this hypothesis was an 'economical' one," Mach grudgingly replied.

It was not a full acceptance, but it was enough for Einstein. "For the moment Einstein was satisfied," his friend Philipp Frank noted. Nevertheless, Einstein began edging away from Mach's skepticism about any theories of reality not built on directly observable data. He developed, said Frank, "a certain aversion to the Machist philosophy."[13] It was the beginning of an important conversion.

Just before moving to Prague, Einstein went to the Dutch town of Leiden to meet Lorentz. Marić accompanied him, and they accepted an invitation to stay with Lorentz and his wife. Einstein wrote that he was looking forward to having a conversation on "the radiation problem," adding, "I wish to assure you in advance that I am not the orthodox light-quantizer for whom you take me."[14]

Einstein had long idolized Lorentz from afar. Just before he went to visit, he wrote a friend: "I admire this man like no other; I might say, I love him." The feeling was reinforced when they finally met. They stayed up late on Saturday night discussing such issues as the relationship between temperature and electrical conductivity.

Lorentz thought he had caught Einstein in a small mathematical mistake in one of his papers on light quanta, but in fact, as Einstein noted, it was simply "a one-time writing error" where he had left out a "$\frac{1}{2}$" that was included later in the paper.[15] Both the hospitality and "scientific stimulus" made Einstein effusive in his next letter. "You radiate so much goodness and benevolence," he wrote, "that the troubling conviction that I did not deserve the great kindness and honors could not even enter my mind during my stay at your house."[16]

Lorentz became, in the words of Abraham Pais, "the one father figure in Einstein's life." After his pleasant visit to Lorentz's study in Leiden, he would return whenever he could find an excuse. The atmosphere of such meetings was captured by their colleague Paul Ehrenfest:

> The best easy chair was carefully pushed in place next to the large work table for his esteemed guest. A cigar was given to him, and then Lorentz quietly began to formulate questions concerning Einstein's theory of the bending of light in a gravitational field . . . As Lorentz spoke on, Einstein began to puff less frequently on his cigar, and he sat more intently in his armchair. And when Lorentz had finished, Einstein bent over the slip of paper on which Lorentz had written mathematical formulas. The cigar was out, and Einstein pensively twisted his finger in a lock of hair over his right ear. Lorentz sat smiling at an Einstein completely lost in meditation, exactly the way that a father looks at a particularly beloved son—full of confidence that the youngster will crack the nut he has given him, but eager to see how. Suddenly, Einstein's head sat up joyfully; he had it. Still a bit of give and take, interrupting one another, a partial disagreement, very quick clarification and a complete mutual understanding, and then both men with beaming eyes skimming over the shining riches of the new theory.[17]

When Lorentz died in 1928, Einstein would say in his eulogy, "I stand at the grave of the greatest and noblest man of our times." And in 1953, for the celebration of the hundredth anniversary of Lorentz's birth, Einstein wrote an essay on his importance. "Whatever came from this supreme mind was as lucid and beautiful as a good work of art," he wrote. "He meant more to me personally than anybody else I have met in my lifetime."[18]

Marić was unhappy about moving to Prague. "I am not going there gladly and I expect very little pleasure," she wrote a friend. But initially, until the city's dirtiness and snobbishness became oppressive, their life there was nice enough. They had electric lighting in their home for the first time, and both the space and money for a live-in maid. "The people are haughty, shabby-genteel, or subservient, depending on their lot in life," Einstein said. "Many of them possess a certain grace."[19]

From Einstein's office at the university he could look down on a beautiful park with shady trees and manicured gardens. In the morning, it would be filled just with women, and in the afternoon just with men. Some walked alone as if deep in thought, Einstein noticed, while others clustered in groups holding animated arguments. Eventually, Einstein asked what the park was. It belonged, he was told, to an insane asylum. When he showed his friend Philipp Frank the view, Einstein commented ruefully, "Those are the madmen who do not occupy themselves with the quantum theory."[20]

The Einsteins became acquainted with Bertha Fanta, a delightfully cultured woman who hosted at her home a literary and musical salon for Prague's Jewish intelligentsia. Einstein was the ideal catch: a rising scholar who was willing, with equal gusto, to play the violin or discuss Hume and Kant, depending on the spirit of the occasion. Other habitués included the young writer Franz Kafka and his friend Max Brod.

In his book *The Redemption of Tycho Brahe,* Brod seemed to use (though he sometimes denied it) Einstein as the model for the character of Johannes Kepler, the brilliant astronomer who had been Brahe's assistant in Prague in 1600. The character is devoted to his scientific work and is always willing to throw away conventional thinking. But in the realm of the personal, he is protected from "the aberrations of feeling" by his aloof and abstracted air. "He had no heart and therefore nothing to fear from the world," Brod wrote. "He was not capable of emotion or love." When the novel came out, a fellow scientist, Walther Nernst, said to Einstein, "You are this man Kepler."[21]

Not really. Despite the image he sometimes cast as a loner, Einstein continued to establish, as he had back in Zurich and Bern, intimate friendships and emotional bonds, particularly with fellow thinkers and scientists. One such friend was Paul Ehrenfest, a young Jewish physicist from Vienna who was teaching at the University of St. Petersburg but feeling professionally stymied there because of his background. In early 1912, he embarked on a trip through Europe looking for a new job, and on his way toward Prague contacted Einstein, with whom he had been corresponding about gravity and radiation. "Do stay at my house so that we can make good use of the time," Einstein responded.[22]

When Ehrenfest arrived one rainy Friday afternoon in February, a cigar-puffing Einstein and his wife were at the train station to meet him. They all walked to a café, where they compared the great cities of Europe. When Marić left, the discussion turned to science, most notably statistical mechanics, and they continued talking as they walked to Einstein's office. "On the way to the institute, first argument about everything," Ehrenfest recorded in his diary of the seven days he spent in Prague.

Ehrenfest was a mousy and insecure man, but his eagerness for friendship and his love of physics made it easy for him to forge a bond with Einstein.[23] They both seemed to crave arguing about science, and Einstein later said that "within a few hours we were friends as if Nature created us for each other." Their intense discussions continued the next day, as Einstein explained his efforts to generalize his theory of relativity. On Sunday evening, they relaxed a bit by performing Brahms, with Ehrenfest on piano, Einstein on violin, and 7-year-old Hans Albert singing. "Yes we will be friends," Ehrenfest wrote in his diary that night. "Was awfully happy."[24]

Einstein was already thinking of leaving Prague, and he suggested Ehrenfest as a possible successor. But he "adamantly refuses to profess any religious affiliation," Einstein lamented. Unlike Einstein, who was willing to relent and write "Mosaic" on his official forms, Ehrenfest had abandoned Judaism and would not profess otherwise. "Your stubborn refusal to acknowledge any religious affiliation really *bugs* me," Einstein wrote him in April. "Drop it for your children's sake. After all, after becoming a professor here you could revert to this strange hobby horse of yours." [25]

Matters eventually came to a happy resolution when Ehrenfest accepted an offer, which Einstein had earlier received but declined, to replace the revered Lorentz, who was cutting back from full-time teaching at the University of Leiden. Einstein was thrilled, for it meant he would now have two friends there to visit regularly. It became, for Einstein, almost a second academic home and a way to escape the oppressive atmosphere he later found in Berlin. Almost every year for the next two decades, until 1933 when Ehrenfest committed suicide and Einstein moved to America, Einstein would make regular pilgrimages to see him and Lorentz in Leiden or at the seaside resorts nearby. [26]

The 1911 Solvay Conference

Ernest Solvay was a Belgian chemist and industrialist who reaped a fortune by inventing a method for making soda. Because he wanted to do something unusual yet useful with his money, and also because he had some odd theories of gravity that he wanted scientists to listen to, he decided to fund an elite gathering of Europe's top physicists. Scheduled for the end of October 1911, it eventually spawned a series of influential meetings, known as Solvay Conferences, that were held sporadically over the ensuing years.

Twenty of Europe's most famous scientists showed up at the Grand Hotel Metropole in Brussels. At 32, Einstein was the youngest. There was Max Planck, Henri Poincaré, Marie Curie, Ernest Rutherford, and Wilhelm Wien. The chemist Walther Nernst organized the event and acted as chaperone for the quirky Ernest Solvay. The kindly Hendrik Lorentz served as the chairman, as his fan Einstein put it, "with incomparable tact and unbelievable virtuosity." [27]

The focus of the conference was "the quantum problem," and Einstein was asked to present a paper on that topic, making him one of only eight "particularly competent members" thus honored. He expressed some annoyance, perhaps a bit more feigned than real, about the prestigious assignment. He dubbed the upcoming meeting "the witch's Sabbath" and complained to Besso, "My twaddle for the Brussels conference weighs down on me."[28]

Einstein's talk was titled "The Present State of the Problem of Specific Heats." Specific heat—the quantity of energy required to increase the temperature of a specific amount of substance by a certain amount—had been a specialty of Einstein's former professor and antagonist at the Zurich Polytechnic, Heinrich Weber. Weber had discovered some anomalies, especially at low temperatures, in the laws that were supposed to govern specific heat. Beginning in late 1906, Einstein had come up with what he called a "quantized" approach to the problem by surmising that the atoms in each substance could absorb energy only in discrete packets.

In his 1911 Solvay lecture, Einstein put these issues into the larger context of the so-called quantum problem. Was it possible, he asked, to avoid accepting the physical reality of these atomistic particles of light, which were like bullets aimed at the heart of Maxwell's equations and, indeed, all of classical physics?

Planck, who had pioneered the concept of the quanta, continued to insist that they came into play only when light was being emitted or absorbed. They were not a real-world feature of light itself, he argued. Einstein, in his talk to the conference, sorrowfully demurred: "These discontinuities, which we find so distasteful in Planck's theory, seem really to exist in nature."[29]

Really to exist in nature. It was, for Einstein, an odd phrase. To a pure proponent of Mach, or for that matter of Hume, the whole phrase "really to exist in nature" lacked clear meaning. In his special relativity theory, Einstein had avoided assuming the existence of such things as absolute time and absolute distance, because it seemed meaningless to say that they "really" existed in nature when they couldn't be observed. But henceforth, during the more than four decades in which he would express his discomfort with quantum theory, he increasingly sounded like a scientific realist, someone who believed that an underlying reality existed in nature that was independent of our ability to observe or measure it.

When he was finished, Einstein faced a barrage of challenges from Lorentz, Planck, Poincaré, and others. Some of what Einstein said, Lorentz rose to point out, "seems in fact to be totally incompatible with Maxwell's equations."

Einstein agreed, perhaps too readily, that "the quantum hypothesis is provisional" and that it "does not seem compatible with the experimentally verified conclusions of the wave theory." Somehow it was necessary, he told his questioners, to accommodate both wave and particle approaches to the understanding of light. "In addition to Maxwell's electrodynamics, which is essential to us, we must also admit a hypothesis such as that of quanta."[30]

It was unclear, even to Einstein, whether Planck was persuaded of the reality of quanta. "I largely succeeded in convincing Planck that my conception is correct, after he has struggled against it for so many years," Einstein wrote his friend Heinrich Zangger. But a week later, Einstein gave Zangger another report: "Planck stuck stubbornly to some undoubtedly wrong preconceptions."

As for Lorentz, Einstein remained as admiring as ever: "A living work of art! He was in my opinion the most intelligent of the theoreticians present." He dismissed Poincaré, who paid little attention to him, with a brusque stroke: "Poincaré was simply negative in general, and, all his acumen notwithstanding, he showed little grasp of the situation."[31]

Overall he gave low marks to the conference, where most of the time was spent bewailing rather than resolving quantum theory's threat to classical mechanics. "The congress in Brussels resembled the lamentations on the ruins of Jerusalem," he wrote Besso. "Nothing positive has come out of it."[32]

There was one interesting sideshow for Einstein: the romance between the widowed Marie Curie and the married Paul Langevin. Dignified and dedicated, Madame Curie was the first woman to win a Nobel Prize; she shared the 1903 physics prize with her husband and one other scientist for their work on radiation. Three years later, her husband was killed by a horse-drawn wagon. She was bereft, and so was her late husband's protégé, Langevin, who taught physics at the Sorbonne with the Curies. Langevin was trapped in a marriage with a wife who physically abused him, and soon he and Marie Curie were having an affair in a Paris apartment. His wife had someone break into it and steal their love letters.

Just as the Solvay Conference was getting under way, with both Curie and Langevin in attendance, the purloined letters began appearing in a Paris tabloid as a prelude to a sensational divorce case. In addition, at that very moment, it was announced that Curie had won the Nobel Prize in chemistry, for discovering radium and polonium.* A member of the Swedish Academy wrote her to suggest that she not appear to receive it, given the furor raised by her relationship with Langevin, but she coolly responded, "I believe there is no connection between my scientific work and the facts of private life." She headed to Stockholm and accepted the prize.[33]

The whole furor seemed silly to Einstein. "She is an unpretentious, honest person," he said, with "a sparkling intelligence." He also rather bluntly came to the conclusion, not justified, that she was not pretty enough to wreck anyone's marriage. "Despite her passionate nature," he said, "she is not attractive enough to represent a danger to anyone."[34]

More gracious was the sturdy letter of support he sent her later that month:

> Do not laugh at me for writing you without having anything sensible to say. But I am so enraged by the base manner in which the public is presently daring to concern itself with you that I absolutely must give vent to this feeling. I am impelled to tell you how much I have come to admire your intellect, your drive, and your honesty, and that I consider myself lucky to have made your personal acquaintance in Brussels. Anyone who does not number among these reptiles is certainly happy, now as before, that we have such personages among us as you, and Langevin too, real people with whom one feels privileged to be in contact. If the rabble continues to occupy itself with you, then simply don't read that hogwash, but rather leave it to the reptile for whom it has been fabricated.[35]

---

* Added to her 1903 physics prize, she thus became the first person to win Nobels in two different fields. The only other person to do so was Linus Pauling, who won for chemistry in 1954, and then won the 1962 Nobel Peace Prize for his fight against nuclear weapons testing.

Enter Elsa

As Einstein wandered around Europe giving speeches and basking in his rising renown, his wife stayed behind in Prague, a city she hated, and brooded about not being part of the scientific circles that she once struggled to join. "I would like to have been there and listened a little, and seen all these fine people," she wrote him after one of his talks in October 1911. "It is so long since we saw each other that I wonder if you will recognize me." She signed herself, "Deine alte D," your old D, as if she were still his Dollie, albeit a bit older.[36]

Her circumstances, perhaps combined with an innate disposition, caused her to become gloomy, even depressed. When Philipp Frank met her in Prague for the first time, he thought that she might be schizophrenic. Einstein concurred, and he later told a colleague that her gloominess "is doubtless traceable to a schizophrenic genetic disposition coming from her mother's family."[37]

Thus it was that Einstein's marriage was once again in an unstable state when he traveled alone to Berlin during the Easter holidays in 1912. There he became reacquainted with a cousin, three years older, whom he had known as a child.

Elsa Einstein* was the daughter of Rudolf ("the rich") Einstein and Fanny Koch Einstein. She was Einstein's cousin on both sides. Her father was the first cousin of Einstein's father, Hermann, and had helped fund his business. Her mother was the sister of Einstein's mother, Pauline (making Elsa and Albert first cousins). After Hermann's death, Pauline had moved in with Rudolf and Fanny Einstein for a few years, helping them keep house.

As children, Albert and Elsa had played together at the home of Albert's parents in Munich and on one occasion had shared a first artistic experience at the opera.[38] Since then, Elsa had been married, divorced, and now, at age 36, was living with her two daughters, Margot and Ilse, in the same apartment building as her parents.

---

* She was born Elsa Einstein, became Elsa Löwenthal during her brief marriage to a Berlin merchant, and was referred to as Elsa Einstein by Albert Einstein even before they married. For clarity, I refer to her as Elsa throughout.

The contrast with Einstein's wife was stark. Mileva Marić was exotic, intellectual, and complex. Elsa wasn't. Instead, she was conventionally handsome and domestically nurturing. She loved heavy German comfort foods and chocolate, which tended to give her a rather ample, matronly look. Her face was similar to her cousin's, and it would become strikingly more so as they aged.[39]

Einstein was looking for new companionship, and he first flirted with Elsa's sister. But by the end of his Easter visit, he had settled on Elsa as offering the comfort and nurturing that he now craved. The love he was seeking, it seems, was not wild romance but uncomplicated support and affection.

And Elsa, who revered her cousin, was eager to give it. When he returned to Prague, she wrote him right away—sending the letter to his office, not his home, and proposing a way they could correspond in secret. "How dear of you not to be too proud to communicate with me in such a way!" he responded. "I can't even begin to tell you how fond I have become of you during these few days." She asked him to destroy her letters, which he did. She, on the other hand, kept his responses for the rest of her life in a folder that she tied and later labeled "Especially beautiful letters from better days."[40]

Einstein apologized for his flirtation with her sister Paula. "It is hard for me to understand how I could have taken a fancy to her," he declared. "But it is in fact simple. She was young, a girl, and complaisant."

A decade earlier, when he was writing his love letters to Marić that celebrated their own rarefied and bohemian approach to life, Einstein would likely have lumped relatives such as Elsa into the category of "bourgeois philistines." But now, in letters that were almost as effusive as the ones he had written to Marić, he professed his new passion for Elsa. "I have to have someone to love, otherwise life is miserable," he wrote. "And this someone is you."

She knew how to make him defensive: she teased him for being under Marić's thumb and asserted that he was "henpecked." As she may have hoped, Einstein responded by protesting that he would show her otherwise. "Do not think about me in such a way!" he said. "I categorically assure you that I consider myself a full-fledged male. Perhaps I will sometime have the opportunity to prove it to you."

Spurred by this new affection and by the prospect of working in the world's capital of theoretical physics, Einstein developed a desire to move to Berlin. "The chances of getting a call to Berlin are, unfortunately, slight," he admitted to Elsa. But on his visit, he did what he could to increase his chances of someday getting a position there. In his notebook he listed appointments he had been able to get with important academic leaders, including the scientists Fritz Haber, Walther Nernst, and Emil Warburg.[41]

Einstein's son Hans Albert later recalled that it was just after his eighth birthday, in the spring of 1912, when he noticed that his parents' marriage was falling apart. But after returning to Prague from Berlin, Einstein seemed to develop qualms about his affair with his cousin. He tried, in two letters, to put an end to it. "There would only be confusion and misfortune if we were to give into our mutual attraction," he wrote Elsa.

Later that month, he tried to be even more definitive. "It will not be good for the two of us, as well as for the others, if we form a closer attachment. So, I am writing to you today for the last time and am submitting again to the inevitable, and you must do the same. You know that it is not hardness of heart or lack of feeling that makes me talk like this, because you know that, like you, I bear my cross without hope."[42]

Einstein and Marić shared one thing: a feeling that living among the middle-class German community in Prague had become wearisome. "These are not people with natural sentiments," he told Besso. They displayed "a peculiar mixture of snobbery and servility, without any kind of goodwill toward their fellow men." The water was undrinkable, the air was full of soot, and an ostentatious luxury was juxtaposed with misery on the streets. But what offended Einstein most were the artificial class structures. "When I come to the institute," he complained, "a servile man who smells of alcohol bows and says, 'your most humble servant.' "[43]

Marić worried that the bad water, milk, and air were hurting the health of their younger son, Eduard. He had lost his appetite and was not sleeping well. It was also now clear that her husband cared more about his science than his family. "He is tirelessly working on his problems; one can say that he lives only for them," she told her friend Helene Savić. "I must confess with a bit of shame that we are unimportant to him and take second place."[44]

So Einstein and his wife decided to return to the one place they thought could restore their relationship.

Zurich, 1912

The Zurich Polytechnic, where Einstein and Marić had blissfully shared their books and their souls, had been upgraded in June 1911 to a full university, now named the Eidgenössische Technische Hochschule (ETH), or the Swiss Federal Institute of Technology, with the right to grant graduate degrees. At 32 and by now quite famous in the world of theoretical physics, Einstein should have been an easy and obvious choice for one of the new professorships available there.

That possibility had been discussed a year earlier. Before he left for Prague, Einstein had made a deal with officials in Zurich. "I promised in private that I would advise them before accepting another offer from somewhere else, so that the administration of the Polytechnic could also make me an offer if they find it fit to do so," he told a Dutch professor who was trying to recruit him to Utrecht.[45]

By November 1911, Einstein had received such an offer from Zurich, or at least so he thought, and as a result he declined the offer to go to Utrecht. But the matter was not completely settled, because some of Zurich's education officials objected. They argued that a professor in theoretical physics was a "luxury," that there was not enough lab space to accommodate one, and that Einstein personally was not a good teacher.

Heinrich Zangger, a longtime friend who was a medical researcher in Zurich, intervened on Einstein's behalf. "A proper theoretical physicist is a necessity these days," he wrote in a letter to one of the top Swiss councilors. He also pointed out that in such a role Einstein "needs no laboratory." As for Einstein's teaching talents, Zangger provided a wonderfully nuanced and revealing description:

> He is not a good teacher for mentally lazy gentlemen who merely want to fill a notebook and then learn it by heart for an exam; he is not a

smooth talker, but anyone wishing to learn honestly how to develop his ideas in physics in an honest way, from deep within, and how to examine all premises carefully and see the pitfalls and the problems in his reflections, will find Einstein a first-class teacher, because all of this is expressed in his lectures, which force the audience to think along.[46]

Zangger wrote Einstein to express his outrage at the dithering in Zurich, and Einstein replied, "The dear Zurich folks can kiss my . . . [und die lieben Züricher können mich auch . . . (ellipses are in original letter)]." He told Zangger not to push the matter further. "Leave the Polytechnic* to God's inscrutable ways."[47]

Einstein, however, decided not to drop the matter but instead to push the Polytechnic through a light ruse. Officials at the university in Utrecht were just about to offer their open post to someone else, Peter Debye, when Einstein asked them to hold off. "I am turning to you with a strange request," he wrote. The Zurich Polytechnic had initially seemed very eager to recruit him, he said, and it had been proceeding with haste out of fear that he would go to Utrecht. "But if they were to learn in the near future that Debye is going to Utrecht, they would lose their fervor at once and keep me forever in suspense. I ask you therefore to wait a little longer with the official offer to Debye."[48]

Rather oddly, Einstein found himself needing letters of recommendation to secure a post at his own alma mater. Marie Curie wrote one. "In Brussels, where I attended a scientific conference in which Mr. Einstein also participated, I was able to admire the clarity of his intellect, the breadth of his information, and the profundity of his knowledge," she noted.[49]

Adding to the irony was that his other main letter of recommendation came from Henri Poincaré, the man who had almost come up with the special theory of relativity but still had not embraced it. Einstein was "one of the most original minds I have ever come across," he said. Particularly poignant was his description of Einstein's willingness, which Poincaré himself lacked, to make radical conceptual leaps:

---

* Although the school had been renamed, Einstein continued to call it the Polytechnic ("Polytechnikum") and, for clarity, I will continue to use this name.

"What I admire in him in particular is the facility with which he adapts himself to new concepts. He does not remain attached to classical principles, and, when presented with a problem in physics, is prompt to envision all the possibilities." Poincaré, however, could not resist asserting, perhaps with relativity in mind, that Einstein might not be right in all his theories: "Since he seeks in all directions one must expect the majority of the paths on which he embarks to be blind alleys."[50]

Soon it all worked out. Einstein would move back to Zurich in July 1912. He thanked Zangger for helping him to prevail "against all odds," and exulted, "I am enormously happy that we will be together again." Marić was thrilled as well. She thought that the return could help save both her sanity and their marriage. Even the children seemed happy to be out of Prague and back to the city of their birth. As Einstein put it in a postcard to another friend, "Great joy about it among us old folks and the two bear cubs."[51]

His departure caused a minor controversy in Prague. Newspaper articles noted that anti-Semitism at the university may have played a role. Einstein felt compelled to issue a public statement. "Despite all presumptions," he said, "I did not feel and did not notice any religious prejudice." The appointment of Philipp Frank, a Jew, as his successor, he added, confirmed that "such considerations" were not a major problem.[52]

Life in Zurich should have been glorious. The Einsteins were able to afford a modern six-room apartment with grand views. They were reunited with friends such as Zangger and Grossmann, and there was even one fewer adversary. "The fierce Weber has died, so it will be very pleasant from a personal point of view," Einstein wrote of their undergraduate physics professor and nemesis, Heinrich Weber.[53]

Once again there were musical gatherings at the home of math professor Adolf Hurwitz. The programs included not only Mozart, Einstein's favorite, but also Schumann, who was Marić's. On Sunday afternoons, Einstein would arrive with his wife and two little boys at the doorstep and announce, "Here comes the whole Einstein hen house."

Despite being back with such friends and diversions, Marić's depression continued to deepen, and her health to decline. She developed rheumatism, which made it hard for her to go out, especially when the streets became icy in winter. She attended the Hurwitz recitals less frequently, and when she did show up her gloom was increasingly evident. In February 1913, to entice her out, the Hurwitz family planned an all-Schumann recital. She came, but seemed paralyzed by pain, both mental and physical.[54]

Thus the atmosphere was ripe for a catalyst that would disrupt this unstable family situation. It came in the form of a letter. After almost a year of silence, Elsa Einstein wrote to her cousin.

The previous May, when he had declared that he was writing her "for the last time," Einstein had nonetheless given her the address of what would be his new office in Zurich. Now Elsa decided to send him a greeting for his thirty-fourth birthday, and she added a request for a picture of him and a recommendation of a good book she could read on relativity. She knew how to flatter.[55]

"There is no book on relativity that is comprehensible to the layman," he replied. "But what do you have a relativity cousin for? If you ever happen to be in Zurich, then we (without my wife, who is unfortunately very jealous) will take a nice walk, and I will tell you about all of those curious things that I discovered." Then he went a bit further. Instead of sending a picture, wouldn't it be better to see each other in person? "If you wish to make me truly happy, then arrange to spend a few days here sometime."[56]

A few days later, he wrote again, with word that he had instructed a photographer to send her a picture. He had been working on generalizing his theory of relativity, he reported, and it was exhausting. As he had a year earlier, he complained about being married to Marić: "What I wouldn't give to be able to spend a few days with you, but without my cross!" He asked Elsa if she would be in Berlin later that summer. "I would like to come for a short visit."[57]

It was therefore not surprising that Einstein was very receptive, a few months later, when the two towers of Berlin's scientific establishment—Max Planck and Walther Nernst—came to Zurich with an enticing proposal. Having been impressed by Einstein at the Solvay Conference of 1911, they had already been sounding out colleagues about getting him to Berlin.

The offer they brought with them, when they arrived with their wives on the night train from Berlin on July 11, 1913, had three impressive components: Einstein would be elected to a coveted vacancy in the Prussian Academy of Sciences, which would come with a hefty stipend; he would become the director of a new physics institute; and he would be made a professor at the University of Berlin. The package included a lot of money, and it was not nearly as much work as it may have seemed on the surface. Planck and Nernst made it clear that Einstein would have no required teaching duties at the university and no real administrative tasks at the institute. And though he would be required to accept German citizenship once again, he could keep his Swiss citizenship as well.

The visitors made their case during a long visit to Einstein's sunny office at the Polytechnic. He said he needed a few hours to think it over, though it is likely he knew he would accept. So Planck and Nernst took their wives on an excursion by funicular railway up one of the nearby mountains. With puckish amusement, Einstein told them he would be awaiting their return to the station with a signal. If he had decided to decline, he would be carrying a white rose, and if he was going to accept, a red rose (some accounts have the signal being a white handkerchief). When they stepped off the train, they happily discovered that he had accepted.[58]

That meant that Einstein would become, at 34, the youngest member of the Prussian Academy. But first Planck had to get him elected. The letter he wrote, which was also signed by Nernst and others, had the memorable but incorrect concession, quoted earlier, that "he might sometimes have overshot the target in his speculations, as for example in his light quantum hypothesis." But the rest of the letter was suffused with extravagant praise for each of his many scientific contributions. "Among the great problems abundant in modern physics, there is hardly one to which Einstein has not made a remarkable contribution."[59]

The Berliners were taking a risk, Einstein realized. He was being recruited not for his teaching skills (as he would not be teaching), nor for his administrative ones. And even though he had been publishing outlines and papers describing his ongoing efforts to generalize relativity, it was unclear whether he would succeed in that quest. "The Germans are gambling on me as they would on a prize-winning hen," he told a friend as they were leaving a party, "but I don't know if I can still lay eggs."[60]

Einstein, likewise, was taking a risk. He had a secure and lucrative post in a city and society that he, his wife, and his family loved. The Swiss personality agreed with him. His wife had a Slav's revulsion for all things Teutonic, and he had a similar distaste that had been ingrained in childhood. As a boy he had run away from Prussian-accented parades and Germanic rigidity. Only the opportunity to be gloriously coddled in the world capital of science could have compelled him to make such a move.

Einstein found the prospect thrilling and a bit amusing. "I am going to Berlin as an Academy-man without any obligations, rather like a living mummy," he wrote fellow physicist Jakob Laub. "I'm already looking forward to this difficult career!"[61] To Ehrenfest he admitted, "I accepted this odd sinecure because giving lectures gets on my nerves."[62] However, to the venerable Hendrik Lorentz in Holland Einstein displayed more gravitas: "I could not resist the temptation to accept a position in which I am relieved of all responsibilities so that I can give myself over completely to rumination."[63]

There was, of course, another factor that made the new job enticing: the chance to be with his cousin and new love, Elsa. As he would later admit to his friend Zangger, "She was the main reason for my going to Berlin, you know."[64]

The same evening that Planck and Nernst left Zurich, Einstein wrote Elsa an excited letter describing the "colossal honor" they had offered. "Next spring at the latest, I'll come to Berlin for good," he exulted. "I already rejoice at the wonderful times we will spend together!"

During the ensuing week, he sent two more such notes. "I rejoice at the thought that I will soon be coming to you," he wrote in the first. And a few days later: "Now we will be together and rejoice in each other!" It is impossible to know for sure what relative weight to assign to each of the factors enticing him to Berlin: the unsurpassed scientific community there, the glories and perks of the post he was offered, or the chance to be with Elsa. But at least to her he claimed it was primarily the latter. "I look forward keenly to Berlin, mainly because I look forward to *you*."[65]

Elsa had actually tried to help him get the offer. Earlier in the year, on her own initiative, she had dropped in on Fritz Haber, who ran the Kaiser Wilhelm Institute of Chemistry in Berlin, and let him know that her cousin might be open to a position that would bring him to Berlin. When he learned of Elsa's intervention, Einstein was amused. "Haber knows who he is dealing with. He knows how to appreciate the influence of a friendly female cousin . . . The nonchalance with which you dropped in on Haber is pure Elsa. Did you tell anyone about it, or did you consult only with your wicked heart? If only I could have looked on!"[66]

Even before Einstein moved to Berlin, he and Elsa began to correspond as if they were a couple. She worried about his exhaustion and sent him a long letter prescribing more exercise, rest, and a healthier diet. He responded by saying that he planned to "smoke like a chimney, work like a horse, eat without thinking, go for a walk *only* in really pleasant company."

He made clear, however, that she should not expect him to abandon his wife: "You and I can very well be happy with each other without her having to be hurt."[67]

Indeed, even amid his flurry of love letters with Elsa, Einstein was still trying to be a suitable family man. For his August 1913 vacation, he decided to take his wife and two sons hiking with Marie Curie and her two daughters. The plan was to go through the mountains of southeastern Switzerland down to Lake Como, where he and Marić had spent their most passionate and romantic moments twelve years earlier.

As it turned out, the sickly Eduard was unable to make the trip, and Marić stayed behind for a few days to get him settled with friends. Then she joined them as they neared Lake Como. During the hikes, Curie challenged Einstein to name all the peaks. They also talked science, especially when the children ran ahead. At one point Einstein stopped suddenly and grabbed Curie's arm. "You understand, what I need to know is exactly what happens to the passengers in an elevator when it falls into emptiness," he said, referring to his ideas about the equivalence of gravity and acceleration. As Curie's daughter noted later, "Such a touching preoccupation made the younger generation roar with laughter."[68]

Einstein then accompanied Marić and their children to visit her family in Novi Sad and at their summer house in Kać. On their final Sunday in Serbia, Marić took the children, without her husband, to be baptized. Hans Albert remembered later the beautiful singing; his brother, Eduard, only 3, was disruptive. As for their father, he seemed sanguine and bemused afterward. "Do you know what the result is?" he told Hurwitz. "They've turned Catholic. Well, it's all the same to me."[69]

The façade of familial harmony, however, masked the deterioration of the marriage. After his visit to Serbia and a stop in Vienna for his annual appearance at the conference of German-speaking physicists, Einstein continued on to Berlin, alone. There he was reunited with Elsa. "I now have someone I can think about with pure delight and I can live for," he told her.[70]

Elsa's home cooking, a hearty pleasure she lavished on him like a mother, became a theme in their letters. Their correspondence, like their relationship, was a stark contrast to that between Einstein and Marić a dozen years earlier. He and Elsa tended to write to each other about domestic comforts—food, tranquillity, hygiene, fondness—rather than about romantic bliss and planted kisses, or intimacies of the soul and insights of the intellect.

Despite such conventional concerns, Einstein still fancied their relationship could avoid sinking into a mundane pattern. "How nice it would be if one of these days we could share in managing a small bohemian household," he wrote. "You have no idea how charming such a life with very small needs and without grandeur can be!"[71] When Elsa gave him a hairbrush, he initially prided himself on his progress in personal grooming, but then he reverted to more slovenly ways and told her, only half jokingly, that it was to guard against the philistines and the bourgeoisie. Those were words he had used with Marić as well, but more earnestly.

Elsa wanted not only to domesticate Einstein but to marry him. Even before he moved to Berlin, she wrote to urge him to divorce Marić. It would become a running battle for years, until she finally won her way. But for the moment, Einstein was resistant. "Do you think," he asked her, "it is so easy to get a divorce if one does not have any proof of the other party's guilt?" She should accept that he had virtually separated from Marić even if he was not going to divorce her. "I treat my wife as an employee whom I cannot fire. I have my own bedroom and avoid being alone with her." Elsa was upset that Einstein did not want to marry her, and she was fearful of how an illicit relationship would affect her daughters, but Einstein insisted it was for the best.[72]

Marić was understandably depressed by the prospect of moving to Berlin. There she would have to deal with Einstein's mother, who had never liked her, and his cousin, whom she rightly suspected of being a rival. In addition, Berlin had sometimes been less tolerant to Slavs than it was even to Jews. "My wife whines to me incessantly about Berlin and her fear of the relatives," Einstein wrote Elsa. "Well, there is some truth in this." In another letter, when he noted that Marić was afraid of her, he added, "Rightly so I hope!"[73]

Indeed, by this point all of the women in his life—his mother, sister, wife, and kissing cousin—were at war with one another. As Christmas 1913 neared, Einstein's struggle to generalize relativity had the added benefit of being a way to avoid family emotions. The effort produced yet another eloquent restatement of how science could rescue him from the merely personal. "The love of science thrives under these circumstances," he told Elsa, "for it lifts me impersonally from the vale of tears into peaceful spheres."[74]

With the approach of the spring of 1914 and their move to Berlin, Eduard came down with an ear infection that made it necessary for Marić to take him to an Alpine resort to recover. "This has a good side," Einstein told Elsa. He would initially be traveling to Berlin alone, and "in order to savor that," he decided to skip a conference in Paris so that he could arrive earlier.

On one of their last evenings in Zurich, he and Marić went to the Hurwitz house for a farewell musical evening. Once again, the program featured Schumann, in an attempt to cheer her up. It didn't. She instead sat by herself in a corner and did not speak to anyone.[75]

Berlin, 1914

By April 1914, Einstein had settled into a spacious apartment just west of Berlin's city center. Marić had picked it out when she visited Berlin over Christmas vacation, and she arrived in late April, after Eduard's ear infection had subsided.[76]

The tensions in Einstein's domestic life were exacerbated by overwork and mental strain. He was settling into a new job—actually three new jobs—and still struggling with his fitful attempts to generalize his theory of relativity and tie it into a theory of gravity. That first April in Berlin, for example, he engaged in an intense correspondence with Paul Ehrenfest over ways to calculate the forces affecting rotating electrons in a magnetic field. He started writing a theory for such situations, then realized it was wrong. "The angel had unveiled itself halfway in its magnificence," he told Ehrenfest, "then on further unveiling a cloven hoof appeared and I ran away."

Even more revealing, perhaps more than he meant it to be, was his comment to Ehrenfest about his personal life in Berlin. "I really delight in my local relatives," he reported, "especially in a cousin of my age."[77]

When Ehrenfest came for a visit at the end of April, Marić had just arrived, and he found her gloomy and yearning for Zurich. Einstein, on the other hand, had thrown himself into his work. "He had the impression that the family was taking a bit too much of his time, and that he had the duty to concentrate completely on his work," his son Hans Albert later recollected about that fateful spring of 1914.[78]

Personal relationships involve nature's most mysterious forces. Outside judgments are easy to make and hard to verify. Einstein repeatedly and plaintively stressed to all of their mutual friends—especially the Bessos, Habers, and Zanggers—that they should try to see the breakup of his marriage from his perspective, despite his own apparent culpability.

It is probably true that he was not solely to blame. The decline of the marriage was a downward spiral. He had become emotionally withdrawn, Marić had become more depressed and dark, and each action reinforced the other. Einstein tended to avoid painful personal emotions by immersing himself in his work. Marić, for her part, was bitter about the collapse of her own dreams and increasingly resentful of her husband's success. Her jealousy made her hostile toward anyone else who was close to Einstein, including his mother (the feeling was reciprocal) and his friends. Her mistrustful nature was, understandably, to some extent an effect of Einstein's detachment, but it was also a cause.

By the time they moved to Berlin, Marić had developed at least one personal involvement of her own, with a mathematics professor in Zagreb named Vladimir Varićak, who had challenged Einstein's interpretations of how special relativity applied to a rotating disk. Einstein was aware of the situation. "He had a kind of relationship with my wife, which can't be held against either of them," he wrote to Zangger in June. "It only made me feel my sense of isolation doubly painfully."[79]

The end came in July. Amid the turmoil, Marić moved with her two boys into the house of Fritz Haber, the chemist who'd recruited Einstein and who ran the institute where his office was located. Haber had his own experience with domestic discord. His wife, Clara, would end up committing suicide the following year after a fight over Haber's participation in the war. But for the time being, she was Mileva Marić's only friend in Berlin, and Fritz Haber became the intermediary as the Einsteins' battles broke into the open.

Through the Habers, Einstein delivered to Marić in mid-July a brutal cease-fire ultimatum. It was in the form of a proposed contract, one in which Einstein's cold scientific approach combined with his personal hostility and emotional alienation to produce an astonishing document. It read in full:

Conditions.

A. You will make sure
   1. that my clothes and laundry are kept in good order;
   2. that I will receive my three meals regularly *in my room*;
   3. that my bedroom and study are kept neat, and especially that my desk is left for *my use only*.
B. You will renounce all personal relations with me insofar as they are not completely necessary for social reasons. Specifically, you will forego

      1. my sitting at home with you;

      2. my going out or traveling with you.

  C. You will obey the following points in your relations with me:

      1. you will not expect any intimacy from me, nor will you re-
         proach me in any way;

      2. you will stop talking to me if I request it;

      3. you will leave my bedroom or study immediately without
         protest if I request it.

  D. You will undertake not to belittle me in front of our children,
    either through words or behavior.[80]

Marić accepted the terms. When Haber delivered her response, Einstein insisted on writing to her again "so that you are completely clear about the situation." He was prepared to live together again "because I don't want to lose the children and I don't want them to lose me." It was out of the question that he would have a "friendly" relationship with her, but he would aim for a "businesslike" one. "The personal aspects must be reduced to a tiny remnant," he said. "In return, I assure you of proper comportment on my part, such as I would exercise to any woman as a stranger."[81]

Only then did Marić realize that the relationship was not salvageable. They all met at Haber's house on a Friday to work out a separation agreement. It took three hours. Einstein agreed to provide Marić and his children 5,600 marks a year, just under half of his primary salary. Haber and Marić went to a lawyer to have the contract drawn up; Einstein did not accompany them, but instead sent his friend Michele Besso, who had come from Trieste to represent him.[82]

Einstein left the meeting at Haber's house and went directly to the home of Elsa's parents, who were also his aunt and uncle. They arrived home late from dinner to find him there, and they received the news about the situation with "a mild distaste." Nevertheless, he ended up staying at their house. Elsa was on summer vacation in the Bavarian Alps with her two daughters, and Einstein wrote to inform her that he was now sleeping in her bed in the apartment upstairs. "It's peculiar how confusingly sentimental one gets," he told her. "It is just a bed like any other, as though you had never slept in it. And yet I find it comforting." She had invited him to visit her in the Bavarian Alps, but he said he could not, "for fear of damaging your reputation again."[83]

The way to a divorce had now been paved, he assured Elsa, and he called it "a sacrifice" he had made on her behalf. Marić would move back to Zurich and take custody of the two boys, and when they came to visit their father they could meet only on "neutral ground," not in any house he shared with Elsa. "This is justified," Einstein conceded to Elsa, "because it is not right to have the children see their father with a woman other than their own mother."

The prospect of parting with his children was devastating for Einstein. He pretended to be detached from personal sentiments, and sometimes he was. But he became deeply emotional as he imagined life apart from his sons. "I would be a real monster if I felt any other way," he wrote Elsa. "I have carried these children around innumerable times day and night, taken them out in their pram, played with them, romped around and joked with them. They used to shout with joy when I came; the little one cheered even now, because he was still too small to grasp the situation. Now they will be gone forever, and their image of their father is being spoiled."[84]

Marić and the two boys left Berlin, accompanied by Michele Besso, aboard the morning train to Zurich on Wednesday, July 29, 1914. Haber went to the station with Einstein, who "bawled like a little boy" all afternoon and evening. It was the most wrenching personal moment for a man who took perverse pride in avoiding personal moments. For all of his reputation of being inured to deep human attachments, he had been madly in love with Mileva Marić and bonded to his children. For one of the few times in his adult life, he found himself crying.

The next day he went to visit his mother, who cheered him up. She had never liked Marić and was delighted that she was gone. "Oh, if your poor Papa had only lived to see it!" she said about the separation. She even professed herself pleased for Elsa, although they had occasionally clashed. And Elsa's mother and father also seemed happy enough with the resolution, though they did express resentment that Einstein had been too financially generous to Marić, which meant the income left for him and Elsa might be "a bit meager."[85]

The whole ordeal left Einstein so drained that, despite what he had said to Elsa just a week earlier, he decided that he was not prepared to get married again. Thus he would not have to force the issue of a legal divorce, which Marić fiercely resisted. Elsa, still on vacation, was "bitterly disappointed" by the news. Einstein sought to reassure her. "For me there is no other female creature besides you," he wrote. "It is not a lack of true affection which scares me away again and again from marriage! Is it a fear of the comfortable life, of nice furniture, of the odium that I burden myself with or even of becoming some sort of contented bourgeois? I myself don't know; but you will see that my attachment to you will endure."

He insisted that she should not feel ashamed or let people pity her for consorting with a man who would not marry her. They would take walks together and be there for each other. Should she choose to offer even more, he would be grateful. But by not marrying, they would be protecting themselves from lapsing into a "contented bourgeois" existence and preventing their relationship "from becoming banal and from growing pale." To him, marriage was confining, which was a state he instinctively resisted. "I'm glad our delicate relationship does not have to founder on a provincial narrow-minded lifestyle." [86]

In the old days, Marić had been the type of soul mate who responded to such bohemian sentiments. Elsa was not such a person. A comfortable life with comfortable furniture appealed to her. So did marriage. She would accept his decision not to get married for a while, but not forever.

In the meantime, Einstein became embroiled in a long-distance battle with Marić over money, furniture, and the way she was allegedly "poisoning" their children against him. [87] And all around them, a chain reaction was taking Europe into the most incomprehensibly bloody war in its history.

Not surprisingly, Einstein reacted to all of this turmoil by throwing himself into his science.

CHAPTER NINE

GENERAL RELATIVITY

1911–1915

Light and Gravity

After Einstein formulated his special theory of relativity in 1905, he realized that it was incomplete in at least two ways. First, it held that no physical interaction can propagate faster than the speed of light; that conflicted with Newton's theory of gravity, which conceived of gravity as a force that acted instantly between distant objects. Second, it applied only to constant-velocity motion. So for the next ten years, Einstein engaged in an interwoven effort to come up with a new field theory of gravity and to generalize his relativity theory so that it applied to accelerated motion.[1]

His first major conceptual advance had come at the end of 1907, while he was writing about relativity for a science yearbook. As noted earlier, a thought experiment about what a free-falling observer would feel led him to embrace the principle that the local effects of being accelerated and of being in a gravitational field are indistinguishable.* A person in a closed windowless chamber who feels his feet pressed to the floor will not be able to tell whether it's because the chamber is in outer space being accelerated upward or because it is at rest in a gravitational

---

* See chapter 7. For purposes of this discussion, we are referring to a uniformly and rectilinearly accelerated reference frame and a static and homogeneous gravitational field.

field. If he pulls a penny from his pocket and lets it go, it will fall to the floor at an accelerating speed in either case. Likewise, a person who feels she is floating in the closed chamber will not know whether it's because the chamber is in free fall or hovering in a gravity-free region of outer space.[2]

This led Einstein to the formulation of an "equivalence principle" that would guide his quest for a theory of gravity and his attempt to generalize relativity. "I realized that I would be able to extend or generalize the principle of relativity to apply to accelerated systems in addition to those moving at a uniform velocity," he later explained. "And in so doing, I expected that I would be able to resolve the problem of gravitation at the same time."

Just as inertial mass and gravitational mass are equivalent, so too there is an equivalence, he realized, between all inertial effects, such as resistance to acceleration, and gravitational effects, such as weight. His insight was that they are both manifestations of the same structure, which we now sometimes call the inertio-gravitational field.[3]

One consequence of this equivalence is that gravity, as Einstein had noted, should bend a light beam. That is easy to show using the chamber thought experiment. Imagine that the chamber is being accelerated upward. A laser beam comes in through a pinhole on one wall. By the time it reaches the opposite wall, it's a little closer to the floor, because the chamber has shot upward. And if you could plot its trajectory across the chamber, it would be curved because of the upward acceleration. The equivalence principle says that this effect should be the same whether the chamber is accelerating upward or is instead resting still in a gravitational field. Thus, light should appear to bend when going through a gravitational field.

For almost four years after positing this principle, Einstein did little with it. Instead, he focused on light quanta. But in 1911, he confessed to Michele Besso that he was weary of worrying about quanta, and he turned his attention back to coming up with a field theory of gravity that would help him generalize relativity. It was a task that would take him almost four more years, culminating in an eruption of genius in November 1915.

In a paper he sent to the *Annalen der Physik* in June 1911, "On the

Influence of Gravity on the Propagation of Light," he picked up his insight from 1907 and gave it rigorous expression. "In a memoir published four years ago I tried to answer the question whether the propagation of light is influenced by gravitation," he began. "I now see that one of the most important consequences of my former treatment is capable of being tested experimentally." After a series of calculations, Einstein came up with a prediction for light passing through the gravitational field next to the sun: "A ray of light going past the sun would undergo a deflection of 0.83 second of arc."*

Once again, he was deducing a theory from grand principles and postulates, then deriving some predictions that experimenters could proceed to test. As before, he ended his paper by calling for just such a test. "As the stars in the parts of the sky near the sun are visible during total eclipses of the sun, this consequence of the theory may be observed. It would be a most desirable thing if astronomers would take up the question."[4]

Erwin Finlay Freundlich, a young astronomer at the Berlin University observatory, read the paper and became excited by the prospect of doing this test. But it could not be performed until an eclipse, when starlight passing near the sun would be visible, and there would be no suitable one for another three years.

So Freundlich proposed that he try to measure the deflection of starlight caused by the gravitational field of Jupiter. Alas, Jupiter did not prove big enough for the task. "If only we had a truly larger planet than Jupiter!" Einstein joked to Freundlich at the end of that summer. "But nature did not deem it her business to make the discovery of her laws easy for us."[5]

The theory that light beams could be bent led to some interesting questions. Everyday experience shows that light travels in straight lines. Carpenters now use laser levels to mark off straight lines and construct level houses. If a light beam curves as it passes through regions of changing gravitational fields, how can a straight line be determined?

---

* I am using the numbers in Einstein's original calculations. Subsequent data caused it to be revised to about 0.85 second of arc. Also, as we shall see, he later revised his theory to predict twice the bending. An arc-second, or second of arc, is an angle of $\frac{1}{3,600}$ of a degree.

One solution might be to liken the path of the light beam through a changing gravitational field to that of a line drawn on a sphere or on a surface that is warped. In such cases, the shortest line between two points is curved, a geodesic like a great arc or a great circle route on our globe. Perhaps the bending of light meant that the fabric of space, through which the light beam traveled, was curved by gravity. The shortest path through a region of space that is curved by gravity might seem quite different from the straight lines of Euclidean geometry.

There was another clue that a new form of geometry might be needed. It became apparent to Einstein when he considered the case of a rotating disk. As a disk whirled around, its circumference would be contracted in the direction of its motion when observed from the reference frame of a person not rotating with it. The diameter of the circle, however, would not undergo any contraction. Thus, the ratio of the disk's circumference to its diameter would no longer be given by pi. Euclidean geometry wouldn't apply to such cases.

Rotating motion is a form of acceleration, because at every moment a point on the rim is undergoing a change in direction, which means that its velocity (a combination of speed and direction) is undergoing a change. Because non-Euclidean geometry would be necessary to describe this type of acceleration, according to the equivalence principle, it would be needed for gravitation as well.[6]

Unfortunately, as he had proved at the Zurich Polytechnic, non-Euclidean geometry was not a strong suit for Einstein. Fortunately, he had an old friend and classmate in Zurich for whom it was.

The Math

When Einstein moved back to Zurich from Prague in July 1912, one of the first things he did was call on his friend Marcel Grossmann, who had taken the notes Einstein used when he skipped math classes at the Zurich Polytechnic. Einstein had gotten a 4.25 out of 6 in his two geometry courses at the Polytechnic. Grossmann, on the other hand, had scored a perfect 6 in both of his geometry courses, had written his dissertation on non-Euclidean geometry, published seven papers on that topic, and was now the chairman of the math department.[7]

"Grossmann, you've got to help me or I will go crazy," Einstein said. He explained that he needed a mathematical system that would express—and perhaps even help him discover—the laws that governed the gravitational field. "Instantly, he was all afire," Einstein recalled of Grossmann's response.[8]

Until then, Einstein's scientific success had been based on his special talent for sniffing out the underlying physical principles of nature. He had left to others the task, which to him seemed less exalted, of finding the best mathematical expressions of those principles, as his Zurich colleague Minkowski had done for special relativity.

But by 1912, Einstein had come to appreciate that math could be a tool for discovering—and not merely describing—nature's laws. Math was nature's playbook. "The central idea of general relativity is that gravity arises from the curvature of spacetime," says physicist James Hartle. "Gravity *is* geometry."[9]

"I am now working exclusively on the gravitation problem and I believe that, with the help of a mathematician friend here, I will overcome all difficulties," Einstein wrote to the physicist Arnold Sommerfeld. "I have gained enormous respect for mathematics, whose more subtle parts I considered until now, in my ignorance, as pure luxury!"[10]

Grossmann went home to think about the question. After consulting the literature, he came back to Einstein and recommended the non-Euclidean geometry that had been devised by Bernhard Riemann.[11]

Riemann (1826–1866) was a child prodigy who invented a perpetual calendar at age 14 as a gift for his parents and went on to study in the great math center of Göttingen, Germany, under Carl Friedrich Gauss, who had been pioneering the geometry of curved surfaces. This was the topic Gauss assigned to Riemann for a thesis, and the result would transform not only geometry but physics.

Euclidean geometry describes flat surfaces. But it does not hold true on curved surfaces. For example, the sum of the angles of a triangle on a flat page is 180°. But look at the globe and picture a triangle formed by the equator as the base, the line of longitude running from the equator to the North Pole through London (longitude 0°) as one side, and the line of longitude running from the equator to the North Pole through New Orleans (longitude 90°) as the third side. If you look at this on a globe, you will see that all three angles of this triangle are right angles, which of course is impossible in the flat world of Euclid.

Gauss and others had developed different types of geometry that could describe the surface of spheres and other curved surfaces. Riemann took things even further: he developed a way to describe a surface no matter how its geometry changed, even if it varied from spherical to flat to hyperbolic from one point to the next. He also went beyond dealing with the curvature of just two-dimensional surfaces and, building on the work of Gauss, explored the various ways that math could describe the curvature of three-dimensional and even four-dimensional space.

That is a challenging concept. We can visualize a curved line or surface, but it is hard to imagine what curved three-dimensional space would be like, much less a curved four dimensions. But for mathematicians, extending the concept of curvature into different dimensions is easy, or at least doable. This involves using the concept of the *metric,* which specifies how to calculate the distance between two points in space.

On a flat surface with just the normal x and y coordinates, any high school algebra student, with the help of old Pythagoras, can calculate the distance between points. But imagine a flat map (of the world, for example) that represents locations on what is actually a curved globe. Things get stretched out near the poles, and measurement gets more complex. Calculating the actual distance between two points on the map in Greenland is different from doing so for points near the equator. Riemann worked out ways to determine mathematically the distance between points in space no matter how arbitrarily it curved and contorted.[12]

To do so he used something called a tensor. In Euclidean geometry, a vector is a quantity (such as of velocity or force) that has both a magnitude and a direction and thus needs more than a single simple number to describe it. In non-Euclidean geometry, where space is curved, we need something more generalized—sort of a vector on steroids—in order to incorporate, in a mathematically orderly way, more components. These are called tensors.

A *metric tensor* is a mathematical tool that tells us how to calculate the distance between points in a given space. For two-dimensional maps, a metric tensor has three components. For three-dimensional space, it has six independent components. And once you get to that glorious four-dimensional entity known as spacetime, the metric tensor needs ten independent components.*

Riemann helped to develop this concept of the metric tensor, which was denoted as $g_{\mu\nu}$ and pronounced *gee-mu-nu*. It had sixteen components, ten of them independent of one another, that could be used to define and describe a distance in curved four-dimensional spacetime.[13]

The useful thing about Riemann's tensor, as well as other tensors that Einstein and Grossmann adopted from the Italian mathematicians Gregorio Ricci-Curbastro and Tullio Levi-Civita, is that they are *generally covariant.* This was an important concept for Einstein as he tried to generalize a theory of relativity. It meant that the relationships between their components remained the same even when there were arbitrary changes or rotations in the space and time coordinate system. In other words, the information encoded in these tensors could go through a variety of transformations based on a changing frame of reference, but the basic laws governing the relationship of the components to each other remained the same.[14]

Einstein's goal as he pursued his general theory of relativity was to find the mathematical equations describing two complementary processes:

---

* Here's how it works. If you are at some point in curved space and want to know the distance to a neighboring point—infinitesimally close—then things can be complicated if you have just the Pythagorean theorem and some general geometry to use. The distance to a nearby point to the north may need to be computed differently from the distance to one to the east or to one in the up direction. You need something comparable to a little scorecard at each point of space to tell you the distance to each of these points. In four-dimensional spacetime your scorecard will require ten numbers for you to be able to deal with all the questions pertaining to spacetime distances to nearby points. You need such a scorecard for every point in the spacetime. But once you have those scorecards, you can figure out the distance along any curve: just add up the distances along each infinitesimal bit using the scorecards as you pass them. These scorecards form the metric tensor, which is a field in spacetime. In other words, it is something defined at every point, but that can have differing values at every point. I am grateful to Professor John D. Norton for helping with this section.

1. How a gravitational field acts on matter, telling it how to move.
2. And in turn, how matter generates gravitational fields in space-time, telling it how to curve.

His head-snapping insight was that gravity could be defined as the curvature of spacetime, and thus it could be represented by a metric tensor. For more than three years he would fitfully search for the right equations to accomplish his mission.[15]

Years later, when his younger son, Eduard, asked why he was so famous, Einstein replied by using a simple image to describe his great insight that gravity was the curving of the fabric of spacetime. "When a blind beetle crawls over the surface of a curved branch, it doesn't notice that the track it has covered is indeed curved," he said. "I was lucky enough to notice what the beetle didn't notice."[16]

The Zurich Notebook, 1912

Beginning in that summer of 1912, Einstein struggled to develop gravitational field equations using tensors along the lines developed by Riemann, Ricci, and others. His first round of fitful efforts are preserved in a scratchpad notebook. Over the years, this revealing "Zurich Notebook" has been dissected and analyzed by a team of scholars including Jürgen Renn, John D. Norton, Tilman Sauer, Michel Janssen, and John Stachel.[17]

In it Einstein pursued a two-fisted approach. On the one hand, he engaged in what was called a "physical strategy," in which he tried to build the correct equations from a set of requirements dictated by his feel for the physics. At the same time, he pursued a "mathematical strategy," in which he tried to deduce the correct equations from the more formal math requirements using the tensor analysis that Grossmann and others recommended.

Einstein's "physical strategy" began with his mission to generalize the principle of relativity so that it applied to observers who were accelerating or moving in an arbitrary manner. Any gravitational field equation he devised would have to meet the following physical requirements:

- It must revert to Newtonian theory in the special case of weak and static gravitational fields. In other words, under certain normal conditions, his theory would describe Newton's familiar laws of gravitation and motion.
- It should preserve the laws of classical physics, most notably the conservation of energy and momentum.
- It should satisfy the principle of equivalence, which holds that observations made by an observer who is uniformly accelerating would be equivalent to those made by an observer standing in a comparable gravitational field.

Einstein's "mathematical strategy," on the other hand, focused on using generic mathematical knowledge about the metric tensor to find a gravitational field equation that was generally (or at least broadly) covariant.

The process worked both ways: Einstein would examine equations that were abstracted from his physical requirements to check their covariance properties, and he would examine equations that sprang from elegant mathematical formulations to see if they met the requirements of his physics. "On page after page of the notebook, he approached the problem from either side, here writing expressions suggested by the physical requirements of the Newtonian limit and energy-momentum conservation, there writing expressions naturally suggested by the generally covariant quantities supplied by the mathematics of Ricci and Levi-Civita," says John Norton.[18]

But something disappointing happened. The two groups of requirements did not mesh. Or at least Einstein thought not. He could not get the results produced by one strategy to meet the requirements of the other strategy.

Using his mathematical strategy, he derived some very elegant equations. At Grossmann's suggestion, he had begun using a tensor developed by Riemann and then a more suitable one developed by Ricci. Finally, by the end of 1912, he had devised a field equation using a tensor that was, it turned out, pretty close to the one that he would eventually use in his triumphant formulation of late November 1915. In other words, in his Zurich Notebook he had come up with what was quite close to the right solution.[19]

But then he rejected it, and it would stagnate in his discard pile for more than two years. Why? Among other considerations, he thought (somewhat mistakenly) that this solution did not reduce, in a weak and static field, to Newton's laws. When he tried it a different way, it did not meet the requirement of the conservation of energy and momentum. And if he introduced a coordinate condition that allowed the equations to satisfy one of these requirements, it proved incompatible with the conditions needed to satisfy the other requirement.[20]

As a result, Einstein reduced his reliance on the mathematical strategy. It was a decision that he would later regret. Indeed, after he finally returned to the mathematical strategy and it proved spectacularly successful, he would from then on proclaim the virtues—both scientific and philosophical—of mathematical formalism.[21]

The Entwurf and Newton's Bucket, 1913

In May 1913, having discarded the equations derived from the mathematical strategy, Einstein and Grossmann produced a sketchy alternative theory based more on the physical strategy. Its equations were constructed to conform to the requirements of energy-momentum conservation and of being compatible with Newton's laws in a weak static field.

Even though it did not seem that these equations satisfied the goal of being suitably covariant, Einstein and Grossmann felt it was the best they could do for the time being. Their title reflected their tentativeness: "Outline of a Generalized Theory of Relativity and of a Theory of Gravitation." The paper thus became known as the *Entwurf*, which was the German word they had used for "outline."[22]

For a few months after producing the *Entwurf*, Einstein was both pleased and depleted. "I finally solved the problem a few weeks ago," he wrote Elsa. "It is a bold extension of the theory of relativity, together with a theory of gravitation. Now I must give myself some rest, otherwise I will go kaput."[23]

However, he was soon questioning what he had wrought. And the more he reflected on the *Entwurf,* the more he realized that its equations did not satisfy the goal of being generally or even broadly covariant. In other words, the way the equations applied to people in arbitrary accelerated motion might not always be the same.

His confidence in the theory was not strengthened when he sat down with his old friend Michele Besso, who had come to visit him in June 1913, to study the implications of the *Entwurf* theory. They produced more than fifty pages of notes on their deliberations, each writing about half, which analyzed how the *Entwurf* accorded with some curious facts that were known about the orbit of Mercury.[24]

Since the 1840s, scientists had been worrying about a small but unexplained shift in the orbit of Mercury. The perihelion is the spot in a planet's elliptical orbit when it is closest to the sun, and over the years this spot in Mercury's orbit had slipped a tiny amount more—about 43 seconds of an arc each century—than what was explained by Newton's laws. At first it was assumed that some undiscovered planet was tugging at it, similar to the reasoning that had earlier led to the discovery of Neptune. The Frenchman who discovered Mercury's anomaly even calculated where such a planet would be and named it Vulcan. But it was not there.

Einstein hoped that his new theory of relativity, when its gravitational field equations were applied to the sun, would explain Mercury's orbit. Unfortunately, after a lot of calculations and corrected mistakes, he and Besso came up with a value of 18 seconds of an arc per century for how far Mercury's perihelion should stray, which was not even halfway correct. The poor result convinced Einstein not to publish the Mercury calculations. But it did not convince him to discard his *Entwurf* theory, at least not yet.

Einstein and Besso also looked at whether rotation could be considered a form of relative motion under the equations of the *Entwurf* theory. In other words, imagine that an observer is rotating and thus experiencing inertia. Is it possible that this is yet another case of relative motion and is indistinguishable from a case where the observer is at rest and the rest of the universe is rotating around him?

The most famous thought experiment along these lines was that described by Newton in the third book of his *Principia.* Imagine a bucket that begins to rotate as it hangs from a rope. At first the water in the bucket stays rather still and flat. But soon the friction from the bucket causes the water to spin around with it, and it assumes a concave shape. Why? Because inertia causes the spinning water to push outward, and therefore it pushes up the side of the bucket.

Yes, but if we suspect that all motion is relative, we ask: What is the water spinning relative to? Not the bucket, because the water is concave when it is spinning along with the bucket, and also when the bucket stops and the water keeps spinning inside for a while. Perhaps the water is spinning relative to nearby bodies such as the earth that exert gravitational force.

But imagine the bucket spinning in deep space with no gravity and no reference points. Or imagine it spinning alone in an otherwise empty universe. Would there still be inertia? Newton believed so, and said it was because the bucket was spinning relative to absolute space.

When Einstein's early hero Ernst Mach came along in the mid-nineteenth century, he debunked this notion of absolute space and argued that the inertia existed because the water was spinning relative to the rest of the matter in the universe. Indeed, the same effects would be observed if the bucket was still and the rest of the universe was rotating around it, he said.[25]

The general theory of relativity, Einstein hoped, would have what he dubbed "Mach's Principle" as one of its touchstones. Happily, when he analyzed the equations in his *Entwurf* theory, he concluded that they *did* seem to predict that the effects would be the same whether a bucket was spinning or was motionless while the rest of the universe spun around it.

Or so Einstein thought. He and Besso made a series of very clever calculations designed to see if indeed this was the case. In their notebook, Einstein wrote a joyous little exclamation at what appeared to be the successful conclusion of these calculations: "Is correct."

Unfortunately, he and Besso had made some mistakes in this work. Einstein would eventually discover those errors two years later and realize, unhappily, that the *Entwurf* did not in fact satisfy Mach's principle. In all likelihood, Besso had already warned him that this might be the case. In a memo that he apparently wrote in August 1913, Besso suggested that a "rotation metric" was not in fact a solution permitted by the field equations in the *Entwurf.*

But Einstein dismissed these doubts, in letters to Besso as well as to Mach and others, at least for the time being.[26] If experiments upheld the theory, "your brilliant investigations on the foundations of mechanics will have received a splendid confirmation," Einstein wrote to Mach days after the *Entwurf* was published. "For it shows that inertia has its origin in some kind of interaction of the bodies, exactly in accordance with your argument about Newton's bucket experiment."[27]

What worried Einstein most about the *Entwurf,* justifiably, was that its mathematical equations did not prove to be generally covariant, thus deflating his goal of assuring that the laws of nature were the same for an observer in accelerated or arbitrary motion as they were for an observer moving at a constant velocity. "Regrettably, the whole business is still so very tricky that my confidence in the theory is still rather hesitant," he wrote in reply to a warm letter of congratulations from Lorentz. "The gravitational equations themselves unfortunately do not have the property of general covariance."[28]

He was soon able to convince himself, at least for a while, that this was inevitable. In part he did so through a thought experiment, which became known as the "hole argument,"[29] that seemed to suggest that the holy grail of making the gravitational field equations generally covariant was impossible to reach, or at least physically uninteresting. "The fact that the gravitational equations are not generally covariant, something that quite disturbed me for a while, is unavoidable," he wrote a friend. "It can easily be shown that a theory with generally covariant equations cannot exist if the demand is made that the field is mathematically completely determined by matter."[30]

For the time being, very few physicists embraced Einstein's new theory, and many came forth to denounce it.[31] Einstein professed pleasure that the issue of relativity "has at least been taken up with the requisite vigor," as he put it to his friend Zangger. "I enjoy controversies. In the manner of Figaro: 'Would my noble Lord venture a little dance? He should tell me! I will strike up the tune for him.' "[32]

Through it all, Einstein continued to try to salvage his *Entwurf* approach. He was able to find ways, or so he thought, to achieve enough covariance to satisfy most aspects of his principle about the equivalence of gravity and acceleration. "I succeeded in proving that the gravitational equations hold for arbitrarily moving reference systems, and thus that the hypothesis of the equivalence of acceleration and gravitational field is absolutely correct," he wrote Zangger in early 1914. "Nature shows us only the tail of the lion. But I have no doubt that the lion belongs with it even if he cannot reveal himself all at once. We see him only the way a louse that sits upon him would."[33]

Freundlich and the 1914 Eclipse

There was, Einstein knew, one way to quell doubts. He often concluded his papers with suggestions for how future experiments could confirm whatever he had just propounded. In the case of general relativity, this process had begun in 1911, when he specified with some precision how much he thought light from a star would be deflected by the gravity of the sun.

This was something that could, he hoped, be measured by photographing stars whose light passed close to the sun and determining whether there appeared to be a tiny shift in their position compared to when their light did not have to pass right by the sun. But this was an experiment that had to be done during an eclipse, when the starlight would be visible.

So it was not surprising that, with his theory arousing noisy attacks from colleagues and quiet doubts in his own mind, Einstein became keenly interested in what could be discovered during the next suitable total eclipse of the sun, which was due to occur on August 21, 1914. That would require an expedition to the Crimea, in Russia, where the path of the eclipse would fall.

Einstein was so eager to have his theory tested during the eclipse that, when it seemed there might be no money for such an expedition, he offered to pay part of the costs himself. Erwin Freundlich, the young Berlin astronomer who had read the light-bending predictions in Einstein's 1911 paper and become eager to prove him correct, was ready to take the lead. "I am extremely pleased that you have taken up the question of the bending of light with so much zeal," Einstein wrote him in early 1912. In August 1913, he was still bombarding the astronomer with encouragement. "Nothing more can be done by the theorists," he wrote. "In this matter it is only you, the astronomers, who can next year perform a simply invaluable service to theoretical physics."[34]

Freundlich got married in August 1913 and decided to take his honeymoon in the mountains near Zurich, in the hope that he could meet Einstein. It worked. When Freundlich described his honeymoon schedule in a letter, Einstein invited him over for a visit. "This is wonderful because it fits in with our plans," Freundlich wrote his fiancée, whose reaction to the prospect of spending part of her honeymoon with a theoretical physicist she had never met is lost to history.

When the newlyweds pulled into the Zurich train station, there was a disheveled Einstein wearing, as Freundlich's wife recalled, a large straw hat, with the plump chemist Fritz Haber at his side. Einstein brought the group to a nearby town where he was giving a lecture, after which he took them to lunch. Not surprisingly, he had forgotten to bring any money, and an assistant who had come along slipped him a 100 franc note under the table. For most of the day, Freundlich discussed gravity and the bending of light with Einstein, even when the group went on a nature hike, leaving his new wife to admire the scenery in peace.[35]

At his speech that day, which was on general relativity, Einstein pointed out Freundlich to the audience and called him "the man who will be testing the theory next year." The problem, however, was raising the money. At the time, Planck and others were trying to lure Einstein from Zurich to Berlin to become a member of the Prussian Academy, and Einstein used the courtship to write Planck and urge him to provide Freundlich the money to undertake the task.

In fact, on the very day that Einstein formally accepted the Berlin post and election to the Academy—December 7, 1913—he wrote Freundlich with the offer to reach into his own pocket. "If the Academy shies away from it, then we will get that little bit of mammon from private individuals," said Einstein. "Should everything fail, then I will pay for the thing myself out of the little bit that I have saved, at least the first 2,000 marks." The main thing, Einstein stressed, was that

Freundlich should proceed with his preparations. "Just go ahead and order the photographic plates, and do not let the time be squandered because of the money problem."[36]

As it turned out, there were enough private donations, mainly from the Krupp Foundation, to make the expedition possible. "You can imagine how happy I am that the external difficulties of your undertaking have now more or less been overcome," Einstein wrote. He added a note of confidence about what would be found: "I have considered the theory from every angle, and I have every confidence in the thing."[37]

Freundlich and two colleagues left Berlin on July 19 for the Crimea, where they were joined by a group from the Córdoba observatory in Argentina. If all went well, they would have two minutes to make photographs that could be used to analyze whether the starlight was deflected by the sun's gravity.

All did not go well. Twenty days before the eclipse, Europe tumbled into World War I and Germany declared war on Russia. Freundlich and his German colleagues were captured by the Russian army, and their equipment was confiscated. Not surprisingly, they were unable to convince the Russian soldiers that, with all of their powerful cameras and location devices, they were mere astronomers planning to gaze at the stars in order to better understand the secrets of the universe.

Even if they had been granted safe passage, it is likely that the observations would have failed. The skies were cloudy during the minutes of the eclipse, and an American group that was also in the region was unable to get any usable photographs.[38]

Yet the termination of the eclipse mission had a silver lining. Einstein's *Entwurf* equations were not correct. The degree to which gravity would deflect light, according to Einstein's theory at the time, was the same as that predicted by Newton's emission theory of light. But, as Einstein would discover a year later, the correct prediction would end up being twice that. If Freundlich had succeeded in 1914, Einstein might have been publicly proven wrong.

"My good old astronomer Freundlich, instead of experiencing a solar eclipse in Russia, will now be experiencing captivity there," Einstein wrote to his friend Ehrenfest. "I am concerned about him."[39]

There was no need to worry. The young astronomer was released in a prisoner exchange within weeks.

Einstein, however, had other reasons to worry in August 1914. His marriage had just exploded. His masterpiece theory still needed work. And now his native country's nationalism and militarism, traits that he had abhorred since childhood, had plunged it into a war that would cast him as a stranger in a strange land. In Germany, it would turn out, that was a dangerous position to be in.

World War I

The chain reaction that pushed Europe into war in August 1914 inflamed the patriotic pride of the Prussians and, in an equal and opposite reaction, the visceral pacifism of Einstein, a man so gentle and averse to conflict that he even disliked playing chess. "Europe in its madness has now embarked on something incredibly preposterous," he wrote Ehrenfest that month. "At such times one sees to what deplorable breed of brutes we belong."[40]

Ever since he ran away from Germany as a schoolboy and was exposed to the gauzy internationalism of Jost Winteler in Aarau, Einstein had harbored sentiments that disposed him toward pacifism, one-world federalism, and socialism. But he had generally shunned public activism.

World War I changed that. Einstein would never forsake physics, but he would henceforth be unabashedly public, for most of his life, in pushing his political and social ideals.

The irrationality of the war made Einstein believe that scientists in fact had a special duty to engage in public affairs. "We scientists in particular must foster internationalism," he said. "Unfortunately, we have had to suffer serious disappointments even among scientists in this regard."[41] He was especially appalled by the lockstep pro-war mentality of his three closest colleagues, the scientists who had lured him to Berlin: Fritz Haber, Walther Nernst, and Max Planck.[42]

Haber was a short, bald, and dapper chemist who was born Jewish but tried mightily to assimilate by converting, getting baptized, and adopting the dress, manner, and even pince-nez glasses of a proper Prussian. The director of the chemistry institute where Einstein had his office, he had been mediating the war between Einstein and Marić just as the larger war in Europe was breaking out. Although he hoped for a commission as an officer in the army, because he was an academic of Jewish heritage he had to settle for being made a sergeant.[43]

Haber reorganized his institute to develop chemical weapons for Germany. He had already found a way to synthesize ammonia from nitrogen, which permitted the Germans to mass-produce explosives. He then turned his attention to making deadly chlorine gas, which, heavier than air, would flow down into the trenches and painfully asphyxiate soldiers by burning through their throats and lungs. In April 1915, modern chemical warfare was inaugurated when some five thousand French and Belgians met that deadly fate at Ypres, with Haber personally supervising the attack. (In an irony that may have been lost on the inventor of dynamite, who endowed the prize, Haber won the 1918 Nobel in chemistry for his process of synthesizing ammonia.)

His colleague and occasional academic rival Nernst, bespectacled and 50, had his wife inspect his style as he practiced marching and saluting in front of their house. Then he took his private car and showed up at the western front to be a volunteer driver. Upon his return to Berlin, he experimented with tear gas and other irritants that could be used as a humane way to flush the enemy out of the trenches, but the generals decided they preferred the lethal approach that Haber was taking, so Nernst became part of that effort.

Even the revered Planck supported what he called Germany's "just war." As he told his students when they went off to battle, "Germany has drawn its sword against the breeding ground of insidious perfidy."[44]

Einstein was able to avoid letting the war cause a personal rift between him and his three colleagues, and he spent the spring of 1915 tutoring Haber's son in math.[45] But when they signed a petition defending Germany's militarism, he felt compelled to break with them politically.

The petition, published in October 1914, was titled "Appeal to the Cultured World" and became known as the "Manifesto of the 93," after the number of intellectuals who endorsed it. With scant regard for the truth, it denied that the German army had committed any attacks on civilians in Belgium and went on to proclaim that the war was necessary. "Were it not for German militarism, German culture would have been wiped off the face of the earth," it asserted. "We shall wage this fight to the very end as a cultured nation, a nation that holds the legacy of Goethe, Beethoven, and Kant no less sacred than hearth and home."[46]

It was no surprise that among the scientists who signed was the conservative Philipp Lenard, of photoelectric effect fame, who would later become a rabid anti-Semite and Einstein hater. What was distressing was that Haber, Nernst, and Planck also signed. As both citizens and scientists, they had a natural instinct to go along with the sentiments of others. Einstein, on the other hand, often displayed a natural inclination *not* to go along, which sometimes was an advantage both as a scientist and as a citizen.

A charismatic adventurer and occasional physician named Georg Friedrich Nicolai, who had been born Jewish (his original name was Lewinstein) and was a friend of both Elsa and her daughter Ilse, worked with Einstein to write a pacifist response. Their "Manifesto to Europeans" appealed for a culture that transcended nationalism and attacked the authors of the original manifesto. "They have spoken in a hostile spirit," Einstein and Nicolai wrote. "Nationalist passions cannot excuse this attitude, which is unworthy of what the world has heretofore called culture."

Einstein suggested to Nicolai that Max Planck, even though he had been one of the signers of the original manifesto, might also want to participate in their countermanifesto because of his "broad-mindedness and good will." He also gave Zangger's name as a possibility. But neither man, apparently, was willing to get involved. In an indication of the temper of the times, Einstein and Nicolai were able to garner only two other supporters. So they dropped their effort, and it was not published at the time.[47]

Einstein also became an early member of the liberal and cautiously pacifist New Fatherland League, a club that pushed for an early peace and the establishment of a federal structure in Europe to avoid future conflicts. It published a pamphlet titled "The Creation of the United

States of Europe," and it helped get pacifist literature into prisons and other places. Elsa went with Einstein to some of the Monday evening meetings until the group was banned in early 1916.[48]

One of the most prominent pacifists during the war was the French writer Romain Rolland, who had tried to promote friendship between his country and Germany. Einstein visited him in September 1915 near Lake Geneva. Rolland noted in his diary that Einstein, speaking French laboriously, gave "an amusing twist to the most serious of subjects."

As they sat on a hotel terrace amid swarms of bees plundering the flowering vines, Einstein joked about the faculty meetings in Berlin where each of the professors would anguish over the topic "why are we Germans hated in the world" and then would "carefully steer clear of the truth." Daringly, maybe even recklessly, Einstein openly said that he thought Germany could not be reformed and therefore hoped the allies would win, "which would smash the power of Prussia and the dynasty."[49]

The following month, Einstein got into a bitter exchange with Paul Hertz, a noted mathematician in Göttingen who was, or had been, a friend. Hertz was an associate member of the New Fatherland League with Einstein, but he had shied away from becoming a full member when it became controversial. "This type of cautiousness, not standing up for one's rights, is the cause of the entire wretched political situation," Einstein berated. "You have that type of valiant mentality the ruling powers love so much in Germans."

"Had you devoted as much care to understanding people as to understanding science, you would not have written me an insulting letter," Hertz replied. It was a telling point, and true. Einstein was better at fathoming physical equations than personal ones, as his family knew, and he admitted so in his apology. "You *must* forgive me, particularly since—as you yourself rightly say—I have *not* bestowed the same care to understanding people as to understanding science," he wrote.[50]

In November, Einstein published a three-page essay titled "My Opinion of the War" that skirted the border of what was permissible, even for a great scientist, to say in Germany. He speculated that there existed "a biologically determined feature of the male character" that was one of the causes of wars. When the article was published by the Goethe League that month, a few passages were deleted for safety's sake, including an attack on patriotism as potentially containing "the moral requisites of bestial hatred and mass murder."[51]

The idea that war had a biological basis in male aggression was a topic Einstein also explored in a letter to his friend in Zurich, Heinrich Zangger. "What drives people to kill and maim each other so savagely?" Einstein asked. "I think it is the sexual character of the male that leads to such wild explosions."

The only method of containing such aggression, he argued, was a world organization that had the power to police member nations.[52] It was a theme he would pick up again eighteen years later, in the final throes of his pure pacifism, when he engaged in a public exchange of letters with Sigmund Freud on both male psychology and the need for world government.

The Home Front, 1915

The early months of the war in 1915 made Einstein's separation from Hans Albert and Eduard more difficult, both emotionally and logistically. They wanted him to come visit them in Zurich for Easter that year, and Hans Albert, who was just turning 11, wrote him two letters designed to pull at his heart: "I just think: At Easter you're going to be here and we'll have a Papa again."

In his next postcard, he said that his younger brother told him about having a dream "that Papa was here." He also described how well he was doing in math. "Mama assigns me problems; we have a little booklet; I could do the same with you as well."[53]

The war made it impossible for him to come at Easter, but he responded to the postcards by promising Hans Albert that he would come in July for a hiking vacation in the Swiss Alps. "In the summer I will take a trip with just you alone for a fortnight or three weeks," he wrote. "This will happen every year, and Tete [Eduard] may also come along when he is old enough for it."

Einstein also expressed his delight that his son had taken a liking to geometry. It had been his "favorite pastime" when he was about the same age, he said, "but I had no one to demonstrate anything to me, so I had to learn it from books." He wanted to be with his son to help teach him math and "tell you many fine and interesting things about science and much else." But that would not always be possible. Perhaps they could do it by mail? "If you write me each time what you already know, I'll give you a nice little problem to solve." He sent along a toy for each of his sons, along with an admonition to brush their teeth well. "I do the same and am very happy now to have kept enough healthy teeth."[54]

But the tension in the family worsened. Einstein and Marić exchanged letters arguing about both money and vacation timing, and at the end of June a curt postcard came from Hans Albert. "If you're so unfriendly to her," he said of his mother, "I don't want to go with you." So Einstein canceled his planned trip to Zurich and instead went with Elsa and her two daughters to the Baltic sea resort of Sellin.

Einstein was convinced that Marić was turning the children against him. He suspected, probably correctly, that her hand was behind the postcards Hans Albert was sending, both the plaintive ones making him feel guilty for not being in Zurich and the sharper ones rejecting vacation hikes. "My fine boy had been alienated from me for a few years already by my wife, who has a vengeful disposition," he complained to Zangger. "The postcard I received from little Albert had been inspired, if not downright dictated, by her."

He asked Zangger, who was a professor of medicine, to check on young Eduard, who had been suffering ear infections and other ailments. "Please write me what is wrong with my little boy," he pleaded. "I'm particularly fondly attached to him; he was still so sweet to me and innocent."[55]

It was not until the beginning of September that he finally made it to Switzerland. Marić felt it would be proper for him to stay with her and the boys, despite the strain. They were, after all, still married. She had hopes of reconciling. But Einstein showed no interest in being with her. Instead, he stayed in a hotel and spent a lot of time with his friends Michele Besso and Heinrich Zangger.

As it turned out, he got a chance to see his sons only twice during the entire three weeks he was in Switzerland. In a letter to Elsa, he blamed his estranged wife: "The cause was mother's fear of the little ones becoming too dependent on me." Hans Albert let his father know that the whole visit made him feel uncomfortable.[56]

After Einstein returned to Berlin, Hans Albert paid a call on Zangger. The kindly medical professor, friends of all sides in the dispute, tried to work out an accord so that Einstein could visit his sons. Besso also played intermediary. Einstein could see his sons, Besso advised in a formal letter he wrote after consulting with Marić, but not in Berlin nor in the presence of Elsa's family. It would be best to do it at "a good Swiss inn," initially just with Hans Albert, where they could spend some time on their own free of all distractions. Over Christmas, Hans Albert was planning to visit Besso's family, and he suggested that perhaps Einstein could come then.[57]

The Race to General Relativity, 1915

What made the flurry of political and personal turmoil in the fall of 1915 so remarkable was that it highlighted Einstein's ability to concentrate on, and compartmentalize, his scientific endeavors despite all distractions. During that period, with great effort and anxiety, he was engaged in a competitive rush to what he later called the greatest accomplishment of his life.[58]

Back when Einstein had moved to Berlin in the spring of 1914, his colleagues had assumed that he would set up an institute and attract acolytes to work on the most pressing problem in physics: the implications of quantum theory. But Einstein was more of a lone wolf. Unlike Planck, he did not want a coterie of collaborators or protégés, and he preferred to focus on what again had become his personal passion: the generalization of his theory of relativity.[59]

So after his wife and sons left him for Zurich, Einstein moved out of their old apartment and rented one that was nearer to Elsa and the center of Berlin. It was a sparsely furnished bachelor's refuge, but still rather spacious: it had seven rooms on the third floor of a new five-story building.[60]

Einstein's study at home featured a large wooden writing table that was cluttered with piles of papers and journals. Padding around this hermitage, eating and working at whatever hours suited him, sleeping when he had to, he waged his solitary struggle.

Through the spring and summer of 1915, Einstein wrestled with his *Entwurf* theory, refining it and defending it against a variety of challenges. He began calling it "the general theory" rather than merely "a generalized theory" of relativity, but that did not mask its problems, which he kept trying to deflect.

He claimed that his equations had the greatest amount of covariance that was permissible given his hole argument and other strictures of physics, but he began to suspect that this was not correct. He also got into an exhausting debate with the Italian mathematician Tullio Levi-Civita, who pointed out problems with his handling of the tensor calculus. And there was still the puzzle of the incorrect result the theory gave for the shift in Mercury's orbit.

At least his *Entwurf* theory still successfully explained—or so he thought through the summer of 1915—rotation as being a form of *relative* motion, that is, a motion that could be defined only relative to the positions and motions of other objects. His field equations, he thought, were invariant under the transformation to rotating coordinates.[61]

Einstein was confident enough in his theory to show it off at a weeklong series of two-hour lectures, starting at the end of June 1915, at the University of Göttingen, which had become the preeminent center for the mathematical side of theoretical physics. Foremost among the geniuses there was David Hilbert, and Einstein was particularly eager—too eager, it would turn out—to explain all the intricacies of relativity to him.

The visit to Göttingen was a triumph. Einstein exulted to Zangger that he had "the pleasurable experience of convincing the mathematicians there thoroughly." Of Hilbert, a fellow pacifist, he added, "I met him and became quite fond of him." A few weeks later, after again reporting, "I was able to convince Hilbert of the general theory of relativity," Einstein called him "a man of astonishing energy and independence." In a letter to another physicist, Einstein was even more effusive: "In Göttingen I had the great pleasure of seeing that everything was understood down to the details. I am quite enchanted with Hilbert!"[62]

Hilbert was likewise enchanted with Einstein and his theory. So much so that he soon set out to see if he could beat Einstein to the goal of getting the field equations right. Within three months of his Göttingen lectures, Einstein was confronted with two distressing discoveries: that his *Entwurf* theory was indeed flawed, and that Hilbert was racing feverishly to come up with the correct formulations on his own.

Einstein's realization that his *Entwurf* theory was unraveling came from an accumulation of problems. But it culminated with two major blows in early October 1915.

The first was that, upon rechecking, Einstein found that the *Entwurf* equations did not actually account for rotation as he had thought.[63] He hoped to prove that rotation could be conceived of as just another form of relative motion, but it turned out that the *Entwurf* didn't actually prove this. The *Entwurf* equations were not, as he had believed, covariant under a transformation that uniformly rotated the coordinate axes.

Besso had warned him in a memo in 1913 that this seemed to be a problem. But Einstein had ignored him. Now, upon redoing his calculations, he was dismayed to see this pillar knocked away. "This is a blatant contradiction," he lamented to the astronomer Freundlich.

He assumed that the same mistake also accounted for his theory's inability to account fully for the shift in Mercury's orbit. And he despaired that he would not be able to find the problem. "I do not believe I am able to find the mistake myself, for in this matter my mind is too set in a deep rut."[64]

In addition, he realized that he had made a mistake in what was called his "uniqueness" argument: that the sets of conditions required by energy-momentum conservation and other physical restrictions uniquely led to the field equations in the *Entwurf.* He wrote Lorentz explaining in detail his previous "erroneous assertions."[65]

Added to these problems were ones he already knew about: the *Entwurf* equations were not generally covariant, meaning that they did not really make all forms of accelerated and nonuniform motion relative, and they did not fully explain Mercury's anomalous orbit. And now, as this edifice was crumbling, he could hear what seemed to be Hilbert's footsteps gaining on him from Göttingen.

Part of Einstein's genius was his tenacity. He could cling to a set of ideas, even in the face of "apparent contradiction" (as he put it in his 1905 relativity paper). He also had a deep faith in his intuitive feel for the physical world. Working in a more solitary manner than most other scientists, he held true to his own instincts, despite the qualms of others.

But although he was tenacious, he was not mindlessly stubborn. When he finally decided his *Entwurf* approach was untenable, he was willing to abandon it abruptly. That is what he did in October 1915.

To replace his doomed *Entwurf* theory, Einstein shifted his focus from the physical strategy, which emphasized his feel for basic principles of physics, and returned to a greater reliance on a mathematical strategy, which made use of the Riemann and Ricci tensors. It was an approach he had used in his Zurich notebooks and then abandoned, but on returning to it he found that it could provide a way to generate generally covariant gravitational field equations. "Einstein's reversal," writes John Norton, "parted the waters and led him from bondage into the promised land of general relativity." [66]

Of course, as always, his approach remained a mix of both strategies. To pursue a revitalized mathematical strategy, he had to revise the physical postulates that were the foundation for his *Entwurf* theory. "This was exactly the sort of convergence of physical and mathematical considerations that eluded Einstein in the Zurich notebook and in his work on the *Entwurf* theory," write Michel Janssen and Jürgen Renn. [67]

Thus he returned to the tensor analysis that he had used in Zurich, with its greater emphasis on the mathematical goal of finding equations that were generally covariant. "Once every last bit of confidence in the earlier theories had given way," he told a friend, "I saw clearly that it was only through general covariance theory, i.e., with Riemann's covariant, that a satisfactory solution could be found." [68]

The result was an exhausting, four-week frenzy during which Einstein wrestled with a succession of tensors, equations, corrections, and updates that he rushed to the Prussian Academy in a flurry of four Thursday lectures. It climaxed, with the triumphant revision of Newton's universe, at the end of November 1915.

Every week, the fifty or so members of the Prussian Academy gathered in the grand hall of the Prussian State Library in the heart of Berlin to address each other as "Your Excellency" and listen to fellow members pour forth their wisdom. Einstein's series of four lectures had been scheduled weeks earlier, but until they began—and even after they had begun—he was still working furiously on his revised theory.

The first was delivered on November 4. "For the last four years," he began, "I have tried to establish a general theory of relativity on the assumption of the relativity even of non-uniform motion." Referring to his discarded *Entwurf* theory, he said he "actually believed I had discovered the only law of gravitation" that conformed to physical realities.

But then, with great candor, he detailed all of the problems that theory had encountered. "For that reason, I completely lost trust in the field equations" that he had been defending for more than two years. Instead, he said, he had now returned to the approach that he and his mathematical caddy, Marcel Grossmann, had been using in 1912. "Thus I went back to the requirement of a more general covariance of the field equations, which I had left only with a heavy heart when I worked together with my friend Grossmann. In fact, we had then already come quite close to the solution."

Einstein reached back to the Riemann and Ricci tensors that Grossmann had introduced him to in 1912. "Hardly anyone who truly understands it can resist the charm of this theory," he lectured. "It signifies a real triumph of the method of the calculus founded by Gauss, Riemann, Christoffel, Ricci, and Levi-Civita."[69]

This method got him much closer to the correct solution, but his equations on November 4 were still not generally covariant. That would take another three weeks.

Einstein was in the throes of one of the most concentrated frenzies of scientific creativity in history. He was working, he said, "horrendously intensely."[70] In the midst of this ordeal, he was also still dealing with the personal crisis within his family. Letters arrived from both his wife and Michele Besso, who was acting on her behalf, that pressed the issue of his financial obligations and discussed the guidelines for his contact with his sons.

On the very day he turned in his first paper, November 4, he wrote an anguished—and painfully poignant—letter to Hans Albert, who was in Switzerland:

> I will try to be with you for a month every year so that you will have a father who is close to you and can love you. You can learn a lot of good things from me that no one else can offer you. The things I have gained from so much strenuous work should be of value not only to strangers but especially to my own boys. In the last few days I completed one of the finest papers of my life. When you are older, I will tell you about it.

He ended with a small apology for seeming so distracted: "I am often so engrossed in my work that I forget to eat lunch."[71]

Einstein also took time off from furiously revising his equations to engage in an awkward fandango with his erstwhile friend and competitor David Hilbert, who was racing him to find the equations of general relativity. Einstein had been informed that the Göttingen mathematician had figured out the flaws in the *Entwurf* equations. Worried about being scooped, he wrote Hilbert a letter saying that he himself had discovered the flaws four weeks earlier, and he sent along a copy of his November 4 lecture. "I am curious whether you will take kindly to this new solution," Einstein asked with a touch of defensiveness.[72]

Hilbert was not only a better pure mathematician than Einstein, he also had the advantage of not being as good a physicist. He did not get all wrapped up, the way Einstein did, in making sure that any new theory conformed to Newton's old one in a weak static field or that it obeyed the laws of causality. Instead of a dual math-and-physics strategy, Hilbert pursued mainly a math strategy, focusing on finding the equations that were covariant. "Hilbert liked to joke that physics was too complicated to be left to the physicists," notes Dennis Overbye.[73]

Einstein presented his second paper the following Thursday, November 11. In it, he used the Ricci tensor and imposed new coordinate conditions that allowed the equations thus to be generally covariant. As it turned out, that did not greatly improve matters. Einstein was still close to the final answer, but making little headway.[74]

Once again, he sent the paper off to Hilbert. "If my present modification (which does not change the equations) is legitimate, then gravitation must play a fundamental role in the composition of matter," Einstein said. "My own curiosity is interfering with my work!"[75]

The reply that Hilbert sent the next day must have unnerved Einstein. He said he was about ready to oblige with "an axiomatic solution to your great problem." He had planned to hold off discussing it until he explored the physical ramifications further. "But since you are so interested, I would like to lay out my theory in very complete detail this coming Tuesday," which was November 16.

He invited Einstein to come to Göttingen and have the dubious pleasure of personally hearing him lay out the answer. The meeting would begin at 6 p.m., and Hilbert helpfully provided Einstein with the arrival times of the two afternoon trains from Berlin. "My wife and I would be very pleased if you stayed with us."

Then, after signing his name, Hilbert felt compelled to add what must surely have been a tantalizing and disconcerting postscript. "As far as I understand your new paper, the solution given by you is entirely different from mine."

Einstein wrote four letters on November 15, a Monday, that give a glimpse into why he was suffering stomach pains. To his son Hans Albert, he suggested that he would like to travel to Switzerland around Christmas and New Year's to visit him. "Maybe it would be better if we were alone somewhere," such as at a secluded inn, he suggested to his son. "What do you think?"

He also wrote his estranged wife a conciliatory letter that thanked her for her willingness not "to undermine my relations with the boys." And he reported to their mutual friend Zangger, "I have modified the theory of gravity, having realized that my earlier proofs had a gap . . . I shall be glad to come to Switzerland at the turn of the year in order to see my dear boy."[76]

Finally, he replied to Hilbert and declined his invitation to visit Göttingen the next day. His letter did not hide his anxiety: "Your analysis interests me tremendously . . . The hints you gave in your messages awaken the greatest of expectations. Nevertheless, I must refrain from traveling to Göttingen for the moment . . . I am tired out and plagued by stomach pains . . . If possible, please send me a correction proof of your study to mitigate my impatience."[77]

Fortunately for Einstein, his anxiety was partly alleviated that week by a joyous discovery. Even though he knew his equations were not in final form, he decided to see whether the new approach he was taking would yield the correct results for what was known about the shift in Mercury's orbit. Because he and Besso had done the calculations once before (and gotten a disappointing result), it did not take him long to redo the calculations using his revised theory.

The answer, which he triumphantly announced in the third of his four November lectures, came out right: 43 arc-seconds per century.[78] "This discovery was, I believe, by far the strongest emotional experience in Einstein's scientific life, perhaps in all his life," Abraham Pais later said. He was so thrilled he had heart palpitations, as if "something had snapped" inside. "I was beside myself with joyous excitement," he told Ehrenfest. To another physicist he exulted: "The results of Mercury's perihelion movement fills me with great satisfaction. How helpful to us is astronomy's pedantic accuracy, which I used to secretly ridicule!"[79]

In the same lecture, he also reported on another calculation he had made. When he first began formulating general relativity eight years earlier, he had said that one implication was that gravity would bend light. He had previously figured that the bending of light by the gravitational field next to the sun would be approximately 0.83 arc-second, which corresponded to what would be predicted by Newton's theory when light was treated as if a particle. But now, using his newly revised theory, Einstein calculated that the bending of light by gravity would be twice as great, because of the effect produced by the curvature of spacetime. Therefore, the sun's gravity would bend a beam by about 1.7 arc-seconds, he now predicted. It was a prediction that would have to wait for the next suitable eclipse, more than three years away, to be tested.

That very morning, November 18, Einstein received Hilbert's new paper, the one that he had been invited to Göttingen to hear presented. Einstein was surprised, and somewhat dismayed, to see how similar it was to his own work. His response to Hilbert was terse, a bit cold, and clearly designed to assert the priority of his own work:

The system you furnish agrees—as far as I can see—exactly with what I found in the last few weeks and have presented to the Academy. The difficulty was not in finding generally covariant equations . . . for this is easily achieved with Riemann's tensor . . . Three years ago with my friend Grossmann I had already taken into consideration the only co-variant equations, which have now been shown to be the correct ones. We had distanced ourselves from it, reluctantly, because it seemed to me that the physical discussion yielded an incongruity with Newton's law. Today I am presenting to the Academy a paper in which I derive quantitatively out of general relativity, without any guiding hypothesis, the perihelion motion of Mercury. No gravitational theory has achieved this until now.[80]

Hilbert responded kindly and quite generously the following day, claiming no priority for himself. "Cordial congratulations on conquering perihelion motion," he wrote. "If I could calculate as rapidly as you, in my equations the electron would have to capitulate and the hydrogen atom would have to produce its note of apology about why it does not radiate."[81]

Yet the day after, on November 20, Hilbert sent in a paper to a Göttingen science journal proclaiming his own version of the equations for general relativity. The title he picked for his piece was not a modest one. "The Foundations of Physics," he called it.

It is not clear how carefully Einstein read the paper that Hilbert sent him or what in it, if anything, affected his thinking as he busily prepared his climactic fourth lecture at the Prussian Academy. Whatever the case, the calculations he had done the week earlier, on Mercury and on light deflection, helped him realize that he could avoid the constraints and coordinate conditions he had been imposing on his gravitational field equations. And thus he produced in time for his final lecture—"The Field Equations of Gravitation," on November 25, 1915—a set of covariant equations that capped his general theory of relativity.

The result was not nearly as vivid to the layman as, say, $E=mc^2$. Yet using the condensed notations of tensors, in which sprawling complexities can be compressed into little subscripts, the crux of the final Einstein field equations is compact enough to be emblazoned, as it indeed often has been, on T-shirts designed for proud physics students. In one of its many variations,[82] it can be written as:

$$R_{\mu\nu} - \tfrac{1}{2} g_{\mu\nu} R = 8\pi T_{\mu\nu}$$

The left side of the equation starts with the term $R_{\mu\nu}$, which is the Ricci tensor he had embraced earlier. The term $g_{\mu\nu}$ is the all-important metric tensor, and the term R is the trace of the Ricci tensor called the Ricci scalar. Together, this left side of the equation—which is now known as the Einstein tensor and can be written simply as $G_{\mu\nu}$—compresses together all of the information about how the geometry of spacetime is warped and curved by objects.

The right side describes the movement of matter in the gravitational field. The interplay between the two sides shows how objects curve spacetime and how, in turn, this curvature affects the motion of objects. As the physicist John Wheeler has put it, "Matter tells spacetime how to curve, and curved space tells matter how to move."[83]

Thus is staged a cosmic tango, as captured by another physicist, Brian Greene:

> Space and time become players in the evolving cosmos. They come alive. Matter here causes space to warp there, which causes matter over here to move, which causes space way over there to warp even more, and so on. General relativity provides the choreography for an entwined cosmic dance of space, time, matter, and energy.[84]

At last Einstein had equations that were truly covariant and thus a theory that incorporated, at least to his satisfaction, all forms of motion, whether it be inertial, accelerated, rotational, or arbitrary. As he proclaimed in the formal presentation of his theory that he published the following March in the *Annalen der Physik*, "The general laws of nature are to be expressed by equations that hold true for all systems of coordinates, that is they are covariant with respect to any substitutions whatever."[85]

Einstein was thrilled by his success, but at the same time he was worried that Hilbert, who had presented his own version five days earlier in Göttingen, would be accorded some of the credit for the theory. "Only one colleague has really understood it," he wrote to his friend

Heinrich Zangger, "and he is seeking to nostrify it (Abraham's expression) in a clever way." The expression "to nostrify" *(nostrifizieren)*, which had been used by the Göttingen-trained mathematical physicist Max Abraham, referred to the practice of nostrification by which German universities converted degrees granted by other universities into degrees of their own. "In my personal experience I have hardly come to know the wretchedness of mankind better." In a letter to Besso a few days later, he added, "My colleagues are acting hideously in this affair. You will have a good laugh when I tell you about it."[86]

So who actually deserves the primary credit for the final mathematical equations? The Einstein-Hilbert priority issue has generated a small but intense historical debate, some of which seems at times to be driven by passions that go beyond mere scientific curiosity. Hilbert presented a version of his equations in his talk on November 16 and a paper that he dated November 20, before Einstein presented his final equations on November 25. However, a team of Einstein scholars in 1997 found a set of proof pages of Hilbert's article, on which Hilbert had made revisions that he then sent back to the publisher on December 16. In the original version, Hilbert's equations differed in a small but important way from Einstein's final version of the November 25 lecture. They were not actually generally covariant, and he did not include a step that involved contracting the Ricci tensor and putting the resulting trace term, the Ricci scalar, into the equation. Einstein did this in his November 25 lecture. Apparently, Hilbert made a correction in the revised version of his article to match Einstein's version. His revisions, quite generously, also added the phrase "first introduced by Einstein" when he referred to the gravitational potentials.

Hilbert's advocates (and Einstein's detractors) respond with a variety of arguments, including that the page proofs are missing one part and that the trace term at issue was either unnecessary or obvious.

It is fair to say that both men—to some extent independently but each also with knowledge of what the other was doing—derived by November 1915 mathematical equations that gave formal expression to the general theory. Judging from Hilbert's revisions to his own page proofs, Einstein seems to have published the final version of these equations first. And in the end, even Hilbert gave Einstein credit and priority.

Either way, it was, without question, Einstein's theory that was being formalized by these equations, one that he had explained to Hilbert during their time together in Göttingen that summer. Even the physicist Kip Thorne, one of those who give Hilbert credit for producing the correct field equations, nonetheless says that Einstein deserves credit for the theory underlying the equations. "Hilbert carried out the last few mathematical steps to its discovery independently and almost simultaneously with Einstein, but Einstein was responsible for essentially everything that preceded these steps," Thorne notes. "Without Einstein, the general relativistic laws of gravity might not have been discovered until several decades later."[87]

Hilbert, graciously, felt the same way. As he stated clearly in the final published version of his paper, "The differential equations of gravitation that result are, as it seems to me, in agreement with the magnificent theory of general relativity established by Einstein." Henceforth he would always acknowledge (thus undermining those who would use him to diminish Einstein) that Einstein was the sole author of the theory of relativity.[88] "Every boy in the streets of Göttingen understands more about four-dimensional geometry than Einstein," he reportedly said. "Yet, in spite of that, Einstein did the work and not the mathematicians."[89]

Indeed, Einstein and Hilbert were soon friendly again. Hilbert wrote in December, just weeks after their dash for the field equations was finished, to say that with his support Einstein had been elected to the Göttingen Academy. After expressing his thanks, Einstein added, "I feel compelled to say something else to you." He explained:

> There has been a certain ill-feeling between us, the cause of which I do not want to analyze. I have struggled against the feeling of bitterness attached to it, with complete success. I think of you again with unmixed geniality and ask you to try to do the same with me. It is a shame when two real fellows who have extricated themselves somewhat from this shabby world do not afford each other mutual pleasure.[90]

They resumed their regular correspondence, shared ideas, and plotted to get a job for the astronomer Freundlich. By February Einstein was even visiting Göttingen again and staying at Hilbert's home.

Einstein's pride of authorship was understandable. As soon as he got printed copies of his four lectures, he mailed them out to friends. "Be sure you take a good look at them," he told one. "They are the most valuable discovery of my life." To another he noted, "The theory is of incomparable beauty."[91]

Einstein, at age 36, had produced one of history's most imaginative and dramatic revisions of our concepts about the universe. The general theory of relativity was not merely the interpretation of some experimental data or the discovery of a more accurate set of laws. It was a whole new way of regarding reality.

Newton had bequeathed to Einstein a universe in which time had an absolute existence that tick-tocked along independent of objects and observers, and in which space likewise had an absolute existence. Gravity was thought to be a force that masses exerted on one another rather mysteriously across empty space. Within this framework, objects obeyed mechanical laws that had proved remarkably accurate—almost perfect—in explaining everything from the orbits of the planets, to the diffusion of gases, to the jiggling of molecules, to the propagation of sound (though not light) waves.

With his special theory of relativity, Einstein had shown that space and time did not have independent existences, but instead formed a fabric of spacetime. Now, with his general version of the theory, this fabric of spacetime became not merely a container for objects and events. Instead, it had its own dynamics that were determined by, and in turn helped to determine, the motion of objects within it— just as the fabric of a trampoline will curve and ripple as a bowling ball and some billiard balls roll across it, and in turn the dynamic curving and rippling of the trampoline fabric will determine the path of the rolling balls and cause the billiard balls to move toward the bowling ball.

The curving and rippling fabric of spacetime explained gravity, its equivalence to acceleration, and, Einstein asserted, the general relativity of all forms of motion.[92] In the opinion of Paul Dirac, the Nobel laureate pioneer of quantum mechanics, it was "probably the greatest scientific discovery ever made." Another of the great giants of twentieth-century physics, Max Born, called it "the greatest feat of human thinking about nature, the most amazing combination of philosophical penetration, physical intuition and mathematical skill."[93]

The entire process had exhausted Einstein but left him elated. His marriage had collapsed and war was ravaging Europe, but Einstein was as happy as he would ever be. "My boldest dreams have now come true," he exulted to Besso. "*General* covariance. Mercury's perihelion motion wonderfully precise." He signed himself "contented but kaput."[94]

CHAPTER TEN

DIVORCE

1916–1919

With Elsa, June 1922

"The Narrow Whirlpool of Personal Experience"

As a young man, Einstein had predicted, in a letter to the mother of his first girlfriend, that the joys of science would be a refuge from painful personal emotions. And thus it was. His conquest of general relativity proved easier than finding the formulas for the forces swirling within his family.

Those forces were complex. At the very moment he was finalizing his field equations—the last week of November 1915—his son Hans Albert was telling Michele Besso that he wanted to spend time alone with his father over Christmas, preferably on Zugerberg mountain or someplace similarly isolated. But simultaneously, the boy was writing his father a nasty letter saying he did not want him to come to Switzerland at all.[1]

How to explain the contradiction? Hans Albert's mind seemed at times to display a duality—he was, after all, only 11—and he had powerfully conflicted attitudes toward his father. That was no surprise. Einstein was intense and compelling and at times charismatic. He was also aloof and distracted and had distanced himself, physically and emotionally, from the boy, who was guarded by a doting mother who felt humiliated.

The stubborn patience that Einstein displayed when dealing with scientific problems was equaled by his impatience when dealing with personal entanglements. So he informed the boy he was canceling the trip. "The unkind tone of your letter dismays me very much," Einstein wrote just days after finishing his last lecture on general relativity. "I see that my visit would bring you little joy, therefore I think it's wrong to sit in a train for two hours and 20 minutes."

There was also the question of a Christmas present. Hans Albert had become an avid little skier, and Marić gave him a set of equipment that cost 70 francs. "Mama bought them for me on condition that you also contribute," he wrote. "I consider them a Christmas present." This did not please Einstein. He replied that he would send him a gift in cash, "but I do think *that a luxury gift costing 70 francs does not match our modest circumstances*," Einstein wrote, underlining the phrase.[2]

Besso put on what he called his "pastoral manner" to mediate. "You should not take serious offense at the boy," he said. The source of the friction was Marić, Besso believed, but he asked Einstein to remember that she was composed "not only of meanness but of goodness." He should try to understand, Besso urged, how difficult it was for Marić to deal with him. "The role as the wife of a genius is never easy."[3] In the case of Einstein, that was certainly true.

The anxiety surrounding Einstein's proposed visit was partly due to a misunderstanding. Einstein had assumed that the plan to have him and his son meet at the Bessos' had been arranged because Marić and Hans Albert wanted it that way. Instead, the boy had no desire to be a bystander while his father and Besso discussed physics. Just the opposite: he wanted his father to himself.

Marić ended up writing to clear up the matter, which Einstein appreciated. "I was likewise a bit disappointed that I would not get Albert to myself but only under Besso's protection," he said.

So Einstein reinstated his plan to visit Zurich, and he promised it would be one of many such trips to see his son. "[Hans] Albert* is now entering the age at which I can mean very much to him," he said. "I want mainly to teach him to think, judge and appreciate things objectively." A week later, in another letter to Marić, he reaffirmed that he was happy to make the trip, "for there is a faint chance that I'll please Albert by coming." He did, however, add rather pointedly, "See to it that he receives me fairly cheerfully. I am quite tired and overworked, and not capable of enduring new agitations and disappointments."[4]

It was not to be. Einstein's exhaustion lingered, and the war made the border crossing from Germany difficult. Two days before Christmas of 1915, when he was supposed to be departing for Switzerland, Einstein instead wrote his son a letter. "I have been working so hard in the last few months that I urgently need a rest during the Christmas holidays," he said. "Aside from this, coming across the border is very uncertain at present, since it has been almost constantly closed recently. That is why I must unfortunately deprive myself of visiting you now."

Einstein spent Christmas at home. That day, he took out of his satchel some of the drawings that Hans Albert had sent him and wrote the boy a postcard saying how much they pleased him. He would come for Easter, he promised, and he expressed delight that his son enjoyed playing piano. "Maybe you can practice something to accompany a violin, and then we can play at Easter when we are together."[5]

After he and Marić separated, Einstein had initially decided not to seek a divorce. One reason was that he had no desire to marry Elsa. Companionship without commitment suited him just fine. "The at-

---

* For clarity, I refer to the boy by both of his given names, Hans Albert, although his father invariably referred to him simply as Albert. At one point, Einstein wrote a letter to his son and signed it "Albert" instead of "Papa." In his next letter, he awkwardly began, "The explanation for the curious signature on my last letter is that, in my absent-mindedness, instead of signing my own name, I frequently sign for the person to whom the letter is addressed" (Einstein to Hans Albert Einstein, March 11 and 16, 1916).

tempts to force me into marriage come from my cousin's parents and is mainly attributable to vanity, though moral prejudice, which is still very much alive in the old generation, plays a part," Einstein wrote Zangger the day after presenting his climactic November 1915 lecture. "If I let myself become trapped, my life would become complicated, and above all it would probably be a heavy blow for my boys. Therefore, I must allow myself not to be moved either by my inclination or by tears, but must remain as I am." It was a resolution he repeated to Besso as well.[6]

Besso and Zangger agreed that he should not seek a divorce. "It is important that Einstein knows that his truest friends," Besso wrote Zangger, "would regard a divorce and subsequent remarriage as a great evil."[7]

But Elsa and her family kept pushing. So in February 1916, Einstein wrote Marić to propose—indeed, beg—that she agree to a divorce, "so that we can arrange the rest of our lives independently." The separation agreement they had worked out with the help of Fritz Haber, he suggested, could serve as the basis for a divorce. "It will surely be possible to have the details settled to your satisfaction," he promised. His letter also included instructions on how to keep their boys from suffering from calcium deficiency.[8]

When Marić resisted, Einstein became more insistent. "For you it involves a mere formality," he said. "For me, however, it is an imperative duty." He informed Marić that Elsa had two daughters whose reputations and chances of marriage were being compromised by "the rumors" that were circulating about the illicit relationship their mother was having with Einstein. "This weighs on me and ought to be redressed by a formal marriage," he told Marić. "Try to imagine yourself in my position for once."

As an enticement, he offered more money. "You would gain from this change," he told Marić. "I wish to do more than I had obligated myself to before." He would transfer 6,000 marks into a fund for the children and increase her payments to 5,600 marks annually. "By making myself such a frugal bed of straw, I am proving to you that my boys' well-being is closest to my heart, above all else in the world."

In return, he wanted the right to have his sons visit him in Berlin. They would not come into contact with Elsa, he pledged. He even added a somewhat surprising promise: he would not be living with Elsa even if they got married. Instead, he would keep his own apartment. "For I shall never give up the state of living alone, which has manifested itself as an indescribable blessing."

Marić did not consent to give him the right to have the boys visit him in Berlin. But she did tentatively agree—or at least so Einstein thought—to allow the start of divorce discussions.[9]

As he had promised Hans Albert, Einstein arrived in Switzerland in early April 1916 for a three-week Easter vacation, moving into a hotel near the Zurich train station. Initially, things went very well. The boys came to see him and greeted him joyously. From his hotel, he sent Marić a note of thanks:

> My compliments on the good condition of our boys. They are in such excellent physical and mental shape that I could not have wished for more. And I know that this is for the most part due to the proper up-bringing you provide them. I am likewise thankful that you have not alienated me from the children. They came to meet me spontaneously and sweetly.

Marić sent word that she wanted to see Einstein herself. Her goal was to be assured that he truly wanted a divorce and was not merely being pressured by Elsa. Both Besso and Zangger tried to arrange such a meeting, but Einstein declined. "There would be no point in a conversation between us and it could serve only to reopen old wounds," he wrote in a note to Marić.[10]

Einstein took Hans Albert off alone, as the boy wished, for what was planned as a ten-day hiking excursion in a mountain resort overlooking Lake Lucerne. There they were caught in a late-season snowstorm that kept them confined to the inn, which initially pleased them both. "We are snowed in at Seelisberg but are enjoying ourselves immensely," Einstein wrote Elsa. "The boy delights me, especially with his clever questions and his undemanding way. No discord exists between us." Unfortunately, soon the weather, and perhaps also their enforced togetherness, became oppressive, and they returned to Zurich a few days early.[11]

Back in Zurich, the tensions revived. One morning, Hans Albert came to visit his father at the physics institute to watch an experiment. It was a pleasant enough activity, but as the boy was leaving for lunch, he urged his father to come by the house and at least pay a courtesy call on Marić.

Einstein refused. Hans Albert, who was just about to turn 12, became angry and said he would not come back for the completion of the experiment that afternoon unless his father relented. Einstein would not. "That's how it remained," he reported to Elsa a week later, on the day he left Zurich. "And I have seen neither of the children since." [12]

Marić subsequently went into an emotional and physical meltdown. She had a series of minor heart incidents in July 1916, accompanied by extreme anxiety, and her doctors told her to remain in bed. The children moved in with the Bessos, and then to Lausanne, where they stayed with Marić's friend Helene Savić, who was riding out the war there.

Besso and Zangger tried to get Einstein to come down from Berlin to be with his sons. But Einstein demurred. "If I go to Zurich, my wife will demand to see me," he wrote Besso. "This I would have to refuse, partly on an inalterable resolve partly also to spare her the agitation. Besides, you know that the personal relations between the children and me deteriorated so much during my stay at Easter (after a very promising start) that I doubt very much whether my presence would be reassuring for them."

Einstein assumed that his wife's illness was largely psychological and even, perhaps, partly faked. "Isn't it possible that nerves are behind it all?" he asked Zangger. To Besso, he was more blunt: "I have the suspicion that the woman is leading both of you kind-hearted men down the garden path. She is not afraid to use all means when she wants to achieve something. You have no idea of the natural craftiness of such a woman." [13] Einstein's mother agreed. "Mileva was never as sick as you seem to think," she told Elsa. [14]

Einstein asked Besso to keep him informed of the situation and made a stab at scientific humor by saying that his reports did not need to have logical "continuity" because "this is permissible in the age of quantum theory." Besso was not sympathetic; he wrote Einstein a sharp letter saying Marić's condition was not "a deception" but was instead caused by emotional stress. Besso's wife, Anna, was even harsher, adding a postscript to the letter that addressed Einstein with the formal *Sie*.[15]

Einstein backed down from his charge that Marić was faking illness, but railed that her emotional distress was unwarranted. "She leads a worry-free life, has her two precious boys with her, lives in a fabulous neighborhood, does what she likes with her time, and innocently stands by as the guiltless party," he wrote Besso.

Einstein was especially stung by the cold postscript, which he mistakenly thought came from Michele rather than Anna Besso. So he added his own postscript: "We have understood each other well for 20 years," he said. "And now I see you developing a bitterness toward me for the sake of a woman who has nothing to do with you. Resist it!" Later that day he realized he had mistaken Anna's harsh postscript for something her husband had written, and he quickly sent along another note apologizing to him.[16]

On Zangger's advice, Marić checked into a sanatorium. Einstein still resisted going to Zurich, even though his boys were at home alone with a maid, but he told Zangger he would change his mind "if you think it's appropriate." Zangger didn't. "The tension on both sides is too great," Zangger explained to Besso, who agreed.[17]

Despite his detached attitude, Einstein loved his sons and would always take care of them. Please let them know, he instructed Zangger, that he would take them under his wing if their mother died. "I would raise the two boys myself," he said. "They would be taught at home, as far as possible by me personally." In various letters over the next few months, Einstein described his different ideas and fantasies for home-schooling his sons, what he would teach, and even the type of walks they would take. He wrote Hans Albert to assure him that he was "constantly thinking of you both."[18]

But Hans Albert was so angry, or hurt, that he had stopped answering his father's letters. "I believe that his attitude toward me has fallen below the freezing point," Einstein lamented to Besso. "Under the given circumstances, I would have reacted in the same way." After three letters to his son went unanswered in three months, Einstein plaintively wrote him: "Don't you remember your father anymore? Are we never going to see each other again?"[19]

Finally, the boy replied by sending a picture of a boat he was constructing out of wood carvings. He also described his mother's return from the sanatorium. "When Mama came home, we had a celebration. I had practiced a sonata by Mozart, and Tete had learned a song."[20]

Einstein did make one concession to the sad situation: he decided to give up asking Marić for a divorce, at least for the time being. That seemed to aid her recovery. "I'll take care that she doesn't get any more disturbance from me," he told Besso. "I have abandoned proceeding with the divorce. Now on to scientific matters!"[21]

Indeed, whenever personal issues began to weigh on him, he took refuge in his work. It shielded him, allowed him to escape. As he told Helene Savić, likely with the intent that it get back to her friend Marić, he planned to retreat into scientific reflection. "I resemble a farsighted man who is charmed by the vast horizon and whom the foreground bothers only when an opaque object prevents him from taking in the long view."[22]

So even as the personal battle was raging, his science provided solace. In 1916, he began writing again about the quantum. He also wrote a formal exposition of his general theory of relativity, which was far more comprehensive, and slightly more comprehensible, than what had poured forth in the weekly lectures during his race with Hilbert the previous November.[23]

In addition, he produced an even more understandable version: a book for the lay reader, *Relativity: The Special and the General Theory,* that remains popular to this day. To make sure that the average person would fathom it, he read every page out loud to Elsa's daughter Margot, pausing frequently to ask whether she indeed got it. "Yes, Albert," she invariably replied, even though (as she confided to others) she found the whole thing totally baffling.[24]

This ability of science to be used as a refuge from painful personal emotions was a theme of a talk he gave at a celebration of Max Planck's sixtieth birthday. Putatively about Planck, it seemed to convey more about Einstein himself. "One of the strongest motives that leads men to art and science is escape from everyday life with its painful crudity and hopeless dreariness," Einstein said. "Such men make this cosmos and its construction the pivot of their emotional life, in order to find the peace and security which they cannot find in the narrow whirlpool of personal experience."[25]

The Treaty

In early 1917, it was Einstein's turn to fall ill. He came down with stomach pains that he initially thought were caused by cancer. Now that his mission was complete, death did not frighten him. He told the astronomer Freundlich that he was not worried about dying because now he had completed his theory of relativity.

Freundlich, on the other hand, did worry about his friend, who was still only 38. He sent Einstein to a doctor, who diagnosed the problem as a chronic stomach malady, one that was exacerbated by wartime food shortages. He put him on a four-week diet of rice, macaroni, and zwieback bread.

These stomach ailments would lay him low for the next four years, then linger for the rest of his life. He was living alone and having trouble getting proper meals. From Zurich, Zangger sent packages to help satisfy the prescribed diet, but within two months Einstein had lost close to fifty pounds. Finally, by the summer of 1917, Elsa was able to rent a second apartment in her building, and she moved him in there to be her neighbor, charge, and companion.[26]

Elsa took great joy in foraging for the food he found comforting. She was resourceful and wealthy enough to commandeer the eggs and butter and bread he liked, even though the war made such staples hard to come by. Every day she cooked for him, doted on him, even found him cigars. Her parents helped as well by having them both over for comforting meals.[27]

The health of his younger son, Eduard, also was precarious. Once again he had fevers, and in early 1917 his lungs became inflamed. After receiving a pessimistic medical prognosis, Einstein lamented to

Besso, "My little boy's condition depresses me greatly. It is impossible that he will become a fully developed person. Who knows if it wouldn't be better for him if he could depart before coming to know life properly."

To Zangger, he ruminated about the "Spartan's method"—leaving sickly children out on a mountain to die—but then said he could not accept that approach. Instead, he promised to pay whatever it took to get Eduard care, and he told Zangger to send him to whatever treatment facility he thought best. "Even if you silently say to yourself that every effort is futile, send him anyway, so that my wife and my Albert think that something is being done." [28]

That summer, Einstein traveled back to Switzerland to take Eduard to a sanatorium in the Swiss village of Arosa. His ability to use science to rise above personal travails was illustrated in a letter he sent to his physicist friend Paul Ehrenfest: "The little one is very sickly and must go to Arosa for a year. My wife is also ailing. Worries and more worries. Nevertheless, I have found a nice generalization of the Sommerfeld-Epstein quantum law." [29]

Hans Albert joined his father on the journey to take Eduard to Arosa, and he then visited when Einstein was staying with his sister, Maja, and her husband, Paul Winteler, in Lucerne. There he found his father bedridden with stomach pains, but his uncle Paul took him hiking. Gradually, with a few rough patches, Einstein's relationship with his older son was being restored. "The letter from my Albert was the greatest joy I've had for the past year," he told Zangger. "I sense with bliss the intimate tie between us." Financial worries were also easing. "I received a prize of 1,500 crowns from the Viennese Academy, which we can use for Tete's cure." [30]

Now that he had moved into the same building as Elsa and she was nursing him back to health, it was inevitable that the issue of a divorce from Marić would arise again. In early 1918, it did. "My desire to put my private affairs in some state of order prompts me to suggest a divorce to you for a second time," he wrote. "I am resolved to do everything to make this step possible." This time his financial offer was even more generous. He would pay her 9,000 marks rather than what had now become a 6,000 annual stipend, with the provision that 2,000 would go into a fund for their children.*

Then he added an amazing new inducement. He was convinced, with good reason, that he would someday win the Nobel Prize. Even though the scientific community had not yet fully come to grips with special relativity, much less his new and unproven theory of general relativity, eventually it would. Or his groundbreaking insights into light quanta and the photoelectric effect would be recognized. And so he made a striking offer to Marić: "The Nobel Prize—in the event of the divorce and the event that it is bestowed upon me—would be ceded to you in full."[31]

It was a financially enticing wager. The Nobel Prize was then, as it is now, very lucrative, indeed huge. In 1918, it was worth about 135,000 Swedish kronor, or 225,000 German marks—more than 37 times what Marić was getting annually. In addition, the German mark was starting to collapse, but the Nobel would be paid in stable Swedish currency. Most poignantly, there would be some symbolic justice: she had helped Einstein with the math and proofreading and domestic support for his 1905 papers, and now she could reap some of the reward.

At first she was furious. "Exactly two years ago, such letters pushed me over the brink into misery, which I still can't get over," she replied. "Why do you torment me so endlessly? I really don't deserve this from you."[32]

But within a few days, she began to assess the situation more clinically. Her life had reached a low point. She suffered pains, anxieties, and depression. Her younger son was in a sanatorium. The sister who had come to help her succumbed to depression and had been committed to an asylum. And her brother, who was serving as a medic in the

---

* Einstein's salary after tax was 13,000 marks. Inflation was beginning to set in, and the value of the German mark had fallen from 24 cents in 1914 to 19 cents in January 1918. One mark at the time would buy two dozen eggs or four loaves of bread. (A year later, the mark would be worth only 12 cents, and when hyperinflation began to rage in January 1920 only 2 cents.) Marić's stipend of 6,000 marks in January 1918 was thus worth about $1,140, or just under $15,000 in inflation-adjusted 2006 dollars. His proposal was to increase this by 50 percent.

Austrian army, had been captured by the Russians. Perhaps an end to the battles with her husband and the chance of financial security might, in fact, be best for her. So she discussed the option with her neighbor Emil Zürcher, who was a lawyer and a friend.

A few days later she decided to take the deal. "Have your lawyer write Dr. Zürcher about how he envisions it, how the contract should be," she replied. "I must leave upsetting things to objective persons. I do not want to stand in the way of your happiness, if you are so resolved."[33]

The negotiations proceeded through letters and third parties through April. "I am curious what will last longer, the world war or our divorce proceedings," he complained lightly at one point. But as things were progressing the way he wanted, he merrily added, "In comparison, this little matter of ours is still much the more pleasant. Amiable greetings to you and kisses to the boys."

The main issue was money. Marić complained to a friend that Einstein was being stingy (in fact he wasn't) because of Elsa. "Elsa's very greedy," Marić charged. "Her two sisters are very rich, and she's always envious of them." Letters went back and forth over exactly how the prospective Nobel Prize money would be paid, what right the children would have to it, what would happen to it if she remarried, and even what compensation he would offer in the unlikely event that the prize was never awarded to him.[34]

Another contentious issue was whether his sons could visit him in Berlin. On barring that, Marić held firm.[35] Finally, at the end of April, he surrendered this final point. "I'm giving in about the children because I now believe you want to handle matters in a conciliatory manner," he said. "Maybe you will later take the view that the boys can come here without reservation. For the time being, I will see them in Switzerland."[36]

Given Marić's poor health, Einstein had tried to work out another option for the two boys: having them live in nearby Lucerne with his sister, Maja, and her husband, Paul Winteler. The Wintelers were willing to take custody of their nephews, and they took the train to Bern one day to see if this could be arranged. But when they arrived, Zanger was away, and they wanted his help before discussing things with

Marić. So Paul went over to see his feisty sister Anna, who was married to Michele Besso, to see if they could have a room for the night.

He had planned not to tell Anna the purpose of their mission, as she had a protective attitude toward Marić and a hair-trigger sense of righteous indignation. "But she guessed the purpose of our coming," Maja reported to Einstein, "and when Paul confirmed her suspicions a torrent of accusations, scoldings, and threats poured forth."[37]

So Einstein wrote a letter to Anna to try to enlist her support. Marić, he argued, was "incapable of running a household" given her condition. It would be best if Hans Albert went to live with Maja and Paul, he argued. Eduard could either do the same or stay in a mountain-air clinic until his health improved. Einstein would pay for it all, including Marić's costs in a sanatorium in Lucerne, where she could see her sons every day.

Unfortunately, Einstein made the mistake of ending the letter by pleading with Anna to help resolve the situation so that he could marry Elsa and end the shame that their relationship was causing her daughters. "Think of the two young girls, whose prospects of getting married are being hampered," he said. "Do put in a good word for me sometime to Miza [Marić] and make it clear to her how unkind it is to complicate the lives of others pointlessly."[38]

Anna shot back that Elsa was the one being selfish. "If Elsa had not wanted to make herself so vulnerable, she should not have run after you so conspicuously."[39]

In truth, Anna was quite difficult, and she soon had a falling out with Marić as well. "She tried to meddle in my affairs in a way that reveals potential human malice," Marić complained to Einstein. At the very least, this helped improve relations between the Einsteins. "I see from your letter that you also have had problems with Anna Besso," he wrote Marić just after they had agreed to the divorce terms. "She has written me such impertinent letters that I've put an end to further correspondence."[40]

It would be a few more months before the divorce decree could become final, but now that the negotiations were complete, everyone seemed relieved that there would be closure. Marić's health improved enough so that the children would remain with her,[41] and the letters back and forth from Berlin and Zurich became friendlier. "A satisfactory relationship has formed between me and my wife through the correspondence about the divorce!" he told Zangger. "A funny opportunity indeed for reconciliation."[42]

This détente meant that Einstein had an option for his summer vacation of 1918: visit his children in Zurich, or have a less stressful holiday with Elsa. He chose the latter, partly because his doctor recommended against the altitude, and for seven weeks he and Elsa stayed in the Baltic Sea resort of Aarenshoop. He brought along some light beach reading, Immanuel Kant's *Prolegomena,* spent "countless hours pondering the quantum problem," and gloried in relaxing and recovering from his stomach ailments. "No telephones, no responsibilities, absolute tranquility," he wrote to a friend. "I am lying on the shore like a crocodile, allowing myself to be roasted by the sun, never see a newspaper, and do not give a hoot about the so-called world."[43]

From this unlikely vacation, he sought to mollify Hans Albert, who had written to say he missed his father. "Write me please why you aren't coming, at least," he asked.[44] Einstein's explanation was sad and very defensive:

> You can easily imagine why I could not come. This winter I was so sick that I had to lie in bed for over two months. Every meal must be cooked separately for me. I may not make any abrupt movements. So I'd have been allowed neither to go on a walk with you nor to eat at the hotel . . . Added to this is that I had quarreled with Anna Besso, and that I did not want to become a burden to Mr. Zangger again, and finally, that I doubted whether my coming mattered much to you.[45]

His son was understanding. He wrote him letters filled with news and ideas, including a description and sketch of an idea he had for a pendulum inside a monorail that would swing and break the electric circuit whenever the train tilted too much.

Einstein had rebuked Hans Albert, unfairly, for not finding some way to visit him in Germany during the vacation. That would have required Marić to waive the provision in their separation agreement that barred such trips, and it would also have been sadly impractical. "My coming to Germany would be almost more impossible than your coming here," Hans Albert wrote, "because in the end I am the only one in the family who can shop for anything." [46]

So Einstein, yearning to be nearer to his boys, found himself briefly tempted to move back to Zurich. During his Baltic vacation that summer of 1918, he considered a combined offer from the University of Zurich and his old Zurich Polytechnic. "You can design your position here exactly as you wish," the physicist Edgar Meyer wrote. As Einstein jokingly noted to Besso, "How happy I would have been 18 years ago with a measly assistantship." [47]

Einstein admitted that he was tormented by the decision. Zurich was his "true home," and Switzerland was the only country for which he felt any affinity. Plus, he would be near his sons.

But there was one rub. If he moved close to his sons he would be moving close to their mother. Even for Einstein, who was good at shielding himself from personal emotions, it would be hard to set up household with Elsa in the same town as his first wife. "My major personal difficulties would persist if I pitched my tent in Zurich again," he told Besso, "although it does seem tempting to be close to my children." [48]

Elsa was also adamantly opposed to the prospect, even appalled. She begged Einstein to promise it would not happen. Einstein could be quite solicitous about Elsa's desires, and so he backed away from a full-time move to Zurich.

Instead, he did something he usually avoided: he compromised. He retained his position in Berlin but agreed to be a guest lecturer in Zurich, making month-long visits there twice a year. That, he thought, could give him the best of both worlds.

In what seemed like an excess of Swiss caution, the Zurich authorities approved the lecture contract, which paid Einstein his expenses but no fee, "by way of experiment." They were in fact wise; Einstein's lectures were initially very popular, but eventually attendance dwindled and they would be canceled after two years.

The Social Democrat

Which would finish first, Einstein had wondered half-jokingly to Marić, the world war or their divorce proceedings? As it turned out, both came to a messy resolution at the end of 1918. As the German Reich was crumbling that November, a revolt by sailors in Kiel mushroomed into a general strike and popular uprising. "Class canceled because of Revolution," Einstein noted in his lecture diary on November 9, the day that protestors occupied the Reichstag and the kaiser abdicated. Four days later, a worker-student revolutionary council took over the University of Berlin and jailed its deans and rector.

With the outbreak of war, Einstein had become, for the first time, an outspoken public figure, advocating internationalism, European federalism, and resistance to militarism. Now, the coming of the peace turned Einstein's political thinking toward more domestic and social issues.

From his youth as an admirer of Jost Winteler and a friend of Friedrich Adler, Einstein had been attracted to the ideal of socialism as well as that of individual freedom. The revolution in Berlin—led by a collection of socialists, workers' councils, communists, and others on the left—caused him to confront cases when these two ideals conflicted.

For the rest of his life Einstein would expound a democratic socialism that had a liberal, anti-authoritarian underpinning. He advocated equality, social justice, and the taming of capitalism. He was a fierce defender of the underdog. But to the extent that any revolutionaries edged over toward a Bolshevik desire to impose centralized control, or to the extent that a regime such as Russia's struck him as authoritarian, Einstein's instinctive love of individual liberty usually provoked a disdainful reaction.

"Socialism to him reflects the ethical desire to remove the appalling chasm between the classes and to produce a more just economic system," his stepson-in-law wrote of Einstein's attitudes during the 1920s. "And yet he cannot accept a socialist program. He appreciates the adventure of solitude and the happiness of freedom too much to welcome a system that threatens completely to eliminate the individual."[49]

It was an attitude that remained constant. "Einstein's basic political philosophy did not undergo any significant changes during his lifetime," said Otto Nathan, a socialist, who became a close friend and then literary executor after Einstein moved to America. "He welcomed the revolutionary development of Germany in 1918 because of his interest in socialism and particularly because of his profound and unqualified devotion to democracy. Basic to his political thinking was the recognition of the dignity of the individual and the protection of political and intellectual freedom."[50]

When the student revolutionaries in Berlin jailed their rector and deans, Einstein got to put this philosophy into practice. The physicist Max Born was in bed that day with the flu when his telephone rang. It was Einstein. He was heading over to the university to see what he could do to get the rector and deans released, and he insisted that Born get out of bed and join him. They also enlisted a third friend, the pioneering Gestalt psychologist Max Wertheimer, perhaps in the belief that his specialty might be more useful than theoretical physics in accomplishing the task.

The three took the tram from Einstein's apartment to the Reichstag, where the students were meeting. At first their way was blocked by a dense mob, but the crowd parted once Einstein was recognized, and they were ushered to a conference room where the student soviet was meeting.

The chairman greeted them and asked them to wait while the group finished hammering out their new statutes for governing the university. Then he turned to Einstein. "Before we come to your request to speak, Professor Einstein, may I be permitted to ask what you think of the new regulations?"

Einstein paused for a moment. Some people are innately conditioned to hedge their words, try to please their listeners, and enjoy the comfort that comes from conforming. Not Einstein. Instead, he responded critically. "I have always thought that the German university's most valuable institution is academic freedom, whereby the lecturers are in no way told what to teach, and the students are able to choose what lectures to attend, without much supervision and control," he said. "Your new statutes seem to abolish all of this. I would be very sorry if the old freedom were to come to an end." At that point, Born recalled, "the high and mighty young gentlemen sat in perplexed silence."

That did not help his mission. After some discussion, the students decided that they did not have the authority to release the rector and deans. So Einstein and company went off to the Reich chancellor's palace to seek out someone who did. They were able to find the new German president, who seemed harried and baffled and perfectly willing to scribble a note ordering the release.

It worked. The trio succeeded in springing their colleagues, and, as Born recalled, "We left the Chancellor's palace in high spirits, feeling that we had taken part in a historical event and hoping to have seen the last of Prussian arrogance."[51]

Einstein then went down the street to a mass meeting of the revived New Fatherland League, where he delivered a two-page speech that he had carried with him to his confrontation with the students. Calling himself "an old-time believer in democracy," he again made clear that his socialist sentiments did not make him sympathetic to Soviet-style controls. "All true democrats must stand guard lest the old class tyranny of the Right be replaced by a new class tyranny of the Left," he said.

Some on the left insisted that democracy, or at least multiparty liberal democracy, needed to be put aside until the masses could be educated and a new revolutionary consciousness take hold. Einstein disagreed. "Do not be seduced by feelings that a dictatorship of the proletariat is temporarily needed in order to hammer the concept of freedom into the heads of our fellow countrymen," he told the rally. Instead, he decried Germany's new left-wing government as "dictatorial," and he demanded that it immediately call open elections, "thereby eliminating all fears of a new tyranny as soon as possible."[52]

Years later, when Adolf Hitler and his Nazis were in power, Einstein would ruefully look back on that day in Berlin. "Do you still remember the occasion some 25 years ago when we went together to the Reichstag building, convinced that we could turn the people there into honest democrats?" he wrote Born. "How naïve we were for men of forty."[53]

Marrying Elsa

Just after the war ended, so did Einstein's divorce proceedings. As part of the process, he had to give a deposition admitting adultery. On December 23, 1918, he appeared before a court in Berlin, stood before a magistrate, and declared, "I have been living together with my cousin, the widow Elsa Einstein, divorced Löwenthal, for about 4½ years and have been continuing these intimate relations since then."[54]

As if to prove it, he brought Elsa when he traveled to Zurich the following month to deliver his first set of lectures there. His opening talks, unlike his later ones, were so well attended that, to Einstein's annoyance, an official was posted at the door to prevent unauthorized auditors from getting in. Hans Albert came to visit him at his hotel, presumably when Elsa was not there, and Einstein spent a few days in Arosa, where Eduard was still recuperating in a sanatorium.[55]

Einstein stayed in Zurich through February 14, when he stood before three local magistrates who granted his final divorce decree. It included the provisions regarding his prospective Nobel Prize award. In his deposition, Einstein had given his religion as "dissenter," but in the divorce decree the clerk designated him "Mosaic." Marić was also designated "Mosaic," even though she had been born and remained a Serbian Orthodox Christian.

As was customary, the decree included the order that "the Defendant [Einstein] is restrained from entering into a new marriage for the period of two years."[56] Einstein had no intention of obeying that provision. He had decided that he would marry Elsa, and he would end up doing so within four months.

His decision to remarry was accompanied by a drama that was, if true, weird even by the standards of his unusual family dynamics. It involved Elsa Einstein's daughter Ilse and the pacifist physician and adventurer Georg Nicolai.

Ilse, then 21, was the elder of Elsa's two daughters. Einstein had hired her as the secretary for the unbuilt Kaiser Wilhelm Institute of Physics that he was supposed to be creating (the only scientist who had been hired so far was his faithful astronomer Freundlich). A spirited, idealistic, swanlike beauty, Ilse's mystique was enhanced by the fact that as a child she had lost the use of an eye in an accident. Like a moth to flame, she was attracted to radical politics and fascinating men.

Thus it was not surprising that she fell for Georg Nicolai, who had collaborated with Einstein in 1914 on the pacifist response to the German intellectuals' "Appeal to the Cultured World." Among other things, Nicolai was a doctor specializing in electrocardiograms who had occasionally treated Elsa. A brilliant egomaniac with a serious sexual appetite, he had been born in Germany and had lived in Paris and Russia. During one visit to Russia, he kept a list of the women he had sex with, totaling sixteen in all, including two mother-daughter pairs.

Ilse fell in love with Nicolai and with his politics. In addition to being, at least briefly, his lover, she helped type and distribute his protest letters. She also helped persuade Einstein to support the publication of Nicolai's pacifist tome, *The Biology of War,* which included their ill-fated 1914 manifesto and a collection of liberal writings by Kant and other classical German authors.[57]

Einstein had initially supported this publishing project, but in early 1917 had labeled the idea "entirely hopeless." Nicolai, who had been drafted as a lowly medical orderly for the German army, somehow thought that Einstein would fund the endeavor, and he kept badgering him. "Nothing is more difficult than turning Nicolai down," Einstein wrote him, addressing him in the third person. "The man, who in other things is so sensitive that even grass growing is a considerable din to him, seems almost deaf when the sound involves a refusal."[58]

On one of Ilse's visits to see Nicolai, she told him that Einstein was now planning to marry her mother. Nicolai, an aficionado of the art of dating both mother and daughter, told Ilse that Einstein had it wrong. He should marry Ilse rather than her mother.

It is unclear what psychological game he was playing with his young lover's mind. And it is likewise unclear what psychological game she was playing with his mind, or her own mind, when she wrote him a detailed letter saying that the Ilse-or-Elsa question had suddenly become a real one for Einstein. The letter is so striking and curious it bears being quoted at length:

You are the only person to whom I can entrust the following and the only one who can give me advice . . . You remember that we recently spoke about Albert's and Mama's marriage and you told me that you thought a marriage between Albert and me would be more proper. I never thought seriously about it until yesterday. Yesterday, the question was suddenly raised about whether Albert wished to marry Mama or me. This question, initially posed half in jest, became within a few minutes a serious matter which must now be considered and discussed fully and completely. Albert himself is refusing to take any decision, he is prepared to marry either me or Mama. I know that Albert loves me very much, perhaps more than any other man ever will. He told me so himself yesterday. On the one hand, he might even prefer me as his wife, since I am young and he could have children with me, which naturally does not apply at all in Mama's case; but he is far too decent and loves Mama too much ever to mention it. You know how I stand with Albert. I love him very much; I have the greatest respect for him as a person. If ever there was true friendship and camaraderie between two beings of different types, those are quite certainly my feelings for Albert. I have never wished nor felt the least desire to be close to him physically. This is otherwise in his case—recently at least. He admitted to me once how difficult it is for him to keep himself in check. But now I do believe that my feelings for him are not sufficient for conjugal life . . . The third person still to be mentioned in this odd and certainly also highly comical affair would be Mother. For the present—because she does not yet firmly believe that I am really serious. She has allowed me to choose completely freely. If she saw that I could really be happy only with Albert, she would surely step aside out of love for me. But it would certainly be bitterly hard for her. And then I do not know whether it really would be fair if—after all her years of struggle—I were to compete with her over the place she had won for herself, now that she is finally at the goal. Philistines like the grandparents are naturally appalled about these new plans. Mother would supposedly be disgraced and other such unpleasant things . . . Albert also thought that if I did not wish to have a child of his it would be nicer for me not to be married to him. And I truly do not have this wish. It will seem peculiar to you that I, a silly little thing of a 20-year-old, should have to decide on such a serious matter; I can hardly believe it myself and feel very unhappy doing so as well. Help me! Yours, Ilse.[59]

She wrote a big note on top of the first page: "Please destroy this letter immediately after reading it!" Nicolai didn't.

Was it true? Was it half-true? Was the truth relative to the observer? The only evidence we have of Einstein's mother-daughter dithering is this one letter. No one else, then or in recollections, ever mentioned the issue. The letter was written by an intense and love-struck young woman to a dashing philanderer whose attentions she craved. Perhaps it was merely her fantasy, or her ploy to provoke Nicolai's jealousy. As with much of nature, especially human nature, the underlying reality, if there is such a thing, may not be knowable.

As it turned out, Einstein married Elsa in June 1919, and Ilse ended up remaining close to both of them.

Einstein's family relations seemed to be improving on all fronts. The very next month, he went to Zurich to see his boys, and he stayed with Hans Albert at his first wife's apartment while she was away. Elsa seemed worried about that arrangement, but he reassured her in at least two letters that Marić would not be around much. "Camping in the lioness's den is proving very worthwhile," he said in one, "and there's no fear of any incident happening." Together he and Hans Albert went sailing, played music, and built a model airplane together. "The boy gives me indescribable joy," he wrote Elsa. "He is very diligent and persistent in everything he does. He also plays piano very nicely." [60]

His relations with his first family were now so calm that, during his July 1919 visit, he once again thought that maybe he should move there with Elsa and her daughters. This completely flummoxed Elsa, who made her feelings very clear. Einstein backed down. "We're going to stay in Berlin, all right," he reassured her. "So calm down and never fear!" [61]

Einstein's new marriage was different from his first. It was not romantic or passionate. From the start, he and Elsa had separate bedrooms at opposite ends of their rambling Berlin apartment. Nor was it intellectual. Understanding relativity, she later said, "is not necessary for my happiness." [62]

She was, on the other hand, talented in practical ways that often eluded her husband. She spoke French and English well, which allowed her to serve as his translator as well as manager when he traveled. "I am not talented in any direction except perhaps as wife and mother," she said. "My interest in mathematics is mainly in the household bills." [63]

That comment reflects her humility and a simmering insecurity, but it sells her short. It was no simple task to play the role of wife and mother to Einstein, who required both, nor to manage their finances and logistics. She did it with good sense and warmth. Even though, every now and then, she succumbed to a few pretenses that came with their standing, she generally displayed an unaffected manner and self-aware humor, and in doing so she thus helped make sure that her husband retained those traits as well.

The marriage was, in fact, a solid symbiosis, and it served adequately, for the most part, the needs and desires of both partners. Elsa was an efficient and lively woman, who was eager to serve and protect him. She liked his fame, and (unlike him) did not try to hide that fact. She also appreciated the social standing it gave them, even if it meant she had to merrily shoo away reporters and other invaders of her husband's privacy.

He was as pleased to be looked after as she was to look after him. She told him when to eat and where to go. She packed his suitcases and doled out his pocket money. In public, she was protective of the man she called "the Professor" or even simply "Einstein."

That allowed him to spend hours in a rather dreamy state, focusing more on the cosmos than on the world around him. All of which gave her excitement and satisfaction. "The Lord has put into him so much that's beautiful, and I find him wonderful, even though life at his side is enervating and difficult," she once said.[64]

When Einstein was in one of his periods of intense work, as was often the case, Elsa "recognized the need for keeping all disturbing elements away from him," a relative noted. She would make his favorite meal of lentil soup and sausages, summon him down from his study, and then would leave him alone as he mechanically ate his meal. But when he would mutter or protest, she would remind him that it was important for him to eat. "People have centuries to find things out," she would say, "but your stomach, no, it will not wait for centuries."[65]

She came to know, from a faraway look in his eyes, when he was "seized with a problem," as she called it, and thus should not be disturbed. He would pace up and down in his study, and she would have food sent up. When his intense concentration was over, he would finally come down to the table for a meal and, sometimes, ask to go on a walk with Elsa and her daughters. They always complied, but they never initiated such a request. "It is he who has to do the asking," a newspaper reported after interviewing her, "and when he asks them for a walk they know that his mind is relieved of work."[66]

Elsa's daughter Ilse would eventually marry Rudolf Kayser, editor of the premier literary magazine in Germany, and they set up a house filled with art and artists and writers. Margot, who liked sculpting, was so shy that she would sometimes hide under the table when guests of her father arrived. She lived at home even after she married, in 1930, a Russian named Dimitri Marianoff. Both of these sons-in-law, it turned out, would end up writing florid but undistinguished books about the Einstein family.

For the time being, Einstein and Elsa and her two daughters lived together in a spacious and somberly furnished apartment near the center of Berlin. The wallpaper was dark green, the tablecloths white linen with lace embroidery. "One felt that Einstein would always remain a stranger in such a household," said his friend and colleague Philipp Frank, "a Bohemian as a guest in a bourgeois home."

In defiance of building codes, they converted three attic rooms into a garret study with a big new window. It was occasionally dusted, never tidied, and papers piled up under the benign gazes of Newton, Maxwell, and Faraday. There Einstein would sit in an old armchair, pad on his knee. Occasionally he would get up to pace, then he would sit back down to scribble the equations that would, he hoped, extend his theory of relativity into an explanation of the cosmos.[67]

CHAPTER ELEVEN

EINSTEIN'S UNIVERSE

1916–1919

In his Berlin home study

Cosmology and Black Holes, 1917

Cosmology is the study of the universe as a whole, including its size and shape, its history and destiny, from one end to the other, from the beginning to the end of time. That's a big topic. And it's not a simple one. It's not even simple to define what those concepts mean, or even if they have meaning. With the gravitational field equations in his general theory of relativity, Einstein laid the foundations for studying the nature of the universe, thereby becoming the primary founder of modern cosmology.

Helping him in this endeavor, at least in the early stages, was a profound mathematician and even more distinguished astrophysicist, Karl Schwarzschild, who directed the Potsdam Observatory. He read Einstein's new formulation of general relativity and, at the beginning of 1916, set about trying to apply it to objects in space.

One thing made Schwarzschild's work very difficult. He had volunteered for the German military during the war, and when he read Einstein's papers he was stationed in Russia, projecting the trajectory of artillery shells. Nevertheless, he was also able to find time to calculate what the gravitational field would be, according to Einstein's theory, around an object in space. It was the wartime counterpart to Einstein's ability to come up with the special theory of relativity while examining patent applications for the synchronization of clocks.

In January 1916, Schwarzschild mailed his result to Einstein with the declaration that it permitted his theory "to shine with increased purity." Among other things, it reconfirmed, with greater rigor, the success of Einstein's equations in explaining Mercury's orbit. Einstein was thrilled. "I would not have expected that the exact solution to the problem could be formulated so simply," he replied. The following Thursday, he personally delivered the paper at the Prussian Academy's weekly meeting.[1]

Schwarzschild's first calculations focused on the curvature of space-time *outside* a spherical, nonspinning star. A few weeks later, he sent Einstein another paper on what it would be like *inside* such a star.

In both cases, something unusual seemed possible, indeed inevitable. If all the mass of a star (or any object) was compressed into a tiny enough space—defined by what became known as the Schwarzschild radius—then all of the calculations seemed to break down. At the center, spacetime would infinitely curve in on itself. For our sun, that would happen if all of its mass were compressed into a radius of less than two miles. For the earth, it would happen if all the mass were compressed into a radius of about one-third of an inch.

What would that mean? In such a situation, nothing within the Schwarzschild radius would be able to escape the gravitational pull, not even light or any other form of radiation. Time would also be part of the warpage as well, dilated to zero. In other words, a traveler nearing the Schwarzschild radius would appear, to someone on the outside, to freeze to a halt.

Einstein did not believe, then or later, that these results actually corresponded to anything real. In 1939, for example, he produced a paper that provided, he said, "a clear understanding as to why these 'Schwarzschild singularities' do not exist in physical reality." A few months later, however, J. Robert Oppenheimer and his student Hartland Snyder argued the opposite, predicting that stars could undergo a gravitational collapse.[2]

As for Schwarzschild, he never had the chance to study the issue further. Weeks after writing his papers, he contracted a horrible autoimmune disease while on the front, which ate away at his skin cells, and he died that May at age 42.

As scientists would discover after Einstein's death, Schwarzschild's odd theory was right. Stars *could* collapse and create such a phenomenon, and in fact they often did. In the 1960s, physicists such as Stephen Hawking, Roger Penrose, John Wheeler, Freeman Dyson, and Kip Thorne showed that this was indeed a feature of Einstein's general theory of relativity, one that was very real. Wheeler dubbed them "black holes," and they have been a feature of cosmology, as well as *Star Trek* episodes, ever since.[3]

Black holes have now been discovered all over the universe, including one at the center of our galaxy that is a few million times more massive than our sun. "Black holes are not rare, and they are not an accidental embellishment of our universe," says Dyson. "They are the only places in the universe where Einstein's theory of relativity shows its full power and glory. Here, and nowhere else, space and time lose their individuality and merge together in a sharply curved four-dimensional structure precisely delineated by Einstein's equations."[4]

Einstein believed that his general theory solved Newton's bucket issue in a way that Mach would have liked: inertia (or centrifugal forces) would not exist for something spinning in a completely empty universe.* Instead, inertia was caused only by rotation *relative* to all the other objects in the universe. "According to my theory, inertia is simply an interaction between masses, not an effect in which 'space' of itself is involved, separate from the observed mass," Einstein told Schwarzschild. "It can be put this way. If I allow all things to vanish, then ac-

---

* Chapter 14 describes Einstein's revision of this view in a 1920 lecture in Leiden.

cording to Newton the Galilean inertial space remains; following my interpretation, however, *nothing* remains."[5]

The issue of inertia got Einstein into a debate with one of the great astronomers of the time, Willem de Sitter of Leiden. Throughout 1916, Einstein struggled to preserve the relativity of inertia and Mach's principle by using all sorts of constructs, including assuming various "border conditions" such as distant masses along the fringes of space that were, by necessity, unable to be observed. As de Sitter noted, that in itself would have been anathema to Mach, who railed against postulating things that could not possibly be observed.[6]

By February 1917, Einstein had come up with a new approach. "I have completely abandoned my views, rightly contested by you," he wrote de Sitter. "I am curious to hear what you will have to say about the somewhat crazy idea I am considering now."[7] It was an idea that initially struck him as so wacky that he told his friend Paul Ehrenfest in Leiden, "It exposes me to the danger of being confined to a madhouse." He jokingly asked Ehrenfest for assurances, before he came to visit, that there were no such asylums in Leiden.[8]

His new idea was published that month in what became yet another seminal Einstein paper, "Cosmological Considerations in the General Theory of Relativity."[9] On the surface, it did indeed seem to be based on a crazy notion: space has no borders because gravity bends it back on itself.

Einstein began by noting that an absolutely infinite universe filled with stars and other objects was not plausible. There would be an infinite amount of gravity tugging at every point and an infinite amount of light shining from every direction. On the other hand, a finite universe floating at some random location in space was inconceivable as well. Among other things, what would keep the stars and energy from flying off, escaping, and depleting the universe?

So he developed a third option: a finite universe, but one without boundaries. The masses in the universe caused space to curve, and over the expanse of the universe they caused space (indeed, the whole four-dimensional fabric of spacetime) to curve completely in on itself. The system is closed and finite, but there is no end or edge to it.

One method that Einstein employed to help people visualize this notion was to begin by imagining two-dimensional explorers on a two-dimensional universe, like a flat surface. These "flatlanders" can wander in any direction on this flat surface, but the concept of going up or down has no meaning to them.

Now, imagine this variation: What if these flatlanders' two dimensions were still on a surface, but this surface was (in a way very subtle to them) gently curved? What if they and their world were still confined to two dimensions, but their flat surface was like the surface of a globe? As Einstein put it, "Let us consider now a two-dimensional existence, but this time on a spherical surface instead of on a plane." An arrow shot by these flatlanders would still seem to travel in a straight line, but eventually it would curve around and come back—just as a sailor on the surface of our planet heading straight off over the seas would eventually return from the other horizon.

The curvature of the flatlanders' two-dimensional space makes their surface finite, and yet they can find no boundaries. No matter what direction they travel, they reach no end or edge of their universe, but they eventually get back to the same place. As Einstein put it, "The great charm resulting from this consideration lies in the recognition that *the universe of these beings is finite and yet has no limits*." And if the flatlanders' surface was like that of an inflating balloon, their whole universe could be expanding, yet there would still be no boundaries to it.[10]

By extension, we can try to imagine, as Einstein has us do, how three-dimensional space can be similarly curved to create a closed and finite system that has no edge. It's not easy for us three-dimensional creatures to visualize, but it is easily described mathematically by the non-Euclidean geometries pioneered by Gauss and Riemann. It can work for four dimensions of spacetime as well.

In such a curved universe, a beam of light starting out in any direction could travel what seems to be a straight line and yet still curve back on itself. "This suggestion of a finite but unbounded space is one of the greatest ideas about the nature of the world which has ever been conceived," the physicist Max Born has declared.[11]

Yes, but what is *outside* this curved universe? What's on the other side of the curve? That's not merely an unanswerable question, it's a meaningless one, just as it would be meaningless for a flatlander to ask what's outside her surface. One could speculate, imaginatively or mathematically, about what things are like in a fourth spatial dimension, but other than in science fiction it is not very meaningful to ask what's in a realm that exists outside of the three spatial dimensions of our curved universe.[12]

This concept of the cosmos that Einstein derived from his general theory of relativity was elegant and magical. But there seemed to be one hitch, a flaw that needed to be fixed or fudged. His theory indicated that the universe would have to be either expanding or contracting, not staying static. According to his field equations, a static universe was impossible because the gravitational forces would pull all the matter together.

This did not accord with what most astronomers thought they had observed. As far as they knew, the universe consisted only of our Milky Way galaxy, and it all seemed pretty stable and static. The stars appeared to be meandering gently, but not receding rapidly as part of an expanding universe. Other galaxies, such as Andromeda, were merely unexplained blurs in the sky. (A few Americans working at the Lowell Observatory in Arizona had noticed that the spectra of some mysterious spiral nebulae were shifted to the red end of the spectrum, but scientists had not yet determined that these were distant galaxies all speeding away from our own.)

When the conventional wisdom of physics seemed to conflict with an elegant theory of his, Einstein was inclined to question that wisdom rather than his theory, often to have his stubbornness rewarded. In this case, his gravitational field equations seemed to imply—indeed, screamed out—that the conventional thinking about a stable universe was wrong and should be tossed aside, just as Newton's concept of absolute time was.[13]

Instead, this time he made what he called a "slight modification" to his theory. To keep the matter in the universe from imploding, Einstein added a "repulsive" force: a little addition to his general relativity equations to counterbalance gravity in the overall scheme.

In his revised equations, this modification was signified by the Greek letter *lambda*, λ, which he used to multiply his metric tensor $g_{\mu\nu}$ in a way that produced a stable, static universe. In his 1917 paper, he was almost apologetic: "We admittedly had to introduce an extension of the field equations that is not justified by our actual knowledge of gravitation."

He dubbed the new element the "cosmological term" or the "cosmological constant" (*kosmologische Glied* was the phrase he used). Later,* when it was discovered that the universe was in fact expanding, Einstein would call it his "biggest blunder." But even today, in light of evidence that the expansion of the universe is accelerating, it is considered a useful concept, indeed a necessary one after all.[14]

During five months in 1905, Einstein had upended physics by conceiving light quanta, special relativity, and statistical methods for showing the existence of atoms. Now he had just completed a more prolonged creative slog, from the fall of 1915 to the spring of 1917, which Dennis Overbye has called "arguably the most prodigious effort of sustained brilliance on the part of one man in the history of physics." His first burst of creativity as a patent clerk had appeared to involve remarkably little anguish. But this later one was an arduous and intense effort, one that left him exhausted and wracked with stomach pains.[15]

During this period he generalized relativity, found the field equations for gravity, found a physical explanation for light quanta, hinted at how the quanta involved probability rather than certainty,† and came up with a concept for the structure of the universe as a whole. From the smallest thing conceivable, the quantum, to the largest, the cosmos itself, Einstein had proven a master.

The Eclipse, 1919

For general relativity, there was a dramatic experimental test that was possible, one that had the potential to dazzle and help heal a war-weary world. It was based on a concept so simple that everyone could understand it: gravity would bend light's trajectory. Specifically, Einstein predicted the degree to which light from a distant star would be

---

* See chapter 14 for Einstein's decision to renounce the term when he discovered the universe was expanding.

† Described in chapter 14.

observed to curve as it went through the strong gravitational field close to the sun.

To test this, astronomers would have to plot precisely the position of a star in normal conditions. Then they would wait until the alignments were such that the path of light from that star passed right next to the sun. Did the star's position seem to shift?

There was one exciting challenge. This observation required a total eclipse, so that the stars would be visible and could be photographed. Fortunately, nature happened to make the size of the sun and moon just properly proportional so that every few years there are full eclipses observable at times and places that make them ideally suited for such an experiment.

Einstein's 1911 paper, "On the Influence of Gravity on the Propagation of Light," and his *Entwurf* equations the following year, had calculated that light would undergo a deflection of approximately (allowing for some data corrections subsequently made) 0.85 arc-second when it passed near the sun, which was the same as would be predicted by an emission theory such as Newton's that treated light as particles. As previously noted, the attempt to test this during the August 1914 eclipse in the Crimea had been aborted by the war, so Einstein was saved the potential embarrassment of being proved wrong.

Now, according to the field equations he formulated at the end of 1915, which accounted for the curvature of spacetime caused by gravity, he had come up with *twice* that deflection. Light passing next to the sun should be bent, he said, by about 1.7 arc-seconds.

In his 1916 popular book on relativity, Einstein issued yet another call for scientists to test this conclusion. "Stars ought to appear to be displaced outwards from the sun by 1.7 seconds of arc, as compared with their apparent position in the sky when the sun is situated at another part of the heavens," he said. "The examination of the correctness or otherwise of this deduction is a problem of the greatest importance, the early solution of which is to be expected of astronomers."[16]

Willem de Sitter, the Dutch astrophysicist, had managed to send a copy of Einstein's general relativity paper across the English Channel in 1916 in the midst of the war and get it to Arthur Eddington, who was the director of the Cambridge Observatory. Einstein was not well-known in England, where scientists then took pride in either ignoring or denigrating their German counterparts. Eddington became an exception. He embraced relativity enthusiastically and wrote an account in English that popularized the theory, at least among scholars.

Eddington consulted with the Astronomer Royal, Sir Frank Dyson, and came up with the audacious idea that a team of English scientists should prove the theory of a German, even as the two nations were at war. In addition, it would help solve a personal problem for Eddington. He was a Quaker and, because of his pacifist faith, faced imprisonment for refusing military service in England. (In 1918, he was 35 years old, still subject to conscription.) Dyson was able to convince the British Admiralty that Eddington could best serve his nation by leading an expedition to test the theory of relativity during the next full solar eclipse.

That eclipse would occur on May 29, 1919, and Dyson pointed out that it would be a unique opportunity. The sun would then be amid the rich star cluster known as the Hyades, which we ordinary stargazers recognize as the center of the constellation Taurus. But it would not be convenient. The eclipse would be most visible in a path that stretched across the Atlantic near the equator from the coast of Brazil to Equatorial Africa. Nor would it be easy. As the expedition was being considered in 1918, there were German U-boats in the region, and their commanders were more interested in the control of the seas than in the curvature of the cosmos.

Fortunately, the war ended before the expeditions began. In early March 1919, Eddington sailed from Liverpool with two teams. One group split off to set up their cameras in the isolated town of Sobral in the Amazon jungle of northern Brazil. The second group, which included Eddington, sailed for the tiny island of Principe, a Portuguese colony a degree north of the equator just off the Atlantic coast of Africa. Eddington set up his equipment on a 500-foot bluff on the island's north tip.[17]

The eclipse was due to begin just after 3:13 p.m. local time on Principe and last about five minutes. That morning it rained heavily. But as the time of the eclipse approached, the sky started to clear. The heavens insisted on teasing and tantalizing Eddington at the most important minutes of his career, with the remaining clouds cloaking and then revealing the elusive sun.

"I did not see the eclipse, being too busy changing plates, except for one glance to make sure it had begun and another halfway through to see how much cloud there was," Eddington noted in his diary. He took sixteen photographs. "They are all good of the sun, showing a very remarkable prominence; but the cloud has interfered with the star images." In his telegram back to London that day, he was more telegraphic: "Through cloud, hopeful. Eddington."[18]

The team in Brazil had better weather, but the final results had to wait until all of the photographic plates from both places could be shipped back to England, developed, measured, and compared. That took until September, with Europe's scientific cognoscenti waiting eagerly. To some spectators, it took on the postwar political coloration of a contest between the English theory of Newton, predicting about 0.85 arc-second deflection, and the German theory of Einstein, predicting a 1.7 arc-seconds deflection.

The photo finish did not produce an immediately clear result. One set of particularly good pictures taken in Brazil showed a deflection of 1.98 arc-seconds. Another instrument, also at the Brazil location, produced photographs that were a bit blurrier, because heat had affected its mirror; they indicated a 0.86 deflection, but with a higher margin of error. And then there were Eddington's own plates from Principe. These showed fewer stars, so a series of complex calculations were used to extract some data. They seemed to indicate a deflection of about 1.6 arc-seconds.

The predictive power of Einstein's theory—the fact that it offered up a testable prediction—perhaps exercised a power over Eddington, whose admiration for the mathematical elegance of the theory caused him to believe in it deeply. He discarded the lower value coming out of Brazil, contending that the equipment was faulty, and with a slight bias toward his own fuzzy results from Africa got an average of just over 1.7 arc-seconds, matching Einstein's predictions. It wasn't the cleanest confirmation, but it was enough for Eddington, and it turned out to be valid. He later referred to getting these results as the greatest moment of his life.[19]

In Berlin, Einstein put on an appearance of nonchalance, but he could not completely hide his eagerness as he awaited word. The downward spiral of the German economy in 1919 meant that the elevator in his apartment building had been shut down, and he was preparing for a winter with little heat. "Much shivering lies ahead for the winter," he wrote his ailing mother on September 5. "There is still no news about the eclipse." In a letter a week later to his friend Paul Ehrenfest in Holland, Einstein ended with an affected casual question: "Have you by any chance heard anything over there about the English solar-eclipse observation?"[20]

Just by asking the question Einstein showed he was not quite as sanguine as he tried to appear, because his friends in Holland would certainly have already sent him such news if they had it. Finally they did. On September 22, 1919, Lorentz sent a cable based on what he had just heard from a fellow astronomer who had talked to Eddington at a meeting: "Eddington found stellar shift at solar limb, tentative value between nine-tenths of a second and twice that." It was wonderfully ambiguous. Was it a shift of 0.85 arc-second, as Newton's emission theory and Einstein's discarded 1912 theory would have it? Or twice that, as he now predicted?

Einstein had no doubts. "Today some happy news," he wrote his mother. "Lorentz telegraphed me that the British expeditions have verified the deflection of light by the sun."[21] Perhaps his confidence was partly an attempt to cheer up his mother, who was suffering from stomach cancer. But it is more likely that it was because he knew his theory was correct.

Einstein was with a graduate student, Ilse Schneider, shortly after Lorentz's news arrived. "He suddenly interrupted the discussion," she later recalled, and reached for the telegram that was lying on a window sill. "Perhaps this will interest you," he said, handing it to her.

Naturally she was overjoyed and excited, but Einstein was quite calm. "I *knew* the theory was correct," he told her.

But, she asked, what if the experiments had shown his theory to be wrong?

He replied, "Then I would have been sorry for the dear Lord; the theory is correct."[22]

As more precise news of the eclipse results spread, Max Planck was among those who gently noted to Einstein that it was good to have his own confidence confirmed by some actual facts. "You have already said many times that you never personally doubted what the result would be," Planck wrote, "but it is beneficial, nonetheless, if now this fact is indubitably established for others as well." For Einstein's stolid patron, the triumph had a transcendent aspect. "The intimate union between the beautiful, the true and the real has again been proved." Einstein replied to Planck with a veneer of humility: "It is a gift from gracious destiny that I have been allowed to experience this."[23]

Einstein's celebratory exchange with his closer friends in Zurich was more lighthearted. The physics colloquium there sent him a piece of doggerel:

> *All doubts have now been spent*
> *At last it has been found:*
> *Light is naturally bent*
> *To Einstein's great renown!*[24]

To which Einstein replied a few days later, referring to the eclipse:

> *Light and heat Mrs. Sun us tenders*
> *Yet loves not he who broods and ponders.*
> *So she contrives many a year*
> *How she may hold her secret dear!*
> *Now came the lunar visitor kind;*
> *For joy, she almost forgot to shine.*
> *Her deepest secrets too she lost*
> *Eddington, you know, has snapped a shot.*[25]

In defense of Einstein's poetic prowess, it should be noted that his verse works better in German, in which the last two lines end with "gekommen" and "aufgenommen."

The first unofficial announcement came at a meeting of the Dutch Royal Academy. Einstein sat proudly onstage as Lorentz described Eddington's findings to an audience of close to a thousand cheering students and scholars. But it was a closed meeting with no press, so the leaks about the results merely added to the great public anticipation leading up to the official announcement scheduled for two weeks later in London.

The distinguished members of the Royal Society, Britain's most venerable scientific institution, met along with colleagues from the Royal Astronomical Society on the afternoon of November 6, 1919, at Burlington House in Piccadilly, for what they knew was likely to be a historic event. There was only one item on the agenda: the report on the eclipse observations.

Sir J. J. Thomson, the Royal Society's president and discoverer of the electron, was in the chair. Alfred North Whitehead, the philosopher, had come down from Cambridge and was in the audience, taking notes. Gazing down on them from an imposing portrait in the great hall was Isaac Newton. "The whole atmosphere of tense interest was exactly that of the Greek drama," Whitehead recorded. "We were the chorus commenting on the decree of destiny . . . and in the background the picture of Newton to remind us that the greatest of scientific generalizations was, now, after more than two centuries, to receive its first modification." [26]

The Astronomer Royal, Sir Frank Dyson, had the honor of presenting the findings. He described in detail the equipment, the photographs, and the complexities of the calculations. His conclusion, however, was simple. "After a careful study of the plates, I am prepared to say that there can be no doubt that they confirm Einstein's prediction," he announced. "The results of the expeditions to Sobral and Principe leave little doubt that a deflection of light takes place in the neighborhood of the sun and that it is of the amount demanded by Einstein's generalized theory of relativity." [27]

There was some skepticism in the room. "We owe it to that great man to proceed very carefully in modifying or retouching his law of gravitation," cautioned Ludwig Silberstein, gesturing at Newton's portrait. But it was the commanding giant J. J. Thomson who set the tone. "The result is one of the greatest achievements of human thought," he declared. [28]

Einstein was back in Berlin, so he missed the excitement. He celebrated by buying a new violin. But he understood the historic impact of the announcement that the laws of Sir Isaac Newton no longer fully governed all aspects of the universe. "Newton, forgive me," Einstein later wrote, noting the moment. "You found the only way which, in your age, was just about possible for a man of highest thought and creative power."[29]

It was a grand triumph, but not one easily understood. The skeptical Silberstein came up to Eddington and said that people believed that only three scientists in the world understood general relativity. He had been told that Eddington was one of them.

The shy Quaker said nothing. "Don't be so modest, Eddington!" said Silberstein.

Replied Eddington, "On the contrary. I'm just wondering who the third might be."[30]

CHAPTER TWELVE

FAME

1919

With Charlie Chaplin and Elsa at the Hollywood
premiere of *City Lights*, January 1931

"Lights All Askew"

Einstein's theory of relativity burst into the consciousness of a
world that was weary of war and yearning for a triumph of human
transcendence. Almost a year to the day after the end of the brutal
fighting, here was an announcement that the theory of a German Jew
had been proven correct by an English Quaker. "Scientists belonging
to two warring nations had collaborated again!" exulted the physicist
Leopold Infeld. "It seemed the beginning of a new era."[1]

The Times of London carried stories on November 7 about the defeated Germans being summoned to Paris to face treaty demands from
the British and French. But it also carried the following triple-decked
headline:

REVOLUTION IN SCIENCE

New Theory of the Universe

NEWTONIAN IDEAS OVERTHROWN

"The scientific concept of the fabric of the Universe must be changed," the paper proclaimed. Einstein's newly confirmed theory will "require a new philosophy of the universe, a philosophy that will sweep away nearly all that has hitherto been accepted."[2]

The *New York Times* caught up with the story two days later.[3] Not having a science correspondent in London, the paper assigned the story to its golf expert, Henry Crouch, who at first decided to skip the Royal Society announcement, then changed his mind, but then couldn't get in. So he telephoned Eddington to get a summary and, somewhat baffled, asked him to repeat it in simpler words.[4]

Perhaps due to Eddington's enthusiasm in the retelling, or due to Crouch's enthusiasm in the reporting, Eddington's appraisal of Einstein's theory was enhanced to read "one of the greatest—perhaps the greatest—of achievements in the history of human thought."[5] But given the frenzy about to ensue, the headline was rather restrained:

ECLIPSE SHOWED GRAVITY VARIATION

Diversion of Light Rays Accepted as Affecting Newton's Principles.

HAILED AS EPOCHMAKING

British Scientist Calls the Discovery One of the Greatest of Human Achievements.

The following day, the *New York Times* apparently decided that it had been too restrained. So it followed up with an even more excited story, its six-deck headline a classic from the days when newspapers knew how to write classic headlines:

LIGHTS ALL ASKEW IN THE HEAVENS

Men of Science More or Less Agog Over Results of Eclipse Observations.

EINSTEIN THEORY TRIUMPHS

Stars Not Where They Seemed or Were Calculated to be, but Nobody Need Worry.

A BOOK FOR 12 WISE MEN

No More in All the World Could Comprehend It, Said Einstein When His Daring Publishers Accepted It.

For days the *New York Times*, with a bygone touch of merry populism, played up the complexity of the theory as an affront to common sense. "This news is distinctly shocking, and apprehensions for confidence even in the multiplication table will arise," it editorialized on November 11. The idea that "space has limits" was most assuredly silly, the paper decided. "It just doesn't, by definition, and that's the end of it—for common folk, however it may be for higher mathematicians." It returned to the theme five days later: "Scientists who proclaim that space comes to an end somewhere are under some obligation to tell us what lies beyond it."

Finally, a week after its first story, the paper decided that some words of calm, more amused than bemused, might be useful. "British scientists seem to have been seized with something like an intellectual panic when they heard of photographic verification of the Einstein theory," the paper pointed out, "but they are slowly recovering as they realize that the sun still rises—apparently—in the east and will continue to do so for some time to come."[6]

An intrepid correspondent for the newspaper in Berlin was able to get an interview with Einstein in his apartment on December 2, and in the process launched one of the apocryphal tales about relativity. After describing Einstein's top-floor study, the reporter asserted, "It was from this lofty library that he observed years ago a man dropping from a neighboring roof—luckily on a pile of soft rubbish—and escaping almost without injury. The man told Dr. Einstein that in falling he experienced no sensation commonly considered as the effect of gravity." That was how, the article said, Einstein developed a "sublimation or supplement" of Newton's law of gravity. As one of the stacked headlines of the article put it, "Inspired as Newton Was, But by the Fall of a Man from a Roof Instead of the Fall of an Apple."[7]

This was, in fact, as the newspaper would say, "a pile of soft rubbish." Einstein had done his thought experiment while working in the Bern patent office in 1907, not in Berlin, and it had not involved a person actually falling. "The newspaper drivel about me is pathetic," he wrote Zangger when the article came out. But he understood, and accepted, how journalism worked. "This kind of exaggeration meets a certain need among the public."[8]

There was, indeed, an astonishing public craving to understand relativity. Why? The theory seemed somewhat baffling, yes, but also very enticing in its mystery. Warped space? The bending of light rays? Time and space not absolute? The theory had the wondrous mix of *Huh?* and *Wow!* that can capture the public imagination.

This was lampooned in a Rea Irvin cartoon in the *New Yorker*, which showed a baffled janitor, fur-clad matron, doorman, kids, and others scratching their heads with wild surmise as they wandered down the street. The caption was a quote from Einstein: "People slowly accustomed themselves to the idea that the physical states of space itself were the final physical reality." As Einstein put it to Grossmann, "Now every coachman and waiter argues about whether or not relativity theory is correct."[9]

Einstein's friends found themselves besieged whenever they lectured on it. Leopold Infeld, who later worked with Einstein, was then a young schoolteacher in a small Polish town. "At the time, I did what hundreds of others did all over the world," he recalled. "I gave a public lecture on the theory of relativity, and the crowd that lined up on a cold winter night was so great that it could not be accommodated in the largest hall in town."[10]

The same thing happened to Eddington when he spoke at Trinity College, Cambridge. Hundreds jammed the hall, and hundreds more were turned away. In his attempt to make the subject comprehensible, Eddington said that if he was traveling at nearly the speed of light he would be only three feet tall. That made newspaper headlines. Lorentz likewise gave a speech to an overflow audience. He compared the earth to a moving vehicle as a way to illustrate some examples of relativity.[11]

Soon many of the greatest physicists and thinkers began writing their own books explaining the theory, including Eddington, von Laue, Freundlich, Lorentz, Planck, Born, Pauli, and even the philosopher and mathematician Bertrand Russell. In all, more than six hundred books and articles on relativity were published in the first six years after the eclipse observations.

Einstein himself had the opportunity to explain it in his own words in *The Times* of London, which commissioned him to write an article called "What Is the Theory of Relativity?"[12] The result was actually quite comprehensible. His own popular book on the subject, *Relativity: The Special and General Theory*, had first appeared in German in 1916. Now, in the wake of the eclipse observation, Einstein published it in English as well. Filled with many thought experiments that could be easily visualized, it became a best seller, with updated editions appearing over the ensuing years.

The Publicity Paradox

Einstein had just the right ingredients to be transformed into a star. Reporters, knowing that the public was yearning for a refreshing international celebrity, were thrilled that the newly discovered genius was not a drab or reserved academic. Instead, he was a charming 40-year-old, just passing from handsome to distinctive, with a wild burst of hair, rumpled informality, twinkling eyes, and a willingness to dispense wisdom in bite-sized quips and quotes.

His friend Paul Ehrenfest found the press attention rather ridiculous. "The startled newspaper ducks flutter up in a hefty bout of quacking," he joked. To Einstein's sister, Maja, who grew up at a time before people actually liked publicity, the attention was astonishing, and she assumed that he found it completely distasteful. "An article was published about you in a Lucerne paper!" she marveled, not fully appreciating that he had made front pages around the world. "I imagine this causes you much unpleasantness that so much is being written about you." [13]

Einstein indeed bemoaned his newfound fame, repeatedly. He was being "hounded by the press and other riff-raff," he complained to Max Born. "It's so dreadful that I can barely breathe anymore, not to mention getting around to any sensible work." To another friend, he painted an even more vivid picture of the perils of publicity: "Since the flood of newspaper articles, I've been so deluged with questions, invitations, and requests that I dream I'm burning in Hell and the postman is the Devil eternally roaring at me, hurling new bundles of letters at my head because I have not yet answered the old ones." [14]

Einstein's aversion to publicity, however, existed a bit more in theory than in reality. It would have been possible, indeed easy, for him to have shunned all interviews, pronouncements, pictures, and public appearances. Those who truly dislike the public spotlight do not turn up, as the Einsteins eventually would, with Charlie Chaplin on a red carpet at one of his movie premieres.

"There was a streak in him that enjoyed the photographers and the crowds," the essayist C. P. Snow said after getting to know him. "He had an element of the exhibitionist and the ham. If there had not been that element, there would have been no photographers and no crowds. Nothing is easier to avoid than publicity. If one genuinely doesn't want it, one doesn't get it." [15]

Einstein's response to adulation was as complex as that of the cosmos to gravity. He was attracted and repelled by the cameras, loved publicity and loved to complain about it. His love-hate relationship with fame and reporters might seem unusual until one reflects on how similar it was to the mix of enjoyment, amusement, aversion, and annoyance that so many other famous people have felt.

One reason that Einstein—unlike Planck or Lorentz or Bohr—became such an icon was because he looked the part and because he could, and would, play the role. "Scientists who become icons must not only be geniuses but also performers, playing to the crowd and enjoying public acclaim," the physicist Freeman Dyson (no relation to the Astronomer Royal) has noted. [16] Einstein performed. He gave interviews readily, peppered them with delightful aphorisms, and knew exactly what made for a good story.

Even Elsa, or perhaps *especially* Elsa, enjoyed the attention. She served as her husband's protector, fearsome in her bark and withering in her near-sighted gaze when unwanted intruders barged into his orbit. But even more than her husband, she reveled in the stature and deference that came with fame. She began charging a fee to photograph him, and she donated the money to charities that fed hungry children in Vienna and elsewhere. [17]

In the current celebrity-soaked age, it is hard to recall the extent to which, a century ago, proper people recoiled from publicity and disdained those who garnered it. Especially in the realm of science, focusing on the personal seemed discordant. When Einstein's friend Max Born published a book on relativity right after the eclipse observations, he included, in his first edition, a frontispiece picture of Einstein and a short biography of him. Max von Laue and other friends of both men were appalled. Such things did not belong in a scientific book, even a popular one, von Laue wrote Born. Chastened, Born left these elements out of the next edition. [18]

As a result, Born was dismayed when it was announced in 1920 that Einstein had cooperated on a forthcoming biography by a Jewish jour-

nalist, Alexander Moszkowski, who had mainly written humor and oc-
cult books. The book advertised itself, in the title, as being based on
conversations with Einstein, and in fact it was. During the war, the
gregarious Moszkowski had befriended Einstein, been solicitous of his
needs, and brought him into a semiliterary circle that hung around at a
Berlin café.

Born was a nonpracticing Jew eager to assimilate into German society, and he feared that the book would stoke the simmering anti-
Semitism. "Einstein's theories had been stamped as 'Jewish physics' by
colleagues," Born recalled, referring to the growing number of German
nationalists who had begun decrying the abstract nature and supposed
moral "relativism" inherent in Einstein's theories. "And now a Jewish
author, who had already published several books with frivolous titles,
came along and wanted to write a similar book on Einstein." So Born
and his wife, Hedwig, who never shied from berating Einstein,
launched a crusade with their friends to stop its publication.

"You must withdraw permission," Hedwig hectored, "at once and
by registered letter." She warned him that the "gutter press" would use
it to tarnish his image and portray him as a self-promoting Jew. "A
completely new and far worse wave of persecution will be unleashed."
The sin, she emphasized, was not what he said but the fact that he was
permitting any publicity for himself:

> If I did not know you well, I would certainly not concede innocent mo-
> tives under these circumstances. I would put it down to vanity. This
> book will constitute your moral death sentence for all but four or five of
> your friends. It could subsequently be the best *confirmation of the accusa-
> tion of self-advertisement.*[19]

Her husband weighed in a week later with a warning that all
of Einstein's anti-Semitic antagonists "will triumph" if he did not
block publication. "Your Jewish 'friends' [i.e., Moszkowski] will have
achieved what a pack of anti-Semites have failed to do."

If Moszkowski refused to back off, Born advised Einstein to get a
restraining order from the public prosecutor's office. "Make sure this is
reported in the newspapers," he said. "I shall send you the details of

where to apply." Like many of their friends, Born worried that Elsa was the one who was more susceptible to the lures of publicity. As he told Einstein, "In these matters you are a little child. We all love you, and you must obey judicious people (not your wife)."[20]

Einstein took the advice of his friends, up to a point, by sending Moszkowski a registered letter demanding that his "splendid" work not appear in print. But when Moszkowski refused to back down, Einstein did not invoke legal measures. Both Ehrenfest and Lorentz agreed that going to court would serve only to inflame the issue and make matters worse, but Born disagreed. "You can flee to Holland," he said, referring to the ongoing effort by Ehrenfest and Lorentz to lure him there, but his Jewish friends who remained in Germany "would be affected by the stench."[21]

Einstein's detachment allowed him to affect an air of amusement rather than anxiety. "The whole affair is a matter of indifference to me, as is all the commotion, and the opinion of *each and every* human being," he said. "I will live through all that is in store for me like an unconcerned spectator."[22]

When the book came out, it made Einstein an easier target for anti-Semites, who used it to bolster their contention that he was a self-promoter trying to turn his science into a business.[23] But it did not cause much of a public commotion. There were, as Einstein noted to Born, no "earth tremors."

In retrospect, the controversy over publicity seems quaint and the book harmless fluff. "I have browsed through it a little, and find it not quite as bad as I had expected," Born later admitted. "It contains many rather amusing stories and anecdotes which are characteristic of Einstein."[24]

Einstein was able to resist letting his fame destroy his simple approach to life. On an overnight trip to Prague, he was afraid that dignitaries or curiosity-seekers would want to celebrate him, so he decided to stay with his friend Philipp Frank and his wife. The problem was that they actually lived in Frank's office suite at the physics laboratory, where Einstein had once worked himself. So Einstein slept on the sofa there. "This was probably not good enough for such a famous man,"

Frank recalled, "but it suited his liking for simple living habits and situations that contravened social conventions."

Einstein insisted that, on the way back from a coffeehouse, they buy food for dinner so that Frank's wife need not go shopping. They chose some calf's liver, which Mrs. Frank proceeded to cook on the Bunsen burner in the office laboratory. Suddenly Einstein jumped up. "What are you doing?" he demanded. "Are you boiling the liver in water?" Mrs. Frank allowed that was indeed what she was doing. "The boiling-point of water is too low," Einstein declared. "You must use a substance with a higher boiling-point such as butter or fat." From then on, Mrs. Frank referred to the necessity of frying liver as "Einstein's theory."

After Einstein's lecture that evening, there was a small reception given by the physics department at which several effusive speeches were made. When it was Einstein's turn to respond, he instead declared, "It will perhaps be pleasanter and more understandable if instead of making a speech I play a piece for you on the violin." He proceeded to perform a sonata by Mozart with, according to Frank, "his simple, precise and therefore doubly moving manner."

The next morning, before he could depart, a young man tracked him down at Frank's office and insisted on showing him a manuscript. On the basis of his $E=mc^2$ equation, the man insisted, it would be possible "to use the energy contained within the atom for the production of frightening explosives." Einstein brushed away the discussion, calling the concept foolish.[25]

From Prague, Einstein took the train to Vienna, where three thousand scientists and excited onlookers were waiting to hear him speak. At the station, his host waited for him to disembark from the first-class car but didn't find him. He looked to the second-class car down the platform, and could not find him there either. Finally, strolling from the third-class car at the far end of the platform was Einstein, carrying his violin case like an itinerant musician. "You know, I like traveling first, but my face is becoming too well known," he told his host. "I am less bothered in third class."[26]

"With fame I become more and more stupid, which of course is a very common phenomenon," Einstein told Zangger.[27] But he soon developed a theory that his fame was, for all of its annoyances, at least a welcome sign of the priority that society placed on people like himself:

> The cult of individual personalities is always, in my view, unjustified . . . It strikes me as unfair, and even in bad taste, to select a few for boundless admiration, attributing superhuman powers of mind and character to them. This has been my fate, and the contrast between the popular estimate of my achievements and the reality is simply grotesque. This extraordinary state of affairs would be unbearable but for one great consoling thought: it is a welcome symptom in an age, which is commonly denounced as materialistic, that it makes heroes of men whose ambitions lie wholly in the intellectual and moral sphere.[28]

One problem with fame is that it can engender resentment. Especially in academic and scientific circles, self-promotion was regarded as a sin. There was a distaste for those who garnered personal publicity, a sentiment that may have been exacerbated by the fact that Einstein was a Jew.

In the piece explaining relativity that he had written for *The Times* of London, Einstein humorously hinted at the issues that could arise. "By an application of the theory of relativity, today in Germany I am called a German man of science, and in England I am represented as a Swiss Jew," he wrote. "If I come to be regarded as a bête noire, the descriptions will be reversed, and I shall become a Swiss Jew for the Germans and a German man of science for the English!"[29]

It was not entirely facetious. Just months after he became world famous, the latter phenomenon occurred. He was told that he was to be given the prestigious gold medal of Britain's Royal Astronomical Society at the beginning of 1920, but a rebellion by a chauvinistic group of English purists forced the honor to be withheld.[30] Far more ominously, a small but growing group in his native country soon began vocally portraying him as a Jew rather than as a German.

"Lone Traveler"

Einstein liked to cast himself as a loner. Although he had an infectious laugh, like the barking of a seal, it could sometimes be wounding rather than warm. He loved being in a group playing music, discussing ideas, drinking strong coffee, and smoking pungent cigars. Yet there was a faintly visible wall that separated him from even family and close friends.[31] Starting with the Olympia Academy, he frequented many parlors of the mind. But he shied away from the inner chambers of the heart.

He did not like to be constricted, and he could be cold to members of his family. Yet he loved the collegiality of intellectual companions, and he had friendships that lasted throughout his life. He was sweet toward people of all ages and classes who floated into his ken, got along well with staffers and colleagues, and tended to be genial toward humanity in general. As long as someone put no strong demands or emotional burdens on him, Einstein could readily forge friendships and even affections.

This mix of coldness and warmth produced in Einstein a wry detachment as he floated through the human aspects of his world. "My passionate sense of social justice and social responsibility has always contrasted oddly with my pronounced lack of need for direct contact with other human beings and communities," he reflected. "I am truly a 'lone traveler' and have never belonged to my country, my home, my friends, or even my immediate family, with my whole heart; in the face of all these ties, I have never lost a sense of distance and a need for solitude."[32]

Even his scientific colleagues marveled at the disconnect between the genial smiles he bestowed on humanity in general and the detachment he displayed to the people close to him. "I do not know anyone as lonely and detached as Einstein," said his collaborator Leopold Infeld. "His heart never bleeds, and he moves through life with mild enjoyment and emotional indifference. His extreme kindness and decency are thoroughly impersonal and seem to come from another planet."[33]

Max Born, another personal and professional friend, noted the same trait, and it seemed to explain Einstein's ability to remain somewhat oblivious to the tribulations afflicting Europe during World War I. "For all his kindness, sociability and love of humanity, he was nevertheless totally detached from his environment and the human beings in it."[34]

Einstein's personal detachment and scientific creativity seemed to be subtly linked. According to his colleague Abraham Pais, this detachment sprang from Einstein's salient trait of "apartness," which led him to reject scientific conventional wisdom as well as emotional intimacies. It is easier to be a nonconformist and rebel, both in science and in a militaristic culture like Germany's, when you can detach yourself easily from others. "The detachment enabled him to walk through life immersed in thought," Pais said. It also allowed him—or compelled him—to pursue his theories in both a "single-minded and single-handed" manner.[35]

Einstein understood the conflicting forces in his own soul, and he seemed to think it was true for all people. "Man is, at one and the same time, a solitary being and a social being," he said.[36] His own desire for detachment conflicted with his desire for companionship, mirroring the struggle between his attraction and his aversion to fame. Using the jargon of psychoanalysis, the pioneering therapist Erik Erikson once pronounced of Einstein, "A certain alternation of isolation and outgoingness seems to have retained the character of a dynamic polarization."[37]

Einstein's desire for detachment was reflected in his extramarital relationships. As long as women did not make any claims on him and he felt free to approach them or not according to his own moods, he was able to sustain a romance. But the fear that he might have to surrender some of his independence led him to erect a shield.[38]

This was even more evident in his relationship with his family. He was not always merely cold, for there were times, especially when it came to Mileva Marić, that the forces of both attraction and repulsion raged inside him with a fiery heat. His problem, especially with his family, was that he was resistant to such strong feelings in others. "He had no gift for empathy," writes historian Thomas Levenson, "no ability to imagine himself into the emotional life of anyone else."[39] When confronted with the emotional needs of others, Einstein tended to retreat into the objectivity of his science.

The collapse of the German currency had caused him to urge Marić to move there, since it had become hard for him to afford her cost of living in Switzerland using depreciated German marks. But once the eclipse observations made him famous and more financially secure, he was willing to let his family stay in Zurich.

To support them, he had the fees from his European lecture trips sent directly to Ehrenfest in Holland, so that the money would not be converted into Germany's sinking currency. Einstein wrote Ehrenfest cryptic letters referring to his hard currency reserves as "results which you and I obtained here on Au ions" (i.e., gold).[40] The money was then disbursed by Ehrenfest to Marić and the children.

Shortly after his remarriage, Einstein visited Zurich to see his sons. Hans Albert, then 15, announced that he had decided to become an engineer.

"I think it's a disgusting idea," said Einstein, whose father and uncle had been engineers.

"I'm still going to become an engineer," replied the boy.

Einstein stormed away angry, and once again their relationship deteriorated, especially after he received a nasty letter from Hans Albert. "He wrote me as no decent person has ever written their father," he explained in a pained letter to his other son, Eduard. "It's doubtful I'll ever be able to take up a relationship with him again."[41]

But Marić by then was intent on improving rather than undermining his relationship with his sons. So she emphasized to the boys that Einstein was "a strange man in many ways," but he was still their father and wanted their love. He could be cold, she said, but also "good and kind." According to an account provided by Hans Albert, "Mileva knew that for all his bluff, Albert could be hurt in personal matters—and hurt deeply."[42]

By later that year, Einstein and his older son were again corresponding regularly about everything from politics to science. He also expressed his appreciation to Marić, joking that she should be happier now that she did not have to put up with him. "I plan on coming to Zurich soon, and we should put all the bad things behind us. You should enjoy what life has given you—like the wonderful children, the house, and that you are not married to me anymore."[43]

Hans Albert went on to enroll at his parents' alma mater, the Zurich Polytechnic, and became an engineer. He took a job at a steel company and then as a research assistant at the Polytechnic, studying hydraulics and rivers. Especially after he scored first in his exams, his father not only became reconciled, but proud. "My Albert has become a sound, strong chap," Einstein wrote Besso in 1924. "He is a total picture of a man, a first-rate sailor, unpretentious and dependable."

Einstein eventually said the same to Hans Albert, adding that he may have been right to become an engineer. "Science is a difficult profession," he wrote. "Sometimes I am glad that you have chosen a practical field, where one does not have to look for a four-leaf clover." [44]

One person who elicited strong and sustained personal emotions in Einstein was his mother. Dying from stomach cancer, she had moved in with him and Elsa at the end of 1919, and watching her suffer overwhelmed whatever human detachment he usually felt or feigned. When she died in February 1920, Einstein was exhausted by the emotions. "One feels right into one's bones what ties of blood mean," he wrote Zangger. Käthe Freundlich had heard him boast to her husband, the astronomer, that no death would affect him, and she was relieved that his mother's death proved that untrue. "Einstein wept like other men," she said, "and I knew that he could really care for someone." [45]

The Ripples from Relativity

For nearly three centuries, the mechanical universe of Isaac Newton, based on absolute certainties and laws, had formed the psychological foundation of the Enlightenment and the social order, with a belief in causes and effects, order, even *duty*. Now came a view of the universe, known as relativity, in which space and time were dependent on frames of reference. This apparent dismissal of certainties, an abandonment of faith in the absolute, seemed vaguely heretical to some people, perhaps even godless. "It formed a knife," historian Paul Johnson wrote in his sweeping history of the twentieth century, *Modern Times*, "to help cut society adrift from its traditional moorings." [46]

The horrors of the great war, the breakdown of social hierarchies, the advent of relativity and its apparent undermining of classical physics all seemed to combine to produce uncertainty. "For some years past, the entire world has been in a state of unrest, mental as well as physical," a Columbia University astronomer, Charles Poor, told the

New York Times the week after the confirmation of Einstein's theory was announced. "It may well be that the physical aspects of the unrest, the war, the strikes, the Bolshevist uprisings, are in reality the visible objects of some underlying deeper disturbance, worldwide in character. This same spirit of unrest has invaded science."[47]

Indirectly, driven by popular misunderstandings rather than a fealty to Einstein's thinking, *relativity* became associated with a new *relativism* in morality and art and politics. There was less faith in absolutes, not only of time and space, but also of truth and morality. In a December 1919 editorial about Einstein's relativity theory, titled "Assaulting the Absolute," the *New York Times* fretted that "the foundations of all human thought have been undermined."[48]

Einstein would have been, and later was, appalled at the conflation of relativity with relativism. As noted, he had considered calling his theory "invariance," because the physical laws of combined spacetime, according to his theory, were indeed invariant rather than relative.

Moreover, he was not a relativist in his own morality or even in his taste. "The word relativity has been widely misinterpreted as relativism, the denial of, or doubt about, the objectivity of truth or moral values," the philosopher Isaiah Berlin later lamented. "This was the opposite of what Einstein believed. He was a man of simple and absolute moral convictions, which were expressed in all he was and did."[49]

In both his science and his moral philosophy, Einstein was driven by a quest for certainty and deterministic laws. If his theory of relativity produced ripples that unsettled the realms of morality and culture, this was caused not by what Einstein believed but by how he was popularly interpreted.

One of those popular interpreters, for example, was the British statesman Lord Haldane, who fancied himself a philosopher and scientific scholar. In 1921, he published a book called *The Reign of Relativity,* which enlisted Einstein's theory to support his own political views on the need to avoid dogmatism in order to have a dynamic society. "Einstein's principle of the relativity of our measurements of space and time cannot be taken in isolation," he wrote. "When its import is considered it may well be found to have its counterpart in other domains of nature and of knowledge generally."[50]

Relativity theory would have profound consequences for theology, Haldane warned the archbishop of Canterbury, who immediately tried to comprehend the theory with only modest success. "The Archbishop," one minister reported to the dean of English science, J. J. Thomson, "can make neither head nor tail of Einstein, and protests that the more he listens to Haldane, and the more newspaper articles he reads on the subject, the less he understands."

Haldane persuaded Einstein to come to England in 1921. He and Elsa stayed at Haldane's grand London townhouse, where they found themselves completely intimidated by their assigned footman and butler. The dinner that Haldane hosted in Einstein's honor convened a pride of English intellectuals leonine enough to awe an Oxford senior common room. Among those present were George Bernard Shaw, Arthur Eddington, J. J. Thomson, Harold Laski, and of course the baffled archbishop of Canterbury, who got a personal briefing from Thomson in preparation.

Haldane seated the archbishop next to Einstein, so he got to pose his burning question directly to the source. What ramifications, His Grace inquired, did the theory of relativity have for religion?

The answer probably disappointed both the archbishop and their host. "None," Einstein said. "Relativity is a purely scientific matter and has nothing to do with religion."[51]

That was no doubt true. However, there was a more complex relationship between Einstein's theories and the whole witch's brew of ideas and emotions in the early twentieth century that bubbled up from the highly charged cauldron of modernism. In his novel *Balthazar*, Lawrence Durrell had his character declare, "The Relativity proposition was directly responsible for abstract painting, atonal music, and formless literature."

The relativity proposition, of course, was *not* directly responsible for any of this. Instead, its relationship with modernism was more mysteriously interactive. There are historical moments when an alignment of forces causes a shift in human outlook. It happened to art and philosophy and science at the beginning of the Renaissance, and again at the beginning of the Enlightenment. Now, in the early twentieth century, modernism was born by the breaking of the old strictures and verities. A spontaneous combustion occurred that included the works of Einstein, Picasso, Matisse, Stravinsky, Schoenberg, Joyce, Eliot, Proust, Diaghilev, Freud, Wittgenstein, and dozens of other path-breakers who seemed to break the bonds of classical thinking.[52]

In his book *Einstein, Picasso: Space, Time, and the Beauty That Causes Havoc,* the historian of science and philosophy Arthur I. Miller explored the common wellsprings that produced, for example, the 1905 special theory of relativity and Picasso's 1907 modernist masterpiece *Les Demoiselles d'Avignon.* Miller noted that both were men of great charm "yet who preferred emotional detachment." Each in his own way felt that something was amiss in the strictures that defined his field, and they were both intrigued by discussions of simultaneity, space, time, and specifically the writings of Poincaré.[53]

Einstein served as a source of inspiration for many of the modernist artists and thinkers, even when they did not understand him. This was especially true when artists celebrated such concepts as being "free from the order of time," as Proust put it in the closing of *Remembrance of Things Past.* "How I would love to speak to you about Einstein," Proust wrote to a physicist friend in 1921. "I do not understand a single word of his theories, not knowing algebra. [Nevertheless] it seems we have analogous ways of deforming Time."[54]

A pinnacle of the modernist revolution came in 1922, the year Einstein's Nobel Prize was announced. James Joyce's *Ulysses* was published that year, as was T. S. Eliot's *The Waste Land.* There was a midnight dinner party in May at the Majestic Hotel in Paris for the opening of *Renard,* composed by Stravinsky and performed by Diaghilev's *Ballets Russes.* Stravinsky and Diaghilev were both there, as was Picasso. So, too, were both Joyce and Proust, who "were destroying 19th century literary certainties as surely as Einstein was revolutionizing physics." The mechanical order and Newtonian laws that had defined classical physics, music, and art no longer ruled.[55]

Whatever the causes of the new relativism and modernism, the un-tethering of the world from its classical moorings would soon produce some unnerving reverberations and reactions. And nowhere was that mood more troubling than in Germany in the 1920s.

CHAPTER THIRTEEN

THE WANDERING ZIONIST

1920–1921

The motorcade in New York City, April 4, 1921

Kinship

In the article he wrote for *The Times* of London after the confirmation of his relativity theory, Einstein quipped that if things went bad the Germans would no longer consider him a compatriot but instead a Swiss Jew. It was a clever remark, made more so because Einstein knew, even then, that there was an odious smell of truth to it. That very week, in a letter to his friend Paul Ehrenfest, he described the mood in Germany. "Anti-Semitism is very strong here," he wrote. "Where is this all supposed to lead?"[1]

The rise of German anti-Semitism after World War I produced a counterreaction in Einstein: it made him identify more strongly with his Jewish heritage and community. At one extreme were German Jews such as Fritz Haber, who did everything they could, including converting to Christianity, to assimilate, and they urged Einstein to do the same. But Einstein took the opposite approach. Just when he was becoming famous, he embraced the Zionist cause. He did not officially join any Zionist organization, nor for that matter did he belong to or worship at any synagogue. But he cast his lot in favor of Jewish settlements in Palestine, a national identity among Jews everywhere, and the rejection of assimilationist desires.

He was recruited by the pioneering Zionist leader Kurt Blumenfeld, who paid a call on Einstein in Berlin in early 1919. "With extreme naïveté he asked questions," Blumenfeld recalled. Among Einstein's queries: With their spiritual and intellectual gifts, why should Jews be called on to create an agricultural nation-state? Wasn't nationalism the problem rather than the solution?

Eventually, Einstein came around to the cause. "I am, as a human being, an opponent of nationalism," he declared. "But as a Jew, I am from today a supporter of the Zionist effort."[2] He also became, more specifically, an advocate for the creation of a new Jewish university in Palestine, which eventually became Hebrew University in Jerusalem.

Once he decided to abandon the postulate that all forms of nationalism were bad, he found it easy to embrace Zionism with greater enthusiasm. "One can be an internationalist without being indifferent to members of one's tribe," he wrote a friend in October 1919. "The Zionist cause is very close to my heart . . . I am glad that there should be a little patch of earth on which our kindred brethren are not considered aliens."[3]

His support for Zionism put Einstein at odds with assimilationists. In April 1920, he was invited to address a meeting of one such group that emphasized its members' loyalty to Germany, the German Citizens of the Jewish Faith. He replied by accusing them of trying to separate themselves from the poorer and less polished eastern European Jews. "Can the 'Aryan' respect such pussyfooters?" he chided.[4]

Privately declining the invitation was not enough. Einstein also felt compelled to write a public attack on those who tried to fit in by talking "about religious faith instead of tribal affiliation."* In particular, he scorned what he called "the assimilatory" approach that sought "to overcome anti-Semitism by dropping nearly everything Jewish." This never worked; indeed, it "appears somewhat comical to a non-Jew," because the Jews are a people set apart from others. "The psychological root of anti-Semitism lies in the fact that the Jews are a group of people unto themselves," he wrote. "Their Jewishness is visible in their physical appearance, and one notices their Jewish heritage in their intellectual work."[5]

The Jews who practiced and preached assimilation tended to be those who took pride in their German or western European heritage. At the time (and through much of the twentieth century), they tended to look down on Jews from eastern Europe, such as Russia and Poland, who seemed less polished, refined, and assimilated. Although Einstein was German Jewish, he was appalled by those from his background who would "draw a sharp dividing line between eastern European Jews and western European Jews." The approach was doomed to backfire against all Jews, he argued, and it was not based on any true distinction. "Eastern European Jewry contains a rich potential of human talents and productive forces that can well stand the comparison to the higher civilization of western European Jews."[6]

Einstein was acutely aware, even more than the assimilationists, that anti-Semitism was not the result of rational causes. "In Germany today hatred of the Jews has taken on horrible expressions," he wrote in early 1920. Part of the problem was that inflation was out of control. The German mark had been worth about 12 cents at the beginning of 1919, which was half of its value from before the war but still manageable. But by the beginning of 1920, the mark was worth a mere 2 cents, and collapsing further each month.

In addition, the loss of the war had been humiliating. Germany had lost 6 million men and then was forced into surrendering land con-

---

* The word Einstein used was *Stammesgenossen*. Although *Stamm* generally means tribe, that translation can have some racial overtones. Some Einstein scholars have said that translations such as "kindred" or "clan" or "lineage" might be clearer.

taining half of its natural resources, plus all of its overseas colonies. Many proud Germans believed it must have been the result of betrayal. The Weimar Republic that had emerged after the war, though supported by liberals and pacifists and Jews such as Einstein, was disdained by much of the old order and even the middle class.

There was one group that could be easily cast as the alien and dark force most responsible for the humiliation facing a proud culture. "People need a scapegoat and make the Jews responsible," Einstein noted. "They are a target of instinctive resentment because they are of a different tribe."[7]

Weyland, Lenard, and the Antirelativists

The explosion of great art and ideas in Germany at the time, as Amos Elon wrote in his book *The Pity of It All*, was largely due to Jewish patrons and pioneers in a variety of fields. This was particularly true in science. As Sigmund Freud pointed out, part of the success of Jewish scientists was their "creative skepticism," which arose from their essential nature as outsiders.[8] What the Jewish assimilationists underestimated was the virulence with which many Germans, whom they considered to be their fellow countrymen, in fact saw them as essentially outsiders or, as Einstein put it, "a different tribe."

Einstein's first public collision with this anti-Semitism came in the summer of 1920. A shady German nationalist named Paul Weyland, an engineer by training, had turned himself into a polemicist with political aspirations. He was an active member of a right-wing nationalistic political party that pledged, in its 1920 official program, to "diminish the dominant Jewish influence showing up increasingly in government and in public."[9]

Weyland realized that Einstein, as a highly publicized Jew, had engendered resentment and jealousy. Likewise, his relativity theory was easy to turn into a target, because many people, including some scientists, were unnerved by the way it seemed to undermine absolutes and be built on abstract hypotheses rather than grounded in solid experiment. So Weyland published articles denouncing relativity as "a big hoax" and formed a ragtag (but mysteriously well-funded) organization grandly dubbed the Study Group of German Scientists for the Preservation of a Pure Science.

Joining with Weyland was an experimental physicist of modest reputation named Ernst Gehrcke, who for years had been assailing relativity with more vehemence than comprehension. Their group lobbed a few personal attacks at Einstein and the "Jewish nature" of relativity theory, then called a series of meetings around Germany, including a large rally at Berlin's Philharmonic Hall on August 24.

Weyland spoke first and, with the orotund rhetoric of a demagogue, accused Einstein of engaging in a "businesslike booming of his theory and his name." Einstein's penchant for publicity, wanted or not, was being used against him, as his assimilationist friends had warned. Relativity was a hoax, Weyland said, and plagiarized to boot. Gehrcke said much the same with a more technical gloss, reading from a written text. The meeting, reported the *New York Times,* "had a decidedly anti-Semitic complexion."[10]

In the middle of Gehrcke's talk, there arose from the audience a quiet murmur: *Einstein, Einstein.* He had come to see the circus and, averse neither to publicity nor controversy, laugh at the spectacle. As his friend Philipp Frank noted, "He always liked to regard events in the world around him as if he were a spectator in a theater." Sitting in the audience with his friend the chemist Walther Nernst, he cackled loudly at times and at the end pronounced the entire event "most amusing."[11]

But he was not truly amused, and he even briefly considered moving away from Berlin.[12] His anger aroused, he made the tactical mistake of responding with a highly charged diatribe that was published three days later on the front page of the *Berliner Tageblatt,* a liberal daily owned by Jewish friends. "I am well aware that the two speakers are unworthy of reply by my pen," he said, but then proceeded not to be restrained by that awareness. Gehrcke and Weyland had not been explicitly anti-Semitic, nor did they overtly criticize Jews in their speeches. But Einstein alleged that they would not have attacked his theory "if I were a German nationalist, with or without a swastika, instead of a Jew."[13]

Einstein spent most of his piece refuting Weyland and Gehrcke.

But he also attacked a more reputable physicist who was not at the meeting but had given support to the antirelativity cause: Philipp Lenard.

Winner of the 1905 Nobel Prize, Lenard had been a pioneer experimenter who described the photoelectric effect. Einstein had once admired him. "I have just read a wonderful paper by Lenard," Einstein had gushed to Marić back in 1901. "Under this beautiful piece I am filled with such happiness and joy that I absolutely must share some of it with you." After Einstein had published his first spate of seminal papers in 1905, citing Lenard by name in the one on light quanta, the two scientists had exchanged flattering letters.[14]

But as an ardent German nationalist, Lenard had become increasingly bitter about the British and the Jews, contemptuous of the publicity Einstein's theory was garnering, and vocal in his attacks on the "absurd" aspects of relativity. He had allowed his name to be used on brochures that were distributed at Weyland's meeting, and as a Nobel laureate he had worked behind the scenes to make sure that Einstein was not awarded the prize.

Because Lenard had refrained from showing up at the Philharmonic Hall rally, and because his published critiques of relativity had been academic in tone, Einstein did not need to attack him in his newspaper piece. But he did. "I admire Lenard as a master of experimental physics, but he has not yet produced anything outstanding in theoretical physics, and his objections to the general theory of relativity are of such superficiality that, up until now, I did not think it necessary to answer them," he wrote. "I intend to make up for this."[15]

Einstein's friends publicly supported him. A group that included von Laue and Nernst published a letter claiming, not altogether accurately, "Whoever is fortunate enough to be close to Einstein knows that he will never be surpassed in his . . . dislike of all publicity."[16]

Privately, however, his friends were appalled. He had been provoked into a display of public anger against those who should have remained unworthy of a reply by his pen, thus stirring up even more distasteful publicity. Max Born's wife, Hedwig, who had freely scolded Einstein about his treatment of his family, now lectured, "[You should] not have allowed yourself to be goaded into that rather unfortunate reply." He should show more respect, she said, for "the secluded temple of science." [17]

Paul Ehrenfest was even harsher. "My wife and I absolutely cannot believe that you yourself wrote some of the phrases in the article," he said. "If you really did write them down with your own hand, it proves that these damn pigs have finally succeeded in touching your soul. I urge you as strongly as I can not to throw one more word on this subject to that voracious beast, the public." [18]

Einstein was somewhat contrite. "Don't be too severe with me," he replied to the Borns. "Everyone must, from time to time, make a sacrifice on the altar of stupidity, to please the deity and mankind. And I did so thoroughly with my article." [19] But he made no apologies for flunking their standards of publicity avoidance. "I had to do this if I wanted to stay in Berlin, where every child recognizes me from photographs," he told Ehrenfest. "If one believes in democracy, then one must grant the public this much right as well." [20]

Not surprisingly, Lenard was outraged by Einstein's article. He insisted on an apology, as he had not even been part of the antirelativity rally. Arnold Sommerfeld, chairman of the German Physical Society, tried to mediate, and he urged Einstein "to write some conciliatory words to Lenard." [21] It was not to be. Einstein refused to back down, and Lenard ended up edging ever closer to being an outright anti-Semite and later a Nazi.

(There was one odd coda to this event. In 1953, according to declassified documents in Einstein's FBI file, a well-dressed German walked into the FBI field office in Miami and told the receptionist he had information that Einstein had admitted to being a communist in an article in *Berliner Tageblatt* in August 1920. The aspiring informer was none other than Paul Weyland, who had landed in Miami and was trying to emigrate after years of being a con man and swindler all over the world. J. Edgar Hoover's FBI was eagerly trying to prove, with no success, that Einstein was a communist, and took up the cause. After three months, the Bureau finally found the article and translated it. There was nothing about being a communist in it. Weyland was, nevertheless, granted American citizenship.) [22]

The public crossfire coming out of the antirelativity rally heightened interest in the upcoming annual meeting of German scientists, scheduled for late September in the spa town of Bad Nauheim. Both Einstein and Lenard were to attend, and Einstein had ended his newspaper response by proclaiming that, at his suggestion, a public discussion of relativity would occur there. "Anyone who can dare face a scientific forum can present his objections there," he said, tossing a gauntlet in Lenard's direction.

During the weeklong gathering in Bad Nauheim, Einstein stayed with Max Born in Frankfurt, twenty miles away, and the two men commuted to the resort town by train each day. The big showdown over relativity, at which both Einstein and Lenard were expected to participate, was on the afternoon of September 23. Einstein had forgotten to bring anything to write with, so he borrowed the pencil of the person next to him in order to take notes while Lenard talked.

Planck was in the chair, and by both his commanding presence and soothing words he was able to prevent any personal attacks. Lenard's objections to relativity were similar to those of many nontheorists. The theory was built on equations rather than observations, he said, and it "offends against the simple common sense of a scientist." Einstein replied that what "seems obvious" changes over time. That was true even of Galileo's mechanics.

It was the first time that Einstein and Lenard had met, but they did not shake hands or speak to each other. And though the official minutes of the meeting do not record it, Einstein apparently lost his equanimity at one point. "Einstein was provoked into making a caustic reply," Born recalled. And a few weeks later, Einstein wrote Born to assure him that he would "not allow myself to get excited again as in Nauheim."[23]

Finally, Planck was able to end the session, before any blood was drawn, with a limp joke. "Since the theory of relativity unfortunately has not so far been able to extend the absolute time available for this meeting," he said, " it must now be adjourned." The papers the next day were left without headlines, and the antirelativity movement subsided for the time being.[24]

As for Lenard, he distanced himself from the weird group of original antirelativists. "Unfortunately Weyland turned out to be a crook," he later said. But he did not let go of his own antipathy toward Einstein. After the Bad Nauheim meeting he became increasingly vitriolic and anti-Semitic in his attacks on Einstein and "Jewish science." He became a proponent of creating a "Deutsche Physik" that purged German physics of Jewish influences, which to him was exemplified by Einstein's relativity theory with its abstract, theoretical, and nonexperimental approach and its odor (at least to him) of a relativism that rejected absolutes, order, and certainties.

A few months later, at the beginning of January 1921, an obscure Munich party functionary picked up the theme. "Science, once our greatest pride, is today being taught by Hebrews," Adolf Hitler wrote in a newspaper polemic.[25] There were even ripples that made it across the Atlantic. That April, the *Dearborn Independent*, a weekly owned by automaker Henry Ford, a strong anti-Semite, blared a banner headline across the top of its front page. "Is Einstein a Plagiarist?" it accusingly asked.[26]

Einstein in America, 1921

Albert Einstein's exploding global fame and budding Zionism came together in the spring of 1921 for an event that was unique in the history of science, and indeed remarkable for any realm: a grand two-month processional through the eastern and midwestern United States that evoked the sort of mass frenzy and press adulation that would thrill a touring rock star. The world had never before seen, and perhaps never will again, such a scientific celebrity superstar, one who also happened to be a gentle icon of humanist values and a living patron saint for Jews.

Einstein had initially thought that his first visit to America might be a way to make some money in a stable currency in order to provide for his family in Switzerland. "I have demanded $15,000 from Princeton and Wisconsin," he told Ehrenfest. "It will probably scare them off. But if they do bite, I will be buying economic independence for myself—and that's not a thing to sniff at."

The American universities did not bite. "My demands were too high," he reported back to Ehrenfest.[27] So by February 1921, he had made other plans for the spring: he would present a paper at the third Solvay Conference in Brussels and give some lectures in Leiden at the behest of Ehrenfest.

It was then that Kurt Blumenfeld, leader of the Zionist movement in Germany, came by Einstein's apartment once again. Exactly two years earlier, Blumenfeld had visited Einstein and enlisted his support for the cause of creating a Jewish homeland in Palestine. Now he was coming with an invitation—or perhaps an instruction—in the form of a telegram from the president of the World Zionist Organization, Chaim Weizmann.

Weizmann was a brilliant biochemist who had emigrated from Russia to England, where he helped his adopted nation in the First World War by coming up with a bacterial method for more efficiently manufacturing the explosive cordite. During that war he worked under former prime minister Arthur Balfour, who was then first lord of the Admiralty. He subsequently helped to persuade Balfour, after he became foreign secretary, to issue the famous 1917 declaration in which Britain pledged to support "the establishment in Palestine of a national home for the Jewish people."

Weizmann's telegram invited Einstein to accompany him on a trip to America to raise funds to help settle Palestine and, in particular, to create Hebrew University in Jerusalem. When Blumenfeld read it to him, Einstein initially balked. He was not an orator, he said, and the role of simply using his celebrity to draw crowds to the cause was "an unworthy one."

Blumenfeld did not argue. Instead, he simply read Weizmann's telegram aloud again. "He is the president of our organization," Blumenfeld said, "and if you take your conversion to Zionism seriously, then I have the right to ask you, in Dr. Weizmann's name, to go with him to the United States."

"What you say is right and convincing," Einstein replied, to the "boundless astonishment" of Blumenfeld. "I realize that I myself am now part of the situation and that I must accept the invitation."[28]

Einstein's reply was indeed a cause for astonishment. He was already committed to the Solvay Conference and other lectures in Europe, he professed to dislike the public spotlight, and his fragile stomach had made him reluctant to travel. He was not a faithful Jew, and his allergy to nationalism kept him from being a pure and unalloyed Zionist.

Yet now he was doing something that went against his nature: accepting an implied command from a figure of authority, one that was based on his perceived bonds and commitments to other people. Why?

Einstein's decision reflected a major transformation in his life. Until the completion and confirmation of his general theory of relativity, he had dedicated himself almost totally to science, to the exclusion even of his personal, familial, and societal relationships. But his time in Berlin had made him increasingly aware of his identity as a Jew. His reaction to the pervasive anti-Semitism was to feel even more connected—indeed, inextricably connected—to the culture and community of his people.

Thus in 1921, he made a leap not of faith but of commitment. "I am really doing whatever I can for the brothers of my race who are treated so badly everywhere," he wrote Maurice Solovine.[29] Next to his science, this would become his most important defining connection. As he would note near the end of his life, after declining the presidency of Israel, "My relationship to the Jewish people has become my strongest human tie."[30]

One person who was not only astonished but dismayed by Einstein's decision was his friend and colleague in Berlin, the chemist Fritz Haber, who had converted from Judaism and assiduously assimilated in order to appear a proper Prussian. Like other assimilationists, he was worried (understandably) that a visit by Einstein to the great wartime enemy at the behest of a Zionist organization would reinforce the belief that Jews had dual loyalties and were not good Germans.

In addition, Haber had been thrilled that Einstein was planning to attend the Solvay Conference in Brussels, the first since the war. No other Germans had been invited, and his attendance was seen as a crucial step for the return of Germany to the larger scientific community.

"People in this country will see this as evidence of the disloyalty of the Jews," Haber wrote when he heard of Einstein's decision to visit America. "You will certainly sacrifice the narrow basis upon which the existence of professors and students of the Jewish faith at German universities rests."[31]

Haber apparently had the letter delivered by hand, and Einstein replied the same day. He took issue with Haber's way of regarding Jews as being people "of the Jewish faith" and instead, once again, cast the identity as being inextricably a matter of ethnic kinship. "Despite my emphatic internationalist beliefs, I have always felt an obligation to stand up for my persecuted and morally oppressed tribal companions," he said. "The prospect of establishing a Jewish university fills me with particular joy, having recently seen countless instances of perfidious and uncharitable treatment of splendid young Jews with attempts to deny their chances of education."[32]

And so it was that the Einsteins sailed from Holland on March 21, 1921, for their first visit to America. To keep things unpretentious and inexpensive, Einstein had said he was willing to travel steerage. The request was not granted, and he was given a nice stateroom. He also asked that he and Elsa be given separate rooms, both aboard the ship and at the hotels, so that he could work while on the trip. That request was granted.

It was, by all accounts, a pleasant Atlantic crossing, during which Einstein tried to explain relativity to Weizmann. Asked upon their arrival whether he understood the theory, Weizmann gave a delightful reply: "During the crossing, Einstein explained his theory to me every day, and by the time we arrived I was fully convinced that he really understands it."[33]

When the ship pulled up to the Battery in lower Manhattan on the afternoon of April 2, Einstein was standing on the deck wearing a faded gray wool coat and a black felt hat that concealed some but not all of his now graying shock of hair. In one hand was a shiny briar pipe; the other clutched a worn violin case. "He looked like an artist," the *New York Times* reported. "But underneath his shaggy locks was a scientific mind whose deductions have staggered the ablest intellects of Europe."[34]

As soon as they were permitted, dozens of reporters and cameramen rushed aboard. The press officer of the Zionist organization told

Einstein that he would have to attend a press conference. "I can't do that," he protested. "It's like undressing in public."[35] But he could, of course, and did.

First he obediently followed directions for almost a half hour as the photographers and newsreel men ordered him and Elsa to strike a variety of poses. Then, in the captain's cabin, he displayed more joy than reluctance as he conducted his first press briefing with all the wit and charm of a merry big-city mayor. "One could tell from his chuckling," the reporter from the *Philadelphia Public Ledger* wrote, "that he enjoyed it."[36] His questioners enjoyed it as well. The whole performance, sprinkled with quips and pithy answers, showed why Einstein was destined to become such a wildly popular celebrity.

Speaking through an interpreter, Einstein began with a statement about his hope "to secure the support, both material and moral, of American Jewry for the Hebrew University of Jerusalem." But the reporters were more interested in relativity, and the first questioner requested a one-sentence description of the theory, a request that Einstein would face at almost every stop on his trip. "All of my life I have been trying to get it into one book," he replied, "and *he* wants me to get it into one sentence!" Pressed to try, he provided a simple overview: "It is a theory of space and time as far as physics is concerned, which leads to a theory of gravitation."

What about those, especially in Germany, who attacked his theory? "No one of knowledge opposes my theory," he answered. "Those physicists who do oppose the theory are animated by political motives."

What political motives? "Their attitude is largely due to anti-Semitism," he replied.

The interpreter finally called the session to a close. "Well, I hope I have passed my examination," Einstein concluded with a smile.

As they were leaving, Elsa was asked if she understood relativity. "Oh, no, although he has explained it to me many times," she replied. "But it is not necessary to my happiness."[37]

Thousands of spectators, along with the fife and drum corps of the Jewish Legion, were waiting in Battery Park when the mayor and other dignitaries brought Einstein ashore on a police tugboat. As blue-and-white flags were waved, the crowd sang the *Star-Spangled Banner* and then the Zionist anthem *Hatikvah*.

The Einsteins and Weizmanns intended to head directly to the Hotel Commodore in Midtown. Instead, their motorcade wound through the Jewish neighborhoods of the Lower East Side late into the evening. "Every car had its horn, and every horn was put in action," Weizmann recalled. "We reached the Commodore at about 11:30, tired, hungry, thirsty and completely dazed." [38]

The following day Einstein entertained a steady procession of visitors and, with what the *Times* called "an unusual impression of geniality," he even held another press gathering. Why, he was asked, had he attracted such an unprecedented explosion of public interest? He professed to being puzzled himself. Perhaps a psychologist could determine why people who generally did not care for science had taken such an interest in him. "It seems psycho-pathological," he said with a laugh. [39]

Weizmann and Einstein were officially welcomed later in the week at City Hall, where ten thousand excited spectators gathered in the park to hear the speeches. Weizmann got polite applause. But Einstein, who said nothing, got a "tumultuous greeting" when he was introduced. "As Dr. Einstein left," the New York *Evening Post* reported, "he was lifted onto the shoulders of his colleagues and into the automobile, which passed in triumphal procession through a mass of waving banners and a roar of cheering voices." [40]

One of Einstein's visitors at the Commodore Hotel was a German immigrant physician named Max Talmey, whose name had been Max Talmud back when he was a poor student in Munich. This was the family friend who had first exposed the young Einstein to math and philosophy, and he was unsure whether the now famous scientist would remember him.

Einstein did. "He had not seen me or corresponded with me for nineteen years," Talmey later noted. "Yet as soon as I entered his room in the hotel, he exclaimed: 'You distinguish yourself through eternal youth!' " [41] They chatted about their days in Munich and their paths since. Einstein invited Talmey back various times during the course of his visit, and before he left even went to Talmey's apartment to meet his young daughters.

Even though he spoke in German about abstruse theories or stood silent as Weizmann tried to cajole money for Jewish settlements in Palestine, Einstein drew packed crowds wherever he went in New York. "Every seat in the Metropolitan Opera House, from the pit to the last row under the roof, was filled, and hundreds stood," reported the *Times* one day. About another lecture that week it likewise reported, "He spoke in German, but those anxious to see and hear the man who has contributed a new theory of space and time and motion to scientific conceptions of the universe filled every seat and stood in the aisles."[42]

After three weeks of lectures and receptions in New York, Einstein paid a visit to Washington. For reasons fathomable only by those who live in that capital, the Senate decided to debate the theory of relativity. Among the leaders asserting that it was incomprehensible were Pennsylvania Republican Boies Penrose, famous for once uttering that "public office is the last refuge of a scoundrel," and Mississippi Democrat John Sharp Williams, who retired a year later, saying, "I'd rather be a dog and bay at the moon than stay in the Senate another six years."

On the House side of the Capitol, Representative J. J. Kindred of New York proposed placing an explanation of Einstein's theories in the *Congressional Record*. David Walsh of Massachusetts rose to object. Did Kindred understand the theory? "I have been earnestly busy with this theory for three weeks," he replied, "and am beginning to see some light." But what relevance, he was asked, did it have to the business of Congress? "It may bear upon the legislation of the future as to general relations with the cosmos."

Such discourse made it inevitable that, when Einstein went with a group to the White House on April 25, President Warren G. Harding would be faced with the question of whether *he* understood relativity. As the group posed for cameras, President Harding smiled and confessed that he did not comprehend the theory at all. The *Washington Post* carried a cartoon showing him puzzling over a paper titled "Theory of Relativity" while Einstein puzzled over one on the "Theory of Normalcy," which was the name Harding gave to his governing philosophy. The *New York Times* ran a page 1 headline: "Einstein Idea Puzzles Harding, He Admits."

At a reception in the National Academy of Sciences on Constitution Avenue (which now boasts the world's most interesting statue of Einstein, a twelve-foot-high full-length bronze figure of him reclining),[43] he listened to long speeches from various honorees, including Prince Albert I of Monaco, who was an avid oceanographer, a North Carolina scholar of hookworms, and a man who had invented a solar stove. As the evening droned on, Einstein turned to a Dutch diplomat seated next to him and said, "I've just developed a new theory of eternity."[44]

By the time Einstein reached Chicago, where he gave three lectures and played violin at a dinner party, he had become more adept at answering irksome questions, particularly the most frequent one, which was sparked by the fanciful *New York Times* headline after the 1919 eclipse that only twelve people could understand his theory.

"Is it true only twelve great minds can understand your theory?" the reporter from the *Chicago Herald and Examiner* asked.

"No, no," Einstein replied with a smile. "I think the majority of scientists who have studied it can understand it."

He then proceeded to try to explain it to the reporter by using his metaphor about how the universe would look to a two-dimensional creature who spent its life moving on a surface of what turned out to be a globe. "It could travel for millions of years and would always return to its starting point," said Einstein. "It would never be conscious of what was above it or beneath it."

The reporter, being a good Chicago newspaperman, was able to spin a glorious tale, written in the third person, about the depths of his own confusion. "When the reporter came to he was vainly trying to light a three-dimensional cigarette with a three-dimensional match," the story concluded. "It began to trickle into his brain that the two-dimensional organism referred to was himself, and far from being the 13th Great Mind to comprehend the theory he was condemned henceforth to be one of the Vast Majority who live on Main Street and ride in Fords."[45]

When a reporter from the rival *Tribune* asked him the same question about only twelve people being able to understand his theory, Einstein again denied it. "Everywhere I go, someone asks me that question," he said. "It's absurd. Anyone who has had sufficient training in science can readily understand the theory." But this time Einstein made no attempt to explain it, nor did the reporter. "The *Tribune* regrets to inform its readers that it will be unable to present to them Einstein's theory of relativity," the article began. "After the professor explained that the most incidental discussion of the question would take from three to four hours, it was decided to confine the interview to other things." [46]

Einstein went on to Princeton, where he delivered a weeklong series of scientific lectures and received an honorary degree "for voyaging through strange seas of thought." Not only did he get a nice fee for the lectures (though apparently not the $15,000 he had originally sought), he also negotiated a deal while there that Princeton could publish his lectures as a book from which he would get a 15 percent royalty. [47]

At the behest of Princeton's president, all of Einstein's lectures were very technical. They included more than 125 complex equations that he scribbled on the blackboard while speaking in German. As one student admitted to a reporter, "I sat in the balcony, but he talked right over my head anyway." [48]

At a party following one of these lectures, Einstein uttered one of his most memorable and self-revealing quotes. Someone excitedly informed him that word had just arrived of a new set of experiments improving on the Michelson-Morley technique that seemed to show that the ether existed and the speed of light was variable. Einstein simply refused to accept it. He knew that his theory was correct. And so he calmly responded, "Subtle is the Lord, but malicious he is not."*

The mathematics professor Oswald Veblen, who was standing there, heard the remark and, when a new math building was built a decade later, asked Einstein for the right to carve the words on the stone mantel of the fireplace in the common room. Einstein happily sent back his approval and further explained to Veblen what he had meant:

---

* I have used the translation preferred by Abraham Pais. Einstein's words in German were, "Raffiniert ist der Herr Gott, aber boshaft ist er nicht."

"Nature hides her secret because of her essential loftiness, but not by means of ruse."[49]

The building, neatly enough, later became the temporary home of the Institute for Advanced Study, and Einstein would have an office there when he immigrated to Princeton in 1933. Near the end of his life, he was in front of the fireplace at a retirement party for the mathematician Hermann Weyl, a friend who had followed him from Germany to Princeton when the Nazis took power. Alluding to his frustration with the uncertainties of quantum mechanics, Einstein nodded to the quote and lamented to Weyl, "Who knows, perhaps He *is* a little malicious."[50]

Einstein seemed to like Princeton. "Young and fresh," he called it. "A pipe as yet unsmoked."[51] For a man who was invariably fondling new briar pipes, this was a compliment. It would not be a surprise, a dozen years hence, that he would decide to move there permanently.

Harvard, where Einstein went next, did not endear itself quite as well. Perhaps it was because Princeton President John Hibben had introduced him in German, whereas Harvard President A. Lawrence Lowell spoke to him in French. In addition, Harvard had invited Einstein to visit, but it did not invite him to give lectures.

Some charged that this slight was due to the influence of a rival Zionist group in America led by Louis Brandeis, a graduate of Harvard Law School, who had become the first Jewish Supreme Court justice. The allegation was so widespread that Brandeis's protégé Felix Frankfurter had to issue a public denial. That prompted an amused letter about the perils of assimilationism from Einstein to Frankfurter. It was "a Jewish weakness," he wrote, "always and eagerly to try to keep the Gentiles in good humor."[52]

The very assimilated Brandeis, who had been born in Kentucky and had turned himself into a proper Bostonian, was an example of the Jews from Germany whose families had arrived in the nineteenth century and tended to look down on the more recent immigrants from eastern Europe and Russia. For both political and personal reasons, Brandeis had clashed with Weizmann, a Russian Jew who had a more assertive and political approach toward Zionism.[53] The enthusiastic crowds that greeted Einstein and Weizmann on their trip were mainly made up of the eastern European Jews, while Brandeis and his ilk remained more aloof.

Most of Einstein's time during the two days he spent in Boston was devoted to appearances, rallies, and dinners (including a kosher banquet for five hundred) with Weizmann to drum up contributions for their Zionist cause. The *Boston Herald* reported on the reaction at one fund-raising event at a synagogue in Roxbury:

> The response was electrifying. Young girl ushers worked their way with difficulty through the crowded aisles, carrying long boxes. Bills of various denominations were rained into these receptacles. A prominent Jewess cried out ecstatically that she had eight sons who had been in the army and wanted to make some donation in proportion to their sacrifices. She held up her watch, a valuable imported timepiece, and slipped the rings from her hands. Others followed her example, and soon baskets and boxes filled with diamonds and other precious ornaments.[54]

While in Boston, Einstein was subjected to a pop quiz known as the Edison test. The inventor Thomas Edison was a practical man, getting crankier with age (he was then 74), who disparaged American colleges as too theoretical and felt the same about Einstein. He had devised a test he gave job applicants that, depending on the position being sought, included about 150 factual questions. How is leather tanned? What country consumes the most tea? What was Gutenberg's type made of?*

The *Times* called it "the ever-present Edison questionnaire controversy," and of course Einstein ran into it. A reporter asked him a question from the test. "What is the speed of sound?" If anyone understood the propagation of sound waves, it was Einstein. But he admitted that he did not "carry such information in my mind since it is readily available in books." Then he made a larger point designed to disparage Edison's view of education. "The value of a college education is not the learning of many facts but the training of the mind to think," he said.[55]

One remarkable feature of most stops on Einstein's grand tour was

---

* Governor Channing Cox had been thrust a version of the test earlier that week, and his first three responses were: Where does shellac come from? "From a can." What is a monsoon? "A funny-sounding word." Where do we get prunes? "Breakfast."

a noisy parade, which was rather unusual for a theoretical physicist. In Hartford, Connecticut, for example, the procession included more than a hundred automobiles headed by a band, a coterie of war veterans, and standard-bearers with the American and Zionist flags. More than fifteen thousand spectators lined the route. "North Main Street was jammed by crowds that struggled to get close to shake hands," the newspaper reported. "The crowds cheered wildly as Dr. Weizmann and Prof. Einstein stood up in the car to receive flowers." [56]

It was an astonishing scene, but it was exceeded in Cleveland. Several thousands thronged Union train depot to meet the visiting delegation, and the parade included two hundred honking and flag-draped cars. Einstein and Weizmann rode in an open car, preceded by a National Guard marching band and a cadre of Jewish war veterans in uniform. Admirers along the way grabbed on to Einstein's car and jumped on the running board, while police tried to pull them away. [57]

While in Cleveland, Einstein spoke at the Case School of Applied Science (now Case Western Reserve), where the famous Michelson-Morley experiments had been conducted. There he met privately, for more than an hour, with Professor Dayton Miller, whose new version of that experiment had provoked Einstein's skeptical response at the Princeton cocktail party. Einstein drew sketches of Miller's ether-drift models and urged him to continue refining his experiments. Miller remained dubious about relativity and partial to the ether, but other experiments eventually affirmed Einstein's faith that the Lord was indeed more subtle than malicious. [58]

The excitement, public outpouring, and dizzying superstar status conferred upon Einstein were unprecedented. But in financial terms, the tour was only a modest success for the Zionist movement. The poorer Jews and recent immigrants had poured out to see him and donated with enthusiasm. But few of the eminent and old-line Jews with great personal fortunes became part of the frenzy. They were, on the whole, more assimilated and less ardently Zionist. Weizmann had hoped to raise at least $4 million. By the end of the year, only $750,000 had actually been collected. [59]

Even after his trip to America, Einstein did not become a full-fledged member of the Zionist movement. He supported the general idea of Jewish settlements in Palestine, and especially Hebrew University in Jerusalem, but he never had a desire to relocate there himself nor to press for the creation of a Jewish nation-state. Instead, his connection was more visceral. He came to feel even more associated with the Jewish people, and he resented even more those who would forsake their roots in order to assimilate.

In this regard, he was part of a momentous trend that was reshaping Jewish identity, by choice and by imposition, in Europe. "Until a generation ago, Jews in Germany did not consider themselves as members of the Jewish people," he told a reporter on the day he was leaving America. "They merely considered themselves as members of a religious community." But anti-Semitism changed that, and there was a silver lining to that cloud, he thought. "The undignified mania of trying to adapt and conform and assimilate, which happens among many of my social standing, has always been very repulsive to me," he said.[60]

The Bad German

Einstein's trip to America indelibly cast him as he wanted to be: a citizen of the world, an internationalist, not a German. That image was reinforced by his trips to Germany's other two Great War enemies. On a visit to England, he spoke at the Royal Society and laid flowers on the grave of Isaac Newton in Westminster Abbey. In France, he charmed the public by lecturing in French and taking a mournful tour of the graves on the famous battlefields.

It was also a time of reconciliation with his family. That summer of 1921, he vacationed on the Baltic with his two boys, instilled in young Eduard a love of math, and then took Hans Albert to Florence. They had such a pleasant time that it helped further restore his relations with Marić. "I'm grateful that you've raised them to have a friendly regard for me," he wrote her. "In fact you've done an exemplary job all around." Most astonishingly, on his way home from Italy he visited Zurich and not only called on Marić but even considered staying in "the little upstairs room," as he called it, at her house there. They all got together with the Hurwitz family and had a musical evening as in the old days.[61]

But the mood was soon sullied by the continued collapse of the German mark, which made it harder for Einstein to support a family whose consumption was in Swiss currency. Before the war the mark had been worth 24 cents, but it had fallen to 2 cents by the beginning of 1920. At that time a mark could buy a loaf of bread. But then the bottom fell out of the currency. By the beginning of 1923, the price of a loaf went to 700 marks and by the end of that year cost 1 billion marks. Yes, 1 billion. In November 1923, a new currency, the Renten-mark, was introduced, backed by the government property; 1 trillion old marks equaled 1 new Rentenmark.

The German people increasingly cast around for scapegoats. They blamed internationalists and pacifists who had forced a surrender in the war. They blamed the French and English for imposing what was in fact an onerous peace. And, no surprise, they blamed the Jews. So Germany in the 1920s was not a good place or time to be an interna-tionalist, pacifist, intellectual Jew.

The milestone that marked the passage of German anti-Semitism from being a nasty undercurrent to a public danger was the assassina-tion of Walther Rathenau. From a wealthy Jewish family in Berlin (his father founded AEG, an electricity firm that competed with that of Einstein's father and then became a huge corporation), he served as a senior official in the war ministry, then reconstruction minister and fi-nally foreign minister.

Einstein had read Rathenau's politics book in 1917, and over dinner told him, "I saw with astonishment and joy how extensive a meeting of minds there is between our outlooks on life." Rathenau returned the compliment by reading Einstein's popular explanation of relativity. "I do not say it comes easily to me, but certainly relatively easily," he joked. Then he peppered Einstein with some very insightful questions: "How does a gyroscope know that it is rotating? How does it distin-guish the direction in space toward which it does not want to be tilted?"[62]

Although they became close friends, there was one issue that di-vided them. Rathenau opposed Zionism and thought, mistakenly, that Jews like himself could reduce anti-Semitism by thoroughly assimilat-ing as good Germans.

In the hope that Rathenau could warm to the Zionist cause, Einstein introduced him to Weizmann and Blumenfeld. They met for discussions, both at Einstein's apartment and at Rathenau's grand manor in Berlin's Grunewald, but Rathenau remained unmoved.[63] The best course, he thought, was for Jews to take public roles and become part of Germany's power structure.

Blumenfeld argued that it was wrong for a Jew to presume to run the foreign affairs of another people, but Rathenau kept insisting that he was a German. It was an attitude that was "all too typical of assimilated German Jews," said Weizmann, who was contemptuous of German Jews who tried to assimilate, and especially of those courtiers who became what he dismissed as *Kaiserjuden*. "They seemed to have no idea that they were sitting on a volcano."[64]

As foreign minister in 1922, Rathenau supported German compliance with the Treaty of Versailles and negotiated the Treaty of Rappallo with the Soviet Union, which caused him to be among the first to be labeled by the fledgling Nazi Party as a member of a Jewish-communist conspiracy. On the morning of June 24, 1922, some young nationalists pulled alongside the open car in which Rathenau was riding to work, sprayed him with machine-gun fire, lobbed in a hand grenade, and then sped away.

Einstein was devastated by the brutal assassination, and most of Germany mourned. Schools, universities, and theaters were closed out of respect on the day of his funeral. A million people, Einstein included, paid tribute in front of the Parliament building.

But not everyone felt sympathy. Adolf Hitler called the killers German heroes. Likewise, at the University of Heidelberg, Einstein's antagonist Philipp Lenard decided to defy the day of mourning and give his regular lecture. A number of students showed up to cheer him, but a group of passing workers were so enraged that they dragged the professor from the class and were about to drop him in the Neckar River when police intervened.[65]

For Einstein, the assassination of Rathenau provided a bitter lesson: assimilation did not bring safety. "I regretted the fact that he became a government minister," Einstein wrote in a tribute he sent to a German magazine. "In view of the attitude that large numbers of educated Germans have towards Jews, I have always thought that the proper conduct of the Jews in public life should be one of proud reserve."[66]

Police warned Einstein that he might be next. His name appeared on the target lists prepared by Nazi sympathizers. He should leave Berlin, officials said, or at least avoid any public lectures.

Einstein moved temporarily to Kiel, took a leave of absence from his teaching duties, and wrote to Planck, backing out of the speech he was scheduled to give to the annual convention of German scientists. Lenard and Gehrcke had led a group of nineteen scientists who published a "Declaration of Protest" aimed at barring him from that convention, and Einstein realized that his fame had come back to haunt him. "The newspapers have mentioned my name too often, thus mobilizing the rabble against me," he explained in his note of apology to Planck.[67]

The months after Rathenau's assassination were "nerve-wracking," Einstein lamented to his friend Maurice Solovine. "I am always on the alert."[68] To Marie Curie he confided that he would probably quit his positions in Berlin and find someplace else to live. She urged him to stay and fight instead: "I think that your friend Rathenau would have encouraged you to make an effort."[69]

One option he considered briefly was a move to Kiel, on Germany's Baltic coast, to work at an engineering firm there run by a friend. He had already developed for the firm a new design for a navigational gyroscope, which it patented in 1922 and for which he was paid 20,000 marks in cash.

The firm's owner was surprised but thrilled when Einstein suggested that he might be willing to move there, buy a villa, and become an engineer rather than a theoretical physicist. "The prospect of a downright normal human existence in quietude, combined with the welcome chance of practical work in the factory, delights me," Einstein said. "Plus the wonderful scenery, sailing—enviable!"

But he quickly abandoned the idea, blaming it on Elsa's "horror" of any change. Elsa, for her part, pointed out, no doubt correctly, that it was really Einstein's own decision. "This business of quietude is an illusion," she wrote.[70]

Why didn't he leave Berlin? He had lived there for eight years, longer than anywhere since running away from Munich as a schoolboy. Anti-Semitism was rising, the economy collapsing, and Kiel was certainly not his only option. The light from his star was causing his friends in both Leiden and Zurich to try repeatedly to recruit him with lucrative job offers.

His inertia is hard to explain, but it is indicative of a change that became evident in both his personal life and his scientific work during the 1920s. He had once been a restless rebel who hopped from job to job, insight to insight, resisting anything that smacked of restraint. He had been repelled by conventional respectability. But now he personified it. From being a romantic youth who fancied himself a footloose bohemian he had settled, with but a few stabs at ironic detachment, into a bourgeois life with a doting hausfrau and a richly wallpapered home filled with heavy Biedermeier furniture. He was no longer restless. He was comfortable.

Despite his qualms about publicity and resolve to lie low, it was not in Einstein's nature to shy away from saying what he thought. Nor was he always able to resist demands that he play a public role. Thus he showed up at a huge pacifist rally in a Berlin public park on August 1, just five weeks after Rathenau's assassination. Although he did not speak, he agreed to be paraded around the rally in a car.[71]

Earlier that year, he had joined the League of Nations' International Committee on Intellectual Cooperation, which sought to promote a pacifist spirit among scholars, and he had persuaded Marie Curie to join as well. Its name and mission was sure to inflame German nationalists. So in the wake of the Rathenau assassination, Einstein declared that he wished to resign. "The situation here is that a Jew would do well to exercise restraint as regards his participation in political affairs," he wrote a League official. "In addition, I must say that I have no desire to represent people who would certainly not choose me as their representative."[72]

Even that small act of public reticence did not hold. Curie and the Oxford professor Gilbert Murray, a leader of the committee, begged him to stay a member, and Einstein promptly withdrew his resignation. For the next two years, he remained peripherally involved, but eventually he broke with the League, partly because it supported France's seizure of the Ruhr region after Germany was unable to make reparation payments.

He treated the League, as he did so many parts of life, with a slightly detached and amused air. Each member was supposed to give an address to Geneva University students, but Einstein gave a violin recital instead. One evening at a dinner, Murray's wife asked him why he remained so cheerful given the depravity of the world. "We must remember that this is a very small star," he responded, "and probably some of the larger and more important stars may be very virtuous and happy." [73]

Asia and Palestine, 1922–1923

The unpleasant atmosphere in Germany made Einstein willing to take the most extensive tour of his life, a six-month excursion beginning in October 1922 that would be the only time he would travel either to Asia or what is now Israel. Wherever he went, he was treated as a celebrity, arousing within him the usual mixed emotions. Upon arrival in Ceylon, the Einsteins were whisked away by a waiting rickshaw. "We rode in small one-man carriages drawn at a trot by men of Herculean strength yet delicate build," he noted in his travel diary. "I was bitterly ashamed to share responsibility for the abominable treatment accorded fellow human beings but was unable to do anything about it." [74]

In Singapore, almost the entire Jewish community of more than six hundred turned up at the dock, fortunately trailing no rickshaws. Einstein's target was the richest of them all, Sir Menasseh Meyer, who was born in Baghdad and made his fortune in the opium and real estate markets. "Our sons are refused admission to the universities of other nations," he declared in his speech seeking donations for Hebrew University. Not many of his listeners understood German, and Einstein called the event a "desperate calamity of language with good tasting cake." But it paid off. Meyer gave a sizable donation. [75]

Einstein's own take was even greater. His Japanese publisher and hosts paid him 2,000 pounds for his lecture series there. It was a huge success. Close to twenty-five hundred paying customers showed up for the first talk in Tokyo, which lasted four hours with translation, and more thronged the Imperial Palace to watch his arrival there to meet the emperor and empress.

Einstein was typically amused by it all. "No living person deserves this sort of reception," he told Elsa as they stood on the balcony of their hotel room at dawn listening to the cheers of a thousand people who had kept an all-night vigil hoping to glimpse him. "I'm afraid we're swindlers. We'll end up in prison yet." The German ambassador, with a bit of edge to his pen, reported that "the entire journey of the famous man has been mounted and executed as a commercial enterprise."[76]

Feeling sorry for his listeners, Einstein shortened his subsequent lecture to under three hours. But as he rode to the next city by train (passing along the way through Hiroshima), he could sense that something was amiss with his hosts. Upon asking what the problem was, he was politely told, "The persons who arranged the second lecture were insulted because it did not last four hours like the first one." Thenceforth, he lectured long to the patient Japanese audiences.

The Japanese people struck him as gentle and unpretentious, with a deep appreciation for beauty and ideas. "Of all the people I have met, I like the Japanese most, as they are modest, intelligent, considerate, and have a feel for art," he wrote his two sons.[77]

On his voyage back west, Einstein made his only visit to Palestine, a memorable twelve-day stay that included stops in Lod, Tel Aviv, Jerusalem, and Haifa. He was greeted with great British pomp, as if he were a head of state rather than a theoretical physicist. A cannon salute announced his arrival at the palatial residence of the British high commissioner, Sir Herbert Samuel.

Einstein, on the other hand, was typically unpretentious; he and Elsa arrived tired because he had insisted that they travel in the coach-class car of the overnight train from the coast rather than the first-class sleeping car that had been prepared for them. Elsa was so unnerved by the British formality that she went to bed early some nights to avoid ceremonial events. "When my husband commits a breach of etiquette, it is said it's because he's a man of genius," she complained. "In my case, however, it is attributed to lack of culture."[78]

Like Lord Haldane, Commissioner Samuel was a serious amateur in philosophy and science. Together he and Einstein walked the Old City of Jerusalem to that holiest shrine for religious Jews, the Western Wall (or Wailing Wall) that flanks Temple Mount. But Einstein's deepening love for his Jewish heritage did not instill any new appreciation for the Jewish religion. "Dull-minded tribal companions are praying, faces turned to the wall, rocking their bodies forward and back," he recorded in his diary. "A pitiful sight of men with a past but without a future."[79]

The sights of industrious Jewish people building a new land evinced a more positive reaction. One day he went to a reception for a Zionist organization, and the gates of the building were stormed by throngs who wanted to hear him. "I consider this the greatest day of my life," Einstein proclaimed in the excitement of the moment. "Before, I have always found something to regret in the Jewish soul, and that is the forgetfulness of its own people. Today, I have been made happy by the sight of the Jewish people learning to recognize themselves and to make themselves recognized as a force in the world."

The most frequent question Einstein was asked was whether he would someday return to Jerusalem to stay. He was unusually discreet in his replies, saying nothing quotable. But he knew, as he confided to one of his hosts, that if he came back he would be "an ornament" with no chance of peace or privacy. As he noted in his diary, "My heart says yes, but my reason says no."[80]

CHAPTER FOURTEEN

NOBEL LAUREATE

1921–1927

Einstein in Paris, 1922

The 1921 Prize

It seemed obvious that Einstein would someday win the Nobel Prize for Physics. He had, in fact, already agreed to transfer the money to his first wife, Mileva Marić, when that occurred. The questions were: When would it happen? and, For what?

Once it was announced—in November 1922, awarding him the prize for 1921—the questions were: What took so long? and, Why "especially for his discovery of the law of the photoelectric effect"?

It has been part of the popular lore that Einstein learned that he had finally won while on his way to Japan. "Nobel Prize for physics awarded to you. More by letter," read the telegram sent on November 10. In fact, he had been alerted as soon as the Swedish Academy made the decision in September, well before he left on his trip.

The chairman of the physics award committee, Svante Arrhenius, had heard that Einstein was planning to go to Japan in October, which meant that he would be away for the ceremony unless he postponed the trip. So he wrote Einstein directly and explicitly: "It will probably be very desirable for you to come to Stockholm in December." Expressing a principle of pre–jet travel physics, he added, "And if you are then in Japan that will be impossible."[1] Coming from the head of a Nobel Prize committee, it was clear what that meant. There are not a lot of other reasons for physicists to be summoned to Stockholm in December.

Despite knowing that he would finally win, Einstein did not see fit to postpone his trip. Partly it was because he had been passed over so often that it had begun to annoy him.

He had first been nominated for the prize in 1910 by the chemistry laureate Wilhelm Ostwald, who had rejected Einstein's pleas for a job nine years earlier. Ostwald cited special relativity, emphasizing that the theory involved fundamental physics and not, as some Einstein detractors argued, mere philosophy. It was a point that he reiterated over the next few years as he resubmitted the nomination.

The Swedish committee was mindful of the charge in Alfred Nobel's will that the prize should go to "the most important discovery or invention," and it felt that relativity theory was not exactly either of those. So it reported that it needed to wait for more experimental evidence "before one can accept the principle and in particular award it a Nobel prize."[2]

Einstein continued to be nominated for his work on relativity during most of the ensuing ten years, gaining support from distinguished theorists such as Wilhelm Wien, although not yet from a still-skeptical Lorentz. His greatest obstacle was that the committee at the time was leery of pure theorists. Three out of the committee's five members throughout the period from 1910 to 1922 were experimentalists from Sweden's Uppsala University, known for its fervent devotion to perfecting experimental and measuring techniques. "Swedish physicists with a strong experimentalist bias dominated the committee," notes Robert Marc Friedman, a historian of science in Oslo. "They held precision measurement as the highest goal for their discipline." That is one reason Max Planck had to wait until 1919 (when he was awarded the delayed prize for 1918) and why Henri Poincaré never won at all.[3]

The dramatic announcement in November 1919 that the eclipse observations had confirmed parts of Einstein's theory should have made 1920 his year. By then Lorentz was no longer such a skeptic. He along with Bohr and six other official nominators wrote in support of Einstein, mostly focusing on his completed theory of relativity. (Planck wrote in support as well, but his letter arrived after the deadline for consideration.) As Lorentz's letter declared, Einstein "has placed himself in the first rank of physicists of all time." Bohr's letter was equally clear: "One faces here an advance of decisive significance."[4]

Politics intervened. Up until then, the primary justifications for denying Einstein a Nobel had been scientific: his work was purely theoretical, it lacked experimental grounding, and it putatively did not involve the "discovery" of any new laws. After the eclipse observations, the explanation of the shift in Mercury's orbit, and other experimental confirmations, these arguments against Einstein were still made, but they were now tinged with more cultural and personal bias. To his critics, the fact that he had suddenly achieved superstar status as the most internationally celebrated scientist since the lightning-tamer Benjamin Franklin was paraded through the streets of Paris was evidence of his self-promotion rather than his worthiness of a Nobel.

This subtext was evident in the internal seven-page report prepared by Arrhenius, the committee chairman, explaining why Einstein should not win the prize in 1920. He noted that the eclipse results had been criticized as ambiguous and that scientists had not yet confirmed the theory's prediction that light coming from the sun would be shifted toward the red end of the spectrum by the sun's gravity. He also cited the discredited argument of Ernst Gehrcke, one of the anti-Semitic antirelativists who led the notorious 1920 rally against Einstein that summer in Berlin, that the shift in Mercury's orbit could be explained by other theories.

Behind the scenes, Einstein's other leading anti-Semitic critic, Philipp Lenard, was waging a crusade against him. (The following year, Lenard would propose Gehrcke for the prize!) Sven Hedin, a

Swedish explorer who was a prominent member of the Academy, later recalled that Lenard worked hard to persuade him and others that "relativity was really not a discovery" and that it had not been proven.[5]

Arrhenius's report cited Lenard's "strong critique of the oddities in Einstein's generalized theory of relativity." Lenard's views were couched as a criticism of physics that was not grounded in experiments and concrete discoveries. But there was a strong undercurrent in the report of Lenard's animosity to the type of "philosophical conjecturing" that he often dismissed as being a feature of "Jewish science."[6]

So the 1920 prize instead went to another Zurich Polytechnic graduate who was Einstein's scientific opposite: Charles-Edouard Guillaume, the director of the International Bureau of Weights and Measures, who had made his modest mark on science by assuring that standard measures were more precise and discovering metal alloys that had practical uses, including making good measuring rods. "When the world of physics had entered upon an intellectual adventure of extraordinary proportions, it was remarkable to find Guillaume's accomplishment, based on routine study and modest theoretical finesse, recognized as a beacon of achievement," says Friedman. "Even those who opposed relativity theory found Guillaume a bizarre choice."[7]

By 1921, the public's Einstein mania was in full force, for better or worse, and there was a groundswell of support for him from both theoreticians and experimentalists, Germans such as Planck and non-Germans such as Eddington. He garnered fourteen official nominations, far more than any other contender. "Einstein stands above his contemporaries even as Newton did," wrote Eddington, offering the highest praise a member of the Royal Society could muster.[8]

This time the prize committee assigned the task of doing a report on relativity to Allvar Gullstrand, a professor of ophthalmology at the University of Uppsala, who had won the prize for medicine in 1911. With little expertise in either the math or the physics of relativity, he criticized Einstein's theory in a sharp but unknowing manner. Clearly determined to undermine Einstein by any means, Gullstrand's fifty-page report declared, for example, that the bending of light was not a true test of Einstein's theory, that the results were not experimentally valid, and that even if they were there were still other ways to explain the phenomenon using classical mechanics. As for Mercury's orbit, he declared, "It remains unknown until further notice whether the Einstein theory can at all be brought into agreement with the perihelion experiment." And the effects of special relativity, he said, "lay below the limits of experimental error." As one who had made his name by devising precision optical measuring instruments, Gullstrand seemed particularly appalled by Einstein's theory that the length of rigid measuring rods could vary relative to moving observers.[9]

Even though some members of the full Academy realized that Gullstrand's opposition was unsophisticated, it was hard to overcome. He was a respected and popular Swedish professor, and he insisted both publicly and privately that the great honor of a Nobel should not be given to a highly speculative theory that was the subject of an inexplicable mass hysteria that would soon deflate. Instead of choosing someone else, the Academy did something that was less (or more?) of a public slap at Einstein: it voted to choose nobody and tentatively bank the 1921 award for another year.

The great impasse threatened to become embarrassing. His lack of a prize had begun to reflect more negatively on the Nobel than on Einstein. "Imagine for a moment what the general opinion will be fifty years from now if the name Einstein does not appear on the list of Nobel laureates," wrote the French physicist Marcel Brillouin in his 1922 nominating letter.[10]

To the rescue rode a theoretical physicist from the University of Uppsala, Carl Wilhelm Oseen, who joined the committee in 1922. He was a colleague and friend of Gullstrand, which helped him gently overcome some of the ophthalmologist's ill-conceived but stubborn objections. And he realized that the whole issue of relativity theory was so encrusted with controversy that it would be better to try a different tack. So Oseen pushed hard to give the prize to Einstein for "the discovery of the law of the photoelectric effect."

Each part of that phrase was carefully calculated. It was not a nomination for relativity, of course. In fact, despite the way it has been phrased by some historians, it was not for Einstein's theory of light quanta, even though that was the primary focus of the relevant 1905 paper. Nor was it for any *theory* at all. Instead, it was for the *discovery* of a *law.*

A report from the previous year had discussed Einstein's "*theory* of the photoelectric effect," but Oseen made clear his different approach with the title of his report: "Einstein's *Law* of the Photoelectric Effect" (emphasis added). In it, Oseen did not focus on the theoretical aspects of Einstein's work. He specified instead what he called a fundamental natural law, fully proven by experiment, that Einstein propounded: the mathematical description of how the photoelectric effect was explained by assuming that light was absorbed and emitted in discrete quanta, and the way this related to the frequency of the light.

Oseen also proposed that giving Einstein the prize delayed from 1921 would allow the Academy to use that as a basis for simultaneously giving Niels Bohr the 1922 prize, because his model of the atom built on the laws that explained the photoelectric effect. It was a clever coupled-entry ticket for making sure that the two greatest theoretical physicists of the time became Nobel laureates without offending the Academy's old-line establishment. Gullstrand went along. Arrhenius, who had met Einstein in Berlin and been charmed, was now also willing to accept the inevitable. On September 6, 1922, the Academy voted accordingly, and Einstein and Bohr were awarded the 1921 and 1922 prizes, respectively.

Thus it was that Einstein became the recipient of the 1921 Nobel Prize, in the words of the official citation, "for his services to theoretical physics, and especially for his discovery of the law of the photoelectric effect." In both the citation and the letter from the Academy's secretary officially informing Einstein, an unusual caveat was explicitly inserted. Both documents specified that the award was given "without taking into account the value that will be accorded your relativity and gravitation theories after these are confirmed in the future."[11] Einstein would not, as it turned out, ever win a Nobel for his work on relativity and gravitation, nor for anything other than the photoelectric effect.

There was a dark irony in using the photoelectric effect as a path to get Einstein the prize. His "law" was based primarily on observations made by Philipp Lenard, who had been the most fervent campaigner to have him blackballed. In his 1905 paper, Einstein had credited Lenard's "pioneering" work. But after the 1920 anti-Semitic rally in Berlin, they had become bitter enemies. So Lenard was doubly outraged that, despite his opposition, Einstein had won the prize and, worse yet, done so in a field that Lenard pioneered. He wrote an angry letter to the Academy, the only official protest it received, in which he said that Einstein misunderstood the true nature of light and was, in addition, a publicity-seeking Jew whose approach was alien to the true spirit of German physics.[12]

Einstein was traveling by train through Japan and missed the official award ceremony on December 10. After much controversy over whether he should be considered German or Swiss, the prize was accepted by the German ambassador, but he was listed as both nationalities in the official record.

The formal presentation speech by Arrhenius, the committee chair, was carefully crafted. "There is probably no physicist living today whose name has become so widely known as that of Albert Einstein," he began. "Most discussion centers on his theory of relativity." He then went on to say, almost dismissively, that "this pertains essentially to epistemology and has therefore been the subject of lively debate in philosophical circles."

After touching briefly on Einstein's other work, Arrhenius explained the Academy's position on why he had won. "Einstein's law of the photoelectrical effect has been extremely rigorously tested by the American Millikan* and his pupils and passed the test brilliantly," he said. "Einstein's law has become the basis of quantitative photochemistry in the same way as Faraday's law is the basis of electrochemistry."[13]

Einstein gave his official acceptance speech the following July at a Swedish science conference with King Gustav Adolf V in attendance. He spoke not about the photoelectric effect, but about relativity, and he

---

* Robert Andrews Millikan would win the Nobel Prize the following year, 1923, for experimental work on the photoelectric effect he had done at the University of Chicago. By then he had become director of the physics lab at the California Institute of Technology, and in the early 1930s he would bring Einstein there as a visiting scientist.

concluded by emphasizing the importance of his new passion, finding a unified field theory that would reconcile general relativity with electromagnetic theory and, if possible, with quantum mechanics.[14]

The prize money that year amounted to 121,572 Swedish kronor, or $32,250, which was more than ten times the annual salary of the average professor at the time. As per his divorce agreement with Marić, Einstein had part of it sent directly to Zurich to reside in a trust for her and their sons, and the rest went into an American account with the interest directed for her use.

This prompted another row. Hans Albert complained that the trust arrangement, which had previously been agreed to, made only the interest on the money accessible to the family. Once again, Zangger intervened and calmed the dispute. Einstein jokingly wrote to his sons, "You all will be so rich that some fine day I may ask you for a loan." The money was eventually used by Marić to buy three homes with rental apartments in Zurich.[15]

Newton's Bucket and the Ether Reincarnated

"Anything truly novel is invented only during one's youth," Einstein lamented to a friend after finishing his work on general relativity and cosmology. "Later one becomes more experienced, more famous— and more *blockheaded.*"[16]

Einstein turned 40 in 1919, the year that the eclipse observations made him world-famous. For the next six years, he continued to make important contributions to quantum theory. But after that, as we shall see, he would begin to seem, if not blockheaded, at least a bit stubborn as he resisted quantum mechanics and embarked on a long, lonely, and unsuccessful effort to devise a unified theory that would subsume it into a more deterministic framework.

Over the ensuing years, researchers would discover new forces in nature, besides electromagnetism and gravity, and also new particles. These would make Einstein's attempts at unification all the more complex. But he would find himself less familiar with the latest data in experimental physics, and he thus would no longer have the same intuitive feel for how to wrest from nature her fundamental principles.

If Einstein had retired after the eclipse observations and devoted himself to sailing for the remaining thirty-six years of his life, would science have suffered? Yes, for even though most of his attacks on quantum mechanics did not prove to be warranted, he did serve to strengthen the theory by coming up with a few advances and also, less intentionally, by his ingenious but futile efforts to poke holes in it.

That raises another question: Why was Einstein so much more creative before the age of 40 than after? Partly, it is an occupational hazard of mathematicians and theoretical physicists to have their great breakthroughs before turning 40.[17] "The intellect gets crippled," Einstein explained to a friend, "but glittering renown is still draped around the calcified shell."[18]

More specifically, Einstein's scientific successes had come in part from his rebelliousness. There was a link between his creativity and his willingness to defy authority. He had no sentimental attachment to the old order, thus was energized by upending it. His stubbornness had worked to his advantage.

But now, just as he had traded his youthful bohemian attitudes for the comforts of a bourgeois home, he had become wedded to the faith that field theories could preserve the certainties and determinism of classical science. His stubbornness henceforth would work to his disadvantage.

It was a fate that he had begun fearing years before, not long after he finished his famous flurry of 1905 papers. "Soon I will reach the age of stagnation and sterility when one laments the revolutionary spirit of the young," he had worried to his colleague from the Olympia Academy, Maurice Solovine.[19]

Now, many triumphs later, there were young revolutionaries who felt this fate had indeed befallen him. In one of his most revealing remarks about himself, Einstein lamented, "To punish me for my contempt of authority, Fate has made me an authority myself."[20]

Thus it is not surprising that, during the 1920s, Einstein found himself scaling back on some of his bolder earlier ideas. For example, in his 1905 special relativity paper he had famously dismissed the concept of the ether as "superfluous." But after he finished his theory of general relativity, he concluded that the gravitational potentials in that theory characterized the physical qualities of empty space and served as a medium that could transmit disturbances. He began referring to this as a new way to conceive of an ether. "I agree with you that the general relativity theory admits of an ether hypothesis," he wrote Lorentz in 1916.[21]

In a lecture in Leiden in May 1920, Einstein publicly proposed a reincarnation, though not a rebirth, of the ether. "More careful reflection teaches us, however, that the special theory of relativity does not compel us to deny ether," he said. "We may assume the existence of an ether, only we must give up ascribing a definite state of motion to it."

This revised view was justified, he said, by the results of the general theory of relativity. He made clear that his new ether was different from the old one, which had been conceived as a medium that could ripple and thus explain how light waves moved through space. Instead, he was reintroducing the idea in order to explain rotation and inertia.

Perhaps he could have saved some confusion if he had chosen a different term. But in his speech he made clear that he was reintroducing the word intentionally:

> To deny the ether is ultimately to assume that empty space has no physical qualities whatever. The fundamental facts of mechanics do not harmonize with this view . . . Besides observable objects, another thing, which is not perceptible, must be looked upon as real, to enable acceleration or rotation to be looked upon as something real . . . The conception of the ether has again acquired an intelligible content, although this content differs widely from that of the ether of the mechanical wave theory of light . . . According to the general theory of relativity, space is endowed with physical qualities; in this sense, there exists an ether. Space without ether is unthinkable; for in such space there not only would be no propagation of light, but also no possibility of existence for standards of space and time (measuring-rods and clocks), nor therefore any spacetime intervals in the physical sense. But this ether may not be thought of as endowed with the qualities of ponderable media, as consisting of parts which may be tracked through time. The idea of motion may not be applied to it.[22]

So what was this reincarnated ether, and what did it mean for Mach's principle and for the question raised by Newton's bucket?* Einstein had initially enthused that general relativity explained rotation as being simply a motion *relative* to other objects in space, just as Mach had argued. In other words, if you were inside a bucket that was dangling in empty space, with no other objects in the universe, there would be no way to tell if you were spinning or not. Einstein even wrote to Mach saying he should be pleased that his principle was supported by general relativity.

Einstein had asserted this claim in a letter to Schwarzschild, the brilliant young scientist who had written to him from Germany's Russian front during the war about the cosmological implications of general relativity. "Inertia is simply an interaction between masses, not an effect in which 'space' of itself is involved, separate from the observed mass," Einstein had declared.[23] But Schwarzschild disagreed with that assessment.

And now, four years later, Einstein had changed his mind. In his Leiden speech, unlike in his 1916 interpretation of general relativity, Einstein accepted that his gravitational field theory implied that empty space had physical qualities. The mechanical behavior of an object hovering in empty space, like Newton's bucket, "depends not only on relative velocities but also on its state of rotation." And that meant "space is endowed with physical qualities."

As he admitted outright, this meant that he was now abandoning Mach's principle. Among other things, Mach's idea that inertia is caused by the presence of all of the distant bodies in the universe implied that these bodies could *instantly* have an effect on an object, even though they were far apart. Einstein's theory of relativity did not accept instant actions at a distance. Even gravity did not exert its force instantly, but only through changes in the gravitational field that obeyed the speed limit of light. "Inertial resistance to acceleration in relation to

---

* See page 119 for Newton's thought experiment about whether water rotating in a bucket in empty space would be subject to inertial pressure and thus press against the sides of the bucket. See page 251 for Einstein's 1916 view, which he was now revising, that an empty universe would have no inertia or fabric of spacetime.

distant masses supposes action at a distance," Einstein lectured. "Because the modern physicist does not accept such a thing as action at a distance, he comes back to the ether, which has to serve as medium for the effects of inertia."[24]

It is an issue that still causes dispute, but Einstein seemed to believe, at least when he gave his Leiden lecture, that according to general relativity as he now saw it, the water in Newton's bucket would be pushed up the walls even if it were spinning in a universe devoid of any other objects. "In contradiction to what Mach would have predicted," Brian Greene writes, "even in an otherwise empty universe, you *will* feel pressed against the inner wall of the spinning bucket . . . In general relativity, empty spacetime provides a benchmark for accelerated motion."[25]

The inertia pushing the water up the wall was caused by its rotation with respect to the metric field, which Einstein now reincarnated as an ether. As a result, he had to face the possibility that general relativity did not necessarily eliminate the concept of absolute motion, at least with respect to the metric of spacetime.[26]

It was not exactly a retreat, nor was it a return to the nineteenth-century concept of the ether. But it was a more conservative way of looking at the universe, and it represented a break from the radicalism of Mach that Einstein had once embraced.

This clearly made Einstein uncomfortable. The best way to eliminate the need for an ether that existed separately from matter, he concluded, would be to find his elusive unified field theory. What a glory that would be! "The contrast between ether and matter would fade away," he said, "and, through the general theory of relativity, the whole of physics would become a complete system of thought."[27]

Niels Bohr, Lasers, and "Chance"

By far the most important manifestation of Einstein's midlife transition from a revolutionary to a conservative was his hardening attitude toward quantum theory, which in the mid-1920s produced a radical new system of mechanics. His qualms about this new quantum mechanics, and his search for a unifying theory that would reconcile it with relativity and restore certainty to nature, would dominate—and to some extent diminish—the second half of his scientific career.

He had once been a fearless quantum pioneer. Together with Max Planck, he launched the revolution at the beginning of the century; unlike Planck, he had been one of the few scientists who truly believed in the physical reality of quanta—that light *actually* came in packets of energy. These quanta behaved at times like particles. They were indivisible units, not part of a continuum.

In his 1909 Salzburg address, he had predicted that physics would have to reconcile itself to a duality in which light could be regarded as both wave and particle. And at the first Solvay Conference in 1911, he had declared that "these discontinuities, which we find so distasteful in Planck's theory, seem really to exist in nature."[28]

This caused Planck, who resisted the notion that his quanta actually had a physical reality, to say of Einstein, in his recommendation that he be elected to the Prussian Academy, "His hypothesis of light quanta may have gone overboard." Other scientists likewise resisted Einstein's quantum hypothesis. Walther Nernst called it "probably the strangest thing ever thought up," and Robert Millikan called it "wholly untenable," even after confirming its predictive power in his lab.[29]

A new phase of the quantum revolution was launched in 1913, when Niels Bohr came up with a revised model for the structure of the atom. Six years younger than Einstein, brilliant yet rather shy and inarticulate, Bohr was Danish and thus able to draw from the work on quantum theory being done by Germans such as Planck and Einstein and also from the work on the structure of the atom being done by the Englishmen J. J. Thomson and Ernest Rutherford. "At the time, quantum theory was a German invention which had scarcely penetrated to England at all," recalled Arthur Eddington.[30]

Bohr had gone to study with Thomson in Cambridge. But the mumbling Dane and brusque Brit had trouble communicating. So Bohr migrated up to Manchester to work with the more gregarious Rutherford, who had devised a model of the atom that featured a positively charged nucleus around which tiny negatively charged electrons orbited.[31]

Bohr made a refinement based on the fact that these electrons did not collapse into the nucleus and emit a continuous spectrum of radiation, as classical physics would suggest. In Bohr's new model, which was based on studying the hydrogen atom, an electron circled a nucleus at certain permitted orbits in states with discrete energies. The atom could absorb energy from radiation (such as light) only in increments that would kick the electron up a notch to another permitted orbit. Likewise, the atom could emit radiation only in increments that would drop the electron down to another permitted orbit.

When an electron moved from one orbit to the next, it was a quantum leap. In other words, it was a disconnected and discontinuous shift from one level to another, with no meandering in between. Bohr went on to show how this model accounted for the lines in the spectrum of light emitted by the hydrogen atom.

Einstein was both impressed and a little jealous when he heard of Bohr's theory. As one scientist reported to Rutherford, "He told me that he had once similar ideas but he did not dare to publish them." Einstein later declared of Bohr's discovery, "This is the highest form of musicality in the sphere of thought."[32]

Einstein used Bohr's model as the foundation for a series of papers in 1916, the most important of which, "On the Quantum Theory of Radiation," was also formally published in a journal in 1917.[33]

Einstein began with a thought experiment in which a chamber is filled with a cloud of atoms. They are being bathed by light (or any form of electromagnetic radiation). Einstein then combined Bohr's model of the atom with Max Planck's theory of the quanta. If each change in an electron orbit corresponded to the absorption or emission of one light quantum, then—presto!—it resulted in a new and better way to derive Planck's formula for explaining blackbody radiation. As Einstein boasted to Michele Besso, "A brilliant idea dawned on me about radiation absorption and emission. It will interest you. An astonishingly simple derivation, I should say *the* derivation of Planck's formula. A thoroughly quantized affair."[34]

Atoms emit radiation in a spontaneous fashion, but Einstein theorized that this process could also be stimulated. A roughly simplified way to picture this is to suppose that an atom is already in a high-energy state from having absorbed a photon. If another photon with a particular wavelength is then fired into it, two photons of the same wavelength and direction can be emitted.

What Einstein discovered was slightly more complex. Suppose there is a gas of atoms with energy being pumped into it, say by pulses of electricity or light. Many of the atoms will absorb energy and go into a higher energy state, and they will begin to emit photons. Einstein argued that the presence of this cloud of photons made it even more likely that a photon of the same wavelength and direction as the other photons in the cloud would be emitted.[35] This process of stimulated emission would, almost forty years later, be the basis for the invention of the laser, an acronym for "light amplification by the stimulated emission of radiation."

There was one part of Einstein's quantum theory of radiation that had strange ramifications. "It can be demonstrated convincingly," he told Besso, "that the elementary processes of emission and absorption are directed processes."[36] In other words, when a photon pulses out of an atom, it does not do so (as the classical wave theory would have it) in all directions at once. Instead, a photon has momentum. In other words, the equations work only if each quantum of radiation is emitted in some particular direction.

That was not necessarily a problem. But here was the rub: *there was no way to determine which direction an emitted photon might go.* In addition, *there was no way to determine when it would happen.* If an atom was in a state of higher energy, it was possible to calculate the *probability* that it would emit a photon at any specific moment. But it was not possible to determine the moment of emission precisely. Nor was it possible to determine the direction. No matter how much information you had. It was all a matter of *chance*, like the roll of dice.

That was a problem. It threatened the strict determinism of Newton's mechanics. It undermined the certainty of classical physics and the faith that if you knew all the positions and velocities in a system you could determine its future. Relativity may have seemed like a radical idea, but at least it preserved rigid cause-and-effect rules. The quirky and unpredictable behavior of pesky quanta, however, was messing with this causality.

"It is a weakness of the theory," Einstein conceded, "that it leaves the time and direction of the elementary process to 'chance.'" The whole concept of chance—"*Zufall*" was the word he used—was so disconcerting to him, so odd, that he put the word in quotation marks, as if to distance himself from it.[37]

For Einstein, and indeed for most classical physicists, the idea that there could be a fundamental randomness in the universe—that events could just happen without a cause—was not only a cause of discomfort, it undermined the entire program of physics. Indeed, he never would become reconciled to it. "The thing about causality plagues me very much," he wrote Max Born in 1920. "Is the quantumlike absorption and emission of light ever conceivable in terms of complete causality?"[38]

For the rest of his life, Einstein would remain resistant to the notion that probabilities and uncertainties ruled nature in the realm of quantum mechanics. "I find the idea quite intolerable that an electron exposed to radiation should choose *of its own free will* not only its moment to jump off but also its direction," he despaired to Born a few years later. "In that case, I would rather be a cobbler, or even an employee of a gaming house, than a physicist."[39]

Philosophically, Einstein's reaction seemed to be an echo of the attitude displayed by the antirelativists, who interpreted (or misinterpreted) Einstein's relativity theory as meaning an end to the certainties and absolutes in nature. In fact, Einstein saw relativity theory as leading to a deeper description of certainties and absolutes—what he called invariances—based on the combination of space and time into one four-dimensional fabric. Quantum mechanics, on the other hand, would be based on true underlying uncertainties in nature, events that could be described only in terms of probabilities.

On a visit to Berlin in 1920, Niels Bohr, who had become the Copenhagen-based ringleader of the quantum mechanics movement, met Einstein for the first time. Bohr arrived at Einstein's apartment bearing Danish cheese and butter, and then he launched into a discussion of the role that chance and probability played in quantum mechanics. Einstein expressed his wariness of "abandoning continuity and causality." Bohr was bolder about going into that misty realm.

Abandoning strict causality, he countered to Einstein, was "the only way open" given the evidence.

Einstein admitted that he was impressed, but also worried, by Bohr's breakthroughs on the structure of the atom and the randomness it implied for the quantum nature of radiation. "I could probably have arrived at something like this myself," Einstein lamented, "but if all this is true then it means the end of physics." [40]

Although Einstein found Bohr's ideas disconcerting, he found the gangly and informal Dane personally endearing. "Not often in life has a human being caused me such joy by his mere presence as you did," he wrote Bohr right after the visit, adding that he took pleasure in picturing "your cheerful boyish face." He was equally effusive behind Bohr's back. "Bohr was here, and I am just as keen on him as you are," he wrote their mutual friend Ehrenfest in Leiden. "He is an extremely sensitive lad and moves around in this world as if in a trance." [41]

Bohr, for his part, revered Einstein. When it was announced in 1922 that they had won sequential Nobel Prizes, Bohr wrote that his own joy had been heightened by the fact that Einstein had been recognized first for "the fundamental contribution that you made to the special field in which I am working." [42]

On his journey home from delivering his acceptance speech in Sweden the following summer, Einstein stopped in Copenhagen to see Bohr, who met him at the train station to take him home by streetcar. On the ride, they got into a debate. "We took the streetcar and talked so animatedly that we went much too far," Bohr recalled. "We got off and traveled back, but again rode too far." Neither seemed to mind, for the conversation was so engrossing. "We rode to and fro," according to Bohr, "and I can well imagine what the people thought about us." [43]

More than just a friendship, their relationship became an intellectual entanglement that began with divergent views about quantum mechanics but then expanded into related issues of science, knowledge, and philosophy. "In all the history of human thought, there is no greater dialogue than that which took place over the years between Niels Bohr and Albert Einstein about the meaning of the quantum," says the physicist John Wheeler, who studied under Bohr. The social philosopher C. P. Snow went further. "No more profound intellectual debate has ever been conducted," he proclaimed.[44]

Their dispute went to the fundamental heart of the design of the cosmos: Was there an objective reality that existed whether or not we could ever observe it? Were there laws that restored strict causality to phenomena that seemed inherently random? Was everything in the universe predetermined?

For the rest of their lives, Bohr would sputter and fret at his repeated failures to convert Einstein to quantum mechanics. *Einstein, Einstein, Einstein,* he would mutter after each infuriating encounter. But it was a discussion that was conducted with deep affection and even great humor. On one of the many occasions when Einstein declared that God would not play dice, it was Bohr who countered with the famous rejoinder: Einstein, stop telling God what to do![45]

Quantum Leaps

Unlike the development of relativity theory, which was largely the product of one man working in near solitary splendor, the development of quantum mechanics from 1924 to 1927 came from a burst of activity by a clamorous congregation of young Turks who worked both in parallel and in collaboration. They built on the foundations laid by Planck and Einstein, who continued to resist the radical ramifications of the quanta, and on the breakthroughs by Bohr, who served as a mentor for the new generation.

Louis de Broglie, who carried the title of prince by virtue of being related to the deposed French royal family, studied history in hopes of being a civil servant. But after college, he became fascinated by physics. His doctoral dissertation in 1924 helped transform the field. If a wave can behave like a particle, he asked, shouldn't a particle also behave like a wave?

In other words, Einstein had said that light should be regarded not only as a wave but also as a particle. Likewise, according to de Broglie, a particle such as an electron could also be regarded as a wave. "I had a sudden inspiration," de Broglie later recalled. "Einstein's wave-particle dualism was an absolutely general phenomenon extending to all of physical nature, and that being the case the motion of all particles—photons, electrons, protons or any other—must be associated with the propagation of a wave."[46]

Using Einstein's law of the photoelectric affect, de Broglie showed that the wavelength associated with an electron (or any particle) would be related to Planck's constant divided by the particle's momentum. It turns out to be an incredibly tiny wavelength, which means that it's usually relevant only to particles in the subatomic realm, not to such things as pebbles or planets or baseballs.*

In Bohr's model of the atom, electrons could change their orbits (or, more precisely, their stable standing wave patterns) only by certain quantum leaps. De Broglie's thesis helped explain this by conceiving of electrons not just as particles but also as waves. Those waves are strung out over the circular path around the nucleus. This works only if the circle accommodates a whole number—such as 2 or 3 or 4—of the particle's wavelengths; it won't neatly fit in the prescribed circle if there's a fraction of a wavelength left over.

De Broglie made three typed copies of his thesis and sent one to his adviser, Paul Langevin, who was Einstein's friend (and Madame Curie's). Langevin, somewhat baffled, asked for another copy to send along to Einstein, who praised the work effusively. It had, Einstein said, "lifted a corner of the great veil." As de Broglie proudly noted, "This made Langevin accept my work."[47]

Einstein made his own contribution when he received in June of that year a paper in English from a young physicist from India named Satyendra Nath Bose. It derived Planck's blackbody radiation law by treating radiation as if it were a cloud of gas and then applying a statistical method of analyzing it. But there was a twist: Bose said that any two photons that had the same energy state were absolutely indistinguishable, in theory as well as fact, and should not be treated separately in the statistical calculations.

Bose's creative use of statistical analysis was reminiscent of Ein-

---

* The de Broglie wavelength of a baseball thrown at 90 mph would be about 10^{-34} meters, incredibly smaller than the size of an atom or even a proton, so infinitessimal as to be unobservable.

stein's youthful enthusiasm for that approach. He not only got Bose's paper published, he also extended it with three papers of his own. In them, he applied Bose's counting method, later called "Bose-Einstein statistics," to actual gas molecules, thus becoming the primary inventor of quantum-statistical mechanics.

Bose's paper dealt with photons, which have no mass. Einstein extended the idea by treating quantum particles *with mass* as being indistinguishable from one another for statistical purposes in certain cases. "The quanta or molecules are not treated as structures statistically independent of one another," he wrote.[48]

The key insight, which Einstein extracted from Bose's initial paper, has to do with how you calculate the probabilities for each possible state of multiple quantum particles. To use an analogy suggested by the Yale physicist Douglas Stone, imagine how this calculation is done for dice. In calculating the odds that the roll of two dice (A and B) will produce a lucky 7, we treat the possibility that A comes up 4 and B comes up 3 as one outcome, and we treat the possibility that A comes up 3 and B comes up 4 as a different outcome—thus counting each of these combinations as different ways to produce a 7. Einstein realized that the new way of calculating the odds of quantum states involved treating these not as two different possibilities, but only as one. A 4-3 combination was indistinguishable from a 3-4 combination; likewise, a 5-2 combination was indistinguishable from a 2-5.

That cuts in half the number of ways two dice can roll a 7. But it does not affect the number of ways they could turn up a 2 or a 12 (using either counting method, there is only one way to roll each of these totals), and it only reduces from five to three the number of ways the two dice could total 6. A few minutes of jotting down possible outcomes shows how this system changes the overall odds of rolling any particular number. The changes wrought by this new calculating method are even greater if we are applying it to dozens of dice. And if we are dealing with billions of particles, the change in probabilities becomes huge.

When he applied this approach to a gas of quantum particles, Einstein discovered an amazing property: unlike a gas of classical particles, which will remain a gas unless the particles attract one another, a gas of quantum particles can condense into some kind of liquid even without a force of attraction between them.

This phenomenon, now called Bose-Einstein condensation,* was a brilliant and important discovery in quantum mechanics, and Einstein deserves most of the credit for it. Bose had not quite realized that the statistical mathematics he used represented a fundamentally new approach. As with the case of Planck's constant, Einstein recognized the physical reality, and the significance, of a contrivance that someone else had devised.[49]

Einstein's method had the effect of treating particles as if they had wavelike traits, as both he and de Broglie had suggested. Einstein even predicted that if you did Thomas Young's old double-slit experiment (showing that light behaved like a wave by shining a beam through two slits and noting the interference pattern) by using a beam of gas molecules, they would interfere with one another as if they were waves. "A beam of gas molecules which passes through an aperture," he wrote, "must undergo a diffraction analogous to that of a light ray."[50]

Amazingly, experiments soon showed that to be true. Despite his discomfort with the direction quantum theory was heading, Einstein was still helping, at least for the time being, to push it ahead. "Einstein is thereby clearly involved in the foundation of wave mechanics," his friend Max Born later said, "and no alibi can disprove it."[51]

Einstein admitted that he found this "mutual influence" of particles to be "quite mysterious," for they seemed as if they should behave independently. "The quanta or molecules are not treated as independent of one another," he wrote another physicist who expressed bafflement. In a postscript he admitted that it all worked well mathematically, but "the physical nature remains veiled."[52]

On the surface, this assumption that two particles could be treated as indistinguishable violated a principle that Einstein would nevertheless try to cling to in the future: the principle of separability, which as-

---

* In 1995, Bose-Einstein condensation was finally achieved experimentally by Eric A. Cornell, Wolfgang Ketterle, and Carl E. Wieman, who were awarded the 2001 Nobel Prize for this work.

serts that particles with different locations in space have separate, independent realities. One aim of general relativity's theory of gravity had been to avoid any "spooky action at a distance," as Einstein famously called it later, in which something happening to one body could instantly affect another distant body.

Once again, Einstein was at the forefront of discovering an aspect of quantum theory that would cause him discomfort in the future. And once again, younger colleagues would embrace his ideas more readily than he would—just as he had once embraced the implications of the ideas of Planck, Poincaré, and Lorentz more readily than they had.[53]

An additional step was taken by another unlikely player, Erwin Schrödinger, an Austrian theoretical physicist who despaired of discovering anything significant and thus decided to concentrate on being a philosopher instead. But the world apparently already had enough Austrian philosophers, and he couldn't find work in that field. So he stuck with physics and, inspired by Einstein's praise of de Broglie, came up with a theory called "wave mechanics." It led to a set of equations that governed de Broglie's wavelike behavior of electrons, which Schrödinger (giving half credit where he thought it was due) called "Einstein–de Broglie waves."[54]

Einstein expressed enthusiasm at first, but he soon became troubled by some of the ramifications of Schrödinger's waves, most notably that over time they can spread over an enormous area. An electron could not, in reality, be waving thus, Einstein thought. So what, in the real world, did the wave equation really represent?

The person who helped answer that question was Max Born, Einstein's close friend and (along with his wife, Hedwig) frequent correspondent, who was then teaching at Göttingen. Born proposed that the wave did not describe the behavior of the particle. Instead, he said that it described the *probability* of its location at any moment.[55] It was an approach that revealed quantum mechanics as being, even more than previously thought, fundamentally based on chance rather than causal certainties, and it made Einstein even more squeamish.[56]

Meanwhile, another approach to quantum mechanics had been developed in the summer of 1925 by a bright-faced 23-year-old hiking enthusiast, Werner Heisenberg, who was a student of Niels Bohr in

His parents, Pauline and
Hermann Einstein

1

2

In a Munich photo studio at age 14

3

Bottom left at the Aarau school, 1896

4

With Mileva Marić, ca. 1905

5

With Mileva and Hans Albert, 1905

6

Eduard, Mileva, and Hans Albert, 1914

7

With Conrad Habicht, left, and Maurice Solovine of the "Olympia Academy," ca. 1902

8

Anna Winteler Besso and Michele Besso

9

At the patent office in Bern during the miracle year, 1905

10

In Prague, 1912

11

Marcel Grossmann, who helped with
math at college and for general relativity

12

Hiking in Switzerland with Madame Curie, 1913

13

With the chemist Fritz Haber, assimilationist
and marriage mediator, July 1914

14

Watched over by Zionist leader Chaim
Weizmann in New York, April 1921

15

Meeting the press in New York, 1930

16

With Elsa at the Grand Canyon, February 1931

17

The 1911 Solvay Conference

18

The 1927 Solvay Conference

19

Receiving the Max Planck medal from its
namesake, 1929

20

In Leiden: Einstein, Ehrenfest, de Sitter in back;
Eddington and Lorentz in front; September 1923

21

With Paul Ehrenfest and Ehrenfest's son in Leiden

Niels Bohr and Einstein discussing quantum mechanics at Ehrenfest's home in Leiden, 1925, in a photo taken by Ehrenfest

23

Werner Heisenberg

24

Erwin Schrödinger

25

Max Born

26

Philipp Lenard

27

Vacationing on the Baltic Sea, 1928

28

Connecting to the cosmos

With Elsa and her daughter Margot, Berlin 1929

30

Margot and Ilse Einstein at the house in Caputh, 1929

31

In Caputh with his son Hans Albert and grandson
Bernhard, 1932

At the Mt. Wilson Observatory near Caltech, discovering that the universe is expanding, January 1931

32

Sailing against the prevailing currents, Long Island Sound, 1936

33

34

Welcoming Hans Albert to America, 1937

35

Margot, Einstein, and Helen Dukas being sworn in as U.S. citizens, October 1940

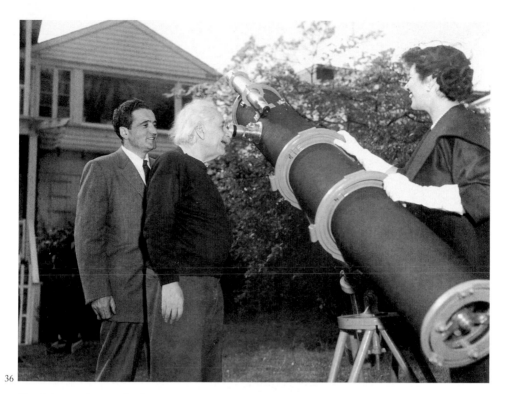

36

Receiving a telescope in the
backyard of 112 Mercer Street,
underneath the picture window
built for his study

37

With Kurt Gödel in Princeton, 1950

38

Princeton, 1953

Copenhagen and then of Max Born in Göttingen. As Einstein had done in his more radical youth, Heisenberg started by embracing Ernst Mach's dictum that theories should avoid any concepts that cannot be observed, measured, or verified. For Heisenberg this meant avoiding the concept of electron orbits, which could not be observed.

He relied instead on a mathematical approach that would account for something that *could* be observed: the wavelengths of the spectral lines of the radiation from these electrons as they lost energy. The result was so complex that Heisenberg gave his paper to Born and left on a camping trip with fellow members of his youth group, hoping that his mentor could figure it out. Born did. The math involved what are known as matrices, and Born sorted it all out and got the paper published.[57] In collaboration with Born and others in Göttingen, Heisenberg went on to perfect a matrix mechanics that was later shown to be equivalent to Schrödinger's wave mechanics.

Einstein politely wrote Born's wife, Hedwig, "The Heisenberg-Born concepts leave us breathless." Those carefully couched words can be read in a variety of ways. Writing to Ehrenfest in Leiden, Einstein was more blunt. "Heisenberg has laid a big quantum egg," he wrote. "In Göttingen they believe in it. I don't."[58]

Heisenberg's more famous and disruptive contribution came two years later, in 1927. It is, to the general public, one of the best known and most baffling aspects of quantum physics: the uncertainty principle.

It is impossible to know, Heisenberg declared, the precise *position* of a particle, such as a moving electron, and its precise *momentum* (its velocity times its mass) at the same instant. The more precisely the position of the particle is measured, the less precisely it is possible to measure its momentum. And the formula that describes the trade-off involves (no surprise) Planck's constant.

The very act of observing something—of allowing photons or electrons or any other particles or waves of energy to strike the object—affects the observation. But Heisenberg's theory went beyond that. An electron does not have a definite position or path until we observe it. This is a feature of our universe, he said, not merely some defect in our observing or measuring abilities.

The uncertainty principle, so simple and yet so startling, was a stake in the heart of classical physics. It asserts that there is no objective reality—not even an objective position of a particle—outside of our observations. In addition, Heisenberg's principle and other aspects of quantum mechanics undermine the notion that the universe obeys strict causal laws. Chance, indeterminacy, and probability took the place of certainty. When Einstein wrote him a note objecting to these features, Heisenberg replied bluntly, "I believe that indeterminism, that is, the nonvalidity of rigorous causality, is necessary."[59]

When Heisenberg came to give a lecture in Berlin in 1926, he met Einstein for the first time. Einstein invited him over to his house one evening, and there they engaged in a friendly argument. It was the mirror of the type of argument Einstein might have had in 1905 with conservatives who resisted his dismissal of the ether.

"We cannot observe electron orbits inside the atom," Heisenberg said. "A good theory must be based on directly observable magnitudes."

"But you don't seriously believe," Einstein protested, "that none but observable magnitudes must go into a physical theory?"

"Isn't that precisely what you have done with relativity?" Heisenberg asked with some surprise.

"Possibly I did use this kind of reasoning," Einstein admitted, "but it is nonsense all the same."[60]

In other words, Einstein's approach had evolved.

Einstein had a similar conversation with his friend in Prague, Philipp Frank. "A new fashion has arisen in physics," Einstein complained, which declares that certain things cannot be observed and therefore should not be ascribed reality.

"But the fashion you speak of," Frank protested, "was invented by you in 1905!"

Replied Einstein: "A good joke should not be repeated too often."[61]

The theoretical advances that occurred in the mid-1920s were shaped by Niels Bohr and his colleagues, including Heisenberg, into what became known as the Copenhagen interpretation of quantum mechanics. A property of an object can be discussed only in the context of how that property is observed or measured, and these observations are not simply aspects of a single picture but are complementary to one another.

In other words, there is no single underlying reality that is independent of our observations. "It is wrong to think that the task of physics is to find out how nature *is*," Bohr declared. "Physics concerns what we can *say* about nature." [62]

This inability to know a so-called "underlying reality" meant that there was no strict determinism in the classical sense. "When one wishes to calculate 'the future' from 'the present' one can only get statistical results," Heisenberg said, "since one can never discover every detail of the present." [63]

As this revolution climaxed in the spring of 1927, Einstein used the 200th anniversary of Newton's death to defend the classical system of mechanics based on causality and certainty. Two decades earlier, Einstein had, with youthful insouciance, toppled many of the pillars of Newton's universe, including absolute space and time. But now he was a defender of the established order, and of Newton.

In the new quantum mechanics, he said, strict causality seemed to disappear. "But the last word has not been said," Einstein argued. "May the spirit of Newton's method give us the power to restore union between physical reality and the profoundest characteristic of Newton's teaching—strict causality." [64]

Einstein never fully came around, even as experiments repeatedly showed quantum mechanics to be valid. He remained a realist, one who made it his creed to believe in an objective reality, rooted in certainty, that existed whether or not we could observe it.

"He does not play dice"

So what made Einstein cede the revolutionary road to younger radicals and spin into a defensive crouch?

As a young empiricist, excited by his readings of Ernst Mach, Einstein had been willing to reject any concepts that could not be observed, such as the ether and absolute time and space and simultaneity. But the success of his general theory convinced him that Mach's skepticism, even though it might be useful for weeding out superfluous concepts, did not provide much help in constructing new theories.

"He rides Mach's poor horse to exhaustion," Einstein complained to Michele Besso about a paper written by a mutual friend.

"We should not insult Mach's poor horse," Besso replied. "Didn't it make possible the tortuous journey through the relativities? And who knows, in the case of the nasty quanta, it may also carry Don Quixote de la Einsteina through it all!"

"You know what I think about Mach's little horse," Einstein wrote Besso in return. "It cannot give birth to anything living. It can only exterminate harmful vermin."[65]

In his maturity, Einstein more firmly believed that there was an objective "reality" that existed whether or not we could observe it. The belief in an external world independent of the person observing it, he repeatedly said, was the basis of all science.[66]

In addition, Einstein resisted quantum mechanics because it abandoned strict causality and instead defined reality in terms of indeterminacy, uncertainty, and probability. A true disciple of Hume would not have been troubled by this. There is no real reason—other than either a metaphysical faith or a habit ingrained in the mind—to believe that nature must operate with absolute certainty. It is just as reasonable, though perhaps less satisfying, to believe that some things simply happen by chance. Certainly, there was mounting evidence that on the subatomic level this was the case.

But for Einstein, this simply did not smell true. The ultimate goal of physics, he repeatedly said, was to discover the laws that strictly determine causes and effects. "I am very, very reluctant to give up complete causality," he told Max Born.[67]

His faith in determinism and causality reflected that of his favorite religious philosopher, Baruch Spinoza. "He was utterly convinced," Einstein wrote of Spinoza, "of the causal dependence of all phenomena, at a time when the success of efforts to achieve a knowledge of the causal relationship of natural phenomena was still quite modest."[68] It was a sentence that Einstein could have written about himself, emphasizing the temporariness implied by the word "still," after the advent of quantum mechanics.

Like Spinoza, Einstein did not believe in a personal God who interacted with man. But they both believed that a divine design was reflected in the elegant laws that governed the way the universe worked.

This was not merely some expression of faith. It was a principle that Einstein elevated (as he had the relativity principle) to the level of a postulate, one that guided him in his work. "When I am judging a theory," he told his friend Banesh Hoffmann, "I ask myself whether, if I were God, I would have arranged the world in such a way."

When he posed that question, there was one possibility that he simply could not believe: that the good Lord would have created beautiful and subtle rules that determined *most* of what happened in the universe, while leaving a few things completely to chance. It felt wrong. "If the Lord had wanted to do that, he would have done it thoroughly, and not kept to a pattern . . . He would have gone the whole hog. In that case, we wouldn't have to look for laws at all."[69]

This led to one of Einstein's most famous quotes, written to Max Born, the friend and physicist who would spar with him over three decades on this topic. "Quantum mechanics is certainly imposing," Einstein said. "But an inner voice tells me that it is not yet the real thing. The theory says a lot, but it does not really bring us any closer to the secrets of the Old One. I, at any rate, am convinced that He does not play dice."[70]

Thus it was that Einstein ended up deciding that quantum mechanics, though it may not be *wrong*, was at least *incomplete*. There must be a fuller explanation of how the universe operates, one that would incorporate both relativity theory and quantum mechanics. In doing so, it would not leave things to chance.

CHAPTER FIFTEEN

UNIFIED FIELD THEORIES

1923–1931

With Bohr at the 1930 Solvay Conference

The Quest

While others continued to develop quantum mechanics, undaunted by the uncertainties at its core, Einstein persevered in his lonelier quest for a more complete explanation of the universe—a unified field theory that would tie together electricity and magnetism and gravity and quantum mechanics. In the past, his genius had been in finding missing links between different theories. The opening sentences of his 1905 general relativity and light quanta papers were such examples.*

---

* From his 1905 special relativity paper: "It is well known that Maxwell's electrodynamics—as usually understood now—when applied to moving bodies leads to asymmetries that do not seem inherent in the phenomena. Take, for example, the electrodynamic

He hoped to extend the gravitational field equations of general relativity so that they would describe the electromagnetic field as well. "The mind striving after unification cannot be satisfied that two fields should exist which, by their nature, are quite independent," Einstein explained in his Nobel lecture. "We seek a mathematically unified field theory in which the gravitational field and the electromagnetic field are interpreted only as different components or manifestations of the same uniform field."[1]

Such a unified theory, he hoped, might make quantum mechanics compatible with relativity. He publicly enlisted Planck in this task with a toast at his mentor's sixtieth birthday celebration in 1918: "May he succeed in uniting quantum theory with electrodynamics and mechanics in a single logical system."[2]

Einstein's quest was primarily a procession of false steps, marked by increasing mathematical complexity, that began with his reacting to the false steps of others. The first was by the mathematical physicist Hermann Weyl, who in 1918 proposed a way to extend the geometry of general relativity that would, so it seemed, serve as a geometrization of the electromagnetic field as well.

Einstein was initially impressed. "It is a first-class stroke of genius," he told Weyl. But he had one problem with it: "I have not been able to settle my measuring-rod objection yet."[3]

Under Weyl's theory, measuring rods and clocks would vary depending on the path they took through space. But experimental observations showed no such phenomenon. In his next letter, after two more days of reflection, Einstein pricked his bubbles of praise with a wry putdown. "Your chain of reasoning is so wonderfully self-contained," he wrote Weyl. "Except for agreeing with reality, it is certainly a grand intellectual achievement."[4]

Next came a proposal in 1919 by Theodor Kaluza, a mathematics professor in Königsberg, that a fifth dimension be added to the four di-

---

interaction between a magnet and a conductor." From the 1905 light quanta paper: "A profound formal difference exists between the theories that physicists have formed about gases and other ponderable bodies, and Maxwell's theory of electromagnetic processes in so-called empty space."

mensions of spacetime. Kaluza further posited that this added spatial dimension was circular, meaning that if you head in its direction you get back to where you started, just like walking around the circumference of a cylinder.

Kaluza did not try to describe the physical reality or location of this added spatial dimension. He was, after all, a mathematician, so he didn't have to. Instead, he devised it as a mathematical device. The metric of Einstein's four-dimensional spacetime required ten quantities to describe all the possible coordinate relationships for any point. Kaluza knew that fifteen such quantities are needed to specify the geometry for a five-dimensional realm.[5]

When he played with the math of this complex construction, Kaluza found that four of the extra five quantities could be used to produce Maxwell's electromagnetic equations. At least mathematically, this might be a way to produce a field theory unifying gravity and electromagnetism.

Once again, Einstein was both impressed and critical. "A five-dimensional cylinder world never dawned on me," he wrote Kaluza. "At first glance I like your idea enormously."[6] Unfortunately, there was no reason to believe that most of this math actually had any basis in physical reality. With the luxury of being a pure mathematician, Kaluza admitted this and challenged the physicists to figure it out. "It is still hard to believe that all of these relations in their virtually unsurpassed formal unity should amount to the mere alluring play of a capricious accident," he wrote. "Should more than an empty mathematical formalism be found to reside behind these presumed connections, we would then face a new triumph of Einstein's general relativity."

By then Einstein had become a convert to the faith in mathematical formalism, which had proven so useful in his final push toward general relativity. Once a few issues were sorted out, he helped Kaluza get his paper published in 1921, and followed up later with his own pieces.

The next contribution came from the physicist Oskar Klein, son of Sweden's first rabbi and a student of Niels Bohr. Klein saw a unified field theory not only as a way to unite gravity and electromagnetism, but he also hoped it might explain some of the mysteries lurking in quantum mechanics. Perhaps it could even come up with a way to find "hidden variables" that could eliminate the uncertainty.

Klein was more a physicist than a mathematician, so he focused more than Kaluza had on what the physical reality of a fourth spatial dimension might be. His idea was that it might be coiled up in a circle, too tiny to detect, projecting out into a new dimension from every point in our observable three-dimensional space.

It was all quite ingenious, but it didn't turn out to explain much about the weird but increasingly well-confirmed insights of quantum mechanics or the new advances in particle physics. The Kaluza-Klein theories were put aside, although Einstein over the years would return to some of the concepts. In fact, physicists still do today. Echoes of these ideas, particularly in the form of extra compact dimensions, exist in string theory.

Next into the fray came Arthur Eddington, the British astronomer and physicist responsible for the famous eclipse observations. He refined Weyl's math by using a geometric concept known as an affine connection. Einstein read Eddington's ideas while on his way to Japan, and he adopted them as the basis for a new theory of his own. "I believe I have finally understood the connection between electricity and gravitation," he wrote Bohr excitedly. "Eddington has come closer to the truth than Weyl."[7]

By now the siren song of a unified theory had come to mesmerize Einstein. "Over it lingers the marble smile of nature," he told Weyl.[8] On his steamer ride through Asia, he polished a new paper and, upon arriving in Egypt in February 1923, immediately mailed it to Planck in Berlin for publication. His goal, he declared, was "to understand the gravitational and electromagnetic field as one."[9]

Once again, Einstein's pronouncements made headlines around the world. "Einstein Describes His Newest Theory," proclaimed the *New York Times*. And once again, the complexity of his approach was played up. As one of the subheads warned: "Unintelligible to Laymen."

But Einstein told the newspaper it was not all that complicated. "I can tell you in one sentence what it is about," the reporter quoted him as saying. "It concerns the relation between electricity and gravitation."

He also gave credit to Eddington, saying, "It is grounded on the theories of the English astronomer."[10]

In his follow-up articles that year, Einstein made explicit that his goal was not merely unification but finding a way to overcome the uncertainties and probabilities in quantum theory. The title of one 1923 paper stated the quest clearly: "Does the Field Theory Offer Possibilities for the Solution of Quanta Problems?"[11]

The paper began by describing how electromagnetic and gravitational field theories provide causal determinations based on partial differential equations combined with initial conditions. In the realm of the quanta, it may not be possible to choose or apply the initial conditions freely. Can we nevertheless have a causal theory based on field equations?

"Quite certainly," Einstein answered himself optimistically. What was needed, he said, was a method to "overdetermine" the field variables in the appropriate equations. That path of overdetermination became yet another proposed tool that he would employ, to no avail, in fixing what he persisted in calling the "problem" of quantum uncertainty.

Within two years, Einstein had concluded that these approaches were flawed. "My article published [in 1923]," he wrote, "does not reflect the true solution of this problem." But for better or worse, he had come up with yet another method. "After searching ceaselessly in the past two years, I think I have now found the true solution."

His new approach was to find the simplest formal expression he could of the law of gravitation in the absence of any electromagnetic field and then generalize it. Maxwell's theory of electromagnetism, he thought, resulted in a first approximation.[12]

He now was relying more on math than on physics. The metric tensor that he had featured in his general relativity equations had ten independent quantities, but if it were made nonsymmetrical there would be sixteen of them, enough to accommodate electromagnetism.

But this approach led nowhere, just like the others. "The trouble with this idea, as Einstein became painfully aware, is that there really is nothing in it that ties the 6 components of the electric and magnetic fields to the 10 components of the ordinary metric tensor that describes gravitation," says University of Texas physicist Steven Weinberg. "A Lorentz transformation or any other coordinate transformation will convert electric or magnetic fields into mixtures of electric and magnetic fields, but no transformation mixes them with the gravitational field."[13]

Undaunted, Einstein went back to work, this time trying an approach he called "distant parallelism." It permitted vectors in different parts of curved space to be related, and from that sprang new forms of tensors. Most wondrously (so he thought), he was able to come up with equations that did not require that pesky Planck constant representing quanta.[14]

"This looks old-fashioned, and my dear colleagues, and also you, will stick their tongues out because Planck's constant is not in the equations," he wrote Besso in January 1929. "But when they have reached the limit of their mania for the statistical fad, they will return full of repentance to the spacetime picture, and then these equations will form a starting point."[15]

What a wonderful dream! A unified theory without that rambunctious quantum. Statistical approaches turning out to be a passing mania. A return to the field theories of relativity. Tongue-sticking colleagues repenting!

In the world of physics, where quantum mechanics was now accepted, Einstein and his fitful quest for a unified theory were beginning to be seen as quaint. But in the popular imagination, he was still a superstar. The frenzy that surrounded the publication of his January 1929 five-page paper, which was merely the latest in a string of theoretical stabs that missed the mark, was astonishing. Journalists from around the world crowded around his apartment building, and Einstein was barely able to escape them to go into hiding at his doctor's villa on the Havel River outside of town. The *New York Times* had started the drumbeat weeks earlier with an article headlined "Einstein on Verge of Great Discovery: Resents Intrusion."[16]

Einstein's paper was not made public until January 30, 1929, but for the entire preceding month the newspapers printed a litany of leaks and speculation. A sampling of the headlines in the *New York Times*, for example, include these:

January 12: "Einstein Extends Relativity Theory / New Work Seeks to Unite Laws of Field of Gravitation and Electro-Magnetism / He Calls It His Greatest 'Book' / Took Berlin Scientist Ten Years to Prepare"

January 19: "Einstein Is Amazed at Stir Over Theory / Holds 100 Journalists at Bay for a Week / BERLIN—For the past week the entire press as represented here has concentrated efforts on procuring the five-page manuscript of Dr. Albert Einstein's 'New Field of Theory.' Furthermore, hundreds of cables from all parts of the world, with prepaid answers and innumerable letters asking for a detailed description or a copy of the manuscript have arrived."

January 25 (page 1): "Einstein Reduces All Physics to One Law / The New Electro-Gravitational Theory Links All Phenomena, Says Berlin Interpreter / Only One Substance Also / Hypothesis Opens Visions of Persons Being Able to Float in Air, Says N.Y.U. Professor / BERLIN—Professor Albert Einstein's newest work, 'A New Field Theory,' which will leave the press soon, reduces to one formula the basic laws of relativistic mechanics and of electricity, according to the person who has interpreted it into English."

Einstein got into the act from his Havel River hideaway. Even before his little paper was published, he gave an interview about it to a British newspaper. "It has been my greatest ambition to resolve the duality of natural laws into unity," he said. "The purpose of my work is to further this simplification, and particularly to reduce to one formula the explanation of the gravitational and electromagnetic fields. For this reason I call it a contribution to 'a unified field theory' . . . Now, but only now, we know that the force that moves electrons in their ellipses about the nuclei of atoms is the same force that moves our earth in its annual course around the sun."[17] Of course, it turned out that he did not know that, nor do we know that even now.

He also gave an interview to *Time,* which put him on its cover, the first of five such appearances. The magazine reported that, while the world waited for his "abstruse coherent field theory" to be made public, Einstein was plodding around his country hideaway looking "haggard, nervous, irritable." His sickly demeanor, the magazine explained, was due to stomach ailments and a constant parade of visitors. In addition, it noted, "Dr. Einstein, like so many other Jews and scholars, takes no physical exercise at all."[18]

The Prussian Academy printed a thousand copies of Einstein's paper, an unusually large number. When it was released on January 30, all were promptly sold, and the Academy went back to the printer for three thousand more. One set of pages was pasted in the window of a London department store, where crowds pushed forward to try to comprehend the complex mathematical treatise with its thirty-three arcane equations not tailored for window shoppers. Wesleyan University in Connecticut paid a significant sum for the handwritten manuscript to be deposited as a treasure in its library.

American newspapers were somewhat at a loss. The *New York Herald Tribune* decided to print the entire paper verbatim, but it had trouble figuring out how to cable all the Greek letters and symbols over telegraph machines. So it hired some Columbia physics professors to devise a coding system and then reconstruct the paper in New York, which they did. The *Tribune's* colorful article about how they transmitted the paper was a lot more comprehensible to most readers than Einstein's paper itself.[19]

The *New York Times,* for its part, raised the unified theory to a religious level by sending reporters that Sunday to churches around the city to report on the sermons about it. "Einstein Viewed as Near Mystic," the headline declared. The Rev. Henry Howard was quoted as saying that Einstein's unified theory supported St. Paul's synthesis and the world's "oneness." A Christian Scientist said it provided scientific backing for Mary Baker Eddy's theory of illusive matter. Others hailed it as "freedom advanced" and a "step to universal freedom."[20]

Theologians and journalists may have been wowed, but physicists were not. Eddington, usually a fan, expressed doubts. Over the next year, Einstein kept refining his theory and insisting to friends that the equations were "beautiful." But he admitted to his dear sister that his work had elicited "the lively mistrust and passionate rejection of my colleagues."[21]

Among those who were dismayed was Wolfgang Pauli. Einstein's new approaches "betrayed" his general theory of relativity, Pauli sharply told him, and relied on mathematical formalism that had no relation to physical realities. He accused Einstein of "having gone over to the pure mathematicians," and he predicted that "within a year, if not before, you will have abandoned that whole distant parallelism, just as earlier you gave up the affine theory."[22]

Pauli was right. Einstein gave up the theory within a year. But he did not give up the quest. Instead, he turned his attention to yet another revised approach that would make more headlines but not more headway in solving the great riddle he had set for himself. "Einstein Completes Unified Field Theory," the *New York Times* reported on January 23, 1931, with little intimation that it was neither the first nor would it be the last time there would be such an announcement. And then again, on October 26 of that year: "Einstein Announces a New Field Theory."

Finally, the following January, he admitted to Pauli, "So you were right after all, you rascal."[23]

And so it went, for another two decades. None of Einstein's offerings ever resulted in a successful unified field theory. Indeed, with the discoveries of new particles and forces, physics was becoming *less* unified. At best, Einstein's effort was justified by the faint praise from the French mathematician Elie Joseph Cartan in 1931: "Even if his attempt does not succeed, it will have forced us to think about the great questions at the foundation of science."[24]

The Great Solvay Debates, 1927 and 1930

The tenacious rearguard action that Einstein waged against the onslaught of quantum mechanics came to a climax at two memorable Solvay Conferences in Brussels. At both he played the provocateur, trying to poke holes in the prevailing new wisdom.

Present at the first, in October 1927, were the three grand masters who had helped launch the new era of physics but were now skeptical of the weird realm of quantum mechanics it had spawned: Hendrik Lorentz, 74, just a few months from death, the winner of the Nobel for his work on electromagnetic radiation; Max Planck, 69, winner of the Nobel for his theory of the quantum; and Albert Einstein, 48, winner of the Nobel for discovering the law of the photoelectric effect.

Of the remaining twenty-six attendees, more than half had won or would win Nobel Prizes as well. The boy wonders of the new quantum mechanics were all there, hoping to convert or conquer Einstein: Werner Heisenberg, 25; Paul Dirac, 25; Wolfgang Pauli, 27; Louis de Broglie, 35; and from America, Arthur Compton, 35. Also there was Erwin Schrödinger, 40, caught between the young Turks and the older skeptics. And, of course, there was the old Turk, Niels Bohr, 42, who had helped spawn quantum mechanics with his model of the atom and become the staunch defender of its counterintuitive ramifications.[25]

Lorentz had asked Einstein to present the conference's report on the state of quantum mechanics. Einstein accepted, then balked. "After much back and forth, I have concluded that I am not competent to give such a report in a way that would match the current state of affairs," he replied. "In part it is because I do not approve of the purely statistical method of thinking on which the new theories are based." He then added rather plaintively, "I beg you not to be angry with me."[26]

Instead, Niels Bohr gave the opening presentation. He was unsparing in his description of what quantum mechanics had wrought. Certainty and strict causality did not exist in the subatomic realm, he said. There were no deterministic laws, only probabilities and chance. It made no sense to speak of a "reality" that was independent of our observations and measurements. Depending on the type of experiment chosen, light could be waves or particles.

Einstein said little at the formal sessions. "I must apologize for not having penetrated quantum mechanics deeply enough," he admitted at the very outset. But over dinners and late-night discussions, resuming again at breakfast, he would engage Bohr and his supporters in animated discourse that was leavened by affectionate banter about dice-playing deities. "One can't make a theory out of a lot of 'maybes,' " Pauli recalls Einstein arguing. "Deep down it is wrong, even if it is empirically and logically right."[27]

"The discussions were soon focused to a duel between Einstein and Bohr about whether atomic theory in its present form could be considered to be the ultimate solution," Heisenberg recalled.[28] As Ehrenfest told his students afterward, "Oh, it was delightful."[29]

Einstein kept lobbing up clever thought experiments, both in sessions and in the informal discussions, designed to prove that quantum mechanics did not give a complete description of reality. He tried to show how, through some imagined contraption, it would be possible, at least in concept, to measure all of the characteristics of a moving particle, with certainty.

For example, one of Einstein's thought experiments involved a beam of electrons that is sent through a slit in a screen, and then the positions of the electrons are recorded as they hit a photographic plate. Various other elements, such as a shutter to open and close the slit instantaneously, were posited by Einstein in his ingenious efforts to show that position and momentum could in theory be known with precision.

"Einstein would bring along to breakfast a proposal of this kind," Heisenberg recalled. He did not worry much about Einstein's machinations, nor did Pauli. "It will be all right," they kept saying, "it will be all right." But Bohr would often get worked up into a muttering frenzy.

The group would usually make their way to the Congress hall together, working on ways to refute Einstein's problem. "By dinner-time we could usually prove that his thought experiments did not contradict uncertainty relations," Heisenberg recalled, and Einstein would concede defeat. "But next morning he would bring along to breakfast a new thought experiment, generally more complicated than the previous one." By dinnertime that would be disproved as well.

Back and forth they went, each lob from Einstein volleyed back by Bohr, who was able to show how the uncertainty principle, in each instance, did indeed limit the amount of knowable information about a moving electron. "And so it went for several days," said Heisenberg. "In the end, we—that is, Bohr, Pauli, and I—knew that we could now be sure of our ground."[30]

"Einstein, I'm ashamed of you," Ehrenfest scolded. He was upset that Einstein was displaying the same stubbornness toward quantum mechanics that conservative physicists had once shown toward relativity. "He now behaves toward Bohr exactly as the champions of absolute simultaneity had behaved toward him."[31]

Einstein's own remarks, given on the last day of the conference, show that the uncertainty principle was not the only aspect of quantum mechanics that concerned him. He was also bothered—and later would become even more so—by the way quantum mechanics seemed to permit action at a distance. In other words, something that happened to one object could, according to the Copenhagen interpretation, instantly determine how an object located somewhere else would be observed. Particles separated in space are, according to relativity theory, independent. If an action involving one can immediately affect another some distance away, Einstein noted, "in my opinion it contradicts the relativity postulate." No force, including gravity, can propagate faster than the speed of light, he insisted.[32]

Einstein may have lost the debates, but he was still the star of the event. De Broglie had been looking forward to meeting him for the first time, and he was not disappointed. "I was particularly struck by his mild and thoughtful expression, by his general kindness, by his simplicity and by his friendliness," he recalled.

The two hit it off well, because de Broglie was trying, like Einstein, to see if there were ways that the causality and certainty of classical physics could be saved. He had been working on what he called "the theory of the double solution," which he hoped would provide a classical basis for wave mechanics.

"The indeterminist school, whose adherents were mainly young and intransigent, met my theory with cold disapproval," de Broglie recalled. Einstein, on the other hand, appreciated de Broglie's efforts, and he rode the train with him to Paris on his way back to Berlin.

At the Gare du Nord they had a farewell talk on the platform. Einstein told de Broglie that all scientific theories, leaving aside their mathematical expressions, ought to lend themselves to so simple a description "that even a child could understand them." And what could be *less* simple, Einstein continued, than the purely statistical interpretation of wave mechanics! "Carry on," he told de Broglie as they parted at the station. "You are on the right track!"

But he wasn't. By 1928, a consensus had formed that quantum mechanics was correct, and de Broglie relented and adopted that view. "Einstein, however, stuck to his guns and continued to insist that the purely statistical interpretation of wave mechanics could not possibly be complete," de Broglie recalled, with some reverence, years later.[33]

Indeed, Einstein remained the stubborn contrarian. "I admire to the highest degree the achievements of the younger generation of physicists that goes by the name quantum mechanics, and I believe in the deep level of truth of that theory," he said in 1929 when accepting the Planck medal from Planck himself. "But"—and there was always a *but* in any statement of support Einstein gave to quantum theory—"I believe that the restriction to statistical laws will be a passing one."[34]

The stage was thus set for an even more dramatic Solvay showdown between Einstein and Bohr, this one at the conference of October 1930. Theoretical physics has rarely seen such an interesting engagement.

This time, in his effort to stump the Bohr-Heisenberg group and restore certainty to mechanics, Einstein devised a more clever thought experiment. One aspect of the uncertainty principle, previously mentioned, is that there is a trade-off between measuring precisely the momentum of a particle and its position. In addition, the principle says that a similar uncertainty is inherent in measuring the energy involved in a process and the time duration of that process.

Einstein's thought experiment involved a box with a shutter that could open and shut so rapidly that it would allow only one photon to escape at a time. The shutter is controlled by a precise clock. The box is weighed exactly. Then, at a certain specified moment, the shutter opens and a photon escapes. The box is now weighed again. The relationship between energy and mass (remember, $E=mc^2$) permitted a precise determination of the energy of the particle. And we know, from the clock, its exact time of departing the system. So there!

Of course, physical limitations would make it impossible to actually *do* such an experiment. But in theory, did it refute the uncertainty principle?

Bohr was shaken by the challenge. "He walked from one person to another, trying to persuade them all that this could not be true, that it would mean the end of physics if Einstein was right," a participant recorded. "But he could think of no refutation. I will never forget the sight of the two opponents leaving the university club. Einstein, a majestic figure, walking calmly with a faint ironic smile, and Bohr trotting along by his side, extremely upset."[35] (See picture, page 336.)

It was one of the great ironies of scientific debate that, after a sleepless night, Bohr was able to hoist Einstein by his own petard. The thought experiment had not taken into account Einstein's own beautiful discovery, the theory of relativity. According to that theory, clocks in stronger gravitational fields run more slowly than those in weaker gravity. Einstein forgot this, but Bohr remembered. During the release of the photon, the mass of the box decreases. Because the box is on a spring scale (in order to be weighed), the box will rise a small amount in the earth's gravity. That small amount is precisely the amount needed to restore the energy-time uncertainty relation.

"It was essential to take into account the relationship between the rate of a clock and its position in a gravitational field," Bohr recalled. He gave Einstein credit for graciously helping to perform the calculations that, in the end, won the day for the uncertainty principle. But Einstein was never fully convinced. Even a year later, he was still churning out variations of such thought experiments.[36]

Quantum mechanics ended up proving to be a successful theory, and Einstein subsequently edged into what could be called his own version of uncertainty. He no longer denounced quantum mechanics as incorrect, only as incomplete. In 1931, he nominated Heisenberg and Schrödinger for the Nobel Prize. (They won in 1932 and 1933, along with Dirac.) "I am convinced that this theory undoubtedly contains a part of the ultimate truth," Einstein wrote in his nominating letter.

Part of the ultimate truth. There was still, Einstein felt, more to reality than was accounted for in the Copenhagen interpretation of quantum mechanics.

Its shortcoming was that it "makes no claim to describe physical reality itself, but only the *probabilities* of the occurrence of a physical reality that we view," he wrote that year in a tribute to James Clerk Maxwell, the master of his beloved field theory approach to physics. His piece concluded with a resounding realist credo—a direct denial of Bohr's declaration that physics concerns not what nature *is* but merely "what we can *say* about nature"—that would have raised the eyebrows of Hume, Mach, and possibly even a younger Einstein. He declared, "Belief in an external world independent of the perceiving subject is the basis of all natural science."[37]

Wresting Principles from Nature

In his more radical salad days, Einstein did not emphasize this credo. He had instead cast himself as an empiricist or positivist. In other words, he had accepted the works of Hume and Mach as sacred texts, which led him to shun concepts, like the ether or absolute time, that were not knowable through direct observations.

Now, as his opposition to the concept of an ether became more subtle and his discomfort with quantum mechanics grew, he edged away from this orthodoxy. "What I dislike in this kind of argumentation," the older Einstein reflected, "is the basic positivistic attitude, which from my point of view is untenable, and which seems to me to come to the same thing as Berkeley's principle, *Esse est percipi.*"[*][38]

There was a lot of continuity in Einstein's philosophy of science, so it would be wrong to insist that there was a clean shift from empiricism to realism in his thinking.[39] Nonetheless, it is fair to say that as he struggled against quantum mechanics during the 1920s, he became less faithful to the dogma of Mach and more of a realist, someone who believed, as he said in his tribute to Maxwell, in an underlying reality that exists independently of our observations.

That was reflected in a lecture that Einstein gave at Oxford in June 1933, called "On the Method of Theoretical Physics," which sketched out his philosophy of science.[40] It began with a caveat. To truly understand the methods and philosophy of physicists, he said, "don't listen to their words, fix your attention on their deeds."

If we look at what Einstein did rather than what he was saying, it is clear that he believed (as any true scientist would) that the end product of any theory must be conclusions that can be confirmed by experience and empirical tests. He was famous for ending his papers with calls for these types of suggested experiments.

But how did he come up with the starting blocks for his theoretical

---

* "To be is to be perceived," meaning that it makes no sense to say that unperceived things—most famously Berkeley's example of trees in a forest "and no body by to perceive them"—actually exist (George Berkeley, *Principles of Human Knowledge,* section 23).

thinking—the principles and postulates that would launch his logical deductions? As we've seen, he did not usually start with a set of experimental data that needed some explanation. "No collection of empirical facts, however comprehensive, can ever lead to the formulation of such complicated equations," he said in describing how he had come up with the general theory of relativity.[41] In many of his famous papers, he made a point of insisting that he had not relied much on any specific experimental data—on Brownian motion, or attempts to detect the ether, or the photoelectric effect—to induce his new theories.

Instead, he generally began with postulates that he had abstracted from his understanding of the physical world, such as the equivalence of gravity and acceleration. That equivalence was not something he came up with by studying empirical data. Einstein's great strength as a theorist was that he had a keener ability than other scientists to come up with what he called "the general postulates and principles which serve as the starting point."

It was a process that mixed intuition with a feel for the patterns to be found in experimental data. "The scientist has to worm these general principles out of nature by discerning, when looking at complexes of empirical facts, certain general features."[42] When he was struggling to find a foothold for a unified theory, he captured the essence of this process in a letter to Hermann Weyl: "I believe that, in order to make any real progress, one would again have to find a general principle wrested from Nature."[43]

Once he had wrested a principle from nature, he relied on a byplay of physical intuition and mathematical formalism to march toward some testable conclusions. In his younger days, he sometimes disparaged the role that pure math could play. But during his final push toward a general theory of relativity, it was the mathematical approach that ended up putting him across the goal line.

From then on, he became increasingly dependent on mathematical formalism in his pursuit of a unified field theory. "The development of the general theory of relativity introduced Einstein to the power of abstract mathematical formalisms, notably that of tensor calculus," writes the astrophysicist John Barrow. "A deep physical insight orchestrated the mathematics of general relativity, but in the years that followed the balance tipped the other way. Einstein's search for a unified theory was characterized by a fascination with the abstract formalisms themselves."[44]

In his Oxford lecture, Einstein began with a nod to empiricism: "All knowledge of reality starts from experience and ends in it." But he immediately proceeded to emphasize the role that "pure reason" and logical deductions play. He conceded, without apology, that his success using tensor calculus to come up with the equations of general relativity had converted him to a faith in a mathematical approach, one that emphasized the simplicity and elegance of equations more than the role of experience.

The fact that this method paid off in general relativity, he said, "justifies us in believing that *nature is the realization of the simplest conceivable mathematical ideas."*[45] That is an elegant—and also astonishingly interesting—creed. It captured the essence of Einstein's thought during the decades when mathematical "simplicity" guided him in his search for a unified field theory. And it echoed the great Isaac Newton's declaration in book 3 of the *Principia:* "Nature is pleased with simplicity."

But Einstein offered no proof of this creed, one that seems belied by modern particle physics.[46] Nor did he ever fully explain what, exactly, he meant by mathematical simplicity. Instead, he merely asserted his deep intuition that this is the way God would make the universe. "I am convinced that we can discover by means of purely mathematical constructions the concepts and the laws connecting them with each other," he claimed.

It was a belief—indeed, a faith—that he had expressed during his previous visit to Oxford, when in May 1931 he had been awarded an honorary doctorate there. In his lecture on that occasion, Einstein explained that his ongoing quest for a unified field theory was propelled by the lure of mathematical elegance, rather than the push of experimental data. "I have been guided not by the pressure from behind of experimental facts, but by the attraction in front from mathematical simplicity," he said. "It can only be hoped that experiments will follow the mathematical flag."[47]

Einstein likewise concluded his 1933 Oxford lecture by saying that he had come to believe that the mathematical equations of field theories were the best way to grasp "reality." So far, he admitted, this had not worked at the subatomic level, which seemed ruled by chance and probabilities. But he told his audience that he clung to the belief that this was not the final word. "I still believe in the possibility of a model of reality—that is to say, of a theory that represents things themselves and not merely the probability of their occurrence."[48]

His Greatest Blunder?

Back in 1917, when Einstein had analyzed the "cosmological considerations" arising from his general theory of relativity, most astronomers thought that the universe consisted only of our Milky Way, floating with its 100 billion or so stars in a void of empty space. Moreover, it seemed a rather stable universe, with stars meandering around but not expanding outward or collapsing inward in a noticeable way.

All of this led Einstein to add to his field equations a cosmological constant that represented a "repulsive" force (see page 254). It was invented to counteract the gravitational attraction that would, if the stars were not flying away from one another with enough momentum, pull all of them together.

Then came a series of wondrous discoveries, beginning in 1924, by Edwin Hubble, a colorful and engaging astronomer working with the 100-inch reflector telescope at the Mount Wilson Observatory in the mountains above Pasadena, California. The first was that the blur known as the Andromeda nebula was actually another galaxy, about the size of our own, close to a million light years away (we now know it's more than twice that far). Soon he was able to find at least two dozen even more distant galaxies (we now believe that there are more than 100 billion of them).

Hubble then made an even more amazing discovery. By measuring the red shift of the stars' spectra (which is the light wave counterpart to the Doppler effect for sound waves), he realized that the galaxies were moving away from us. There were at least two possible explanations for the fact that distant stars in all directions seemed to be flying away from us: (1) because we are the center of the universe, something that since the time of Copernicus only our teenage children believe; (2) because the entire metric of the universe was expanding, which meant that everything was stretching out in all directions so that all galaxies were getting farther away from one another.

It became clear that the second explanation was the case when Hubble confirmed that, in general, the galaxies were moving away from us at a speed that was proportional to their distance from us. Those twice as far moved away twice as fast, and those three times as far moved away three times as fast.

One way to understand this is to imagine a grid of dots that are each spaced an inch apart on the elastic surface of a balloon. Then assume that the balloon is inflated so that the surface expands to twice its original dimensions. The dots are now two inches away from each other. So during the expansion, a dot that was originally one inch away moved another one inch away. And during that same time period, a dot that was originally two inches away moved another two inches away, one that was three inches away moved another three inches away, and one that was ten inches away moved another ten inches away. The farther away each dot was originally, the faster it receded from our dot. And that would be true from the vantage point of each and every dot on the balloon.

All of which is a simple way to say that the galaxies are not merely flying away from us, but instead, the entire metric of space, or the fabric of the cosmos, is expanding. To envision this in 3-D, imagine that the dots are raisins in a cake that is baking and expanding in all directions.

On his second visit to America in January 1931, Einstein decided to go to Mount Wilson (conveniently up the road from Caltech, where he was visiting) to see for himself. He and Edwin Hubble rode in a sleek Pierce-Arrow touring car up the winding road. There at the top to meet him was the aging and ailing Albert Michelson, of ether-drift experiment fame.

It was a sunny day, and Einstein merrily played with the telescope's dials and instruments. Elsa came along as well, and it was explained to her that the equipment was used to determine the scope and shape of the universe. She reportedly replied, "Well, my husband does that on the back of an old envelope."[49]

The evidence that the universe was expanding was presented in the popular press as a challenge to Einstein's theories. It was a scientific drama that captured the public imagination. "Great stellar systems," an Associated Press story began, "rushing away from the earth at 7,300 miles a second, offer a problem to Dr. Albert Einstein."[50]

But Einstein welcomed the news. "The people at the Mt. Wilson observatory are outstanding," he wrote Besso. "They have recently found that the spiral nebulae are distributed approximately uniformly in space, and they show a strong Doppler effect, proportional to their distances, that one can readily deduce from general relativity theory without the 'cosmological' term."

In other words, the cosmological constant, which he had reluctantly concocted to account for a static universe, was apparently not necessary, for the universe was in fact expanding.* "The situation is truly exciting," he exulted to Besso.[51]

Of course, it would have been even more exciting if Einstein had trusted his original equations and simply announced that his general theory of relativity predicted that the universe is expanding. If he had done that, then Hubble's confirmation of the expansion more than a decade later would have had as great an impact as when Eddington confirmed his prediction of how the sun's gravity would bend rays of light. The Big Bang might have been named the Einstein Bang, and it would have gone down in history, as well as in the popular imagination, as one of the most fascinating theoretical discoveries of modern physics.[52]

As it was, Einstein merely had the pleasure of renouncing the cosmological constant, which he had never liked.[53] In a new edition of his popular book on relativity published in 1931, he added an appendix explaining why the term he had pasted into his field equations was, thankfully, no longer necessary.[54] "When I was discussing cosmological problems with Einstein," George Gamow later recalled, "he remarked

---

* As Eddington showed, the cosmological term probably would not have worked even if the universe had turned out to be static. Because it required such a delicate balance, any small disturbance would have caused a runaway expansion or contraction of the universe.

that the introduction of the cosmological term was the biggest blunder he ever made in his life." [55]

In fact, Einstein's blunders were more fascinating and complex than even the triumphs of lesser scientists. It was hard simply to banish the term from the field equations. "Unfortunately," says Nobel laureate Steven Weinberg, "it was not so easy just to drop the cosmological constant, because anything that contributes to the energy density of the vacuum acts just like a cosmological constant." [56]

It turns out that the cosmological constant not only was difficult to banish but is still needed by cosmologists, who use it today to explain the accelerating expansion of the universe. [57] The mysterious dark energy that seems to cause this expansion behaves as if it were a manifestation of Einstein's constant. As a result, two or three times each year fresh observations produce reports that lead with sentences along the lines of this one from November 2005: "The genius of Albert Einstein, who added a 'cosmological constant' to his equation for the expansion of the universe but then retracted it, may be vindicated by new research." [58]

CHAPTER SIXTEEN

TURNING FIFTY

1929–1931

Einstein's house in Caputh near Berlin

Caputh

Einstein wanted some solitude for his fiftieth birthday, a refuge from publicity. So in March 1929 he fled once again, as he had during the publication of his unified field theory paper of a few months earlier, to the gardener's cottage of an estate on the Havel River owned by Janos Plesch, a flamboyant and gossipy Hungarian-born celebrity doctor who had added Einstein to his showcase collection of patient-friends.

For days he lived by himself, cooking his own meals, while journalists and official well-wishers searched for him. His whereabouts became a matter of newspaper speculation. Only his family and assistant knew where he was, and they refused to tell even close friends.

Early on the morning of his birthday, he walked from this hideaway, which had no phone, to a nearby house to call Elsa. She started to wish him well on reaching the half-century mark, but he interrupted.

"Such a fuss about a birthday," he laughed. He was phoning about a matter involving physics, not the merely personal. He had made a small mistake in some calculations he had given to his assistant Walther Mayer, he told her, and he wanted her to take down the corrections and pass them along.

Elsa and her daughters came out that afternoon for a small, private celebration. She was dismayed to find him in his oldest suit, which she had hidden. "How did you manage to find it?" she asked.

"Ah," he replied, "I know all about those hiding places."[1]

The *New York Times*, as intrepid as ever, was the only paper that managed to track him down. A family member later recalled that Einstein's angry look drove the reporter away. That was not true. The reporter was smart and Einstein, despite his feigned fury, was as accommodating as usual. "Einstein Is Found Hiding on His Birthday" was the paper's headline. He showed the reporter a microscope he had been given as a gift, and the paper reported that he was like a "delighted boy" with a new toy.[2]

From around the world came other gifts and greetings. The ones that moved him the most were from ordinary people. A seamstress had sent him a poem, and an unemployed man had saved a few coins to get him a small packet of tobacco. The latter gift brought tears to his eyes and was the first for which he wrote a thank-you letter.[3]

Another birthday gift caused more problems. The city of Berlin, at the suggestion of the ever-meddling Dr. Plesch, decided to honor its most famous citizen by giving him lifelong rights to live in a country house that was part of a large lakeside estate that the city had acquired. There he would be able to escape, sail his wooden boat, and scribble his equations in serenity.

It was a generous and gracious gesture. It was also a welcome one. Einstein loved sailing and solitude and simplicity, but he owned no weekend retreat and had to store his sailboat with friends. He was thrilled to accept.

The house, in a classical style, was nestled in a park near the village of Cladow on a lake of the Havel River. Pictures of it appeared in the papers, and a relative called it "the ideal residence for a person of creative intellect and a man fond of sailing." But when Elsa went to inspect it, she found still living there the aristocratic couple who sold the estate to the city. They claimed that they had retained the right to live on the property. A study of the documents proved them right, and they could not be evicted.

So the city decided to give the Einsteins another part of the estate on which they could build their own home. But that, too, violated the city's purchase agreement. Pressure and publicity only hardened the resolve of the original family to block the Einsteins from building on the land, and it became an embarrassing front-page fiasco, especially after a third suggested alternative also proved unsuitable.

Finally it was decided that the Einsteins should simply find their own piece of land, and the city would buy it. So Einstein picked out a parcel, owned by some friends, farther out of town near a village just south of Potsdam called Caputh. It was in a sylvan spot between the Havel and a dense forest, and Einstein loved it. The mayor accordingly asked the assembly of city deputies to approve spending 20,000 marks to buy the property as the fiftieth birthday gift to Einstein.

A young architect drew up plans, and Einstein bought a small garden plot nearby. Then politics intervened. In the assembly, the right-wing German Nationalists objected, delayed the vote, and insisted that the proposal be put on a future agenda for a full debate. It became clear that Einstein personally would become the focus of that debate.

So he wrote a letter, tinged with amusement, declining the gift. "Life is very short," he told the mayor, "while the authorities work slowly. My birthday is already past, and I decline the gift." The headline the next day in the *Berliner Tageblatt* newspaper read, "Public Disgrace Complete / Einstein Declines."[4]

By this point, the Einsteins had fallen in love with the plot of land in Caputh, negotiated its purchase, and had a design for a house to build upon it. So they went ahead and bought it with their own money. "We have spent most of our savings," Elsa complained, "but we have our land."

The house they built was simple, with polished wood panels inside and unvarnished planks showing to the outside. Through a large picture window was a serene view of the Havel. Marcel Breuer, the famed Bauhaus furniture designer, had offered to do the interior design, but

Einstein was a man of conservative tastes. "I am not going to sit on furniture that continually reminds me of a machine shop or a hospital operating room," he said. Some leftover heavy pieces from the Berlin apartment were used instead.

Einstein's room on the ground floor had a spartan wooden table, a bed, and a small portrait of Isaac Newton. Elsa's room was also downstairs, with a shared bathroom between them. Upstairs were small rooms with sleeping niches for her two daughters and their maid. "I like living in the new little wooden house enormously, even though I am broke as a result," he wrote his sister shortly after moving in. "The sailboat, the sweeping view, the solitary fall walks, the relative quiet— it is a paradise."[5]

There he sailed the new twenty-three-foot boat his friends had given him for his birthday, the *Tümmler*, or Dolphin, which was built fat and solid to his specifications. He liked to go out on the water alone, even though he didn't swim. "He was absurdly happy as soon as he reached the water," recalled a visitor.[6] For hours he would let the boat drift and glide aimlessly as he gently toyed with the rudder. "His scientific thinking, which never leaves him even on the water, takes on the nature of a daydream," according to one relative. "Theoretical thinking is rich in imagination."[7]

Companions

Throughout Einstein's life, his relationships with women seemed subject to untamed forces. His magnetic appeal and soulful manner repeatedly attracted women. And even though he usually shielded himself from entangling commitments, he occasionally found himself caught in the swirl of a passionate attraction, just as he had been with Mileva Marić and even Elsa.

In 1923, after marrying Elsa, he had fallen in love with his secretary, Betty Neumann. Their romance was serious and passionate, according to newly revealed letters. That fall, while on a visit to Leiden, he wrote to suggest that he might take a job in New York, and she could come as his secretary. She would live there with him and Elsa, he fantasized. "I will convince my wife to allow this," he said. "We could live together forever. We could get a large house outside New York."

She replied by ridiculing both him and the idea, which prompted him to concede how much of a "crazy ass" he had been. "You have more respect for the difficulties of triangular geometry than I, old mathematicus, have."[8]

He finally terminated their romance with the lament that he "must seek in the stars" the true love that was denied to him on earth. "Dear Betty, laugh at me, the old donkey, and find somebody who is ten years younger than me and loves you just as much as I do."[9]

But the relationship lingered. The following summer, Einstein went to see his sons in southern Germany, and from there he wrote to his wife that he could not visit her and her daughters, who were at a resort nearby, because that would be "too much of a good thing." At the same time, he was writing Betty Neumann saying that he was going secretly to Berlin, but she should not tell anyone because if Elsa found out she "will fly back."[10]

After he built the house in Caputh, a succession of women friends visited him there, with Elsa's grudging acquiescence. Toni Mendel, a wealthy widow with an estate on the Wannsee, sometimes came sailing with him in Caputh, or he would pilot his boat up to her villa and stay late into the night playing the piano. They even went to the theater together in Berlin occasionally. Once when she picked Einstein up in her chauffeured limousine, Elsa got into a furious fight with him and would not give him any pocket money.

He also had a relationship with a Berlin socialite named Ethel Michanowski. She tagged along on one of his trips to Oxford, in May 1931, and apparently stayed in a local hotel. He composed a five-line poem for her one day on a Christ Church college notecard. "Long-branched and delicately strung, Nothing that will escape her gaze," it began. A few days later she sent him an expensive present, which was not appreciated. "The small package really angered me," he wrote. "You have to stop sending me presents incessantly . . . And to send something like that to an English college where we are surrounded by senseless affluence anyway!"[11]

When Elsa found out that Michanowski had visited Einstein in Oxford, she was furious, particularly at Michanowski for misleading her about where she was going. Einstein wrote from Oxford to tell Elsa to calm down. "Your dismay toward Frau M is totally groundless because she behaved completely according to the best Jewish-Christian morality," he said. "Here is the proof: 1) What one enjoys and doesn't harm others, one should do. 2) What one doesn't enjoy and only aggravates others, one should not do. Because of #1, she came with me, and because of #2 she didn't tell you anything about it. Isn't that impeccable behavior?" But in a letter to Elsa's daughter Margot, Einstein claimed that Michanowski's pursuit was unwanted. "Her chasing me is getting out of control," he wrote Margot, who was Michanowski's friend. "I don't care what people are saying about me, but for mother [Elsa] and for Frau M, it is better that not every Tom, Dick and Harry gossip about it."[12]

In his letter to Margot, he insisted that he was not particularly attached to Michanowski nor to most of the other women who flirted with him. "Of all the women, I am actually attached only to Frau L, who is perfectly harmless and respectable," he said, not so reassuringly.[13] That was a reference to a blond Austrian named Margarete Lebach, with whom he had a very public relationship. When Lebach visited Caputh, she brought pastries for Elsa. But Elsa, understandably, could not abide her, and she took to leaving the village to go shopping in Berlin on the days that Lebach came.

On one visit, Lebach left a piece of clothing in Einstein's sailboat, which caused a family row and prompted Elsa's daughter to urge her to force Einstein to end the relationship. But Elsa was afraid that her husband would refuse. He had let it be known that he believed that men and women were not naturally monogamous.[14] In the end, she decided that she was better off preserving what she could of their marriage. In other respects, it suited her aspirations.[15]

Elsa liked her husband, and she also revered him. She realized that she must accept him with all of his complexities, especially since her life as Mrs. Einstein included much that made her happy. "Such a genius should be irreproachable in every respect," she told the artist and etcher Hermann Struck, who did Einstein's portrait around the time of his fiftieth birthday (as he had done a decade earlier). "But nature does not behave this way. Where she gives extravagantly, she takes away extravagantly." The good and the bad had to be accepted as a whole. "You have to see him all of one piece," she explained. "God has given him so much nobility, and I find him wonderful, although life with him is exhausting and complicated, and not only in one way but in others."[16]

The most important other woman in Einstein's life was one who was completely discreet, protective, loyal, and not threatening to Elsa. Helen Dukas came to work as Einstein's secretary in 1928, when he was confined to bed with an inflamed heart. Elsa knew her sister, who ran the Jewish Orphans Organization, of which Elsa was honorary president. Elsa interviewed Dukas before allowing her to meet Einstein, and she felt that Dukas would be trustworthy and, more to the point, safe in all respects. She offered Dukas the job even before she had met Einstein.

When Dukas, then 32, was ushered into Einstein's sickroom in April 1928, he stretched out his hand and smiled, "Here lies an old child's corpse." From that moment until his death in 1955—indeed until her own death in 1982—the never-married Dukas was fiercely protective of his time, his privacy, his reputation, and later his legacy. "Her instincts were as infallible and straightforward as a magnetic compass," George Dyson later declared. Although she could display a pleasant smile and lively directness with those she liked, she was generally austere, hard-boiled, and at times quite prickly.[17]

More than a secretary, she could appear to intrusive outsiders as Einstein's pit bull—or, as he referred to her, his Cerberus, the guard dog at the gates of his own little kingdom of Hades. She would keep journalists at bay, shield him from letters she thought a waste of his time, and cover up any matters that she decreed should remain private. After a while, she became like a member of the family.

Another frequent visitor was a young mathematician from Vienna, Walther Mayer, who became an assistant and, in Einstein's words, "the calculator." Einstein collaborated with him on some unified field theory papers, and he called him "a splendid fellow who would have long had a professorship if he were not a Jew."[18]

Even Mileva Marić, who had gone back to using her maiden name after the divorce, started using the name Einstein again and was able to establish a strained but workable relationship with him. When he visited South America, he brought her back baskets of cactuses. Since she loved the plants, it was presumably meant as an amicable gift. On his visits to Zurich, he stayed at her apartment occasionally.

He even invited her to stay with him and Elsa when she came to Berlin, an arrangement that likely would have made every single person involved uncomfortable. But she wisely stayed with the Habers instead. Their relationship had improved so much, he told her, that he was now surprising his friends by recounting how well they were getting along. "Elsa is also happy that you and the boys are not hostile to her anymore," he added.[19]

Their two sons, he told Marić, were the best part of his inner life, a legacy that would remain after the clock of his own body had worn down. Despite this, or because of it, his relationship with his sons remained fraught with tensions. This was particularly true when Hans Albert decided to get married.

As if the gods wished to extract their revenge, the situation was similar to the one Einstein had put his own parents through when he decided to marry Mileva Marić. Hans Albert had fallen in love, while studying at the Zurich Polytechnic, with a woman nine years his senior named Frieda Knecht. Less than five feet tall, she was plain and had an abrupt manner but was very smart. Both Marić and Einstein, reunited by this cause, agreed that she was scheming, unattractive, and would likely produce physically unsuitable offspring. "I tried my best to convince him that marrying her would be crazy," he wrote Marić. "But it seems like he is totally dependent on her, so it was in vain."[20]

Einstein assumed that his son had been ensnared because he was shy and inexperienced with women. "She was the one to grab you first, and now you consider her to be the embodiment of femininity," he wrote Hans Albert. "That is the well-known way that women take advantage of unworldly people." So he suggested that an attractive woman would remedy such problems.

But Hans Albert was as stubborn as his father had been twenty-five years earlier, and he was determined to marry Frieda. Einstein conceded that he couldn't stop him, but he urged his son to promise not to have children. "And should you ever feel like you have to leave her, you should not be too proud to come talk to me," Einstein wrote. "After all, that day *will* come."[21]

Hans Albert and Frieda married in 1927, had children, and remained married until her death thirty-one years later. As Evelyn Einstein, their adopted daughter, recalled years later, "Albert had such a hell of a time with his parents over his own marriage that you would think he would have had the sense not to interfere with his son's. But no. When my father went to marry my mother, there was explosion after explosion."[22]

Einstein expressed his dismay about Hans Albert's marriage in letters to Eduard. "The deterioration of the race is a serious problem," Einstein wrote. "That is why I cannot forgive [Hans] Albert his sin. I instinctively avoid meeting him, because I cannot show him a happy face."[23]

But within two years, Einstein had begun to accept Frieda. The couple came to visit him in the summer of 1929, and he reported back to Eduard that he had made his peace. "She made a better impression than I had feared," he wrote. "He is really sweet with her. God bless those rose-colored spectacles."[24]

For his part, Eduard was becoming increasingly dreamy in his academic pursuits, and his psychological problems were becoming more apparent. He liked poetry and wrote doggerel and aphorisms that often had an edge to them, especially when the subject was his family. He played the piano, particularly Chopin, with a passion that was initially a welcome contrast to his usual lethargy but eventually became scary.

His letters to his father were equally intense, pouring out his soul about philosophy and the arts. Einstein responded sometimes tenderly, and occasionally with detachment. "I often sent my father rather rapturous letters, and several times got worried afterwards because he was of a cooler disposition," Eduard later recalled. "I learned only a lot later how much he treasured them."

Eduard went to Zurich University, where he studied medicine and planned to become a psychiatrist. He became interested in Sigmund Freud, whose picture he hung in his bedroom, and attempted his own self-analysis. His letters to his father during this period are filled with his efforts, often astute, to use Freud's theories to analyze various realms of life, including movies and music.

Not surprisingly, Eduard was especially interested in relationships between fathers and sons. Some of his comments were simple and poignant. "It's at times difficult to have such an important father, because one feels so unimportant," he wrote at one point. A few months later, he poured out more insecurities: "People who fill their time with intellectual work bring into the world sickly, nervous at times even completely idiotic children (for example, you me)."[25]

Later his comments became more complex, such as when he analyzed his father's famous lament that fate had punished him for his contempt for authority by making him an authority himself. Eduard wrote, "This means psychoanalytically that, because you didn't want to bend in front of your own father and instead fought with him, you had to become an authority in order to step into his place."[26]

Einstein met Freud when he came from Vienna to Berlin for New Year 1927. Freud, then 70, had cancer of the mouth and was deaf in one ear, but the two men had a pleasant talk, partly because they focused on politics rather than on their respective fields of study. "Einstein understands as much about psychology as I do about physics," Freud wrote to a friend.[27]

Einstein never asked Freud to meet or treat his son, nor did he seem impressed by the idea of psychoanalysis. "It may not always be helpful to delve into the subconscious," he once said. "Our legs are controlled by a hundred different muscles. Do you think it would help us to walk if we analyzed our legs and knew the exact purpose of each muscle and the order in which they work?" He certainly never expressed any interest in undergoing therapy himself. "I should like very much to remain in the darkness of not having been analyzed," he declared.[28]

Eventually, however, he did concede to Eduard, perhaps to make him happy, that there might be some merit to Freud's work. "I must admit that, through various little personal experiences, I am convinced at least of his main theses."[29]

While at the university, Eduard fell in love with an older woman, a trait that apparently ran in the family and might have amused Freud.

When the relationship came to a painful conclusion, he fell into a list-less depression. His father suggested he find a dalliance with a younger "plaything." He also suggested that he find a job. "Even a genius like Schopenhauer was crushed by unemployment," he wrote. "Life is like riding a bicycle. To keep your balance you must keep moving."[30]

Eduard was unable to keep his balance. He began cutting classes and staying in his room. As he grew more troubled, Einstein's care and affection for him seemed to increase. There was a painful sweetness in his letters to his troubled son as he engaged with his ideas about psychology and wrestled with his enigmatic aphorisms.

"There is no meaning to life outside of life itself," Eduard declared in one of these aphorisms.

Einstein replied politely that he could accept this, "but that clarifies very little." Life for its own sake, Einstein went on, was hollow. "People who live in a society, enjoy looking into each other's eyes, who share their troubles, who focus their efforts on what is important to them and find this joyful—these people lead a full life."[31]

There was a knowing, self-referential quality in that exhortation. Einstein himself had little inclination or talent for sharing other people's troubles, and he compensated by focusing on what was important to him. "Tete really has a lot of myself in him, but with him it seems more pronounced," Einstein conceded to Marić. "He's an interesting fellow, but things won't be easy for him."[32]

Einstein visited Eduard in October 1930, and together with Marić tried to deal with his downward mental spiral. They played piano together, but to no avail. Eduard continued to slip into a darker realm. Soon after he left, the young man threatened to throw himself out of his bedroom window, but his mother restrained him.

The complex strands of Einstein's family life came together in an odd scene in November 1930. Four years earlier, a conniving Russian writer named Dimitri Marianoff had sought to meet Einstein. With great nerve and tenacity, he presented himself at Einstein's apartment and was able to convince Elsa to let him in. There he proceeded to charm Einstein by talking about Russian theater, and also to turn the head of Elsa's daughter Margot by engaging in a grand show of handwriting analysis.

Margot was so painfully shy that she often hid from strangers, but Marianoff's wiles soon brought her out of her shell. Their wedding occurred a few days after Eduard had tried to commit suicide, and a distraught Marić made an unannounced visit to Berlin to ask her former husband for help. Marianoff later described the scene at the end of his wedding ceremony: "As we came down the steps I noticed a woman standing near the portico. I would not have noticed her, except that she looked at us with such an intensely burning gaze that it impressed me. Margot said under her breath, 'It's Mileva.' "[33]

Einstein was shaken deeply by his son's illness. "This sorrow is eating up Albert," Elsa wrote. "He finds it difficult to cope with."[34]

There was, however, not much he could do. The morning after the wedding, he and Elsa left by train to Antwerp, from which they would sail for their second voyage to the United States. It was a hectic departure. Einstein got separated from Elsa at the Berlin station, then lost their train tickets.[35] But eventually they got everything together and embarked on what would be another triumphal American visit.

America Again

Einstein's second trip to America, beginning in December 1930, was supposed to be different from his first. This time, there would be no public frenzy or odd hoopla. Instead, he was coming for a two-month working visit as a research fellow at the California Institute of Technology. The officials who arranged it were eager to protect his privacy and, like his friends in Germany, they viewed any publicity as undignified.

As usual, Einstein seemed to agree—in theory. Once it was known that he was coming, he was swamped with dozens of telegrams each day with speaking offers and award invitations, all of which he declined. On the way over, he and his mathematical calculator, Walther Mayer, holed up, working on revisions to his unified field theory, in an upper-deck suite with a sailor guarding the door.[36]

He even decided that he would not disembark when his ship docked in New York. "I hate facing cameras and having to answer a crossfire of questions," he claimed. "Why popular fancy should seize on me, a scientist, dealing in abstract things and happy if left alone, is a manifestation of mass psychology that is beyond me."[37]

But by then the world, and especially America, had irrevocably entered the new age of celebrity. Aversion to fame was no longer considered natural. Publicity was still something that many proper people tended to avoid, but its lure had begun to be accepted. The day before his ship docked in New York, Einstein sent word that he had relented to reporters' requests and would hold a press conference and photo opportunity upon his arrival.[38]

It was "worse than the most fantastic expectation," he recorded in his travel diary. Fifty reporters plus fifty more cameramen swarmed aboard, accompanied by the German consul and his fat assistant. "The reporters asked exquisitely inane questions, to which I replied with cheap jokes, which were enthusiastically received."[39]

Asked to define the fourth dimension in a word, Einstein replied, "You will have to ask a spiritualist." Could he define relativity in one sentence? "It would take me three days to give a short definition."

There was, however, one question that he tried to answer seriously, and which he alas got wrong. It was about a politician whose party had risen from obscurity three months earlier to win 18 percent of the vote in the German elections. "What do you think of Adolf Hitler?" Einstein replied, "He is living on the empty stomach of Germany. As soon as economic conditions improve, he will no longer be important."[40]

Time magazine that week featured Elsa on its cover, wearing a sprightly hat and exulting in her role as wife of the world's most famous scientist. The magazine reported, "Because Mathematician Einstein cannot keep his bank account correctly," his wife had to balance his finances and handle the arrangements for the trip. "All these things I must do so that he will think he is free," she told the magazine. "He is all my life. He is worth it. I like being Mrs. Einstein very much."[41] One duty she assigned herself was to charge $1 for her husband's autograph and $5 for his photograph; she kept a ledger and donated the money to charities for children.

Einstein changed his mind about staying secluded aboard ship while it was docked in New York. In fact, he seemed to pop up everywhere. He celebrated Hanukkah with fifteen thousand people in

Madison Square Garden, toured Chinatown by car, lunched with the editorial board of the *New York Times*, was cheered when he arrived at the Metropolitan Opera to hear the sensational soprano Maria Jeritza sing *Carmen*, received the keys to the city (which Mayor Jimmy Walker quipped were given "relatively"), and was introduced by the president of Columbia University as "the ruling monarch of the mind."[42]

He also paid a visit to Riverside Church, a massive structure with a 2,100-seat nave, which had just been completed. It was a Baptist church, but above the west portal, carved in stone amid a dozen other great thinkers in history, was a full-length statue of Einstein. Harry Emerson Fosdick, the noted senior minister, met Einstein and Elsa at the door and gave them a tour. Einstein paused to admire a stained-glass window of Immanuel Kant in his garden, then asked about his own statue. "Am I the only living man among all these figures of the ages?" Dr. Fosdick, with a sense of gravity duly noted by the reporters present, replied, "That is true, Professor Einstein."

"Then I will have to be very careful for the rest of my life as to what I do and say," Einstein answered. Afterward, according to an article in the church bulletin, he joked, "I might have imagined that they could make a Jewish saint of me, but I never thought I'd become a Protestant one!"[43]

The church had been built with donations from John D. Rockefeller Jr., and Einstein arranged to have a meeting with the great capitalist and philanthropist. The purpose was to discuss the complex restrictions the Rockefeller foundations were putting on research grants. "The red tape," Einstein said, "encases the mind like the hands of a mummy."

They also discussed economics and social justice in light of the Great Depression. Einstein suggested that working hours be shortened so that, at least in his understanding of economics, more people would have a chance to be employed. He also said that lengthening the school year would help keep young people out of the workforce.

"Does not such an idea," Rockefeller asked, "impose an unwarranted restriction upon individual freedom?" Einstein replied that the current economic crisis justified measures like those taken during wartime. This gave Einstein the opportunity to propound his pacifist positions, which Rockefeller politely declined to share.[44]

His most memorable speech was a pacifist clarion call that he gave to the New History Society, in which he called for an "uncompromising war resistance and refusal to do military service under any circumstances." Then he issued what became a famous call for a brave 2 percent:

> The timid might say, "What's the use? We shall be sent to prison." To them I would reply: Even if only 2% of those assigned to perform military service should announce their refusal to fight . . . governments would be powerless, they would not dare send such a large number of people to jail.

The speech quickly became a manifesto for war resisters. Buttons that simply said "2%" began sprouting on the lapels of students and pacifists.* The *New York Times* headlined the story on page 1 and reprinted the speech in its entirety. One German paper also headlined it, but with less enthusiasm: "Einstein Begging for Military Service Objectors: Scientist's Unbelievable Publicity Methods in America."[45]

On the day he left New York, Einstein revised slightly one of the statements he had made upon his arrival. Asked again about Hitler, he declared that if the Nazis were ever able to gain control, he would consider leaving Germany.[46]

Einstein's ship headed to California through the Panama Canal. While his wife spent time at the hairdresser, Einstein dictated letters to Helen Dukas and worked on unified field theory equations with Walther Mayer. Although he complained about the "perpetual photographing" he had to endure from his fellow passengers, he did let one young man sketch him, and then he appended his own self-deprecating doggerel to turn it into a collector's item.

In Cuba, where he relished the warm weather, Einstein addressed the local Academy of Sciences. Then it was on to Panama, where a rev-

---

* The pacifists assumed that no other explanation was needed, but some contemporary accounts somehow thought the buttons referred to 2 percent beer.

olution was brewing that would depose a president who, it turned out, was also a graduate of the Zurich Polytechnic. That didn't stop officials from offering Einstein an elaborate welcome ceremony at which he was presented a hat that "an illiterate Ecuadorian Indian worked for six months weaving." On Christmas day, he broadcast holiday greetings to America via the ship's radio.[47]

When his ship docked in San Diego on the last morning of 1930, dozens of newsmen clambered aboard, with two of them falling off the ladder as they rushed their way onto the deck. Five hundred uniformed girls stood on the dock, waiting to serenade him. The gaudy arrival ceremony lasted four hours, filled with speeches and presentations.

Were there men, he was asked, living elsewhere in the universe? "Other beings, perhaps, but not men," he answered. Did science and religion conflict? Not really, he said, "though it depends, of course, on your religious views."[48]

Friends who saw all the arrival hoopla on newsreels back in Germany were astonished and somewhat appalled. "I am always very amused to see and hear you in the weekly newsreel," wrote the sharp-penned Hedwig Born, "being presented with a floral float containing lovely sea-nymphs in San Diego, and that sort of thing. However crazy things must look from the outside, I always have the feeling that the dear Lord knows what he's up to."[49]

It was on this trip, as noted in the previous chapter, that Einstein visited the Mount Wilson Observatory, was shown evidence of the expanding universe, and renounced the cosmological constant he had added to his general relativity equations. He also paid tribute to the aging Albert Michelson, carefully praising his famous experiments that detected no ether drift, without explicitly saying that they were a basis for his special theory of relativity.

Einstein soaked in a variety of the delights that southern California could offer. He attended the Rose Bowl parade, was given a special screening of *All Quiet on the Western Front*, and sunbathed nude in the Mojave desert while at a friend's house for the weekend. At a Hollywood studio, the special effects team filmed him pretending to drive a parked car, and then that evening amused him by showing how they made it seem as if he were zipping through Los Angeles, soaring up into the clouds, flying over the Rockies, and eventually landing in the German countryside. He even was offered some movie roles, which he politely declined.

He went sailing in the Pacific with Robert A. Millikan, Caltech's president, who Einstein noted in his diary "plays the role of God" at the university. Millikan was a physicist who had won the Nobel Prize in 1923 for, as the organization noted, having "verified experimentally Einstein's all-important photoelectric equation." He likewise verified Einstein's interpretation of Brownian motion. So it was understandable that, as he was building Caltech into one of the world's preeminent scientific institutions, he worked diligently to bring Einstein there.

Despite all they had in common, Millikan and Einstein were different enough in their personal outlooks that they were destined to have an awkward relationship. Millikan was so conservative scientifically that he resisted Einstein's interpretation of the photoelectric effect and his dismissal of the ether even after they were apparently verified by his own experiments. And he was even more conservative politically. A robust and athletic son of an Iowa preacher, he had a penchant for patriotic militarism that was as pronounced as Einstein's aversion to it.

Moreover, Millikan was enhancing Caltech through hefty donations from like-minded conservatives. Einstein's pacifist and socialist sentiments unnerved many of them, and they urged Millikan to restrain him from making pronouncements on earthly rather than cosmic issues. As Major General Amos Fried put it, they must avoid "aiding and abetting the teaching of treason to the youth of this country by being hosts to Dr. Albert Einstein." Millikan responded sympathetically by denouncing Einstein's call for military resistance and declaring that "the 2% comment, if he ever made it, is one which no experienced man could possibly have made." [50]

Millikan particularly disdained the crusading writer and union advocate Upton Sinclair, whom he called "the most dangerous man in California," and the actor Charlie Chaplin, who equaled Einstein in global celebrity and surpassed him in left-wing sentiments. Much to Millikan's dismay, Einstein promptly befriended both.

Einstein had corresponded with Sinclair about their shared commitment to social justice, and upon arriving in California was happy to accept his invitations to a variety of dinners, parties, and meetings. He even remained polite, though amused, while attending a farcical séance at Sinclair's home. When Mrs. Sinclair challenged his views on science and spirituality, Elsa chided her for having such presumption. "You know, my husband has the greatest mind in the world," she said. Mrs. Sinclair responded, "Yes, I know, but surely he doesn't know everything."[51]

During a tour of Universal Studios, Einstein mentioned that he had always wanted to meet Charlie Chaplin. So the studio boss called him, and he came right over to join the Einsteins for lunch in the commissary. The result, a few days later, was one of the most memorable scenes in the new era of celebrity: Einstein and Chaplin arriving together, dressed in black tie, with Elsa beaming, for the premiere of *City Lights*. As they were applauded on their way into the theater, Chaplin memorably (and accurately) noted, "They cheer me because they all understand me, and they cheer you because no one understands you."[52]

Einstein struck a more serious pose when he addressed the Caltech student body near the end of his stay. His sermon, grounded in his humanistic outlook, was on how science had not yet been harnessed to do more good than harm. During war it gave people "the means to poison and mutilate one another," and in peacetime it "has made our lives hurried and uncertain." Instead of being a liberating force, "it has enslaved men to machines" by making them work "long wearisome hours mostly without joy in their labor." Concern for making life better for ordinary humans must be the chief object of science. "Never forget this when you are pondering over your diagrams and equations!"[53]

The Einsteins took a train east across America for their return sail from New York. Along the way, they stopped at the Grand Canyon, where they were greeted by a contingent of Hopi Indians (employed by the concession stand at the canyon, though Einstein did not know that), who initiated him into their tribe as "the Great Relative" and gave him a bountiful feathered headdress that resulted in some classic photographs.[54]

When his train reached Chicago, Einstein gave a speech from its rear platform to a rally of pacifists who had come to celebrate him. Millikan must have been appalled. It was similar to the "2%" speech Einstein had given in New York. "The only way to be effective is through the revolutionary method of refusing military service," he declared. "Many who consider themselves good pacifists will not want to participate in such a radical form of pacifism; they will claim that patriotism prevents them from adopting such a policy. But in an emergency, such people cannot be counted on anyhow."[55]

Einstein's train pulled into New York City on the morning of March 1, and for the next sixteen hours Einstein mania reached new heights. "Einstein's personality, for no clear reason, triggers outbursts of a kind of mass hysteria," the German consul reported to Berlin.

Einstein first went to his ship, where four hundred members of the War Resisters' League were waiting to greet him. He invited them all on board and addressed them in a ballroom. "If in time of peace members of pacifist organizations are not ready to make sacrifices by opposing authorities at the risk of imprisonment, they will certainly fail in time of war, when only the most steeled and resolute person can be expected to resist." The crowd erupted in delirium, with overwrought pacifists rushing up to kiss his hand and touch his clothing.[56]

The socialist leader Norman Thomas was at the meeting, and he tried to convince Einstein that pacifism could not occur without radical economic reforms. Einstein disagreed. "It is easier to win over people to pacifism than to socialism," he said. "We should work first for pacifism, and only later for socialism."[57]

That afternoon, the Einsteins were taken to the Waldorf Hotel, where they had a sprawling suite in which they could meet a stream of visitors, such as Helen Keller and various journalists. Actually, it was two full suites connected by a grand private dining room. When one friend arrived that afternoon, he asked Elsa, "Where is Albert?"

"I don't know," she replied with some exasperation. "He always gets lost somewhere in all these rooms."

They finally found him wandering around, trying to find his wife. The ostentatious spread annoyed him. "I'll tell you what to do," the friend suggested. "Lock the second suite entirely off, and you will feel better." Einstein did, and it worked.[58]

That evening, he addressed a sold-out fund-raising dinner on be-half of the Zionist cause, and he finally made it back to his ship just be-fore midnight. But even then his day was not over. A large crowd of young pacifists, chanting "No War Forever," cheered him wildly as he reached the pier. They later formed the Youth Peace Federation, and Einstein sent them a scrawled message of encouragement: "I wish you great progress in the radicalization of pacifism."[59]

Einstein's Pacifism

This radical pacifism had been building in Einstein throughout the 1920s. Even as he was retreating from the fore of physics, he was be-coming, at age 50, more engaged in politics. His primary cause, at least until Adolf Hitler and his Nazis took power, was that of disarmament and resistance to war. "I am not only a pacifist," he told one interviewer on his trip to America. "I am a militant pacifist."[60]

He rejected the more modest approach taken by the League of Na-tions, the international organization formed after World War I, which the United States had declined to join. Instead of calling for complete disarmament, the League was nibbling at the margins by trying to de-fine proper rules of engagement and arms control. When he was asked in January 1928 to attend one of the League's disarmament commis-sions, which was planning to study ways to limit gas warfare, he pub-licly proclaimed his disgust with such half measures:

> It seems to me an utterly futile task to prescribe rules and limitations for the conduct of war. War is not a game; hence one cannot wage war by rules as one would in playing games. Our fight must be against war it-self. The masses of people can most effectively fight the institution of war by establishing an organization for the absolute refusal of military service.[61]

Thus he became one of the spiritual leaders of the growing move-ment led by War Resisters' International. "The international move-ment to refuse participation in any kind of war service is one of the most encouraging developments of our time," he wrote the London branch of that group in November 1928.[62]

Even as the Nazis began their rise to power, Einstein refused to admit, at least initially, that there might be exceptions to his pacifist postulate. What would he do, a Czech journalist asked, if another European war broke out and one side was clearly the aggressor? "I would unconditionally refuse all war service, direct or indirect, and would seek to persuade my friends to adopt the same position, regardless of how I might feel about the causes of any particular war," he answered.[63] The censors in Prague refused to allow the remark to be published, but it was made public elsewhere and enhanced Einstein's status as the standard-bearer of pacifist purists.

Such sentiments were not unusual at the time. The First World War had shocked people by being so astonishingly brutal and apparently unnecessary. Among those who shared Einstein's pacifism were Upton Sinclair, Sigmund Freud, John Dewey, and H. G. Wells. "We believe that everybody who sincerely wants peace should demand the abolition of military training for youth," they declared in a 1930 manifesto, which Einstein signed. "Military training is the education of the mind and body in the technique of killing. It thwarts the growth of man's will for peace."[64]

Einstein's advocacy of war resistance reached its peak in 1932, the year before the Nazis seized power. That year a General Disarmament Conference, organized by the League of Nations plus the United States and Russia, convened in Geneva.

Einstein initially had grand hopes that the conference, as he wrote in an article for the *Nation*, "will be decisive for the fate of the present generation and the one to come." But he warned that it must not merely content itself with feckless arms-limitation rules. "Mere agreements to limit armaments confer no protection," he said. Instead, there should be an international body empowered to arbitrate disputes and enforce the peace. "Compulsory arbitration must be supported by an executive force."[65]

His fears were realized. The conference became mired in such issues as how to calculate the offensive power of aircraft carriers in assessing an arms-control balance. Einstein showed up in Geneva in May, just as that topic was being tackled. When he appeared in the visitors' gallery, the delegates stopped their discussions and rose to applaud him. But Einstein was not pleased. That afternoon, he called a press conference at his hotel to denounce their timidity.

"One does not make war less likely to occur by formulating rules of warfare," he declared to dozens of excited journalists who abandoned the conference to cover his criticism. "We should be standing on rooftops, all of us, and denouncing this conference as a travesty!" He argued that it would be better for the conference to fail outright than to end with an agreement to "humanize war," which he considered a tragic delusion.[66]

"Einstein tended to become impractical once outside the scientific field," his novelist friend and fellow pacifist Romain Rolland commented. It is true that, given what was about to happen in Germany, disarmament was a chimera, and pacifist hopes were, to use a word sometimes flung at Einstein, naïve. Yet it should be noted that there was some merit to his criticisms. The arms-control acolytes in Geneva were no less naïve. They spent five years in futile, arcane debates as Germany rearmed itself.

Political Ideals

"Go One Step Further, Einstein!" the headline exhorted. It was on an essay, published in August 1931 as an open letter to Einstein, by the German socialist leader Kurt Hiller, one of many activists on the left who urged Einstein to expand his pacifism into a more radical politics. Pacifism was only a partial step, Hiller argued. The real goal was to advocate socialist revolution.

Einstein labeled the piece "rather stupid." Pacifism did not require socialism, and socialist revolutions sometimes led to the suppression of freedom. "I am not convinced that those who would gain power through revolutionary actions would act in accord with my ideals," he wrote to Hiller. "I also believe that the fight for peace must be pushed energetically, far ahead of any efforts to bring about social reforms."[67]

Einstein's pacifism, world federalism, and aversion to nationalism were part of a political outlook that also included a passion for social justice, a sympathy for underdogs, an antipathy toward racism, and a predilection toward socialism. But during the 1930s, as in the past, his wariness of authority, his fealty to individualism, and his fondness for personal freedom made him resist the dogmas of Bolshevism and communism. "Einstein was neither Red nor dupe," writes Fred Jerome, who has analyzed both Einstein's politics and the large dossier of material gathered on him by the FBI.[68]

This wariness of authority reflected the most fundamental of all of Einstein's moral principles: Freedom and individualism are necessary for creativity and imagination to flourish. He had demonstrated this as an impertinent young thinker, and he proclaimed the principle clearly in 1931. "I believe that the most important mission of the state is to protect the individual and to make it possible for him to develop into a creative personality," he said.[69]

Thomas Bucky, the son of a doctor who cared for Elsa's daughters, was 13 when he met Einstein in 1932, and they began what would become a longstanding discussion of politics. "Einstein was a humanist, socialist, and a democrat," he recalled. "He was completely anti-totalitarian, no matter whether it was Russian, German or South American. He approved of a combination of capitalism and socialism. And he hated all dictatorships of the right or left."[70]

Einstein's skepticism about communism was evident when he was invited to the 1932 World Antiwar Congress. Though putatively a pacifist group, it had become a front for Soviet communists. The official call for the conference, for example, denounced the "imperialist powers" for encouraging Japan's aggressive attitude toward the Soviet Union. Einstein refused to attend or support its manifesto. "Because of the glorification of Soviet Russia it includes, I cannot bring myself to sign it," he said.

He had come to some somber conclusions about Russia, he added. "At the top there appears to be a personal struggle in which the foulest means are used by power-hungry individuals acting from purely selfish motives. At the bottom there seems to be complete suppression of the individual and freedom of speech. One wonders whether life is worth living under such conditions." Perversely, when the FBI later compiled a secret dossier on Einstein during the Red Scare of the 1950s, one piece of evidence cited against him was that he had *supported*, rather than rejected, the invitation to be active in this world congress.[71]

One of Einstein's friends at the time was Isaac Don Levine, a Russian-born American journalist who had been sympathetic to the communists but had turned strongly against Stalin and his brutal regime as a columnist for the Hearst newspapers. Along with other defenders of civil liberties, including ACLU founder Roger Baldwin and Bertrand Russell, Einstein supported the publication of Levine's exposé of Stalinist horrors, *Letters from Russian Prisons*. He even provided an essay, written in longhand, in which he denounced "the regime of frightfulness in Russia."[72]

Einstein also read Levine's subsequent biography of Stalin, a fiercely critical exposé of the dictator's brutalities, and called it "profound." He saw in it a clear lesson about tyrannical regimes on both the left and the right. "Violence breeds violence," he wrote Levine in a letter of praise. "Liberty is the necessary foundation for the development of all true values."[73]

Eventually, however, Einstein began to break with Levine. Like many former communist sympathizers who swung over to the anticommunist cause, Levine had the zeal of a convert and an intensity that made it hard for him to appreciate any of the middle shades of the spectrum. Einstein, on the other hand, was too willing to accept, Levine felt, some aspects of Soviet repression as being an unfortunate byproduct of revolutionary change.

There were, indeed, many aspects of Russia that Einstein admired, including what he saw as its attempt to eliminate class distinctions and economic hierarchies. "I regard class differences as contrary to justice," he wrote in a personal statement of his credo. "I also consider that plain living is good for everybody, physically and mentally."[74]

These sentiments led Einstein to be critical of what he saw as the excessive consumption and disparities of wealth in America. As a result, he enlisted in a variety of racial and social justice movements. He took up, for example, the cause of the Scottsboro Boys, a group of young black men who were convicted of a gang rape in Alabama after a controversial trial, and of Tom Mooney, a labor activist imprisoned for murder in California.[75]

At Caltech, Millikan was upset with Einstein's activism, and wrote him to say so. Einstein responded diplomatically. "It cannot be my affair," he agreed, "to insist in a matter that concerns only the citizens of your country."[76] Millikan thought Einstein naïve in his politics, as did many people. To some extent he was, but it should be remembered that his qualms about the convictions of the Scottsboro Boys and Mooney proved justified, and his advocacy of racial and social justice turned out to be on the right side of history.

Despite his association with the Zionist cause, Einstein's sympathies extended to the Arabs who were being displaced by the influx of Jews into what would eventually be Israel. His message was a prophetic one. "Should we be unable to find a way to honest cooperation and honest pacts with the Arabs," he wrote Weizmann in 1929, "then we have learned absolutely nothing during our 2,000 years of suffering."[77]

He proposed, both to Weizmann and in an open letter to an Arab, that a "privy council" of four Jews and four Arabs, all independent-minded, be set up to resolve any disputes. "The two great Semitic peoples," he said, "have a great common future." If the Jews did not assure that both sides lived in harmony, he warned friends in the Zionist movement, the struggle would haunt them in decades to come.[78] Once again, he was labeled naïve.

The Einstein-Freud Exchange

When a group known as the Institute for Intellectual Cooperation invited him in 1932 to exchange letters with a thinker of his choice on issues relating to war and politics, Einstein picked as his correspondent Sigmund Freud, the era's other great intellectual and pacifist icon. Einstein began by proposing an idea that he had been refining over the years. The elimination of war, he said, required nations to surrender some of their sovereignty to a "supranational organization competent to render verdicts of incontestable authority and enforce absolute submission to the execution of its verdicts." In other words, some international authority more powerful than the League of Nations must be created.

Ever since he was a teenager rankling at German militarism, Einstein had been repulsed by nationalism. One of the fundamental postulates of his political view, which would remain invariant even after

Hitler's rise made him waver on the principles of pacifism, was his support for an international or "supranational" entity that would transcend the chaos of national sovereignty by imposing the resolution of disputes.

"The quest of international security," he wrote Freud, "involves the unconditional surrender by every nation, in a certain measure, of its liberty of action—its sovereignty that is to say—and it is clear that no other road can lead to such security." Years later, Einstein would become even more committed to this approach as a way to transcend the military dangers of the atomic age that he helped to spawn.

Einstein ended by posing a question to "the expert in the lore of human instincts." Because humans have within them a "lust for hatred and destruction," leaders can manipulate it to stir up militaristic passions. "Is it possible," Einstein asked, "to control man's mental evolution so as to make him secure against the psychosis of hate and destructiveness?"[79]

In a complex and convoluted response, Freud was bleak. "You surmise that man has in him an active instinct for hatred and destruction," he wrote. "I entirely agree." Psychoanalysts had come to the conclusion that two types of human instincts were woven together: "those that conserve and unify, which we call 'erotic' . . . and, secondly, the instincts to destroy and kill, which we assimilate as the aggressive or destructive instincts." Freud cautioned against labeling the first good and the second evil. "Each of these instincts is every whit as indispensable as its opposite, and all the phenomena of life derive from their activity, whether they work in concert or in opposition."

Freud thus came to a pessimistic conclusion:

> The upshot of these observations is that there is no likelihood of our being able to suppress humanity's aggressive tendencies. In some happy corners of the earth, they say, where nature brings forth abundantly whatever man desires, there flourish races whose lives go gently by; unknowing of aggression or constraint. This I can hardly credit; I would like further details about these happy folk. The Bolshevists, too, aspire to do away with human aggressiveness by insuring the satisfaction of material needs and enforcing equality between man and man. To me this hope seems vain. Meanwhile they busily perfect their armaments.[80]

Freud was not pleased with the exchange, and he joked that he doubted it would win either of them the Nobel Peace Prize. In any event, by the time it was ready for publication in 1933, Hitler had come to power. Thus the topic was suddenly moot, and only a few thousand copies were printed. Einstein, like a good scientist, was by then revising his theories based on new facts.

CHAPTER SEVENTEEN

EINSTEIN'S GOD

Santa Barbara beach, 1933

One evening in Berlin, Einstein and his wife were at a dinner party when a guest expressed a belief in astrology. Einstein ridiculed the notion as pure superstition. Another guest stepped in and similarly disparaged religion. Belief in God, he insisted, was likewise a superstition.

At this point the host tried to silence him by invoking the fact that even Einstein harbored religious beliefs.

"It isn't possible!" the skeptical guest said, turning to Einstein to ask if he was, in fact, religious.

"Yes, you can call it that," Einstein replied calmly. "Try and penetrate with our limited means the secrets of nature and you will find that, behind all the discernible laws and connections, there remains something subtle, intangible and inexplicable. Veneration for this force

384

beyond anything that we can comprehend is my religion. To that extent I am, in fact, religious."[1]

As a child, Einstein had gone through an ecstatic religious phase, then rebelled against it. For the next three decades, he tended not to pronounce much on the topic. But around the time he turned 50, he began to articulate more clearly—in various essays, interviews, and letters—his deepening appreciation of his Jewish heritage and, somewhat separately, his belief in God, albeit a rather impersonal, deistic concept of God.

There were probably many reasons for this, in addition to the natural propensity toward reflections about the eternal that can occur at age 50. The kinship he felt with fellow Jews due to their continued oppression reawakened some of his religious sentiments. But mainly, his beliefs seemed to arise from the sense of awe and transcendent order that he discovered through his scientific work.

Whether embracing the beauty of his gravitational field equations or rejecting the uncertainty in quantum mechanics, he displayed a profound faith in the orderliness of the universe. This served as a basis for his scientific outlook—and also his religious outlook. "The highest satisfaction of a scientific person," he wrote in 1929, is to come to the realization "that God Himself could not have arranged these connections any other way than that which does exist, any more than it would have been in His power to make four a prime number."[2]

For Einstein, as for most people, a belief in something larger than himself became a defining sentiment. It produced in him an admixture of confidence and humility that was leavened by a sweet simplicity. Given his proclivity toward being self-centered, these were welcome graces. Along with his humor and self-awareness, they helped him to avoid the pretense and pomposity that could have afflicted the most famous mind in the world.

His religious feelings of awe and humility also informed his sense of social justice. It impelled him to cringe at trappings of hierarchy or class distinction, to eschew excess consumption and materialism, and to dedicate himself to efforts on behalf of refugees and the oppressed.

Shortly after his fiftieth birthday, Einstein gave a remarkable interview in which he was more revealing than he had ever been about his religious thinking. It was with a pompous but ingratiating poet and propagandist named George Sylvester Viereck, who had been born in Germany, moved to America as a child, and then spent his life writing gaudily erotic poetry, interviewing great men, and expressing his complex love for his fatherland.

Having bagged interviews with people ranging from Freud to Hitler to the kaiser, which he would eventually publish as a book called *Glimpses of the Great,* he was able to secure an appointment to talk to Einstein in his Berlin apartment. There Elsa served raspberry juice and fruit salad; then the two men went up to Einstein's hermitage study. For reasons not quite clear, Einstein assumed Viereck was Jewish. In fact, Viereck proudly traced his lineage to the family of the kaiser, and he would later become a Nazi sympathizer who was jailed in America during World War II for being a German propagandist.[3]

Viereck began by asking Einstein whether he considered himself a German or a Jew. "It's possible to be both," replied Einstein. "Nationalism is an infantile disease, the measles of mankind."

Should Jews try to assimilate? "We Jews have been too eager to sacrifice our idiosyncrasies in order to conform."

To what extent are you influenced by Christianity? "As a child I received instruction both in the Bible and in the Talmud. I am a Jew, but I am enthralled by the luminous figure of the Nazarene."

You accept the historical existence of Jesus? "Unquestionably! No one can read the Gospels without feeling the actual presence of Jesus. His personality pulsates in every word. No myth is filled with such life."

Do you believe in God? "I'm not an atheist. The problem involved is too vast for our limited minds. We are in the position of a little child entering a huge library filled with books in many languages. The child knows someone must have written those books. It does not know how. It does not understand the languages in which they are written. The child dimly suspects a mysterious order in the arrangement of the books but doesn't know what it is. That, it seems to me, is the attitude of even the most intelligent human being toward God. We see the universe marvelously arranged and obeying certain laws but only dimly understand these laws."

Is this a Jewish concept of God? "I am a determinist. I do not believe in free will. Jews believe in free will. They believe that man shapes his own life. I reject that doctrine. In that respect I am not a Jew."

Is this Spinoza's God? "I am fascinated by Spinoza's pantheism, but I admire even more his contribution to modern thought because he is the first philosopher to deal with the soul and body as one, and not two separate things."

How did he get his ideas? "I'm enough of an artist to draw freely on my imagination. Imagination is more important than knowledge. Knowledge is limited. Imagination encircles the world."

Do you believe in immortality? "No. And one life is enough for me."[4]

Einstein tried to express these feelings clearly, both for himself and all of those who wanted a simple answer from him about his faith. So in the summer of 1930, amid his sailing and ruminations in Caputh, he composed a credo, "What I Believe." It concluded with an explanation of what he meant when he called himself religious:

> The most beautiful emotion we can experience is the mysterious. It is the fundamental emotion that stands at the cradle of all true art and science. He to whom this emotion is a stranger, who can no longer wonder and stand rapt in awe, is as good as dead, a snuffed-out candle. To sense that behind anything that can be experienced there is something that our minds cannot grasp, whose beauty and sublimity reaches us only indirectly: this is religiousness. In this sense, and in this sense only, I am a devoutly religious man.[5]

People found it evocative, even inspiring, and it was reprinted repeatedly in a variety of translations. But not surprisingly, it did not satisfy those who wanted a simple, direct answer to the question of whether he believed in God. As a result, getting Einstein to answer that question concisely replaced the earlier frenzy of trying to get him to give a one-sentence explanation of relativity.

A Colorado banker wrote that he had already gotten responses from twenty-four Nobel Prize winners to the question of whether they believed in God, and he asked Einstein to reply as well. "I cannot conceive of a personal God who would directly influence the actions of individuals or would sit in judgment on creatures of his own creation,"

Einstein scribbled on the letter. "My religiosity consists of a humble admiration of the infinitely superior spirit that reveals itself in the little that we can comprehend about the knowable world. That deeply emotional conviction of the presence of a superior reasoning power, which is revealed in the incomprehensible universe, forms my idea of God."[6]

A little girl in the sixth grade of a Sunday school in New York posed the question in a slightly different form. "Do scientists pray?" she asked. Einstein took her seriously. "Scientific research is based on the idea that everything that takes place is determined by laws of nature, and this holds for the actions of people," he explained. "For this reason, a scientist will hardly be inclined to believe that events could be influenced by a prayer, i.e. by a wish addressed to a supernatural Being."

That did not mean, however, there was no Almighty, no spirit larger than ourselves. As he went on to explain to the young girl:

> Every one who is seriously involved in the pursuit of science becomes convinced that a spirit is manifest in the laws of the Universe—a spirit vastly superior to that of man, and one in the face of which we with our modest powers must feel humble. In this way the pursuit of science leads to a religious feeling of a special sort, which is indeed quite different from the religiosity of someone more naïve.[7]

For some, only a clear belief in a personal God who controls our daily lives qualified as a satisfactory answer, and Einstein's ideas about an impersonal cosmic spirit, as well as his theories of relativity, deserved to be labeled for what they were. "I very seriously doubt that Einstein himself really knows what he is driving at," Boston's Cardinal William Henry O'Connell said. But one thing seemed clear. It was godless. "The outcome of this doubt and befogged speculation about time and space is a cloak beneath which hides the ghastly apparition of atheism."[8]

This public blast from a cardinal prompted the noted Orthodox Jewish leader in New York, Rabbi Herbert S. Goldstein, to send a very direct telegram: "Do you believe in God? Stop. Answer paid. 50 words." Einstein used only about half his allotted number of words. It became the most famous version of an answer he gave often: "I believe in Spinoza's God, who reveals himself in the lawful harmony of all that exists, but not in a God who concerns himself with the fate and the do-ings of mankind."[9]

Einstein's response was not comforting to everyone. Some religious Jews, for example, noted that Spinoza had been excommunicated from the Jewish community of Amsterdam for holding these beliefs, and he had also been condemned by the Catholic Church for good measure. "Cardinal O'Connell would have done well had he not attacked the Einstein theory," said one Bronx rabbi. "Einstein would have done bet-ter had he not proclaimed his nonbelief in a God who is concerned with fates and actions of individuals. Both have handed down dicta outside their jurisdiction."[10]

Nevertheless, most people were satisfied, whether they fully agreed or not, because they could appreciate what he was saying. The idea of an impersonal God, whose hand is reflected in the glory of creation but who does not meddle in daily existence, is part of a respectable tradi-tion in both Europe and America. It is to be found in some of Ein-stein's favorite philosophers, and it generally accords with the religious beliefs of many of America's founders, such as Jefferson and Franklin.

Some religious believers dismiss Einstein's frequent invocations of God as a mere figure of speech. So do some nonbelievers. There were many phrases he used, some of them playful, ranging from *der Herrgott* (the Lord God) to *der Alte* (the Old One). But it was not Einstein's style to speak disingenuously in order to appear to conform. In fact, just the opposite. So we should do him the honor of taking him at his word when he insists, repeatedly, that these oft-used phrases were not merely a semantic way of disguising that he was actually an atheist.

Throughout his life, he was consistent in deflecting the charge that he was an atheist. "There are people who say there is no God," he told a friend. "But what makes me really angry is that they quote me for support of such views."[11]

Unlike Sigmund Freud or Bertrand Russell or George Bernard Shaw, Einstein never felt the urge to denigrate those who believe in God; instead, he tended to denigrate atheists. "What separates me from most so-called atheists is a feeling of utter humility toward the unattainable secrets of the harmony of the cosmos," he explained.[12]

In fact, Einstein tended to be more critical of the debunkers, who seemed to lack humility or a sense of awe, than of the faithful. "The fanatical atheists," he explained in a letter, "are like slaves who are still feeling the weight of their chains which they have thrown off after hard struggle. They are creatures who—in their grudge against traditional religion as the 'opium of the masses'—cannot hear the music of the spheres."[13]

Einstein would later engage in an exchange on this topic with a U.S. Navy ensign he had never met. Was it true, the sailor asked, that Einstein had been converted by a Jesuit priest into believing in God? That was absurd, Einstein replied. He went on to say that he considered the belief in a God who was a fatherlike figure to be the result of "childish analogies." Would Einstein permit him, the sailor asked, to quote his reply in his debates against his more religious shipmates? Einstein warned him not to oversimplify. "You may call me an agnostic, but I do not share the crusading spirit of the professional atheist whose fervor is mostly due to a painful act of liberation from the fetters of religious indoctrination received in youth," he explained. "I prefer the attitude of humility corresponding to the weakness of our intellectual understanding of nature and of our own being."[14]

How did this religious instinct relate to his science? For Einstein, the beauty of his faith was that it informed and inspired, rather than conflicted with, his scientific work. "The cosmic religious feeling," he said, "is the strongest and noblest motive for scientific research."[15]

Einstein later explained his view of the relationship between science and religion at a conference on that topic at the Union Theological Seminary in New York. The realm of science, he said, was to ascertain what was the case, but not evaluate human thoughts and actions about what *should* be the case. Religion had the reverse mandate. Yet the endeavors worked together at times. "Science can be created only by those who are thoroughly imbued with the aspiration toward truth and understanding," he said. "This source of feeling, however, springs from the sphere of religion."

The talk got front-page news coverage, and his pithy conclusion became famous: "The situation may be expressed by an image: science without religion is lame, religion without science is blind."

But there was one religious concept, Einstein went on to say, that science could not accept: a deity who could meddle at whim in the events of his creation or in the lives of his creatures. "The main source of the present-day conflicts between the spheres of religion and of science lies in this concept of a personal God," he argued. Scientists aim to uncover the immutable laws that govern reality, and in doing so they must reject the notion that divine will, or for that matter human will, plays a role that would violate this cosmic causality.[16]

This belief in causal determinism, which was inherent in Einstein's scientific outlook, conflicted not only with the concept of a personal God. It was also, at least in Einstein's mind, incompatible with human free will. Although he was a deeply moral man, his belief in strict determinism made it difficult for him to accept the idea of moral choice and individual responsibility that is at the heart of most ethical systems.

Jewish as well as Christian theologians have generally believed that people have this free will and are responsible for their actions. They are even free to choose, as happens in the Bible, to defy God's commands, despite the fact that this seems to conflict with a belief that God is all-knowing and all-powerful.

Einstein, on the other hand, believed, as did Spinoza,[17] that a person's actions were just as determined as that of a billiard ball, planet, or star. "Human beings in their thinking, feeling and acting are not free but are as causally bound as the stars in their motions," Einstein declared in a statement to a Spinoza Society in 1932.[18]

Human actions are determined, beyond their control, by both physical and psychological laws, he believed. It was a concept he drew also from his reading of Schopenhauer, to whom he attributed, in his 1930 "What I Believe" credo, a maxim along those lines:

> I do not at all believe in free will in the philosophical sense. Everybody acts not only under external compulsion but also in accordance with inner necessity. Schopenhauer's saying, "A man can do as he wills, but not will as he wills,"[19] has been a real inspiration to me since my youth; it has been a continual consolation in the face of life's hardships, my own and others', and an unfailing wellspring of tolerance.[20]

Do you believe, Einstein was once asked, that humans are free agents? "No, I am a determinist," he replied. "Everything is determined, the beginning as well as the end, by forces over which we have no control. It is determined for the insect as well as for the star. Human beings, vegetables, or cosmic dust, we all dance to a mysterious tune, intoned in the distance by an invisible player."[21]

This attitude appalled some friends, such as Max Born, who thought it completely undermined the foundations of human morality. "I cannot understand how you can combine an entirely mechanistic universe with the freedom of the ethical individual," he wrote Einstein. "To me a deterministic world is quite abhorrent. Maybe you are right, and the world is that way, as you say. But at the moment it does not really look like it in physics—and even less so in the rest of the world."

For Born, quantum uncertainty provided an escape from this dilemma. Like some philosophers of the time, he latched on to the indeterminacy that was inherent in quantum mechanics to resolve "the discrepancy between ethical freedom and strict natural laws."[22] Einstein conceded that quantum mechanics called into question strict determinism, but he told Born he still believed in it, both in the realm of personal actions and physics.

Born explained the issue to his high-strung wife, Hedwig, who was always eager to debate Einstein. She told Einstein that, like him, she was "unable to believe in a 'dice-playing' God." In other words, unlike her husband, she rejected quantum mechanics' view that the universe was based on uncertainties and probabilities. But, she added, "nor am I able to imagine that you believe—as Max has told me—that your 'complete rule of law' means that everything is predetermined, for example whether I am going to have my child inoculated."[23] It would mean, she pointed out, the end of all ethics.

In Einstein's philosophy, the way to resolve this issue was to look upon free will as something that was useful, indeed necessary, for a civilized society, because it caused people to take responsibility for their own actions. Acting *as if* people were responsible for their actions would, psychologically and practically, prompt them to act in a more responsible manner. "I am compelled to act as if free will existed," he explained, "because if I wish to live in a civilized society I must act responsibly." He could even hold people responsible for their good or evil, since that was both a pragmatic and sensible approach to life, while still believing intellectually that everyone's actions were predetermined. "I know that philosophically a murderer is not responsible for his crime," he said, "but I prefer not to take tea with him."[24]

In defense of Einstein, as well as of both Max and Hedwig Born, it should be noted that philosophers through the ages have struggled, sometimes awkwardly and not very successfully, to reconcile free will with determinism and an all-knowing God. Whether Einstein was more or less adept than others at grappling with this knot, there is one salient fact about him that should be noted: he was able to develop, and to practice, a strong personal morality, at least toward humanity in general if not always toward members of his family, that was not hampered by all these irresolvable philosophical speculations. "The most important human endeavor is the striving for morality in our actions," he wrote a Brooklyn minister. "Our inner balance and even our existence depend on it. Only morality in our actions can give beauty and dignity to life."[25]

The foundation of that morality, he believed, was rising above the "merely personal" to live in a way that benefited humanity. There were times when he could be callous to those closest to him, which shows that, like the rest of us humans, he had flaws. Yet more than most people, he dedicated himself honestly and sometimes courageously to actions that he felt transcended selfish desires in order to encourage human progress and the preservation of individual freedoms. He was generally kind, good-natured, gentle, and unpretentious. When he and Elsa left for Japan in 1922, he offered her daughters some advice on how to lead a moral life. "Use for yourself little," he said, "but give to others much."[26]

CHAPTER EIGHTEEN

THE REFUGEE

1932–1933

With Winston Churchill at his home, Chartwell, 1933

"Bird of Passage"

"Today I resolved to give up my Berlin position and shall be a bird of passage for the rest of my life," Einstein wrote in his travel diary. "I am learning English, but it doesn't want to stay in my old brain."[1]

It was December 1931, and he was sailing across the Atlantic for a third visit to America. He was in a reflective mood, aware that the course of science might be proceeding without him and that events in his native land might again make him rootless. When a ferocious storm, far greater than any he had ever witnessed, seized his ship, he recorded his thoughts in his travel diary. "One feels the insignificance of the individual," he wrote, "and it makes one happy."[2]

Yet Einstein was still torn about whether to forsake Berlin for good. It had been his home for seventeen years, Elsa's for even longer. Despite the challenge from Copenhagen, it was still the greatest center for theoretical physics in the world. For all of its dark political undercurrents, it remained a place where he was generally loved and revered, whether he was holding court in Caputh or taking his seat at the Prussian Academy.

In the meantime, his options continued to grow. This trip to America was for another two-month visiting professorship at Caltech, which Millikan was trying to turn into a permanent arrangement. Einstein's friends in Holland had for years also been trying to recruit him, and now so too was Oxford.

Soon after he settled into his rooms at the Athenaeum, the graceful faculty club at Caltech, yet another possibility arose. One morning, he was visited there by the noted American educator Abraham Flexner, who spent more than an hour walking the cloistered courtyard with him. When Elsa found them and summoned her husband to a luncheon engagement, he waved her off.

Flexner, who had helped reshape American higher education as an officer of the Rockefeller Foundation, was in the process of creating a "haven" where scholars could work without any academic pressures or teaching duties and, as he put it, "without being carried off in the maelstrom of the immediate."[3] Funded by a $5 million donation from Louis Bamberger and his sister Caroline Bamberger Fuld, who had the good fortune to sell their department store chain just weeks before the 1929 stock market crash, it would be named the Institute for Advanced Study and located in New Jersey, probably next to (but not formally affiliated with) Princeton University, where Einstein had already spent some enjoyable time.

Flexner had come to Caltech to get some ideas from Millikan, who (to his later regret) insisted he talk to Einstein. When Flexner finally set up such a meeting, he was impressed, he later wrote, with Einstein's "noble bearing, simply charming manner, and his genuine humility."

It was obvious that Einstein would be a perfect anchor and ornament for Flexner's new institute, but it would have been inappropriate for Flexner to make an offer on Millikan's home turf. Instead, they agreed that Flexner would visit Einstein in Europe to discuss matters further. Flexner claimed in his autobiography that, even after their

Caltech meeting, "I had no idea that he [Einstein] would be interested in being connected to the Institute." But that was belied by the letters he wrote to his patrons at the time, in which he referred to Einstein as an "unhatched chicken" whose prospects they needed to treat circumspectly.[4]

By then Einstein had grown slightly disenchanted with life in southern California. When he gave a speech to an international relations group, in which he denounced arms-control compromises and advocated complete disarmament, his audience seemed to treat him as celebrity entertainment. "The propertied classes here seize upon anything that might provide ammunition in the struggle against boredom," he noted in his diary. Elsa reflected his annoyance in a letter to a friend. "The affair was not only lacking in seriousness but was treated as a kind of social entertainment."[5]

As a result, he was dismissive when his friend Ehrenfest in Leiden wrote to ask for his help in getting a job in America. "I must tell you honestly that in the long term I would prefer to be in Holland rather than in America," Einstein replied. "Apart from the handful of really fine scholars, it is a boring and barren society that would soon make you shiver."[6]

Nevertheless, on this and other topics Einstein's mind was not a simple one. He clearly enjoyed America's freedom, excitement, and even (yes) the celebrity status it conferred upon him. Like many others, he could be critical of America yet also attracted to it. He could recoil at its occasional displays of crassness and materialism, yet find himself powerfully drawn to the freedoms and unvarnished individuality that were on the flip side of the same coin.

Soon after returning to Berlin, where the political situation had become even more unnerving, Einstein went to Oxford to give another series of lectures. Once again, he found its refined formality oppressive, especially in contrast to America. At the stultifying sessions of the governing body of Christ Church, his college at Oxford, he sat in the senior common room holding a notepad under the tablecloth so that he could scribble equations. He came to realize, once again, that America, for all of its lapses of taste and excesses of enthusiasm, offered freedoms he might never find again in Europe.[7]

Thus he was pleased when Flexner came, as promised, to continue the conversation they had started at the Athenaeum. Both men knew, from the outset, that it was not merely an abstract discussion but part of an effort to recruit Einstein. So Flexner was a bit disingenuous when he later wrote that it was only while they were pacing around the manicured lawns of Christ Church's Tom Quad that it "dawned on me" that Einstein might be interested in coming to the new institute. "If on reflection you conclude that it would give you the opportunities that you value," Flexner said, "you would be welcome on your own terms."[8]

The arrangement that would bring Einstein to Princeton was concluded the following month, June 1932, when Flexner visited Caputh. It was cool that day, and Flexner wore an overcoat, but Einstein was in summer clothes. He preferred, he joked, to dress "according to the season not according to the weather." They sat on the veranda of Einstein's beloved new cottage and spoke all afternoon and then through dinner, up until Einstein walked Flexner to the Berlin bus at 11 p.m.

Flexner asked Einstein how much he would expect to make. About $3,000, Einstein tentatively suggested. Flexner looked surprised. "Oh," Einstein hastened to add, "could I live on less?"

Flexner was amused. He had more, not less, in mind. "Let Mrs. Einstein and me arrange it," he said. They ended up settling on $10,000 per year. That was soon increased when Louis Bamberger, the primary backer, discovered that mathematician Oswald Veblen, the Institute's other jewel, was making $15,000 a year. Bamberger insisted that Einstein's salary be equal.

There was one additional deal point. Einstein insisted that his assistant, Walther Mayer, be given a job of his own as well. The previous year he had let authorities in Berlin know that he was entertaining offers in America that would provide for Mayer, something Berlin had been unwilling to do. Caltech had balked at this request, as did Flexner initially. But then Flexner relented.[9]

Einstein did not consider his post at the Institute a full-time job, but it was likely to be his primary one. Elsa delicately broached this in her letter to Millikan. "Will you, under the circumstances, still want my husband in Pasadena next winter?" she asked. "I doubt it."[10]

Actually, Millikan did want him, and they agreed that Einstein would come back again in January, before the Institute would be open in Princeton. Millikan was upset, however, that he had not finalized a long-term deal, and he realized that Einstein would end up being, at best, an occasional visitor to Caltech. As it turned out, the upcoming January 1933 trip that Elsa helped arrange would end up being his last trip to California.

Millikan vented his anger at Flexner. Einstein's connection with Caltech "has been laboriously built up during the past ten years," he wrote. As a result of Flexner's pernicious raid, Einstein would be spending his time at some new haven rather than a great center of experimental as well as theoretical physics. "Whether the progress of science in the U.S. will be advanced by such a move, or whether Professor Einstein's productivity will be increased by such a transfer, is at least debatable." He proposed, as a compromise, that Einstein split his time in America between the Institute and Caltech.

Flexner was not magnanimous in victory. He protested, falsely, that it was "altogether by accident" that he ended up in Oxford and speaking to Einstein, a tale that even his own memoirs later contradicted. As for sharing Einstein, Flexner declined. He claimed that he was looking after Einstein's interests. "I cannot believe that annual residence for brief periods at several places is sound or wholesome," he wrote. "Looking at the entire matter from Professor Einstein's point of view, I believe that you and all of his friends will rejoice that it has been possible to create for him a permanent post." [11]

For his part, Einstein was unsure how he would divide his time. He thought that he might be able to juggle visiting professorships in Princeton, Pasadena, and Oxford. In fact, he even hoped that he could keep his position in the Prussian Academy and his beloved cottage in Caputh, if things did not worsen in Germany. "I am not abandoning Germany," he announced when the Princeton post became public in August. "My permanent home will still be in Berlin."

Flexner spun the relationship the other way, telling the *New York Times* that Princeton would be Einstein's primary home. "Einstein will devote his time to the Institute," Flexner said, "and his trips abroad will be vacation periods for rest and meditation at his summer home outside of Berlin." [12]

As it turned out, the issue would be settled by events out of either man's control. Throughout the summer of 1932, the political situation in Germany darkened. As the Nazis continued to lose national elections but increase their share of the vote, the octogenarian president, Paul von Hindenburg, selected as chancellor the bumbling Franz von Papen, who tried to rule through martial authority. When Philipp Frank came to visit him in Caputh that summer, Einstein lamented, "I am convinced that a military regime will not prevent the imminent National Socialist [Nazi] revolution."[13]

As Einstein was preparing to leave for his third visit to Caltech in December 1932, he had to suffer one more indignity. The headlines about his future post in Princeton had aroused the indignation of the Woman Patriot Corporation, a once powerful but fading group of American self-styled guardians against socialists, pacifists, communists, feminists, and undesirable aliens. Although Einstein fit into only the first two of these categories, the women patriots felt sure that he fit into them all, with the possible exception of feminists.

The leader of the group, Mrs. Randolph Frothingham (who, given this context, seemed as if her distinguished family name had been conjured up by Dickens), submitted a sixteen-page typed memo to the U.S. State Department detailing reasons to "refuse and withhold such passport visa to Professor Einstein." He was a militant pacifist and communist who advocated doctrines that "would allow anarchy to stalk in unmolested," the memo charged. "*Not even Stalin himself* is affiliated with so many anarcho-communist international groups to promote this 'preliminary condition' of world revolution and ultimate anarchy as ALBERT EINSTEIN." (Emphasis and capitalization are in the original.)[14]

State Department officials could have ignored the memo. Instead, they put it into a file that would grow over the next twenty-three years into an FBI dossier of 1,427 pages of documents. In addition, they sent the memo to the U.S. consulate in Berlin so that officers there could interview Einstein and see if the charges were true before granting him another visa.

Initially, Einstein was quite amused when he read newspaper accounts of the women's allegations. He called up the Berlin bureau chief of United Press, Louis Lochner, who had become a friend, and gave him a statement that not only ridiculed the charges but also proved conclusively that he could not be accused of feminism:

> Never yet have I experienced from the fair sex such energetic rejection of all advances, or if I have, never from so many at once. But are they not right, these watchful citizenesses? Why should one open one's doors to a person who devours hard-boiled capitalists with as much appetite and gusto as the ogre Minotaur in Crete once devoured luscious Greek maidens—a person who is also so vulgar as to oppose every sort of war, except the inevitable one with his own wife? Therefore, give heed to your clever and patriotic women folk and remember that the capital of mighty Rome was once saved by the cackling of its faithful geese.[15]

The *New York Times* ran the story on its front page with the headline, "Einstein Ridicules Women's Fight on Him Here / Remarks Cackling Geese Once Saved Rome."[16] But Einstein was far less amused two days later when, as he and Elsa were packing to leave, he received a telephone call from the U.S. consular office in Berlin asking him to come by for an interview that afternoon.

The consul general was on vacation, so his hapless deputy conducted the interview, which Elsa promptly recounted to reporters.[17] According to the *New York Times*, which ran three stories the next day on the incident, the session started well enough but then degenerated.

"What is your political creed?" he was asked. Einstein gave a blank stare and then burst out laughing. "Well, I don't know," he replied. "I can't answer that question."

"Are you a member of any organization?" Einstein ran his hand through "his ample hair" and turned to Elsa. "Oh yes!" he exclaimed. "I am a War Resister."

The interview dragged on for forty-five minutes, and Einstein became increasingly impatient. When he was asked whether he was a sympathizer of any communist or anarchist parties, Einstein lost his temper. "Your countrymen invited me," he said. "Yes, begged me. If I am to enter your country as a suspect, I don't want to go at all. If you don't want to give me a visa, please say so."

Then he reached for his coat and hat. "Are you doing this to please yourselves," he asked, "or are you acting on orders from above?" Without waiting for an answer, he left with Elsa in tow.

Elsa let the papers know that Einstein had quit packing and had left Berlin for his cottage in Caputh. If he did not have a visa by noon the next day, he would cancel his trip to America. By late that night, the consulate issued a statement saying that it had reviewed the case and would issue a visa immediately.

As the *Times* correctly reported, "He is not a Communist and has declined invitations to lecture in Russia because he did not want to give the impression that he was in sympathy with the Moscow regime." What none of the papers reported, however, was that Einstein did agree to sign a declaration, requested by the consulate, that he was not a member of the Communist Party or any organization intent on overthrowing the U.S. government.[18]

"Einstein Resumes Packing for America," read the *Times* headline the next day. "From the deluge of cables reaching us last night," Elsa told reporters, "we know Americans of all classes were deeply disturbed over the case." Secretary of State Henry Stimson said that he regretted the incident, but he also noted that Einstein "was treated with every courtesy and consideration." As they left Berlin by train for Bremerhaven to catch their ship, Einstein joked about the incident and said that all had turned out well in the end.[19]

Pasadena, 1933

When the Einsteins left Germany in December 1932, he still thought that he might be able to return, but he wasn't sure. He wrote to his longtime friend Maurice Solovine, now publishing his works in Paris, to send copies "to me next April at my Caputh address." Yet when they left Caputh, Einstein said to Elsa, as if with a premonition, "Take a very good look at it. You will never see it again." With them on the steamer *Oakland* as it headed for California were thirty pieces of luggage, probably more than necessary for a three-month trip.[20]

Thus it was awkward, and painfully ironic, that the one public duty Einstein was scheduled to perform in Pasadena was to give a speech to

celebrate German-American friendship. To finance Einstein's stay at
Caltech, President Millikan had obtained a $7,000 grant from the
Oberlaender Trust, a foundation that sought to promote cultural exchanges with Germany. The sole requirement was that Einstein would
make "one broadcast which will be helpful to German-American relations." Upon Einstein's arrival, Millikan announced that Einstein was
"coming to the United States on a mission of molding public opinion
to better German-American relations,"[21] a view that may have surprised Einstein, with his thirty pieces of luggage.

Millikan usually preferred that his prize visitor avoid speaking on
nonscientific matters. In fact, soon after Einstein arrived, Millikan
forced him to cancel a speech he was scheduled to give to the UCLA
chapter of the War Resisters' League, in which he had planned to denounce compulsory military service again. "There is no power on earth
from which we should be prepared to accept an order to kill," he wrote
in the draft of the speech he never gave.[22]

But as long as Einstein was expressing pro-German rather
than pacifist sentiments, Millikan was happy for him to talk about
politics—especially as there was funding involved. Not only had Millikan been able to secure the $7,000 Oberlaender grant by scheduling
the speech, which was to be broadcast on NBC radio, he also had invited big donors to a black-tie dinner preceding it at the Athenaeum.

Einstein was such a draw that there was a wait list to buy tickets.
Among those seated at Einstein's table was Leon Watters, a wealthy
pharmaceutical manufacturer from New York. Noticing that Einstein
looked bored, he reached across the woman seated between them to
offer him a cigarette, which Einstein consumed in three drags. The two
men subsequently became close friends, and Einstein would later stay
at Watters's Fifth Avenue apartment when he visited New York from
Princeton.

When the dinner was over, Einstein and the other guests went to
the Pasadena Civic Auditorium, where several thousand people waited
to hear his address. His text had been translated for him by a friend,
and he delivered it in halting English.

After making fun of the difficulties of sounding serious while wearing a tuxedo, he proceeded to attack people who used words "laden

with emotion" to intimidate free expression. "Heretic," as used during the Inquisition, was such a case, he said. Then he cited examples that had similar hateful connotations for people in a variety of countries: "the word Communist in America today, or the word bourgeoisie in Russia, or the word Jew for the reactionary group in Germany." Not all of these examples seemed calculated to please Millikan or his anticommunist and pro-German funders.

Nor was his critique of the current world crisis one that would appeal to ardent capitalists. The economic depression, especially in America, seemed to be caused, he said, mainly by technological advances that "decreased the need for human labor" and thereby caused a decline in consumer purchasing power.

As for Germany, he made a couple of attempts to express sympathy and earn Millikan's grant. America would be wise, he said, not to press too hard for continued payment of debts and reparations from the world war. In addition, he could see some justification in Germany's demand for military equality.

That did not mean, however, that Germany should be allowed to reintroduce mandatory military service, he hastened to add. "Universal military service means the training of youth in a warlike spirit," he concluded.[23] Millikan may have gotten his speech about Germany, but the price he paid was swallowing a few thoughts from the war resistance speech he had forced Einstein to cancel.

A week later, all of these items—German-American friendship, debt payments, war resistance, even Einstein's pacifism—were dealt a blow that would render them senseless for more than a decade. On January 30, 1933, while Einstein was safely in Pasadena, Adolf Hitler took power as the new chancellor of Germany.

Einstein initially seemed unsure what this meant for him. During the first week of February, he was writing letters to Berlin about how to calculate his salary for his planned return in April. His sporadic entries in his trip journal that week recorded only serious scientific discussions, such as on cosmic ray experiments, and frivolous social encounters, such as: "Evening Chaplin. Played Mozart quartets there. Fat lady whose occupation consists of making friends with all celebrities."[24]

By the end of February, however, with the Reichstag in flames and brownshirts ransacking the homes of Jews, things had become clearer. "Because of Hitler, I don't dare step on German soil," Einstein wrote one of his women friends.[25]

On March 10, the day before he left Pasadena, Einstein was strolling in the gardens of the Athenaeum. Evelyn Seeley of the *New York World Telegram* found him there in an expansive mood. They talked for forty-five minutes, and one of his declarations made headlines around the world. "As long as I have any choice in the matter, I shall live only in a country where civil liberty, tolerance and equality of all citizens before the law prevail," he said. "These conditions do not exist in Germany at the present time."[26]

Just as Seeley was leaving, Los Angeles was struck by a devastating earthquake—116 people were killed in the area—but Einstein barely seemed to notice. With the acquiescence of an indulgent editor, Seeley was able to end her article with a dramatic metaphor: "As he left for the seminar, walking across campus, Dr. Einstein felt the ground shaking under his feet."

In retrospect, Seeley would be saved from sounding too portentous by a drama that was occurring that very day back in Berlin, although neither she nor Einstein knew it. His apartment there, with Elsa's daughter Margot cowering inside, was raided twice that afternoon by the Nazis. Her husband, Dimitri Marianoff, was out doing errands and was almost trapped by one of the roving mobs of thugs. He sent word for Margot to get Einstein's papers to the French embassy and then meet him in Paris. She was able to do both. Ilse and her husband, Rudolph Kayser, successfully escaped to Holland. During the next two days, the Berlin apartment was ransacked three more times. Einstein would never see it again. But his papers were safe.[27]

On his train ride east from Caltech, Einstein reached Chicago on his fifty-fourth birthday. There he attended a Youth Peace Council rally, where speakers pledged that the pacifist cause should continue despite the events in Germany. Some left with the impression that he was in full agreement. "Einstein will never abandon the peace movement," one noted.

They were wrong. Einstein had begun to mute his pacifist rhetoric. At a birthday luncheon that day in Chicago, he spoke vaguely about the need for international organizations to keep the peace, but he refrained from repeating his calls for war resistance. He was similarly cautious a few days later at a New York reception for an anthology featuring his pacifist writings, *The Fight against War.* He mainly talked about the distressing turn of events in Germany. The world should make its moral disapproval of the Nazis known, he said, but he added that the German population itself should not be demonized.

It was unclear, even as he was about to sail, where he would now live. Paul Schwartz, the German consul in New York who had been Einstein's friend in Berlin, met with him privately to make sure that he did not plan to go back to Germany. "They'll drag you through the streets by the hair," he warned.[28]

His initial destination, where the ship would let him off, was Belgium, and he suggested to friends that he might go to Switzerland after that. When the Institute for Advanced Study opened the following year, he planned to spend four or five months there each year. Perhaps it would turn out to be even more. On the day before he sailed, he and Elsa slipped away to Princeton to look at houses they might buy.

The only place in Germany that he wanted to see again, he told family members, was Caputh. But on the journey across the Atlantic, he received word that the Nazis had raided his cottage under the pretense of looking for a cache of communist weaponry (there was none). Later they came back and confiscated his beloved boat on the pretense it might be used for smuggling. "My summer house was often honored by the presence of many guests," he said in a message from the ship. "They were always welcome. No one had any reason to break in."[29]

The Bonfires

The news of the raid on his Caputh cottage determined Einstein's relationship to his German homeland. He would never go back there.

As soon as his ship docked in Antwerp on March 28, 1933, he had a car drive him to the German consulate in Brussels, where he turned in his passport and (as he had done once before when a teenager) declared that he was renouncing his German citizenship. He also mailed a letter, written during the crossing, in which he submitted his resigna-

tion to the Prussian Academy. "Dependence on the Prussian government," he stated, "is something that, under the present circumstances,
I feel to be intolerable."[30]

Max Planck, who had recruited him to the Academy nineteen years
earlier, was relieved. "This idea of yours seems to be the only way that
would ensure for you an honorable severance of your relations with the
Academy," Planck wrote back with an almost audible sigh. He added
his gracious plea that "despite the deep gulf that divides our political
opinions, our personal amicable relations will never undergo any
change."[31]

What Planck was hoping to avoid, amid the flurry of anti-Semitic
diatribes against Einstein in the Nazi press, were formal disciplinary
hearings against Einstein, which some government ministers were demanding. That would cause Planck personal agony and the Academy
historic embarrassment. "Starting formal exclusion procedures against
Einstein would bring me into gravest conflicts of conscience," he wrote
an Academy secretary. "Even though on political matters a deep gulf
divides me from him, I am, on the other hand, absolutely certain that in
the history of centuries to come, Einstein's name will be celebrated as
one of the brightest stars that ever shone in the Academy."[32]

Alas, the Academy was not content to leave bad enough alone.
The Nazis were furious that he had preempted them by renouncing,
very publicly, with headlines in the papers, his citizenship and Academy membership before they could strip him of both. So a Nazi-
sympathizing secretary of the Academy issued a statement on its
behalf. Referring to the press reports of some of his comments in
America, which in fact had been very cautious, it denounced Einstein's
"participation in atrocity-mongering" and his "activities as an agitator
in foreign countries," concluding, "It has, therefore, no reason to regret
Einstein's withdrawal."[33]

Max von Laue, a longtime colleague and friend, protested. At a
meeting of the Academy later that week, he tried to get members to
disavow the secretary's action. But no other member would go along,
not even Haber, the converted Jew who had been one of Einstein's
closest friends and supporters.

Einstein was not willing to let such a slander pass. "I hereby declare that I have never taken any part in atrocity-mongering," he responded. He had merely spoken the truth about the situation in Germany, without resorting to purveying tales of atrocities. "I described the present state of affairs in Germany as a state of psychic distemper in the masses," he wrote.[34]

By then there was no doubt this was true. Earlier in the week, the Nazis had called for a boycott of all Jewish-owned businesses and stationed storm troopers outside of their stores. Jewish teachers and students were barred from the university in Berlin and their academic identification cards were confiscated. And the Nobel laureate Philipp Lenard, Einstein's longtime antagonist, declared in a Nazi newspaper, "The most important example of the dangerous influence of Jewish circles on the study of nature has been provided by Herr Einstein."[35]

The exchanges between Einstein and the Academy descended into petulance. An official wrote Einstein that, even if he had not actively spread slanders, he had failed to join "the side of the defenders of our nation against the flood of lies that has been let loose upon it . . . A good word from you in particular might have produced a great effect abroad." Einstein thought that absurd. "By giving such testimony in the present circumstances I would have been contributing, if only indirectly, to moral corruption and the destruction of all existing cultural values," he replied.[36]

The entire dispute was becoming moot. Early in April 1933, the German government passed a law declaring that Jews (defined as anyone with a Jewish grandparent) could not hold an official position, including at the Academy or at the universities. Among those forced to flee were fourteen Nobel laureates and twenty-six of the sixty professors of theoretical physics in the country. Fittingly, such refugees from fascism who left Germany or the other countries it came to dominate—Einstein, Edward Teller, Victor Weisskopf, Hans Bethe, Lise Meitner, Niels Bohr, Enrico Fermi, Otto Stern, Eugene Wigner, Leó Szilárd, and others—helped to assure that the Allies rather than the Nazis first developed the atom bomb.

Planck tried to temper the anti-Jewish policies, even to the extent of appealing to Hitler personally. "Our national policies will not be revoked or modified, even for scientists," Hitler thundered back. "If the dismissal of Jewish scientists means the annihilation of contemporary German science, then we shall do without science for a few years!" After that, Planck quietly went along and cautioned other scientists that it was not their role to challenge the political leadership.

Einstein could not bring himself to be angry at Planck, who was like an uncle as well as a patron. Even amid his angry exchanges with the Academy, he agreed to Planck's request that they keep their personal respect intact. "In spite of everything, I am happy that you greet me in old friendship and that even the greatest stresses have failed to cloud our mutual relations," he wrote, using the formal and respectful style he always used when writing to Planck. "These continue in their ancient beauty and purity, regardless of what, in a manner of speaking, is happening further below." [37]

Among those fleeing the Nazi purge was Max Born, who with his tart-tongued wife, Hedwig, ended up in England. "I have never had a particularly favorable opinion of the Germans," Einstein wrote when he received the news. "But I must confess that the degree of their brutality and cowardice came as something of a surprise."

Born took it all rather well, and he developed, like Einstein, a deeper appreciation for his heritage. "As regards my wife and children, they have only become conscious of being Jews or 'non-Aryans' (to use the delightful technical term) during the last few months, and I myself have never felt particularly Jewish," he wrote in his letter back to Einstein. "Now, of course, I am extremely conscious of it, not only because we are considered to be so, but because oppression and injustice provoke me to anger and resistance." [38]

Even more poignant was the case of Fritz Haber, friend to both Einstein and Marić, who thought that he had become German by converting to Christianity, affecting a Prussian air, and pioneering gas warfare for his Fatherland in the First World War. But with the new laws, even he was forced from his position at Berlin University and in the Academy, at age 64, just before he would have been eligible for a pension.

As if to atone for forsaking his heritage, Haber threw himself into organizing Jews who suddenly needed to find jobs outside of Germany. Einstein could not resist gigging him, in the bantering manner they had often used in their letters, about the failure of his theory of assimilation. "I can understand your inner conflicts," he wrote. "It is somewhat like having to give up a theory on which one has worked one's whole life. It is not the same for me because I never believed it in the least."[39]

In the process of helping his newfound tribal companions to emigrate, Haber became friends with the Zionist leader Chaim Weizmann. He even tried to mend a rift that had come between Weizmann and Einstein over Jewish treatment of the Arabs and the management of Hebrew University. "In my whole life I have never felt so Jewish as now!" he exulted, though that was not actually saying much.

Einstein replied by saying how pleased he was that "your former love for the blond beast has cooled off a bit." The Germans were all a bad breed, Einstein insisted, "except a few fine personalities (Planck 60% noble, and Laue 100%)." Now, in this time of adversity, they could at least take comfort that they were thrown together with their true kinsmen. "For me the most beautiful thing is to be in contact with a few fine Jews—a few millennia of a civilized past do mean something after all."[40]

Einstein would never again see Haber, who decided that he would try to make a new life at Hebrew University in Jerusalem, which Einstein had helped to launch. But in Basel, on his way there, Haber's heart gave out and he died.

Close to forty thousand Germans gathered in front of Berlin's opera house on May 10, 1933, as a parade of swastika-wearing students and beer-hall thugs carrying torches tossed books into a huge bonfire. Ordinary citizens poured forth carrying volumes looted from libraries and private homes. "Jewish intellectualism is dead," propaganda minister Joseph Goebbels, his face fiery, yelled from the podium. "The German soul can again express itself."

What happened in Germany in 1933 was not just a brutality perpetrated by thuggish leaders and abetted by ignorant mobs. It was also, as Einstein described, "the utter failure of the so-called intellectual aristocracy." Einstein and other Jews were ousted from what had been among the world's greatest citadels of open-minded inquiry, and those who remained did little to resist. It represented the triumph of the ilk of Philipp Lenard, Einstein's longtime anti-Semitic baiter, who was named by Hitler to be the new chief of Aryan science. "We must recognize that it is unworthy of a German to be the intellectual follower of a Jew," Lenard exulted that May. "Heil Hitler!" It would be a dozen years before Allied troops would fight their way in and oust him from that role.[41]

Le Coq sur Mer, 1933

Having found himself deposited in Belgium, more by the happenstance of ocean liner routes than by conscious choice, Einstein and his entourage—Elsa, Helen Dukas, Walther Mayer—set up household there for the time being. He was not, he realized after a little consideration, quite up for the emotional energy it would take to relocate his new family in Zurich alongside his old one. Nor was he ready to commit to Leiden or Oxford while he awaited his scheduled visit, or perhaps move, to Princeton. So he rented a house on the dunes of Le Coq sur Mer, a resort near Ostend, where he could contemplate, and Mayer could calculate, the universe and its waves in peace.

Peace, however, was elusive. Even by the sea he could not completely escape the threats of the Nazis. The newspapers reported that his name was on a list of assassination targets, and one rumor had it that there was a $5,000 bounty on his head. Upon hearing this, Einstein touched that head and cheerfully proclaimed, "I didn't know it was worth that much!" The Belgians took the danger more seriously and, much to his annoyance, assigned two beefy police officers to stand guard at the house.[42]

Philipp Frank, who still had Einstein's old job and office in Prague, happened to be passing through Ostend that summer and decided to pay a surprise visit. He asked local residents how to find Einstein and, despite all the security injunctions about giving out such information, was promptly directed to the cottage amid the dunes. As he approached, he saw two robust men, who certainly did not look like Einstein's usual visitors, in intense conversation with Elsa. Suddenly, as Frank later recalled, "the two men saw me, threw themselves at me and seized me."

Elsa, her face chalky white with fright, intervened. "They suspected you of being the rumored assassin."

Einstein found the entire situation quite hilarious, including the naïveté of the people in the neighborhood who kindly showed Frank the way to his house. Einstein described his exchange of letters with the Prussian Academy, which he had put into a folder with some lines of humorous verse he had composed for an imaginary response: "Thank you for your note so tender / It's typically German, like the sender."

When Einstein said that leaving Berlin had proved liberating, Elsa defended the city that she had loved for so long. "You often said to me after coming home from the physics colloquium that such a gathering of outstanding physicists is not to be found anywhere else."

"Yes," Einstein replied, "from a purely scientific point of view life in Berlin was often very nice. Nevertheless, I always had a feeling that something was pressing on me, and I always had a premonition that the end would not be good."[43]

With Einstein a free agent, offers flowed in from all over Europe. "I now have more professorships than rational ideas in my head," he told Solovine.[44] Although he had committed to spend at least a few months each year in Princeton, he began accepting these invitations somewhat promiscuously. He was never very good at declining requests.

Partly it was because the offers were enticing and he was flattered. Partly it was because he was still trying to leverage a better deal for his assistant, Walther Mayer. In addition, the offers became a way for him and the various universities to show their defiance of what the Nazis were doing to German academies. "You may feel that it would have been my duty not to accept the Spanish and French offers," he confessed to Paul Langevin in Paris, "however, such a refusal might have been misinterpreted since both invitations were, at least to some extent, political demonstrations that I considered important and did not want to spoil."[45]

His acceptance of a post at the University of Madrid made headlines in April. "Spanish Minister Announces Physicist Has Accepted Professorship," said the *New York Times*. "News Received with Joy." The paper pointed out that this should not affect his annual stints in

Princeton, but Einstein warned Flexner that it could if Mayer was not given a full rather than an associate professorship at the new Institute. "You will by now have learned through the press that I have accepted a chair at Madrid University," he wrote. "The Spanish government has given me the right to recommend to them a mathematician to be appointed as a full professor . . . I therefore find myself in a difficult position: either to recommend him for Spain or to ask you whether you could possibly extend his appointment to a full professorship." In case the threat was not clear enough, Einstein added, "His absence from the Institute might even create some difficulties for my own work." [46]

Flexner compromised. In a four-page letter, he cautioned Einstein about the perils of becoming too attached to one assistant, told tales of how that had worked out badly in other cases, but then relented. Although Mayer's title remained associate professor, he was given tenure, which was enough to secure the deal. [47]

Einstein also accepted or expressed interest in lectureships in Brussels, Paris, and Oxford. He was particularly eager to spend some time at the latter. "Do you think that Christ Church could find a small room for me?" he wrote his friend Professor Frederick Lindemann, a physicist there who would become an important adviser to Winston Churchill. "It need not be so grand as in the two previous years." At the end of the letter, he added a wistful little note: "I shall never see the land of my birth again." [48]

This raised one obvious question: Why did he not consider spending some time at Hebrew University in Jerusalem? After all, it was partly his baby. Einstein spent the spring of 1933 actively talking about starting up a new university, perhaps in England, that could serve as a refuge for displaced Jewish academics. Why wasn't he instead recruiting them for, and committing himself personally to, Hebrew University?

The problem was that for the previous five years, Einstein had been doing battle with administrators there, and it came to an untimely showdown in 1933, just as he and other professors were fleeing the Nazis. The target of his ire was the university's president, Judah Magnes, a former rabbi from New York who felt a duty to please his wealthy American backers, including on faculty appointments, even if this meant compromising on scholarly distinction. Einstein wanted the university to operate more in the European tradition, with the academic departments given great power over curriculum and tenured faculty decisions.[49]

While he was in Le Coq sur Mer, his frustrations with Magnes boiled over. "This ambitious and weak person surrounded himself with other morally inferior men," he wrote Haber in cautioning him about going to Hebrew University. He described it to Born as "a pigsty, complete charlatanism."[50]

Einstein's complaints put him at odds with the Zionist leader Chaim Weizmann. When Weizmann and Magnes sent him a formal invitation to join the Hebrew University faculty, he allowed his distaste to pour forth publicly. He told the press that the university was "unable to satisfy intellectual needs" and declared that he had thus rejected the invitation.[51]

Magnes must go, Einstein declared. He wrote Sir Herbert Samuel, the British high commissioner, who had been appointed to a committee to propose reforms, that Magnes had wrought "enormous damage" and that "if ever people want my collaboration, his immediate resignation is my condition." In June he said the same to Weizmann: "Only a decisive change of personnel would alter things."[52]

Weizmann was an adroit broken-field runner. He decided to turn Einstein's challenge into an opportunity to lessen Magnes's power. If he succeeded, then Einstein should feel compelled to join the faculty. On a trip to America later in June, he was asked why Einstein was not going to Jerusalem, where he surely belonged. He should indeed go there, Weizmann agreed, and he had been invited to do so. If he went to Jerusalem, Weizmann added, "he would cease to be a wanderer among the universities of the world."[53]

Einstein was furious. His reasons for not going to Jerusalem were well known to Weizmann, he said, "and he also knows under what circumstances I would be prepared to undertake work for the Hebrew University." That led Weizmann to appoint a committee that, he knew, would remove Magnes from direct control of the academic side of the university. He then announced, during a visit to Chicago, that Einstein's conditions had been met and therefore he should be coming to Hebrew University after all. "Albert Einstein has definitely decided to accept direction of the physics institute at the Hebrew University," the Jewish Telegraphic Agency reported, based on information from Weizmann.

It was a ruse by Weizmann that was not true and would never come to pass. But in addition to frightening Flexner in Princeton, it allowed the Hebrew University controversy to simmer down and for reforms to be made at the university.[54]

The End of Pacifism

Like a good scientist, Einstein could change his attitudes when confronted with new evidence. Among his deepest personal principles was his pacifism. But in early 1933, with Hitler's ascension, the facts had changed.

So Einstein forthrightly declared that he had come to the conclusion that absolute pacifism and military resistance were, at least for the moment, not warranted. "The time seems inauspicious for further advocacy of certain propositions of the radical pacifist movement," he wrote to a Dutch minister who wanted his support for a peace organization. "For example, is one justified in advising a Frenchman or a Belgian to refuse military service in the face of German rearmament?" Einstein felt the answer was now clear. "Frankly, I do not believe so."

Instead of pushing pacifism, he redoubled his commitment to a world federalist organization, like a League of Nations with real teeth, that would have its own professional army to enforce its decisions. "It seems to me that in the present situation we must support a supranational organization of force rather than advocate the abolition of all forces," he said. "Recent events have taught me a lesson in this respect."[55]

This met resistance from the War Resisters' International, an organization that he had long supported. Its leader, Lord Arthur Ponsonby, denounced the idea, calling it "undesirable because it is an admission that force is the factor that can resolve international disputes." Einstein disagreed. In the wake of the new threat arising in Germany, his new philosophy, he wrote, was "no disarmament without security."[56]

Four years earlier, while visiting Antwerp, Einstein had been invited to the Belgian royal palace by Queen Elisabeth,[57] the daughter of a Bavarian duke who was married to King Albert I. The queen loved music, and Einstein spent the afternoon playing Mozart with her, drinking tea, and attempting to explain relativity. Invited back the following year, he met her husband, the king, and became charmed by the least regal of all royals. "These two simple people are of a purity and goodness that is seldom to be found," he wrote Elsa. Once again he and the queen played Mozart, then Einstein was invited to stay and dine alone with the couple. "No servants, vegetarian, spinach with fried egg and potatoes," he recounted. "I liked it enormously, and I am sure that the feeling is mutual."[58]

Thus began a lifelong friendship with the Belgian queen. Later, his relationship with her would play a minor role in Einstein's involvement with the atomic bomb. But in July 1933, the issue at stake was pacifism and military resistance.

"The husband of the second violinist would like to talk to you on an urgent matter." It was a cryptic way for King Albert to identify himself that Einstein, but few others, would recognize. Einstein headed to the palace. On the king's mind was a case that was roiling his country. Two conscientious objectors were being held in jail for refusing service in the Belgian army, and international pacifists were pressuring Einstein to speak out on their behalf. This, of course, would cause problems.

The king hoped that Einstein would refrain from getting involved. Out of friendship, out of respect for the leader of a country that was hosting him, and also out of his new and sincere beliefs, Einstein agreed. He even went so far as to write a letter that he allowed to be made public.

"In the present threatening situation, created by the events in Germany, Belgium's armed forces can be regarded only as a means of defense, not an instrument of aggression," he declared. "And now, of all times, such defense forces are urgently needed."

Being Einstein, however, he felt compelled to add a few additional thoughts. "Men who, by their religious and moral convictions, are constrained to refuse military service should not be treated as criminals," he argued. "They should be offered the alternative of accepting more onerous and hazardous work than military service." For example, they could be put to work as low-paid conscripts doing "mine labor, stoking furnaces aboard ships, hospital service in infectious disease wards or in certain sections of mental institutions."[59] King Albert sent back a warm note of gratitude, which politely avoided any discussions of alternative service.

When Einstein changed his mind, he did not try to hide the fact. So he also wrote a public letter to the leader of the pacifist group that was encouraging him to intervene in the Belgian case. "Until recently, we in Europe could assume that personal war resistance constituted an effective attack on militarism," he said. "Today we face an altogether different situation. In the heart of Europe lies a power, Germany, that is obviously pushing to war with all available means."

He even went so far as to proclaim the unthinkable: he himself would join the army if he were a young man.

> I must tell you candidly: Under today's conditions, if I were a Belgian, I would not refuse military service, but gladly take it upon me in the knowledge of serving European civilization. This does not mean that I am surrendering the principle for which I have stood heretofore. I have no greater hope than that the time may not be far off when refusal of military service will once again be an effective method of serving the cause of human progress.[60]

For weeks the story reverberated around the world. "Einstein Alters His Pacifist Views / Advises the Belgians to Arm Themselves Against the Threat of Germany," headlined the *New York Times*.[61] Einstein not only held firm, but explained himself more passionately in response to each successive attack.

> *To the French secretary of the War Resisters' International:* "My views have not changed, but the European situation has . . . So long as Germany persists in rearming and systematically indoctrinating its citizens for a war of revenge, the nations of western Europe depend,

unfortunately, on military defense. Indeed, I will go so far as to assert that if they are prudent, they will not wait, unarmed, to be attacked . . . I cannot shut my eyes to realities." [62]

To Lord Ponsonby, his pacifist partner from England: "Can you possibly be unaware of the fact that Germany is feverishly rearming and that the whole population is being indoctrinated with nationalism and drilled for war? . . . What protection, other than organized power, would you suggest?" [63]

To the Belgian War Resisters' Committee: "As long as no international police force exists, these countries must undertake the defense of culture. The situation in Europe has changed sharply within the past year; we should be playing into the hands of our bitterest enemies were we to close our eyes to this fact." [64]

To an American professor: "To prevent the greater evil, it is necessary that the lesser evil—the hated military—be accepted for the time being." [65]

And even a year later, to an upset rabbi from Rochester: "I am the same ardent pacifist I was before. But I believe that we can advocate refusing military service only when the military threat from aggressive dictatorships toward democratic countries has ceased to exist." [66]

After years of being called naïve by his conservative friends, now it was those on the left who felt that his grasp of politics was shaky. "Einstein, a genius in his scientific field, is weak, indecisive and inconsistent outside it," the dedicated pacifist Romain Rolland wrote in his diary. [67] The charge of inconsistency would have amused Einstein. For a scientist, altering your doctrines when the facts change is not a sign of weakness.

Farewell

The previous fall, Einstein had gotten a long, rambling, and, as often was the case, intensely personal letter from Michele Besso, one of his oldest friends. Most of it was about poor Eduard, Einstein's younger son, who had continued to succumb to his mental illness and was now confined to an asylum near Zurich. Einstein was pictured so often with his stepdaughters, but never with his sons, Besso noted. Why didn't he travel with them? Perhaps he could take Eduard on one of his trips to America and get to know him better.

Einstein loved Eduard. Elsa told a friend, "This sorrow is eating up Albert." But he felt that Eduard's schizophrenia was inherited from his mother's side, as to some extent it probably was, and there was little that he could do about it. That was also the reason he resisted psychoanalysis for Eduard. He considered it ineffective, especially in cases of severe mental illness that seemed to have hereditary causes.

Besso, on the other hand, had gone through psychoanalysis, and in his letter he was expansive and disarming, just as he had been back when they used to walk home from the patent office together more than a quarter-century earlier. He had his own problems in marriage, Besso said, referring to Anna Winteler, whom Einstein had introduced him to. But by forging a better relationship with his own son, he had made his marriage work and his life more meaningful.

Einstein replied that he hoped to take Eduard with him to visit Princeton. "Unfortunately, everything indicates that strong heredity manifests itself very definitely," he lamented. "I have seen that coming slowly but inexorably since Tete's youth. Outside influences play only a small part in such cases, compared to internal secretions, about which nobody can do anything."[68]

The tug was there, and Einstein knew that he had to, and wanted to, see Eduard. He was supposed to visit Oxford in late May, but he decided to delay the trip for a week so that he could go to Zurich and be with his son. "I could not wait six weeks before going to see him," he wrote Lindemann, asking his indulgence. "You are not a father, but I know you will understand."[69]

His relationship with Marić had improved so much that, when she heard he could not go back to Germany, she invited both him and Elsa to come to Zurich and live in her apartment building. He was pleasantly surprised, and he stayed with her when he came alone that May. But his visit with Eduard turned out to be more wrenching than he had anticipated.

Einstein had brought with him his violin. Often he and Eduard had played together, expressing emotions with their music in ways they could not with words. The photograph of them on that visit is particularly poignant. They are sitting awkwardly next to each other, wearing suits, in what seems to be the visiting room of the asylum. Einstein is holding his violin and bow, looking away. Eduard is staring down intensely at a pile of papers, the pain seeming to contort his now fleshy face.

When Einstein left Zurich for Oxford, he was still assuming that he would be spending half of each ensuing year in Europe. What he did not know was that, as things would turn out, this would be the last time he would see his first wife and their younger son.

While at Oxford, Einstein gave his Herbert Spencer Lecture, in which he explained his philosophy of science, and then went to Glasgow, where he gave an account of his path toward the discovery of general relativity. He enjoyed the trip so much that, soon after his return to Le Coq sur Mer, he decided to go back to England in late July, this time at the invitation of one of his unlikeliest acquaintances.

British Commander Oliver Locker-Lampson was most things that Einstein was not. The adventurous son of a Victorian poet, he became a World War I aviator, leader of an armored division in Lapland and Russia, an adviser to Grand Duke Nicholas, and potential plotter in the murder of Rasputin. Now he was a barrister, journalist, and member of Parliament. He had studied in Germany, knew the language and the people, and had become, perhaps as a consequence, an early advocate for preparing to fight the Nazis. With an appetite for the interesting, he began writing Einstein, whom he had met only in passing once at Oxford, asking him to be his guest in England.

When Einstein accepted his offer, the dashing commander made the most of it. He took Einstein to see Winston Churchill, then suffering through his wilderness years as an opposition member of Parliament. At lunch in the gardens of Churchill's home, Chartwell, they discussed Germany's rearmament. "He is an eminently wise man," Einstein wrote Elsa that day. "It became clear to me that these people have made preparations and are determined to act resolutely and soon."[70] It sounded like an assessment from someone who had just eaten lunch with Churchill.

Locker-Lampson also brought Einstein to Austen Chamberlain, another advocate of rearmament, and former Prime Minister Lloyd

George. When he arrived at the home of the latter, Einstein was given the guest book to sign. When he got to the space for home address, he paused for a few moments, then wrote *ohne,* without any.

Locker-Lampson recounted the incident the next day when, with great flourish, he introduced a bill in Parliament, as Einstein watched from the visitors' gallery wearing a white linen suit, to "extend opportunities of citizenship for Jews." Germany was in the process of destroying its culture and threatening the safety of its greatest thinkers. "She has turned out her most glorious citizen, Albert Einstein," he said. "When he is asked to put his address in visitors' books he has to write, 'without any.' How proud this country must be to have offered him shelter at Oxford!"[71]

When he returned to his seaside cottage in Belgium, Einstein decided there was one issue he should clear up, or at least try to, before he embarked for America again. The Woman Patriot Corporation and others were still seeking to bar him as a dangerous subversive or communist, and he found their allegations to be both offensive and potentially problematic.

Because of his socialist sentiments, history of pacifism, and opposition to fascism, it was thought then—and throughout his life—that Einstein might be sympathetic to the Russian communists. Nor did it help that he had an earnest willingness to lend his name to almost any worthy-sounding manifesto or masthead that arrived in his mail, without always determining whether the groups involved might be fronts for other agendas.

Fortunately, his willingness to lend his name to sundry organizations was accompanied by an aversion to actually showing up for any meetings or spending time in comradely planning sessions. So there were not many political groups, and certainly no communist ones, in which he actually participated. And he made it a point never to visit Russia, because he knew that he could be used for propaganda purposes.

As his departure date neared, Einstein gave two interviews to make these points clear. "I am a convinced democrat," he told fellow German refugee Leo Lania for the *New York World Telegram.* "It is for this reason that I do not go to Russia, although I have received very cordial invitations. My voyage to Moscow would certainly be exploited by the rulers of the Soviets to profit their own political aims. Now I am an adversary of Bolshevism just as much as of fascism. I am against all dictatorships."[72]

In another interview, which appeared both in the *Times* of London and the *New York Times,* Einstein admitted that occasionally he had been "fooled" by organizations that pretended to be purely pacifist or humanitarian but "are in truth nothing less than camouflaged propaganda in the service of Russian despotism." He emphasized, "I have never favored communism and do not favor it now." The essence of his political belief was to oppose any power that "enslaves the individual by terror and force, whether it arises under a Fascist or Communist flag."[73]

These statements were made, no doubt, to tamp down any controversy in America about his alleged political leanings. But they had the added virtue of being true. He had occasionally been duped by groups whose agendas were not what they seemed, but he had, since childhood, kept as his guiding principle an aversion to authoritarianism, whether of the left or the right.

At the end of the summer, Einstein received some devastating news. Having recently separated from his wife and collaborator, his friend Paul Ehrenfest had gone to visit his 16-year-old son, who was in an Amsterdam institution with Down syndrome. He pulled out a gun, shot the boy in the face, taking out his eye but not killing him. Then he turned the gun on himself and committed suicide.

More than twenty years earlier, Ehrenfest, a wandering young Jewish physicist, had shown up in Prague, where Einstein was working, and asked for help finding a job. After visiting the cafés and talking physics for hours that day, the two men became deeply devoted friends. Ehrenfest's mind was very different from Einstein's in many ways. He had "an almost morbid lack of self-confidence," Einstein said, and was better at critically poking holes in existing theories than at building new ones. That made him a good teacher, "the best I have ever known," but his "sense of inadequacy, objectively unjustified, plagued him incessantly."

But there was one important way in which he was like Einstein. He could never make his peace with quantum mechanics. "To learn and teach things that one cannot fully accept in one's heart is always a difficult matter," Einstein wrote of Ehrenfest, "doubly difficult for a man of fanatical honesty."

Einstein, who knew what it was like to turn 50, followed this with a description that said as much about his own approach to quantum mechanics as it did about Ehrenfest's: "Added to this was the increasing difficulty of adapting to new thoughts which always confronts the man past fifty. I do not know how many readers of these lines will be capable of fully grasping that tragedy."[74] Einstein was.

Ehrenfest's suicide deeply unnerved Einstein, as did the increased intensity of the threats against his own life. His name had been falsely associated with a book attacking Hitler's terror; as was often the case, he had let his name be used as the honorary chair of a committee, which then published the book, but he had not read any of it. German papers headlined "Einstein's infamy" in red letters. One magazine pictured him on a list of enemies of the German regime, listed his "crimes," and concluded with the phrase "not yet hanged."

So Einstein decided to take Locker-Lampson up on his English hospitality yet again for the final month before his scheduled departure for America in October. Elsa, who wanted to stay behind in Belgium to pack, asked a reporter from the Sunday Express to arrange for Einstein to get to England safely. Being a good journalist, he accompanied Einstein on the trip himself and reported that on the channel crossing Einstein pulled out his notebook and went to work on his equations.

In a drama worthy of a James Bond movie, Locker-Lampson had two young female "assistants" take Einstein up to a secluded cottage he owned that was nestled on a coastal moor northeast of London. There he was swept into a slapstick whirl of secrecy and publicity. The two young women posed next to him holding hunting shotguns for a picture that was given to the press agencies, and Locker-Lampson declared, "If any unauthorized person comes near they will get a charge of buckshot." Einstein's own assessment of his security was less intimidating. "The beauty of my bodyguards would disarm a conspirator sooner than their shotguns," he told a visitor.

Among those who penetrated this modest security perimeter were a former foreign minister, who wanted to discuss the crisis in Europe;

Einstein's stepson-in-law, Dimitri Marianoff, who had come to interview him for an article he had sold to a French publication; Walther Mayer, who helped continue the Sisyphean task of finding unified field theory equations; and the noted sculptor Jacob Epstein, who spent three days making a beautiful bust of Einstein.

The only one who ran afoul of the female guards was Epstein, who asked if they would take one of the doors off its hinges so he could get a better angle for his sculpting. "They facetiously asked whether I would like the roof off next," he recalled. "I thought I should have liked that too, but I did not demand it as the attendant angels seemed to resent a little my intrusion into the retreat of their professor." After three days, however, the guardians warmed to Epstein, and everyone began drinking beer together at the end of his sittings.[75]

Einstein's humor stayed intact through it all. Among the letters he received in England was one from a man who had a theory that gravity meant that as the earth rotated people were sometimes upside down or horizontal. Perhaps that led people to do foolish things, he speculated, like falling in love. "Falling in love is not the most stupid thing that people do," Einstein scribbled on the letter, "but gravitation cannot be held responsible for it."[76]

Einstein's main appearance on this trip was a speech on October 3 in London's Royal Albert Hall, which was designed to raise money for displaced German scholars. Some suspected, no doubt with reason, that Locker-Lampson had hyped the security threat and publicity about Einstein's hideaway in order to promote ticket sales. If so, he was successful. All nine thousand seats were filled, and others jammed the aisles and lobbies. A thousand students acted as guides and guards against any pro-Nazi demonstration that might materialize (none did).

Einstein spoke, in English, about the current menace to freedom, but he was careful not to attack the German regime specifically. "If we want to resist the powers that threaten to suppress intellectual and individual freedom, we must be clear what is at stake," he said. "Without such freedom there would have been no Shakespeare, no Goethe, no Newton, no Faraday, no Pasteur, no Lister." Freedom was a foundation for creativity.

He also spoke of the need for solitude. "The monotony of a quiet life stimulates the creative mind," he said, and he repeated a suggestion he had made when younger that scientists might be employed as light-house keepers so they could "devote themselves undisturbed" to think-ing.[77]

It was a revealing remark. For Einstein, science was a solitary pursuit, and he seemed not to realize that for others it could be far more fruitful when pursued collaboratively. In Copenhagen and elsewhere, the quantum mechanics team had been building on one another's ideas with a frenzy. But Einstein's great breakthroughs had been those that could be done, with perhaps just an occasional sounding board and mathematical assistant, by someone in a Bern patent office, the garret of a Berlin apartment, or a lighthouse.

The ocean liner *Westmoreland,* which had sailed from Antwerp with Elsa and Helen Dukas aboard, picked up Einstein and Walther Mayer in Southampton on October 7, 1933. He did not think he would be away for long. In fact, he planned to spend another term at Christ Church, Oxford, the next spring. But although he would live for another twenty-two years, Einstein would never see Europe again.

CHAPTER NINETEEN

AMERICA

1933–1939

112 Mercer Street

Princeton

The ocean liner *Westmoreland,* which carried Einstein, at age 54, to what would become his new home country, arrived in New York Harbor on October 17, 1933. Waiting to meet him in the rain at the Twenty-third Street pier was an official committee led by his friend Samuel Untermyer, a prominent attorney, who carried some orchids he had grown, plus a group of cheerleaders that was scheduled to parade with him to a welcoming pageant.

Einstein and his entourage, however, were nowhere to be found. Abraham Flexner, the director of the Institute for Advanced Study, was obsessed with shielding him from publicity, whatever Einstein's quirky preferences might be. So he had sent a tugboat, with two Institute trustees, to spirit Einstein away from the *Westmoreland* as soon as it cleared quarantine. "Make no statement and give no interviews on any subject," he had cabled. To reiterate the message, he sent a letter with one of the trustees who greeted Einstein's ship. "Your safety in America depends upon silence and refraining from attendance at public functions," it said.[1]

Carrying his violin case, with a profusion of hair poking out from a wide-brimmed black hat, Einstein surreptitiously disembarked onto the tug, which then ferried him and his party to the Battery, where a car was waiting to whisk them to Princeton. "All Dr. Einstein wants is to be left in peace and quiet," Flexner told reporters.[2]

Actually, he also wanted a newspaper and an ice cream cone. So as soon as he had checked into Princeton's Peacock Inn, he changed into casual clothes and, smoking his pipe, went walking to a newsstand, where he bought an afternoon paper and chuckled over the headlines about the mystery of his whereabouts. Then he walked into an ice cream parlor, the Baltimore, pointed his thumb at the cone a young divinity student had just bought, and then pointed at himself. As the waitress made change for him, she announced, "This one goes in my memory book."[3]

Einstein was given a corner office in a university hall that served as the temporary headquarters of the Institute. There were eighteen scholars in residence then, including the mathematicians Oswald Veblen (nephew of the social theorist Thorstein Veblen) and John von Neumann, a pioneer of computer theory. When shown his office, he was asked what equipment he might need. "A desk or table, a chair, paper and pencils," he replied. "Oh yes, and a large wastebasket, so I can throw away all my mistakes."[4]

He and Elsa soon found a house to rent, which they celebrated by hosting a small musical recital featuring the works of Haydn and Mozart. The noted Russian violinist Toscha Seidel played lead, with Einstein as second fiddle. In return for some violin tips, Einstein tried to explain relativity theory to Seidel and made him some drawings of moving rods contracting in length.[5]

Thus began a proliferation of popular tales in town about Einstein's love for music. One involved Einstein playing in a quartet with violin virtuoso Fritz Kreisler. At a certain point they got out of sync. Kreisler stopped playing and turned to Einstein in mock exasperation. "What's the matter, professor, can't you count?"[6] More poignantly, there was an evening where a Christian prayer group gathered to make intercessions for persecuted Jews. Einstein surprised them by asking if he could come. He brought his violin and, as if offering a prayer, played a solo.[7]

Many of his performances were purely impromptu. That first Halloween, he disarmed some astonished trick-or-treaters, a group of 12-year-old girls who had come with the intent of playing a prank, by appearing at the door and serenading them with his violin. And at Christmastime, when members of the First Presbyterian Church came by to sing carols, he stepped out into the snow, borrowed a violin from one of the women, and accompanied them. "He was just a lovely person," one of them recalled.[8]

Einstein soon acquired an image, which grew into a near legend but was nevertheless based on reality, of being a kindly and gentle professor, distracted at times but unfailingly sweet, who wandered about lost in thought, helped children with their homework, and rarely combed his hair or wore socks. With his amused sense of self-awareness, he catered to such perceptions. "I'm a kind of ancient figure known primarily for his non-use of socks and wheeled out on special occasions as a curiosity," he joked. His slightly disheveled appearance was partly an assertion of his simplicity and partly a mild act of rebellion. "I have reached an age when, if someone tells me to wear socks, I don't have to," he told a neighbor.[9]

His baggy, comfortable clothes became a symbol of his lack of pretense. He had a leather jacket that he tended to wear to events both formal and informal. When a friend found out that he had a mild allergy to wool sweaters, she went to a surplus store and bought him some cotton sweatshirts, which he wore all the time. And his dismissive attitude toward haircuts and grooming was so infectious that Elsa, Margot, and his sister, Maja, all sported the same disheveled gray profusion.

He was able to make his rumpled-genius image as famous as Chaplin did the little tramp. He was kindly yet aloof, brilliant yet baffled. He floated around with a distracted air and a wry sensibility. He exuded honesty to a fault, was sometimes but not always as naïve as he seemed, cared passionately about humanity and sometimes about people. He would fix his gaze on cosmic truths and global issues, which allowed him to seem detached from the here and now. This role he played was not far from the truth, but he enjoyed playing it to the hilt, knowing that it was such a great role.

He had also, by then, adapted willingly to the role Elsa played, that of a wife who could be both doting and demanding, protective yet afflicted with occasional social aspirations. They had grown comfortable together, after some rough patches. "I manage him," she said proudly, "but I never let him know that I manage him."[10]

Actually, he knew, and he found it mildly amusing. He surrendered, for example, to Elsa's nagging that he smoked too much and on Thanksgiving bet her that he would be able to abstain from his pipe until the new year. When Elsa boasted of this at a dinner party, Einstein grumbled, "You see, I am no longer a slave to my pipe, but I am a slave to that woman." Einstein kept his word, but "he got up at daylight on New Year's morning, and he hasn't had his pipe out of his mouth since except to eat and sleep," Elsa told neighbors a few days after the deal was over.[11]

The greatest source of friction for Einstein came from Flexner's desire to protect him from publicity. Einstein was, as always, less fastidious about this than were his friends, patrons, and self-appointed protectors. An occasional flash of the limelight made his eyes twinkle. More important, he was willing and even eager to endure such indignities if he could use his fame to raise money and sympathy for the worsening plight of European Jews.

Such political activism made Einstein's penchant for publicity even more disconcerting to Flexner, an old-line and assimilated American Jew. It might provoke anti-Semitism, he thought, especially in Princeton, where the Institute was luring Jewish scholars into an environment that was, to say the least, socially wary of them.[12]

Flexner was particularly upset when Einstein, quite charmingly, agreed one Saturday to meet at his home with a group of boys from a Newark school who had named their science club after him. Elsa baked cookies, and when the discussion turned to Jewish political leaders, she noted, "I don't think there is any anti-Semitism in this country." Einstein agreed. It would have amounted to no more than a sweet visit, except that the adviser who accompanied the boys wrote a colorful account, focusing on Einstein's thoughts about the plight of Jews, that was bannered atop the front page of the Newark *Sunday Ledger*.[13]

Flexner was furious. "I simply want to protect him," he wrote in a sharp letter to Elsa, and he sent the Newark article to her with a stern note attached. "This is exactly the sort of thing that seems to me absolutely unworthy of Professor Einstein," he scolded. "It will hurt him in the esteem of his colleagues, for they will believe that he seeks such publicity, and I do not see how they can be convinced that such is not the case."[14]

Flexner went on to ask Elsa to dissuade her husband from being featured at a scheduled musical recital in Manhattan, which he had already accepted, that was to raise money for Jewish refugees. But like her husband, Elsa was not totally averse to publicity, nor to helping Jewish causes, and she resented Flexner's attempts at control. So she replied with a very frank refusal.

That provoked Flexner to send an astonishingly blunt letter the next day, which he noted he had discussed with the president of Princeton University. Echoing the sentiments of some of Einstein's European friends, including the Borns, Flexner warned Elsa that if Jews got too much publicity it would stoke anti-Semitism:

> It is perfectly possible to create anti-Semitic feeling in the United States. There is no danger that any such feeling would be created except by the Jews themselves. There are already signs which are unmistakable that anti-Semitism has increased in America. It is because I am myself a Jew and because I wish to help oppressed Jews in Germany that my efforts, though continuous and in a measure successful, are absolutely quiet and anonymous . . . The questions involved are the dignity of your husband and the Institute according to the highest American standards and the most effective way of helping the Jewish race in America and in Europe.[15]

That same day, Flexner wrote Einstein directly to make the case that Jews like themselves should keep a low profile because a penchant for publicity could arouse anti-Semitism. "I have felt this from the moment that Hitler began his anti-Jewish policy, and I have acted accordingly," he wrote. "There have been indications in American universities that Jewish students and Jewish professors will suffer unless the utmost caution is used."[16]

Not surprisingly, Einstein went ahead with his planned benefit recital in Manhattan, for which 264 guests paid $25 apiece to attend. It featured Bach's Concerto for Two Violins in D-minor and Mozart's G Major Quartet. It was even opened to the press. "He became so absorbed in the music," *Time* magazine reported, "that with a far-away look he was still plucking at the strings when the performance was all over."[17]

In his attempt to prevent such events, Flexner had begun intercepting Einstein's mail and declining invitations on his behalf. The stage was thus set for a showdown when Rabbi Stephen Wise of New York decided it would be a good idea to get Einstein invited to visit President Franklin Roosevelt, which Wise hoped would focus attention on Germany's treatment of Jews. "F.D.R. has not lifted a finger on behalf of the Jews of Germany, and this would be little enough," Wise wrote a friend.[18]

The result was a telephone call from Roosevelt's social secretary, Colonel Marvin MacIntyre, inviting Einstein to the White House. When Flexner found out, he was furious. He called the White House and gave a stern lecture to the somewhat surprised Colonel MacIntyre. All invitations must go through him, Flexner said, and on Einstein's behalf he declined.

For good measure, Flexner proceeded to write an official letter to the president. "I felt myself compelled this afternoon to explain to your secretary," Flexner said, "that Professor Einstein had come to Princeton for the purpose of carrying out his scientific work in seclusion and that it was absolutely impossible to make any exception which would inevitably bring him into public notice."

Einstein knew none of this until Henry Morgenthau, a prominent Jewish leader who was about to become treasury secretary, inquired about the apparent snub. Dismayed to discover Flexner's presumption, Einstein wrote to Eleanor Roosevelt, his political soul mate. "You can hardly imagine of what great interest it would have been for me to meet the man who is tackling with gigantic energy the greatest and most difficult problem of our time," he wrote. "However, as a matter of fact, no invitation whatever has reached me."

Eleanor Roosevelt answered personally and politely. The confusion came, she explained, because Flexner had been so adamant in his phone call to the White House. "I hope you and Mrs. Einstein will come sometime soon," she added. Elsa responded graciously. "First excuse my poor English please," she wrote. "Dr. Einstein and myself accept with feelings of gratitude your very kind invitation."

He and Elsa arrived at the White House on January 24, 1934, had dinner, and spent the night. The president was able to converse with them in passable German. Among other things, they discussed Roosevelt's marine prints and Einstein's love for sailing. The next morning, Einstein wrote an eight-line piece of doggerel on a White House note card to Queen Elisabeth of the Belgians marking his visit, but he made no public statements.[19]

Flexner's interference infuriated Einstein. He complained about it in a letter to Rabbi Wise—on which he put as his return address "Concentration Camp, Princeton"—and he sent a five-page litany of Flexner's meddling to the Institute's trustees. Either they must assure him that there would be no more "constant interference of the type that no self-respecting person would tolerate," Einstein threatened, or "I would propose that I discuss with you severing my relationship with your institute in a dignified manner."[20]

Einstein prevailed, and Flexner backed off. But as a result, he lost his influence with Flexner, whom he would later refer to as one of his "few enemies" in Princeton.[21] When Erwin Schrödinger, Einstein's fellow traveler in the minefields of quantum mechanics, arrived as a refugee in Princeton that March, he was offered a job at the university. But he wanted instead to be tapped for the Institute for Advanced Study. Einstein lobbied Flexner on his behalf, but to no avail. Flexner was doing him no more favors, even if it meant depriving the Institute of Schrödinger.

During his short stay in Princeton, Schrödinger asked Einstein if he was indeed going to come back to Oxford later that spring, as scheduled. He had called himself a "bird of passage" when heading off to Caltech in 1931, and it was unclear, perhaps even in his own mind, whether he saw this as a liberation or a lament. But now he found himself comfortable in Princeton, with no desire to take wing again.

"Why should an old fellow like me not enjoy peace and quiet for once?" he asked his friend Max Born. So he told Schrödinger to pass along his sincere regrets. "I am sorry to say that he asked me to write you a definite no," Schrödinger informed Lindemann. "The reason for his decision is really that he is frightened of all the ado and the fuss that would be laid upon him if he came to Europe." Einstein also worried that he would be expected to go to Paris and Madrid if he went to Oxford, "and I lack the courage to undertake all this."[22]

The stars had aligned to create for Einstein a sense of inertia, or at least a weariness of further wandering. In addition, Princeton, which he called a "pipe as yet unsmoked" on his first visit in 1921, captured him with its leafy charm and its neo-Gothic echoes of a European university town. "A quaint and ceremonious village of puny demigods strutting on stiff legs," he called it in a letter to Elisabeth, the queen mother of Belgium since the death of the king. "By ignoring certain social conventions, I have been able to create for myself an atmosphere conducive to study and free from distraction."[23]

Einstein particularly liked the fact that America, despite its inequalities of wealth and racial injustices, was more of a meritocracy than Europe. "What makes the new arrival devoted to this country is the democratic trait among the people," he marveled. "No one humbles himself before another person or class."[24]

This was a function of the right of individuals to say and think what they pleased, a trait that had always been important to Einstein. In addition, the lack of stifling traditions encouraged more creativity of the sort he had relished as a student. "American youth has the good fortune not to have its outlook troubled by outworn traditions," he noted.[25]

Elsa likewise loved Princeton, which was important to Einstein. She had taken such good care of him for so long that he had become more solicitous of her desires, particularly her nesting instinct. "The whole of Princeton is one great park with wonderful trees," she wrote a friend. "We might almost believe that we are in Oxford." The architecture and countryside reminded her of England, and she felt somewhat guilty that she was so comfortable while others back in Europe were suffering. "We are very happy here, perhaps too happy. Sometimes one has a bad conscience."[26]

So in April 1934, just six months after his arrival, Einstein announced that he was staying in Princeton indefinitely and assuming full-time status at the Institute. As it turned out, he would never live anywhere else for the remaining twenty-one years of his life. Nevertheless, he made appearances at the "farewell" parties that had been scheduled that month as fund-raisers for various of his favorite charities. These causes had become almost as important to him as his science. As he declared at one of the events, "Striving for social justice is the most valuable thing to do in life."[27]

Sadly, just when they had decided to settle in, Elsa had to travel back to Europe to care for her spirited and adventurous elder daughter, Ilse, who had dallied with the romantic radical Georg Nicolai and married the literary journalist Rudolf Kayser. Ilse was afflicted with what they thought was tuberculosis but what turned out to be leukemia, and her condition had taken a turn for the worse. Now she had gone to Paris to be nursed by her sister, Margot.

Insisting that her problems were mainly psychosomatic, Ilse resisted medications and turned instead to prolonged psychotherapy. Early during her illness, Einstein had tried to persuade her to go to a regular doctor, but she had refused. Now there was little that could be done as the whole family, absent Einstein himself, gathered by her bed in Margot's Paris apartment.

Ilse's death devastated Elsa. She "changed and aged," Margot's husband recalled, "almost beyond recognition." Instead of having Ilse's ashes deposited in a crypt, Elsa had them put in a sealed bag for her. "I cannot be separated," she said. "I have to have them." She then sewed the bag inside a pillow so that she could have them close to her on the trip home to America.[28]

Elsa also carried back cases of her husband's papers, which Margot had earlier smuggled from Berlin to Paris using French diplomatic channels and the anti-Nazi underground. To help get them into America, Elsa enlisted the help of a kindly neighbor from Princeton, Caroline Blackwood, who was on the same ship home.

Elsa had met the Blackwoods a few months earlier in Princeton, and they mentioned that they were going to Palestine and Europe and wished to meet some Zionist leaders.

"I didn't know you were Jews," Elsa said.

Mrs. Blackwood said that they actually were Presbyterian, but there was a deep connection between the Jewish heritage and the Christian, "and besides, Jesus was a Jew."

Elsa hugged her. "No Christian has ever said that to me in my life." She also asked for help in getting a German-language Bible, as they had lost theirs in the move from Berlin. Mrs. Blackwood found her a copy of Martin Luther's translation, which Elsa clasped to her heart. "I wish I had more faith," she told Mrs. Blackwood.

Elsa had taken note of what liner the Blackwoods were traveling on, and she purposely booked passage on it when she returned to America. One morning she brought Mrs. Blackwood into the ship's deserted lounge to ask a favor. Because she was not a citizen, she was afraid that her husband's papers might be held at the border. Would the Blackwoods bring them in?

They agreed, although Mr. Blackwood was careful not to lie on his customs declaration. "Material acquired in Europe for scholarly purposes," he wrote. Later, Einstein came over in the rain to the Blackwoods' shed to collect his papers. "Did I write this drivel?" he joked as he looked at one journal. But the Blackwoods' son, who was there, recalled that Einstein "was obviously deeply moved to have his books and papers in his hands."[29]

Ilse's death, accompanied in the summer of 1934 by Hitler's consolidation of power during the "Night of the Long Knives," severed the Einsteins' remaining bonds with Europe. Margot immigrated that year to Princeton, after she and her odd Russian husband separated. Hans Albert soon followed. She was "not longing for Europe at all," Elsa wrote Caroline Blackwood soon after returning. "I feel such a homelike feeling for this country."[30]

Recreations

When Elsa returned from Europe, she joined Einstein at a summer cottage he had rented in Watch Hill, Rhode Island, a quiet enclave on a peninsula near where Long Island Sound meets the Atlantic. It was perfect for sailing, which is why Einstein, at Elsa's urging, decided to summer there with his friend Gustav Bucky and his family.

Bucky was a physician, engineer, inventor, and pioneer of X-ray technology. A German who had gained American citizenship during the 1920s, he had met the Einsteins in Berlin. When Einstein came to America, his friendship with Bucky deepened; they even took out a joint patent on a device they came up with to control a photographic diaphragm, and Einstein testified as an expert witness for Bucky in a dispute over another invention.[31]

His son Peter Bucky happily spent time driving Einstein around, and he later wrote down some of his recollections in extensive notebooks. They provide a delightful picture of the mildly eccentric but deeply un-affected Einstein in his later years. Peter tells, for example, of driving in his convertible with Einstein when it suddenly started to rain. Einstein pulled off his hat and put it under his coat. When Peter looked quizzi-cal, Einstein explained: "You see, my hair has withstood water many times before, but I don't know how many times my hat can."[32]

Einstein relished the simplicity of life in Watch Hill. He puttered around its lanes and even shopped for groceries with Mrs. Bucky. Most of all, he loved sailing his seventeen-foot wooden boat *Tinef*, which is Yiddish for a piece of junk. He usually went out on his own, aimlessly and often carelessly. "Frequently he would go all day long, just drifting around," remembered a member of the local yacht club who went to re-trieve him on more than one occasion. "He apparently was just out there meditating."

As he had at Caputh, Einstein would drift with the breeze and sometimes scribble equations in his notebook when becalmed. "Once we all waited with growing concern for his return from an afternoon sail," Bucky recalls. "Finally, at 11 pm, we decided to send the Coast Guard out to search for him. The guardsmen found him in the Bay, not in the least concerned about his situation."

At one point a friend gave him an expensive outboard motor for emergency use. Einstein declined. He had a childlike delight about taking small risks—he still never took a life jacket even though he could not swim—and escaping to where he could be by himself. "To the average person, being becalmed for hours might be a terrible trial," said Bucky. "To Einstein, this could simply have provided more time to think."[33]

The sailing rescue sagas continued the following summer, when the Einsteins began renting in Old Lyme, Connecticut, also on Long Island Sound. One such tale even made the *New York Times*. "Relative Tide and Sand Bars Trap Einstein," read the headline. The young boys who saved him were invited to the house for raspberry juice.[34]

Elsa loved the Old Lyme house, although both she and her family found it a bit too imposing. It was set on twenty acres, with a tennis court and swimming pool, and the dining room was so large that they initially were afraid to use it. "Everything is so luxurious here that the first ten days—I swear to you—we ate in the pantry," Elsa wrote a friend. "The dining room was too magnificent for us."[35]

When the summers were over, the Einsteins would visit the Bucky family at their Manhattan home once or twice a month. Einstein would also stay, especially when he was by himself, at the home of the widower Leon Watters, the pharmaceutical company owner he had met in Pasadena. He once surprised Watters by arriving without a dressing gown or pajamas. "When I retire, I sleep as nature made me," he said. Watters recalled that he did, however, ask to borrow a pencil and notepad for his bedside.

Out of both politeness and his touch of vanity, Einstein found it hard to decline requests from artists and photographers who wanted him to pose. One weekend in April 1935, when he was staying with Watters, Einstein sat for two artists in one day. His first session was with the wife of Rabbi Stephen Wise, not known for her artistic ability. Why was he doing it? "Because she's a nice woman," he answered.

Later that day, Watters picked Einstein up to ferry him to Greenwich Village for a session with the Russian sculptor Sergei Konenkov, a practitioner of Soviet realism, who was producing what would be a distinguished bust of Einstein that is now at the Institute for Advanced Study. Einstein had been introduced to Konenkov through Margot, who was also a sculptor. Soon, all of them became friends with his wife, Margarita Konenkova, who, unbeknown to Einstein, was a Soviet spy.

In fact, Einstein would later become, after Elsa's death, romantically involved with her, which would end up creating, as we shall see, more complexities than he ever knew.[36]

Now that they had decided to stay in the United States, it made sense for Einstein to seek citizenship. When Einstein visited the White House, President Roosevelt had suggested that he should accept the offer of some congressmen to have a special bill passed on his behalf, but Einstein instead decided to go through the normal procedures. That meant leaving the country, so that he—and Elsa, Margot, and Helen Dukas—could come in not as visitors but as people seeking citizenship.

So in May 1935 they all sailed on the *Queen Mary* to Bermuda for a few days to satisfy these formalities. The royal governor was there to greet them when they arrived in Hamilton, and he recommended the island's two best hotels. Einstein found them stuffy and pretentious. As they walked through town, he saw a modest guest cottage, and that is where they ended up.

Einstein declined all official invitations from the Bermuda gentry and socialized instead with a German cook he met at a restaurant, who invited him to come sailing on his little boat. They were away for seven hours, and Elsa feared that Nazi agents may have nabbed her husband. But she found him at the cook's home, where he had gone to enjoy a dinner of German dishes.[37]

That summer, a house down the block from the one they were renting in Princeton went on sale. A modest white clapboard structure that peeked through a little front yard onto one of the town's pleasant tree-lined arteries, 112 Mercer Street was destined to become a world-famous landmark not because of its grandeur but because it so perfectly suited and symbolized the man who lived there. Like the public persona that he adopted in later life, the house was unassuming, sweet, charming, and unpretentious. It sat there right on a main street, highly visible yet slightly cloaked behind a veranda.

Its modest living room was a bit overwhelmed by Elsa's heavy German furniture, which had somehow caught up with them after all their wanderings. Helen Dukas commandeered the small library on the first floor as a workroom in which she dealt with Einstein's correspondence and took charge of the only telephone in the house (Princeton 1606 was the unlisted number).

Elsa oversaw the construction of a second-floor office for Einstein. They removed part of the back wall and installed a picture window that looked out on the long and lush backyard garden. Bookcases on both sides went up to the ceiling. A large wooden table, cluttered with papers and pipes and pencils, sat in the center with a view out of the window, and there was an easy chair where Einstein would sit for hours scribbling on a pad of paper in his lap.

The usual pictures of Faraday and Maxwell were tacked on the walls. There was also, of course, one of Newton, although after a while it fell off its hook. To that mix was added a fourth: Mahatma Gandhi, Einstein's new hero now that his passions were as much political as they were scientific. As a small joke, the only award displayed was a framed certificate of Einstein's membership in the Bern Scientific Society.

Besides his menagerie of women, the household was joined, over the years, by various pets. There was a parrot named Bibo, who required an unjustifiable amount of medical care; a cat named Tiger; and a white terrier named Chico that had belonged to the Bucky family. Chico was an occasional problem. "The dog is very smart," Einstein explained. "He feels sorry for me because I receive so much mail. That's why he tries to bite the mailman." [38]

"The professor does not drive," Elsa often said. "It's too complicated for him." Instead, he loved to walk, or, more precisely, shuffle, up Mercer Street each morning to his office at the Institute. People often snapped their heads when he passed, but the sight of him walking lost in thought was soon one of the well-known attractions of the town.

On his walk back home at midday, he would often be joined by three or four professors or students. Einstein would usually walk calmly and quietly, as if in a reverie, while they pranced around him, waved their arms, and tried to make their points. When they got to the house, the others would peel off, but Einstein sometimes just stood there thinking. Every now and then, unwittingly, he even started drifting back to the Institute. Dukas, always watching from her window, would come outside, take his arm, and lead him inside for his macaroni lunch. Then he would nap, dictate some answers to his mail, and pad up to his study for another hour or two of rumination about potential unified field theories.[39]

Occasionally, he would take rambling walks on his own, which could be dicey. One day someone called the Institute and asked to speak to a particular dean. When his secretary said that the dean wasn't available, the caller hesitantly asked for Einstein's home address. That was not possible to give out, he was informed. The caller's voice then dropped to a whisper. "Please don't tell anybody," he said, "but I *am* Dr. Einstein, I'm on my way home, and I've forgotten where my house is."[40]

This incident was recounted by the son of the dean, but like many of the tales about Einstein's distracted behavior it may have been exaggerated. The absentminded professor image fit him so nicely and naturally that it became reinforcing. It was a role that Einstein was happy to play in public and that his neighbors relished recounting. And like most assumed roles, there was a core of truth to it.

At one dinner where Einstein was being honored, for example, he got so distracted that he pulled out his notepad and began scribbling equations. When he was introduced, the crowd burst into a standing ovation, but he was still lost in thought. Dukas caught his attention and told him to get up. He did, but noticing the crowd standing and applauding, he assumed it was for someone else and heartily joined in. Dukas had to come over and inform him that the ovation was for him.[41]

In addition to the tales of the dreamy Einstein, another common theme was that of the kindly Einstein helping a child, usually a little girl, with her homework. The most famous of these involved an 8-year-old neighbor on Mercer Street, Adelaide Delong, who rang his bell and asked for help with a math problem. She carried a plate of homemade fudge as a bribe. "Come in," he said. "I'm sure we can solve it." He helped explain the math to her, but made her do her own homework. In return for the fudge, he gave her a cookie.

After that the girl kept reappearing. When her parents found out, they apologized profusely. Einstein waved them off. "That's quite unnecessary," he said. "I'm learning just as much from your child as she is learning from me." He loved to tell, with a twinkle in his eye, the tale of her visits. "She was a very naughty girl," he would laugh. "Do you know she tried to bribe me with candy?"

A friend of Adelaide's recalled going with her and another girl on one of these visits to Mercer Street. When they got up to his study, Einstein offered them lunch, and they accepted. "So he moved a whole bunch of papers from the table, opened four cans of beans with a can opener, and heated them on a Sterno stove one by one, stuck a spoon in each and that was our lunch," she recalled. "He didn't give us anything to drink." [42]

Later, Einstein famously told another girl who complained about her problems with math, "Do not worry about your difficulties in mathematics; I can assure you that mine are even greater." But lest it be thought he helped only girls, he hosted a group of senior boys from the Princeton Country Day School who were baffled by a problem on their math final exam. [43]

He also helped a 15-year-old boy at Princeton High School, Henry Rosso, who was doing poorly in a journalism course. His teacher had offered an A to anyone who scored an interview with Einstein, so Rosso showed up at Mercer Street but was rebuffed at the door. As he was slinking away, the milkman gave him a tip: Einstein could be found walking a certain route every morning at 9:30. So Rosso snuck out of school one day, positioned himself accordingly, and was able to accost Einstein as he wandered by.

Rosso was so flummoxed that he did not know what to ask, which may have been why he was doing poorly in the course. Einstein took pity on him and suggested questions. No personal topics, he insisted. Ask about math instead. Rosso was smart enough to follow his advice. "I discovered that nature was constructed in a wonderful way, and our task is to find out the mathematical structure of the nature itself," Einstein explained of his own education at age 15. "It is a kind of faith that helped me through my whole life."

The interview earned Rosso an A. But it also caused him a bit of dismay. He had promised Einstein that it would only be used in the school paper, but without his permission it got picked up by the Trenton newspaper and then others around the world, which provided yet another lesson in journalism.[44]

Elsa's Death

Soon after they moved into 112 Mercer Street, Elsa became afflicted with a swollen eye. Tests in Manhattan showed that it was a symptom of heart and kidney problems, and she was ordered to remain immobile in bed.

Einstein sometimes read to her, but mostly he threw himself more intently into his studies. "Strenuous intellectual work and looking at God's nature are the reconciling, fortifying yet relentlessly strict angels that shall lead me through all of life's troubles," he had written to the mother of his first girlfriend. Then as now, he could escape the complexity of human emotions by delving into the mathematical elegance that could describe the cosmos. "My husband sticks fearsomely to his calculations," Elsa wrote Watters. "I have never seen him so engrossed in his work."[45]

Elsa painted a warmer picture of her husband when writing to her friend Antonina Vallentin. "He has been so upset by my illness," she reported. "He wanders around like a lost soul. I never thought he loved me so much. And that comforts me."

Elsa decided that they would be better off if they went away for the summer, as they usually did, and so they rented a cottage on Saranac Lake in the Adirondack Mountains of New York. "I'm certain to get better there," she said. "If my Ilse walked into my room now, I would recover at once."[46]

It turned out to be an enjoyable summer, but by winter Elsa was again bedridden and getting weaker. She died on December 20, 1936.

Einstein was hit harder than he might have expected. In fact, he actually cried, as he had done when his mother died. "I had never seen him shed a tear," Peter Bucky reported, "but he did then as he sighed, 'Oh, I shall really miss her.' "[47]

Their relationship had not been a model romance. Before their marriage, Einstein's letters to her were filled with sweet endearments, but those disappeared over the years. He could be prickly and demanding at times, seemingly inured to her emotional needs, and occasionally a flirt or more with other women.

Yet beneath the surface of many romances that evolve into partnerships, there is a depth not visible to outside observers. Elsa and Albert Einstein liked each other, understood each other, and perhaps most important (for she, too, was actually quite clever in her own way) were amused by each other. So even if it was not the stuff of poetry, the bond between them was a solid one. It was forged by satisfying each other's desires and needs, it was genuine, and it worked in both directions.

Not surprisingly, Einstein found solace in his work. He admitted to Hans Albert that focusing was difficult, but the attempt provided him the means to escape the painfully personal. "As long as I am able to work, I must not and will not complain, because work is the only thing that gives substance to life."[48]

When he came to the office, he was "ashen with grief," his collaborator Banesh Hoffmann noted, but he insisted on delving into their work each day. He needed it more than ever, he said. "At first his attempts to concentrate were pitiful," Hoffmann recalled. "But he had known sorrow before and had learned that work was a precious antidote."[49] Together they worked that month on two major papers: one that explored how the bending of light by the gravitational fields of galaxies could create "cosmic lenses" that would magnify distant stars, and another that explored the existence of gravitational waves.[50]

Max Born learned of Elsa's death in a letter from Einstein in which it was mentioned almost as an afterthought in explaining why he had become less social. "I live like a bear in my cave, and really feel more at home than ever before in my eventful life," he told his old friend. "This bearlike quality has been further enhanced by the death of my woman comrade, who was better with other people than I am." Born later marveled at "the incidental way" in which Einstein broke the news of his wife's death. "For all his kindness, sociability and love of humanity," commented Born, "he was nevertheless totally detached from his environment and the human beings in it."[51]

That was not entirely true. For a self-styled bear in a cave, Einstein attracted a clan wherever he went. Whether it was walking home from the Institute, puttering around 112 Mercer Street, or sharing summer cottages and Manhattan weekends with the Watters or Bucky families, Einstein was rarely alone, except when he trundled up to his study. He could keep an ironic detachment and retreat into his own reveries, but he was a true loner only in his own mind.

After Elsa died, he still lived with Helen Dukas and his stepdaughter Margot, and soon thereafter his sister moved in. Maja had been living near Florence with her husband, Paul Winteler. But when Mussolini enacted laws in 1938 that withdrew resident status from all foreign Jews, Maja moved to Princeton on her own. Einstein, who loved her dearly and liked her immensely, was thrilled.

Einstein also encouraged Hans Albert, now 33, to come to America, at least for a visit. Their relationship had been rocky, but Einstein had come to admire the diligence of his son's engineering work, especially regarding the flow of rivers, a topic he had once studied himself.[52] He had also changed his mind and encouraged his son to have children, and he was now happy to have two young grandsons.

In October 1937, Hans Albert arrived for a three-month stay. Einstein met him at the pier, where they posed for photographs, and Hans Albert playfully lit a long Dutch pipe he had brought his father. "My father would like me to come here with my family," he said. "You know his wife died recently and he is all alone now."[53]

During the visit, young and eager Peter Bucky offered to drive Hans Albert across America so that he could visit universities and seek positions as an engineering professor. The trip, which covered ten thousand miles, took them to Salt Lake City, Los Angeles, Iowa City, Knoxville, Vicksburg, Cleveland, Chicago, Detroit, and Indianapolis.[54] Einstein reported to Mileva Marić how much he had enjoyed being with their son. "He has such a great personality," he wrote. "It is unfortunate that he has this wife, but what can you do if he's happy?"[55]

Einstein had written Frieda a few months earlier and suggested that she not accompany her husband on the trip.[56] But with his affection for Hans Albert fully restored, Einstein urged both of them to return together the following year, with their two children, and stay in America. They did. Hans Albert found a job studying soil conservation at a U.S. Department of Agriculture extension station in Clemson,

South Carolina, where he became an authority on alluvial transport by rivers. Displaying his father's taste, he built a simple wooden house, reminiscent of that in Caputh, in nearby Greenville, where he applied for American citizenship in December 1938.[57]

While his father was becoming more connected to his Jewish heritage, Hans Albert became, under the influence of his wife, a Christian Scientist. The rejection of medical care, as sometimes entailed by that faith, had tragic results. A few months after their arrival, their 6-year-old son, Klaus, contracted diphtheria and died. He was buried at a tiny new cemetery in Greenville. "The deepest sorrow loving parents can experience has come upon you," Einstein wrote in a condolence note. His relationship with his son became increasingly secure and even, at times, affectionate.

During the five years that Hans Albert lived in South Carolina, before moving to Caltech and then Berkeley, Einstein would occasionally take the train down to visit. There they would discuss engineering puzzles that reminded Einstein of his days at the Swiss patent office. In the afternoon, he would sometimes wander the roads and forests, often in dreamy thought, spawning colorful anecdotes from astonished locals who helped him find his way home.[58]

Because he was a mental patient, Eduard was not allowed to immigrate to America. As his illness progressed, his face became bloated, his speech slow. Marić increasingly had trouble allowing him back home, so his stays in the institution became more prolonged. Her sister Zorka, who had come to help care for them, descended into her own hell. After their mother died, she became an alcoholic, accidentally burned all the family money, which had been hidden in an old stove, and died a recluse in 1938 on a straw-covered floor surrounded only by her cats.[59] Marić lived on, through it all, in increasing despair.

Prewar Politics

In retrospect, the rise of the Nazis created a fundamental moral challenge for America. At the time, however, this was not so clear. That was especially true in Princeton, which was a conservative town, and at its university, which harbored a surprising number of students who shared the amorphous anti-Semitic attitude found among some in their social class. A survey of incoming freshmen in 1938 produced a result that is now astonishing, and should have been back then as well: Adolf Hitler polled highest as the "greatest living person." Albert Einstein was second.[60]

"Why do They Hate the Jews?" Einstein wrote in an article for the popular weekly *Collier's* that year. He used the article not just to explore anti-Semitism but also to explain how the social creed inbred in most Jews, which he personally tried to live by, was part of a long and proud tradition. "The bond that has united the Jews for thousands of years and that unites them today is, above all, the democratic ideal of social justice coupled with the ideal of mutual aid and tolerance among all men."[61]

His kinship with his fellow Jews, and his horror at the plight that was befalling them, plunged him into the effort for refugee relief. It was both a public and a private endeavor. He gave dozens of speeches for the cause, was feted at even more dinners, and even gave occasional violin recitals for the American Friends Service Committee or the United Jewish Appeal. One gimmick that organizers used was to have people write their checks to Einstein himself. He would then endorse them to the charity. The donor would thus have as a keepsake a cancelled check with Einstein's autograph.[62] He also quietly backed scores of individuals who needed financial guarantees in order to emigrate, especially as the United States made it harder to get visas.

Einstein also became a supporter of racial tolerance. When Marian Anderson, the black contralto, came to Princeton for a concert in 1937, the Nassau Inn refused her a room. So Einstein invited her to stay at his house on Mercer Street, in what was a deeply personal as well as a publicly symbolic gesture. Two years later, when she was barred from performing in Washington's Constitution Hall, she gave what became a historic free concert on the steps of the Lincoln Memorial. Whenever she returned to Princeton, she stayed with Einstein, her last visit coming just two months before he died.[63]

One problem with Einstein's willingness to sign on to various and sundry movements, appeals, and honorary chairmanships was that, as before, it opened him to charges that he was a dupe for those that were fronts for communists or other subversives. This purported sin was compounded, in the eyes of those who were suspicious about his loyalty, when he declined to sign on to some crusades that attacked Stalin or the Soviets.

For example, when his friend Isaac Don Levine, whose anticommunist writings Einstein had previously endorsed, asked him to sign a petition in 1934 condemning Stalin's murder of political prisoners, this time Einstein balked. "I, too, regret immensely that the Russian political leaders let themselves be carried away," Einstein wrote. "In spite of this, I cannot associate myself with your action. It will have no impact in Russia. The Russians have proved that their only aim is really the improvement of the lot of the Russian people."[64]

It was a gauzy view of the Russians and of Stalin's murderous regime, one that history would prove wrong. Einstein was so intent on fighting the Nazis, and so annoyed that Levine had shifted so radically from left to right, that he reacted strongly against those who would equate the Russian purges with the Nazi holocaust.

An even larger set of trials in Moscow began in 1936, involving supporters of the exiled Leon Trotsky, and again Einstein rebuffed some of his former friends from the left who had now swung to become ardently anticommunist. The philosopher Sidney Hook, a recovering Marxist, wrote Einstein, asking him to speak out in favor of the creation of an international public commission to assure that Trotsky and his supporters would get a fair hearing rather than merely a show trial. "There is no doubt that every accused person should be given an opportunity to establish his innocence," Einstein replied. "This certainly holds true for Trotsky." But how should this be accomplished? Einstein suggested it would best be done privately, without a public commission.[65]

In a very long letter, Hook tried to rebut each of Einstein's concerns, but Einstein lost interest in arguing with Hook and did not respond. So Hook phoned him in Princeton. He reached Helen Dukas, and somehow was able to make it through her defensive shield to set up an appointment.

Einstein received Hook cordially, brought him up to his study lair, smoked his pipe, and spoke in English. After listening to Hook again make his case, Einstein expressed sympathy but said he thought the whole enterprise was unlikely to succeed. "From my point of view," he proclaimed, "both Stalin and Trotsky are political gangsters." Hook later said that even though he disagreed with Einstein, "I could appreciate his reasons," especially because Einstein emphasized that he was "aware of what communists were capable of doing."

Wearing an old sweatshirt and no socks, Einstein walked Hook back to the train station. Along the way, he explained his anger at the Germans. They had raided his house in Caputh searching for communist weapons, he said, and found only a bread knife to confiscate. One remark he made turned out to be very prescient. "If and when war comes," he said, "Hitler will realize the harm he has done Germany by driving out the Jewish scientists."[66]

CHAPTER TWENTY

QUANTUM ENTANGLEMENT

1935

"Spooky Action at a Distance"

The thought experiments that Einstein had lobbed like grenades into the temple of quantum mechanics had done little damage to the edifice. In fact, they helped test it and permit a better understanding of its implications. But Einstein remained a resister, and he continued to conjure up new ways to show that the uncertainties inherent in the interpretations formulated by Niels Bohr, Werner Heisenberg, Max Born, and others meant that something was missing in their explanation of "reality."

Just before he left Europe in 1933, Einstein attended a lecture by Léon Rosenfeld, a Belgian physicist with a philosophical bent. When it was over, Einstein rose from the audience to ask a question. "Suppose two particles are set in motion towards each other with the same, very large, momentum, and they interact with each other for a very short time when they pass at known positions," he posited. When the particles have bounced far apart, an observer measures the momentum of one of them. "Then, from the conditions of the experiment, he will obviously be able to deduce the momentum of the other particle," Einstein said. "If, however, he chooses to measure the position of the first particle, he will be able to tell where the other particle is."

Because the two particles were far apart, Einstein was able to assert, or at least to *assume,* that "all physical interaction has ceased between them." So his challenge to the Copenhagen interpreters of quantum mechanics, posed as a question to Rosenfeld, was simple: "How can the final state of the second particle be influenced by a measurement performed on the first?"[1]

Over the years, Einstein had increasingly come to embrace the concept of realism, the belief that there is, as he put it, "a real factual situation" that exists "independent of our observations."[2] This belief was one aspect of his discomfort with Heisenberg's uncertainty principle and other tenets of quantum mechanics that assert that observations determine realities. With his question to Rosenfeld, Einstein was deploying another concept: locality.* In other words, if two particles are spatially distant from each other, anything that happens to one is independent from what happens to the other, and no signal or force or influence can move between them faster than the speed of light.

Observing or poking one particle, Einstein posited, could not *instantaneously* jostle or jangle another one far away. The only way an action on one system can affect a distant one is if some wave or signal or information traveled between them—a process that would have to obey the speed limit of light. That was even true of gravity. If the sun suddenly disappeared, it would not affect the earth's orbit for about eight minutes, the amount of time it would take the change in the gravitational field to ripple to the earth at the speed of light.

As Einstein said, "There is one supposition we should, in my opinion, absolutely hold fast: the *real factual situation* of the system S_2 is independent of what is done with the system S_1, which is spatially separated from the former."[3] It was so intuitive that it seemed obvious. But as Einstein noted, it was a "supposition." It had never been proven.

---

* There are two related concepts that Einstein uses. *Separability* means that different particles or systems that occupy different regions in space have an independent reality; *locality* means that an action involving one of these particles or systems cannot influence a particle or system in another part of space unless something travels the distance between them, a process limited by the speed of light.

To Einstein, realism and localism were related underpinnings of physics. As he declared to his friend Max Born, coining a memorable phrase, "Physics should represent a reality in time and space, free from spooky action at a distance."[4]

Once he had settled in Princeton, Einstein began to refine this thought experiment. His sidekick, Walther Mayer, less loyal to Einstein than Einstein was to him, had drifted away from the front lines of fighting quantum mechanics, so Einstein enlisted the help of Nathan Rosen, a 26-year-old new fellow at the Institute, and Boris Podolsky, a 49-year-old physicist Einstein had met at Caltech who had since moved to the Institute.

The resulting four-page paper, published in May 1935 and known by the initials of its authors as the EPR paper, was the most important paper Einstein would write after moving to America. "Can the Quantum-Mechanical Description of Physical Reality Be Regarded as Complete?" they asked in their title.

Rosen did a lot of the math, and Podolsky wrote the published English version. Even though they had discussed the content at length, Einstein was displeased that Podolsky had buried the clear conceptual issue under a lot of mathematical formalism. "It did not come out as well as I had originally wanted," Einstein complained to Schrödinger right after it was published. "Rather, the essential thing was, so to speak, smothered by the formalism."[5]

Einstein was also annoyed at Podolsky for leaking the contents to the *New York Times* before it was published. The headline read: "Einstein Attacks Quantum Theory / Scientist and Two Colleagues Find It Not 'Complete' Even though 'Correct.'" Einstein, of course, had occasionally succumbed to giving interviews about upcoming articles, but this time he declared himself dismayed by the practice. "It is my invariable practice to discuss scientific matters only in the appropriate forum," he wrote in a statement to the *Times,* "and I deprecate advance publication of any announcement in regard to such matters in the secular press."[6]

Einstein and his two coauthors began by defining their realist premise: "If without in any way disturbing a system we can predict with certainty the value of a physical quantity, then there exists an element of physical *reality* corresponding to this physical quantity."[7] In other words, if by some process we could learn with absolute certainty the position of a particle, and we have not disturbed the particle by observing it, then we can say the particle's position is real, that it exists in reality totally independent of our observations.

The paper went on to expand Einstein's thought experiment about two particles that have collided (or have flown off in opposite directions from the disintegration of an atom) and therefore have properties that are correlated. We can take measurements of the first particle, the authors asserted, and from that gain knowledge about the second particle "without in any way disturbing the second particle." By measuring the position of the first particle, we can determine precisely the position of the second particle. And we can do the same for the momentum. "In accordance with our criterion for reality, in the first case we must consider the quantity P as being an element of reality, in the second case the quantity Q is an element of reality."

In simpler words: at any moment the second particle, which we have not observed, has a position that is real and a momentum that is real. These two properties are features of reality that quantum mechanics does not account for; thus the answer to the title's question should be no, quantum mechanics' description of reality is not complete.[8]

The only alternative, the authors argued, would be to claim that the process of measuring the first particle affects the reality of the position and momentum of the second particle. "No reasonable definition of reality could be expected to permit this," they concluded.

Wolfgang Pauli wrote Heisenberg a long and angry letter. "Einstein has once again expressed himself publicly on quantum mechanics (together with Podolsky and Rosen—no good company, by the way)," he fumed. "As is well known, every time that happens it is a catastrophe."[9]

When the EPR paper reached Niels Bohr in Copenhagen, he realized that he had once again been cast in the role, which he played so well at the Solvay Conferences, of defending quantum mechanics from yet another Einstein assault. "This onslaught came down on us as a bolt from the blue," a colleague of Bohr's reported. "Its effect on Bohr was remarkable." He had often reacted to such situations by wandering around and muttering, "Einstein . . . Einstein . . . Einstein!" This time he added some collaborative doggerel as well: "Podolsky, Opodolsky, Iopodolsky, Siopodolsky . . ."[10]

"Everything else was abandoned," Bohr's colleague recalled. "We had to clear up such a misunderstanding at once." Even with such intensity, it took Bohr more than six weeks of fretting, writing, revising, dictating, and talking aloud before he finally sent off his response to EPR.

It was longer than the original paper. In it Bohr backed away somewhat from what had been an aspect of the uncertainty principle: that the mechanical disturbance caused by the act of observation was a cause of the uncertainty. He admitted that in Einstein's thought experiment, "there is no question of a mechanical disturbance of the system under investigation."[11]

This was an important admission. Until then, the disturbance caused by a measurement had been part of Bohr's physical explanation of quantum uncertainty. At the Solvay Conferences, he had rebutted Einstein's ingenious thought experiments by showing that the simultaneous knowledge of, say, position and momentum was impossible at least in part because determining one attribute caused a disturbance that made it impossible to then measure the other attribute precisely.

However, using his concept of complementarity, Bohr added a significant caveat. He pointed out that the two particles were part of one whole phenomenon. Because they have interacted, the two particles are therefore "entangled." They are part of one whole phenomenon or one whole system that has one quantum function.

In addition, the EPR paper did not, as Bohr noted, truly dispel the uncertainty principle, which says that it is not possible to know *both* the precise position and momentum of a particle *at the same moment*. Einstein is correct, that if we measure the *position* of particle A, we can indeed know the *position* of its distant twin B. Likewise, if we measure the *momentum* of A, we can know the *momentum* of B. However, even if we can *imagine* measuring the position and then the momentum of particle A, and thus ascribe a "reality" to those attributes in particle B, we cannot *in fact* measure *both* these attributes precisely at any one time for particle A, and thus we cannot know them both precisely for particle B. Brian Greene, discussing Bohr's response, has put it simply: "If you don't have both of these attributes of the right-moving particle in hand, you don't have them for the left-moving particle either. Thus there is no conflict with the uncertainty principle."[12]

Einstein continued to insist, however, that he had pinpointed an important example of the incompleteness of quantum mechanics by showing how it violated the principle of separability, which holds that two systems that are spatially separated have an independent existence. It likewise violated the related principle of locality, which says that an action on one of these systems cannot immediately affect the other. As an adherent of field theory, which defines reality using a spacetime continuum, Einstein believed that separability was a fundamental feature of nature. And as a defender of his own theory of relativity, which rid Newton's cosmos of spooky action at a distance and decreed instead that such actions obey the speed limit of light, he believed in locality as well.[13]

Schrödinger's Cat

Despite his success as a quantum pioneer, Erwin Schrödinger was among those rooting for Einstein to succeed in deflating the Copenhagen consensus. Their alliance had been forged at the Solvay Conferences, where Einstein played God's advocate and Schrödinger looked on with a mix of curiosity and sympathy. It was a lonely struggle, Einstein lamented in a letter to Schrödinger in 1928: "The Heisenberg-Bohr tranquilizing philosophy—or religion?—is so delicately contrived that, for the time being, it provides a gentle pillow for the true believer from which he cannot very easily be aroused."[14]

So it was not surprising that Schrödinger sent Einstein a congratulatory note as soon as he read the EPR paper. "You have publicly caught dogmatic quantum mechanics by its throat," he wrote. A few weeks later, he added happily, "Like a pike in a goldfish pond it has stirred everyone up."[15]

Schrödinger had just visited Princeton, and Einstein was still hoping, in vain, that Flexner might be convinced to hire him for the Institute. In his subsequent flurry of exchanges with Schrödinger, Einstein began conspiring with him on ways to poke holes in quantum mechanics.

"I do not believe in it," Einstein declared flatly. He ridiculed as "spiritualistic" the notion that there could be "spooky action at a distance," and he attacked the idea that there was no reality beyond our ability to observe things. "This epistemology-soaked orgy ought to burn itself out," he said. "No doubt, however, you smile at me and think that, after all, many a young whore turns into an old praying sister, and many a young revolutionary becomes an old reactionary."[16] Schrödinger did smile, he told Einstein in his reply, because he had likewise edged from revolutionary to old reactionary.

On one issue Einstein and Schrödinger diverged. Schrödinger did not feel that the concept of locality was sacred. He even coined the term that we now use, *entanglement*, to describe the correlations that exist between two particles that have interacted but are now distant from each other. The quantum states of two particles that have interacted must subsequently be described together, with any changes to one particle instantly being reflected in the other, no matter how far apart they now are. "Entanglement of predictions arises from the fact that the two bodies at some earlier time formed in a true sense *one* system, that is were interacting, and have left behind *traces* on each other," Schrödinger wrote. "If two separated bodies enter a situation in which they influence each other, and separate again, then there occurs what I have just called *entanglement* of our knowledge of the two bodies."[17]

Einstein and Schrödinger together began exploring another way—one that did not hinge on issues of locality or separation—to raise questions about quantum mechanics. Their new approach was to look at what would occur when an event in the quantum realm, which includes subatomic particles, interacted with objects in the macro world, which includes those things we normally see in our daily lives.

In the quantum realm, there is no definite location of a particle, such as an electron, at any given moment. Instead, a mathematical function, known as a wave function, describes the probability of finding the particle in some particular place. These wave functions also describe quantum states, such as the probability that an atom will, when observed, be decayed or not. In 1925, Schrödinger had come up with his famous equation that described these waves, which spread and smear throughout space. His equation defined the probability that a particle, when observed, will be found in a particular place or state.[18]

According to the Copenhagen interpretation developed by Niels Bohr and his fellow pioneers of quantum mechanics, until such an observation is made, the reality of the particle's position or state consists only of these probabilities. By measuring or observing the system, the observer causes the wave function to collapse and one distinct position or state to snap into place.

In a letter to Schrödinger, Einstein gave a vivid thought experiment showing why all this discussion of wave functions and probabilities, and of particles that have no definite positions until observed, failed his test of completeness. He imagined two boxes, one of which we know contains a ball. As we prepare to look in one of the boxes, there is a 50 percent chance of the ball being there. After we look, there is either a 100 percent or a 0 percent chance it is in there. But all along, *in reality*, the ball was in one of the boxes. Einstein wrote:

> I describe a state of affairs as follows: the probability is ½ that the ball is in the first box. Is that a complete description? NO: A complete statement is: the ball *is* (or is not) in the first box. That is how the characterization of the state of affairs must appear in a complete description. YES: Before I open them, the ball is by no means in *one* of the two boxes. Being in a definite box comes about only when I lift the covers.[19]

Einstein clearly preferred the former explanation, a statement of his realism. He felt that there was something incomplete about the second answer, which was the way quantum mechanics explained things.

Einstein's argument is based on what appears to be common sense. However, sometimes what seems to make sense turns out not to be a good description of nature. Einstein realized this when he developed his relativity theory; he defied the accepted common sense of the time and forced us to change the way we think about nature. Quantum mechanics does something similar. It asserts that particles do not have a definite state except when observed, and two particles can be in an entangled state so that the observation of one determines a property of the other instantly. As soon as any observation is made, the system goes into a fixed state.[20]

Einstein never accepted this as a complete description of reality, and along these lines he proposed another thought experiment to Schrödinger a few weeks later, in early August 1935. It involved a situation in which quantum mechanics would assign only probabilities, even though common sense tells us that there is *obviously* an underlying reality that exists with certainty. Imagine a pile of gunpowder that, due to the instability of some particle, will combust at some point, Einstein said. The quantum mechanical equation for this situation "describes a sort of blend of not-yet and already-exploded systems." But this is not "a *real* state of affairs," Einstein said, "for *in reality* there is just no intermediary between exploded and not-exploded."[21]

Schrödinger came up with a similar thought experiment—involving a soon-to-be-famous fictional feline rather than a pile of gunpowder—to show the weirdness inherent when the indeterminacy of the quantum realm interacts with our normal world of larger objects. "In a lengthy essay that I have just written, I give an example that is very similar to your exploding powder keg," he told Einstein.[22]

In this essay, published that November, Schrödinger gave generous credit to Einstein and the EPR paper for "providing the impetus" for his argument. It poked at a core concept in quantum mechanics, namely that the timing of the emission of a particle from a decaying nucleus is indeterminate until it is actually observed. In the quantum world, a nucleus is in a "superposition," meaning it exists simultaneously as being decayed and undecayed until it is observed, at which point its wave function collapses and it becomes either one or the other.

This may be conceivable for the microscopic quantum realm, but it is baffling when one imagines the intersection between the quantum realm and our observable everyday world. So, Schrödinger asked in his thought experiment, when does the system stop being in a superposition incorporating both states and snap into being one reality?

This question led to the precarious fate of an imaginary creature, which was destined to become immortal whether it was dead or alive, known as Schrödinger's cat:

> One can even set up quite ridiculous cases. A cat is penned up in a steel chamber, along with the following device (which must be secured against direct interference by the cat): in a Geiger counter there is a tiny

bit of radioactive substance, *so* small, that *perhaps* in the course of the hour one of the atoms decays, but also, with equal probability, perhaps none; if it happens, the counter tube discharges and through a relay releases a hammer which shatters a small flask of hydrocyanic acid. If one has left this entire system to itself for an hour, one would say that the cat still lives *if* meanwhile no atom has decayed. The psi-function of the entire system would express this by having in it the living and dead cat (pardon the expression) mixed or smeared out.[23]

Einstein was thrilled. "Your cat shows that we are in complete agreement concerning our assessment of the character of the current theory," he wrote back. "A psi-function that contains the living as well as the dead cat just cannot be taken as a description of a real state of affairs."[24]

The case of Schrödinger's cat has spawned reams of responses that continue to pour forth with varying degrees of comprehensibility. Suffice it to say that in the Copenhagen interpretation of quantum mechanics, a system stops being a superposition of states and snaps into a single reality when it is observed, but there is no clear rule for what constitutes such an observation. Can the cat be an observer? A flea? A computer? A mechanical recording device? There's no set answer. However, we do know that quantum effects generally are not observed in our everyday visible world, which includes cats and even fleas. So most adherents of quantum mechanics would not argue that Schrödinger's cat is sitting in that box somehow being both dead and alive until the lid is opened.[25]

Einstein never lost faith in the ability of Schrödinger's cat and his own gunpowder thought experiments of 1935 to expose the incompleteness of quantum mechanics. Nor has he received proper historical credit for helping give birth to that poor cat. In fact, he would later mistakenly give Schrödinger credit for both of the thought experiments in a letter that exposed the animal to being blown up rather than poisoned. "Contemporary physicists somehow believe that the quantum theory provides a description of reality, and even a *complete* description," Einstein wrote Schrödinger in 1950. "This interpretation is, however, refuted most elegantly by your system of radioactive atom + Geiger counter + amplifier + charge of gunpowder + cat in a box, in which the psi-function of the system contains the cat both alive and blown to bits."[26]

Einstein's so-called mistakes, such as the cosmological constant he added to his gravitational field equations, often turned out to be more intriguing than other people's successes. The same was true of his parries against Bohr and Heisenberg. The EPR paper would not succeed in showing that quantum mechanics was wrong. But it did eventually become clear that quantum mechanics was, as Einstein argued, incompatible with our commonsense understanding of locality—our aversion to spooky action at a distance. The odd thing is that Einstein, apparently, was far more right than he hoped to be.

In the years since he came up with the EPR thought experiment, the idea of entanglement and spooky action at a distance—the quantum weirdness in which an observation of one particle can instantly affect another one far away—has increasingly become part of what experimental physicists study. In 1951, David Bohm, a brilliant assistant professor at Princeton, recast the EPR thought experiment so that it involved the opposite "spins" of two particles flying apart from an interaction.[27] In 1964, John Stewart Bell, who worked at the CERN nuclear research facility near Geneva, wrote a paper that proposed a way to conduct experiments based on this approach.[28]

Bell was less than comfortable with quantum mechanics. "I hesitated to think it was wrong," he once said, "but I knew that it was rotten."[29] That, plus his admiration of Einstein, caused him to express some hope that Einstein rather than Bohr might be proven right. But when the experiments were undertaken in the 1980s by the French physicist Alain Aspect and others, they provided evidence that locality was not a feature of the quantum world. "Spooky action at a distance," or, more precisely, the potential entanglement of distant particles, was.[30]

Even so, Bell ended up appreciating Einstein's efforts. "I felt that Einstein's intellectual superiority over Bohr, in this instance, was enormous, a vast gulf between the man who saw clearly what was needed, and the obscurantist," he said. "So for me, it is a pity that Einstein's idea doesn't work. The reasonable thing just doesn't work."[31]

Quantum entanglement—an idea discussed by Einstein in 1935 as a way of undermining quantum mechanics—is now one of the weirder elements of physics, because it is so counterintuitive. Every year the evidence for it mounts, and public fascination with it grows. At the end of 2005, for example, the *New York Times* published a survey article called "Quantum Trickery: Testing Einstein's Strangest Theory," by Dennis Overbye, in which Cornell physicist N. David Mermin called it "the closest thing we have to magic."[32] And in 2006, the *New Scientist* ran a story titled "Einstein's 'Spooky Action' Seen on a Chip," which began:

> A simple semiconductor chip has been used to generate pairs of entangled photons, a vital step towards making quantum computers a reality. Famously dubbed "spooky action at a distance" by Einstein, entanglement is the mysterious phenomenon of quantum particles whereby two particles such as photons behave as one regardless of how far apart they are.[33]

Might this spooky action at a distance—where something that happens to a particle in one place can be instantly reflected by one that is billions of miles away—violate the speed limit of light? No, the theory of relativity still seems safe. The two particles, though distant, remain part of the same physical entity. By observing one of them, we may affect its attributes, and that is correlated to what would be observed of the second particle. But no information is transmitted, no signal sent, and there is no traditional cause-and-effect relationship. One can show by thought experiments that quantum entanglement cannot be used to send information instantaneously. "In short," says physicist Brian Greene, "special relativity survives by the skin of its teeth."[34]

During the past few decades, a number of theorists, including Murray Gell-Mann and James Hartle, have adopted a view of quantum mechanics that differs in some ways from the Copenhagen interpretation and provides an easier explanation of the EPR thought experiment. Their interpretation is based on alternative histories of the universe, coarse-grained in the sense that they follow only certain variables and ignore (or average over) the rest. These "decoherent" histories form a tree-like structure, with each of the alternatives at one time branching out into alternatives at the next time and so forth.

In the case of the EPR thought experiment, the position of one of the two particles is measured on one branch of history. Because of the common origin of the particles, the position of the other one is determined as well. On a different branch of history, the momentum of one of the particles may be measured, and the momentum of the other one is also determined. On each branch nothing occurs that violates the laws of classical physics. The information about one particle *implies* the corresponding information about the other one, but nothing *happens* to the second particle as a result of the measurement of the first one. So there is no threat to special relativity and its prohibition of instantaneous transmission of information. What is special about quantum mechanics is that the simultaneous determination of the position and the momentum of a particle is impossible, so if these two determinations occur, it must be on different branches of history.[35]

"Physics and Reality"

Einstein's fundamental dispute with the Bohr-Heisenberg crowd over quantum mechanics was not merely about whether God rolled dice or left cats half dead. Nor was it just about causality, locality, or even completeness. It was about reality.[36] Does it exist? More specifically, is it meaningful to speak about a physical reality that exists independently of whatever observations we can make? "At the heart of the problem," Einstein said of quantum mechanics, "is not so much the question of causality but the question of realism."[37]

Bohr and his adherents scoffed at the idea that it made sense to talk about what might be beneath the veil of what we can observe. All we can know are the results of our experiments and observations, not some ultimate reality that lies beyond our perceptions.

Einstein had displayed some elements of this attitude in 1905, back when he was reading Hume and Mach while rejecting such unobservable concepts as absolute space and time. "At that time my mode of thinking was much nearer positivism than it was later on," he recalled.

"My departure from positivism came only when I worked out the general theory of relativity." [38]

From then on, Einstein increasingly adhered to the belief that there *is* an objective classical reality. And though there are some consistencies between his early and late thinking, he admitted freely that, at least in his own mind, his realism represented a move away from his earlier Machian empiricism. "This credo," he said, "does not correspond with the point of view I held in younger years." [39] As the historian Gerald Holton notes, "For a scientist to change his philosophical beliefs so fundamentally is rare." [40]

Einstein's concept of realism had three main components:

1. His belief that a reality exists independent of our ability to observe it. As he put it in his autobiographical notes: "Physics is an attempt conceptually to grasp reality as it is thought independently of its being observed. In this sense one speaks of 'physical reality.'" [41]

2. His belief in separability and locality. In other words, objects are located at certain points in spacetime, and this separability is part of what defines them. "If one abandons the assumption that what exists in different parts of space has its own independent, real existence, then I simply cannot see what it is that physics is supposed to describe," he declared to Max Born. [42]

3. His belief in strict causality, which implies certainty and classical determinism. The idea that probabilities play a role in reality was as disconcerting to him as the idea that our observations might play a role in collapsing those probabilities. "Some physicists, among them myself, cannot believe," he said, "that we must accept the view that events in nature are analogous to a game of chance." [43]

It is possible to imagine a realism that has only two, or even just one, of these three attributes, and on occasion Einstein pondered such a possibility. Scholars have debated which of these three was most fundamental to his thinking. [44] But Einstein kept coming back to the hope, and faith, that all three attributes go together. As he said in a speech to a doctors convention in Cleveland near the end of his life, "Everything should lead back to conceptual objects in the realm of space and time and to lawlike relations that obtain for these objects."[45]

At the heart of this realism was an almost religious, or perhaps childlike, awe at the way all of our sense perceptions—the random sights and sounds that we experience every minute—fit into patterns, follow rules, and make sense. We take it for granted when these perceptions piece together to represent what seem to be external objects, and it does not amaze us when laws seem to govern the behavior of these objects.

But just as he felt awe when first pondering a compass as a child, Einstein was able to feel awe that there are rules ordering our perceptions, rather than pure randomness. Reverence for this astonishing and unexpected comprehensibility of the universe was the foundation for his realism as well as the defining character of what he called his religious faith.

He expressed this in a 1936 essay, "Physics and Reality," written on the heels of his defense of realism in the debates over quantum mechanics. "The very fact that the totality of our sense experiences is such that, by means of thinking, it can be put in order, this fact is one that leaves us in awe," he wrote. "The eternal mystery of the world is its comprehensibility . . . The fact that it is comprehensible is a miracle."[46]

His friend Maurice Solovine, with whom he had read Hume and Mach in the days of the Olympia Academy, told Einstein that he found it "strange" that he considered the comprehensibility of the world to be "a miracle or an eternal mystery." Einstein countered that it would be logical to assume that the opposite was the case. "Well, a priori, one should expect a chaotic world which cannot be grasped by the mind in any way," he wrote. "There lies the weakness of positivists and professional atheists."[47] Einstein was neither.

To Einstein, this belief in the existence of an underlying reality had a religious aura to it. That dismayed Solovine, who wrote to say that he had an "aversion" to such language. Einstein disagreed. "I have no better expression than 'religious' for this confidence in the rational nature of reality and in its being accessible, to some degree, to human reason.

When this feeling is missing, science degenerates into mindless empiricism."[48]

Einstein knew that the new generation viewed him as an out-of-touch conservative clinging to the old certainties of classical physics, and that amused him. "Even the great initial success of the quantum theory does not make me believe in a fundamental dice-game," he told his friend Max Born, "although I am well aware that our younger colleagues interpret this as a consequence of senility."[49]

Born, who loved Einstein dearly, agreed with the Young Turks that Einstein had become as "conservative" as the physicists of a generation earlier who had balked at his relativity theory. "He could no longer take in certain new ideas in physics which contradicted his own firmly held philosophical convictions."[50]

But Einstein preferred to think of himself not as a conservative but as (again) a rebel, a nonconformist, one with the curiosity and stubbornness to buck prevailing fads. "The necessity of conceiving of nature as an *objective reality* is said to be obsolete prejudice while the quantum theoreticians are vaunted," he told Solovine in 1938. "Each period is dominated by a mood, with the result that most men fail to see the tyrant who rules over them."[51]

Einstein pushed his realist approach in a textbook on the history of physics that he coauthored in 1938, *The Evolution of Physics*. Belief in an "objective reality," the book argued, had led to great scientific advances throughout the ages, thus proving that it was a useful concept even if not provable. "Without the belief that it is possible to grasp reality with our theoretical constructions, without the belief in the inner harmony of our world, there could be no science," the book declared. "This belief is and always will remain the fundamental motive for all scientific creation."[52]

In addition, Einstein used the text to defend the utility of field theories amid the advances of quantum mechanics. The best way to do that was to view particles not as independent objects but as a special manifestation of the field itself:

> There is no sense in regarding matter and field as two qualities quite different from each other . . . Could we not reject the concept of matter and build a pure field physics? We could regard matter as the regions in

space where the field is extremely strong. A thrown stone is, from this point of view, a changing field in which the states of the greatest field intensity travel through space with the velocity of the stone.[53]

There was a third reason that Einstein helped to write this textbook, a more personal one. He wanted to help Leopold Infeld, a Jew who had fled Poland, collaborated briefly in Cambridge with Max Born, and then moved to Princeton.[54] Infeld began working on relativity with Banesh Hoffmann, and he proposed that they offer themselves to Einstein. "Let's see if he'd like us to work with him," Infeld suggested.

Einstein was delighted. "We did all the dirty work of calculating the equations and so on," Hoffmann recalled. "We reported the results to Einstein and then it was like having a headquarters conference. Sometimes his ideas seemed to come from left field, to be quite extraordinary."[55] Working with Infeld and Hoffmann, Einstein in 1937 came up with elegant ways to explain more simply the motion of planets and other massive objects that produced their own curvatures of space.

But their work on unified field theory never quite gelled. At times, the situation seemed so hopeless that Infeld and Hoffmann became despondent. "But Einstein's courage never faltered, nor did his inventiveness fail him," Hoffmann recalled. "When excited discussion failed to break the deadlock, Einstein would quietly say in his quaint English, 'I will a little tink.'" The room would become silent, and Einstein would pace slowly up and down or walk around in circles, twirling a lock of his hair around his forefinger. "There was a dreamy, far-away, yet inward look on his face. No sign of stress. No outward indication of intense concentration." After a few minutes, he would suddenly return to the world, "a smile on his face and an answer to the problem on his lips."[56]

Einstein was so pleased with Infeld's help that he tried to get Flexner to give him a post at the Institute. But Flexner, who was annoyed that the Institute had already been forced to hire Walther Mayer, balked. Einstein even went to a fellows meeting in person, which he rarely did, to argue for a mere $600 stipend for Infeld, but to no avail.[57]

So Infeld came up with a plan to write a history of physics with Einstein, which was sure to be successful, and split the royalties. When he went to Einstein to pitch the idea, Infeld became incredibly tongue-tied, but he was finally able to stammer out his proposal. "This is not at all a stupid idea," Einstein said. "Not stupid at all. We shall do it."[58]

In April 1937, Richard Simon and Max Schuster, founders of the house that published this biography, drove out to Einstein's home in Princeton to secure the rights. The gregarious Schuster tried to win Einstein over with jokes. He had discovered something faster than the speed of light, he said: "The speed with which a woman arriving in Paris goes shopping."[59] Einstein was amused, or at least so Schuster recalled. In any event, the trip was successful, and the *Evolution of Physics*, which is in its forty-fourth printing, not only propagandized for the role of field theories and a faith in objective reality, it also made Infeld (and Einstein) more secure financially.

No one could accuse Infeld of being ungrateful. He later called Einstein "perhaps the greatest scientist and kindest man who ever lived." He also wrote a flattering biography of Einstein, while his mentor was still alive, that praised him for his willingness to defy conventional thinking in his quest for a unified theory. "His tenacity in sticking to a problem for years, in returning to the problem again and again—this is the characteristic feature of Einstein's genius," he wrote.[60]

Against the Current

Was Infeld right? Was tenacity the characteristic feature of Einstein's genius? To some extent he had always been blessed by this trait, especially in his long and lonely quest to generalize relativity. There was also ingrained in him, since his school days, a willingness to sail against the current and defy the reigning authorities. All of this was evident in his quest for a unified theory.

But even though he liked to claim that an analysis of empirical data had played a minimal role in the construction of his great theories, he had generally been graced with an intuitive feel for the insights and principles that could be wrested from nature based on current experiments and observations. This trait was now becoming less evident.

By the late 1930s, he was becoming increasingly detached from new experimental discoveries. Instead of the unification of gravity and electromagnetism, there was greater disunity as two new forces, the weak and the strong nuclear forces, were found. "Einstein chose to ignore those new forces, although they were not any less fundamental than the two which have been known about longer," his friend Abraham Pais recalled. "He continued the old search for a unification of gravitation and electromagnetism."[61]

In addition, a menagerie of new fundamental particles were discovered beginning in the 1930s. Currently there are dozens of them, ranging from bosons such as photons and gluons to fermions such as electrons, positrons, up quarks, and down quarks. This did not seem to bode well for Einstein's quest to unify everything. His friend Wolfgang Pauli, who joined him at the Institute in 1940, quipped about the futility of this quest. "What God has put asunder," he said, "let no man join together."[62]

Einstein found the new discoveries to be vaguely disconcerting, but he felt comfortable not putting much emphasis on them. "I can derive only small pleasure from the great discoveries, because for the time being they do not seem to facilitate for me an understanding of the foundations," he wrote Max von Laue. "I feel like a kid who cannot get the hang of the ABCs, even though, strangely enough, I do not abandon hope. After all, one is dealing here with a sphinx, not with a willing streetwalker."[63]

So Einstein beat on against the current, borne back ceaselessly into the past. He realized that he had the luxury to pursue his lonely course, something that would be too risky for younger physicists still trying to make their reputations.[64] But as it turned out, there were usually at least two or three younger physicists, attracted by Einstein's aura, willing to collaborate with him, even if the vast majority of the physics priesthood considered his search for a unified field theory to be quixotic.

One of these young assistants, Ernst Straus, remembers working on an approach that Einstein pursued for almost two years. One evening, Straus found, to his dismay, that their equations led to some conclusions that clearly could not be true. The next day, he and Einstein explored the issue from all angles, but they could not avoid the disappointing result. So they went home early. Straus was dejected, and he assumed that Einstein would be even more so. To his surprise, Einstein was as eager and excited as ever the next day, and he proposed yet another approach they could take. "This was the start of an entirely new theory, also relegated to the trash heap after a half-year's work and mourned no longer than its predecessor," Straus recalls.[65]

Einstein's quest was driven by his intuition that mathematical simplicity, an attribute he never fully defined though he felt he knew it when he saw it, was a feature of nature's handiwork.[66] Every now and then, when a particularly elegant formulation cropped up, he would exult to Straus, "This is so simple God could not have passed it up."

Enthusiastic letters to friends continued to pour forth from Princeton about the progress of his crusade against the quantum theorists who seemed wedded to probabilities and averse to believing in an underlying reality. "I am working with my young people on an extremely interesting theory with which I hope to defeat modern proponents of mysticism and probability and their aversion to the notion of reality in the domain of physics," he wrote Maurice Solovine in 1938.[67]

Likewise, headlines continued to emanate from Princeton on purported breakthroughs. "Soaring over a hitherto unscaled mathematical mountain-top, Dr. Albert Einstein, climber of cosmic Alps, reports having sighted a new pattern in the structure of space and matter," the distinguished *New York Times* science reporter William Laurence reported in a page 1 article in 1935. The same writer and the same paper reported on page 1 in 1939, "Albert Einstein revealed today that after twenty years of unremitting search for a law that would explain the mechanism of the cosmos in its entirety, reaching out from the stars and galaxies in the vastness of infinite space down to the mysteries within the heart of the infinitesimal atom, he has at last arrived within sight of what he hopes may be the 'Promised Land of Knowledge,' holding what may be the master key to the riddle of creation."[68]

The triumphs in his salad days had come partly from having an instinct that could sniff out underlying physical realities. He could intuitively sense the implications of the relativity of all motion, the constancy of the speed of light, and the equivalence of gravitational and inertial mass. From that he could build theories based on a feel for the physics. But he later became more reliant on a mathematical formalism, because it had guided him in his final sprint to complete the field equations of general relativity.

Now, in his quest for a unified theory, there seemed to be a lot of mathematical formalism but very few fundamental physical insights guiding him. "In his earlier search for the general theory, Einstein had been guided by his principle of equivalence linking gravitation with acceleration," said Banesh Hoffmann, a Princeton collaborator. "Where were the comparable guiding principles that could lead to the construction of a unified field theory? No one knew. Not even Einstein. Thus the search was not so much a search as a groping in the gloom of a mathematical jungle inadequately lit by physical intuition." Jeremy Bernstein later called it "like an all but random shuffling of mathematical formulas with no physics in view."[69]

After a while, the optimistic headlines and letters stopped emanating from Princeton, and Einstein publicly admitted that he was, at least for the time being, stymied. "I am not as optimistic," he told the *New York Times*. For years the paper had regularly headlined each of Einstein's purported breakthroughs toward a unified theory, but now its headline read, "Einstein Baffled by Cosmos Riddle."

Nonetheless, Einstein insisted that he still could not "accept the view that events in nature are analogous to a game of chance." And so he pledged to continue his quest. Even if he failed, he felt that the effort would be meaningful. "It is open to every man to choose the direction of his striving," he explained, "and every man may take comfort from the fine saying that the search for truth is more precious than its possession."[70]

Around the time of Einstein's sixtieth birthday, early in the spring of 1939, Niels Bohr came to Princeton for a two-month visit. Einstein remained somewhat aloof toward his old friend and sparring partner. They met at a few receptions, exchanged some small talk, but did not reengage in their old game of volleying thought experiments about quantum weirdness.

Einstein gave only one lecture during that period, which Bohr attended. It dealt with his latest attempts to find a unified field theory. At the end, Einstein fixed his eyes on Bohr and noted that he had long tried to explain quantum mechanics in such a fashion. But he made clear that he would prefer not to discuss the issue further. "Bohr was profoundly unhappy with this," his assistant recalled.[71]

Bohr had arrived in Princeton with a piece of scientific news that was related to Einstein's discovery of the link between energy and mass, $E=mc^2$. In Berlin, Otto Hahn and Fritz Strassman had gotten some interesting experimental results by bombarding heavy uranium with neutrons. These had been sent to their former colleague, Lise Meitner, who had just been forced to flee to Sweden because she was half Jewish. She in turn shared them with her nephew Otto Frisch, and they concluded that the atom had been split, two lighter nuclei created, and a small amount of lost mass turned into energy.

After they substantiated the results, which they dubbed *fission*, Frisch informed his colleague Bohr, who was about to leave for America. Upon his arrival in late January 1939, Bohr described the new discovery to colleagues, and it was discussed at a weekly gathering of physicists in Princeton known as the Monday Evening Club. Within days the results had been replicated, and researchers began churning out papers on the process, including one that Bohr wrote with a young untenured physics professor, John Archibald Wheeler.

Einstein had long been skeptical about the possibility of harnessing atomic energy or unleashing the power implied by $E=mc^2$. On a visit to Pittsburgh in 1934, he had been asked the question and replied that "splitting the atom by bombardment is something akin to shooting birds in the dark in a place where there are only a few birds." That produced a banner headline across the front page of the *Post-Gazette:* "Atom Energy Hope Is Spiked by Einstein / Efforts at Loosing Vast Force Is Called Fruitless / Savant Talked Here."[72]

With the news in early 1939 that it was, apparently, very possible to bombard and split an atomic nucleus, Einstein faced the question again. In an interview for his sixtieth birthday that March, he was asked whether mankind would find some use for the process. "Our results so far concerning the splitting of the atom do not justify the assumption of a practical utilization of the energies released," he replied. This time he was cautious, however, and went on to hedge his answer slightly. "There is no physicist with soul so poor who would allow this to affect his interest in this highly important subject." [73]

Over the next four months, his interest would indeed grow rapidly.

THE BOMB

1939–1945

With Leó Szilárd reenacting (in 1946) their 1939 meeting

The Letter

Leó Szilárd, a charming and slightly eccentric Hungarian physicist, was an old friend of Einstein's. While living in Berlin in the 1920s, they had collaborated on the development of a new type of refrigerator, which they patented but were unable to market successfully.[1] After Szilárd fled the Nazis, he made his way to England and then New York, where he worked at Columbia University on ways to create a nuclear chain reaction, an idea he had conceived while waiting at a stoplight in London a few years earlier. When he heard of the discovery of fission using uranium, Szilárd realized that element might be used to produce this potentially explosive chain reaction.

Szilárd discussed this possibility with his close friend Eugene Wigner, another refugee physicist from Budapest, and they began to worry that the Germans might try to buy up the uranium supplies of the Congo, which was then a colony of Belgium. But how, they asked themselves, could two Hungarian refugees in America find a way to warn the Belgians? Then Szilárd recalled that Einstein happened to be friends with that country's queen mother.

Einstein was spending the summer of 1939 in a rented cottage on the north fork of eastern Long Island, across the Great Peconic Bay from the villages of the Hamptons. There he sailed his small boat *Tinef*, bought sandals from the local department store, and played Bach with the store's owner.[2]

"We knew that Einstein was somewhere on Long Island but we didn't know precisely where," Szilárd recalled. So he phoned Einstein's Princeton office and was told he was renting the house of a Dr. Moore in the village of Peconic. On Sunday, July 16, 1939, they embarked on their mission with Wigner at the wheel (Szilárd, like Einstein, did not drive).

But when they arrived they couldn't find the house, and nobody seemed to know who Dr. Moore was. Just as they were ready to give up, Szilárd saw a young boy standing by the curb. "Do you, by any chance, know where Professor Einstein lives?" Like most people in town, even those who had no idea who Dr. Moore was, the boy did, and he led them up to a cottage near the end of Old Grove Road, where they found Einstein lost in thought.[3]

Sitting at a bare wooden table on the screen porch of the sparsely furnished cottage, Szilárd explained the process of how an explosive chain reaction could be produced in uranium layered with graphite by the neutrons released from nuclear fission. "I never thought of that!" Einstein interjected. He asked a few questions, went over the process for fifteen minutes, and then quickly grasped the implications. Instead of writing to the queen mother, Einstein suggested, perhaps they should write to a Belgian minister he knew.

Wigner, showing some sensible propriety, suggested that perhaps three refugees should not be writing to a foreign government about secret security matters without consulting with the State Department. In which case, they decided, perhaps the proper channel was a letter from Einstein, the only one of them famous enough to be heeded, to the

Belgian ambassador, with a cover letter to the State Department. With that tentative plan in mind, Einstein dictated a draft in German. Wigner translated it, gave it to his secretary to be typed, and then sent it to Szilárd.[4]

A few days later, a friend arranged for Szilárd to talk to Alexander Sachs, an economist at Lehman Brothers and a friend of President Roosevelt. Showing a bit more savvy than the three theoretical physicists, Sachs insisted that the letter should go right to the White House, and he offered to hand-deliver it.

It was the first time Szilárd had met Sachs, but his bold plan was appealing. "It could not do any harm to try this way," he wrote Einstein. Should they talk by phone or meet in person to revise the letter? Einstein replied that he should come back out to Peconic.

By that point Wigner had gone to California for a visit. So Szilárd enlisted, as driver and scientific sidekick, another friend from the amazing group of Hungarian refugees who were theoretical physicists, Edward Teller.[5] "I believe his advice is valuable, but also I think you might enjoy getting to know him," Szilárd told Einstein. "He is particularly nice."[6] Another plus was that Teller had a big 1935 Plymouth. So once again, Szilárd headed out to Peconic.

Szilárd brought with him the original draft from two weeks earlier, but Einstein realized that they were now planning a letter that was far more momentous than one asking Belgian ministers to be careful about Congolese uranium exports. The world's most famous scientist was about to tell the president of the United States that he should begin contemplating a weapon of almost unimaginable impact that could unleash the power of the atom. "Einstein dictated a letter in German," Szilárd recalled, "which Teller took down, and I used this German text as a guide in preparing two drafts of a letter to the President."[7]

According to Teller's notes, Einstein's dictated draft not only raised the question of Congo's uranium, but also explained the possibility of chain reactions, suggested that a new type of bomb could result, and urged the president to set up formal contact with physicists working on this topic. Szilárd then prepared and sent back to Einstein a 45-line version and a 25-line one, both dated August 2, 1939, "and left it up to

Einstein to choose which he liked best." Einstein signed them both in a small scrawl, rather than with the flourish he sometimes used.[8]

The longer version, which is the one that eventually reached Roosevelt, read in part:

> Sir:
>
> Some recent work by E. Fermi and L. Szilárd, which has been communicated to me in a manuscript, leads me to expect that the element uranium may be turned into a new and important source of energy in the immediate future. Certain aspects of this situation which has arisen seem to call for watchfulness and, if necessary, quick action on the part of the Administration. I believe therefore that it is my duty to bring to your attention the following facts and recommendations:
>
> . . . It may become possible to set up a nuclear chain reaction in a large mass of uranium, by which vast amounts of power and large quantities of new radium-like elements would be generated. Now it appears almost certain that this could be achieved in the immediate future.
>
> This new phenomena would also lead to the construction of bombs, and it is conceivable—though much less certain—that extremely powerful bombs of a new type may thus be constructed. A single bomb of this type, carried by boat and exploded in a port, might very well destroy the whole port together with some of the surrounding territory . . .
>
> In view of this situation you may think it desirable to have some permanent contact maintained between the administration and the group of physicists working on chain reactions in America.

It ended with a warning that German scientists might be pursuing a bomb. Once the letter had been written and signed, they still had to figure out who could best get it into the hands of President Roosevelt. Einstein was unsure about Sachs. They considered, instead, financier Bernard Baruch and MIT President Karl Compton.

More amazingly, when Szilárd sent back the typed version of the letter, he suggested that they use as their intermediary Charles Lindbergh, whose solo transatlantic flight twelve years earlier had made him a celebrity. All three of the refugee Jews were apparently unaware that the aviator had been spending time in Germany, was decorated the year before by the Nazi Hermann Göring with that nation's medal of honor, and was becoming an isolationist and Roosevelt antagonist.

Einstein had briefly met Lindbergh a few years earlier in New York, so he wrote a note of introduction, which he included when he returned the signed letters to Szilárd. "I would like to ask you to do me a favor of receiving my friend Dr. Szilárd and think very carefully about what he will tell you," Einstein wrote to Lindbergh. "To one who is outside of science the matter he will bring up may seem fantastic. However, you will certainly become convinced that a possibility is presented here which has to be very carefully watched in the public interest."[9]

Lindbergh did not respond, so Szilárd wrote him a reminder letter on September 13, again asking for a meeting. Two days later, they realized how clueless they had been when Lindbergh gave a nationwide radio address. It was a clarion call for isolationism. "The destiny of this country does not call for our involvement in European wars," Lindbergh began. Interwoven were hints of Lindbergh's pro-German sympathies and even some anti-Semitic implications about Jewish ownership of the media. "We must ask who owns and influences the newspaper, the news picture, and the radio station," he said. "If our people know the truth, our country is not likely to enter the war."[10]

Szilárd's next letter to Einstein stated the obvious: "Lindbergh is not our man."[11]

Their other hope was Alexander Sachs, who had been given the formal letter to Roosevelt that Einstein signed. Even though it was obviously of enormous importance, Sachs was not able to find the opportunity to deliver it for almost two months.

By then, events had turned what was an important letter into an urgent one. At the end of August 1939, the Nazis and Soviets stunned the world by signing their war alliance pact and proceeded to carve up Poland. That prompted Britain and France to declare war, starting the century's second World War. For the time being, America stayed neutral, or at least did not declare war. The country did, however, begin to rearm and to develop whatever new weapons might be necessary for its future involvement.

Szilárd went to see Sachs in late September and was horrified to discover that he still had not been able to schedule an appointment with Roosevelt. "There is a distinct possibility Sachs will be of no use to us," Szilárd wrote Einstein. "Wigner and I have decided to accord him ten days grace."[12] Sachs barely made the deadline. On the afternoon of Wednesday, October 11, he was ushered into the Oval Office carrying Einstein's letter, Szilárd's memo, and an eight-hundred-word summary he had written on his own.

The president greeted him jovially. "Alex, what are you up to?"

Sachs could be loquacious, which may be why the president's handlers made it hard for him to get an appointment, and he tended to tell the president parables. This time it was a tale about an inventor who told Napoleon that he would build him a new type of ship that could travel using steam rather than sails. Napoleon dismissed him as crazy. Sachs then revealed that the visitor was Robert Fulton and, so went the lesson, the emperor should have listened.[13]

Roosevelt responded by scribbling a note to an aide, who hurried off and soon returned with a bottle of very old and rare Napoleon brandy that Roosevelt said had been in his family for a while. He poured two glasses.

Sachs worried that if he left the memos and papers with Roosevelt, they might be glanced at and then pushed aside. The only reliable way to deliver them, he decided, was to read them aloud. Standing in front of the president's desk, he read his summation of Einstein's letter, parts of Szilárd's memo, and some other paragraphs from assorted historical documents.

"Alex, what you are after is to see that the Nazis don't blow us up," the president said.

"Precisely," Sachs replied.

Roosevelt called in his personal assistant. "This requires action," he declared.[14]

That evening, plans were drawn up for an ad hoc committee, coordinated by Dr. Lyman Briggs, director of the Bureau of Standards, the nation's physics laboratory. It met informally for the first time in Washington on October 21. Einstein was not there, nor did he want to be. He was neither a nuclear physicist nor someone who enjoyed proximity to political or military leaders. But his Hungarian émigré trio—Szilárd, Wigner, and Teller—were there to launch the effort.

The following week, Einstein received a polite and formal thank-you letter from the president. "I have convened a board," Roosevelt wrote, "to thoroughly investigate the possibilities of your suggestion regarding the element of uranium."[15]

Work on the atomic project proceeded slowly. Over the next few months, the Roosevelt administration approved only $6,000 for graphite and uranium experiments. Szilárd became impatient. He was becoming more convinced of the feasibility of chain reaction and more worried about reports he was getting from fellow refugees on the activity in Germany.

So in March 1940, he went to Princeton to see Einstein again. They composed another letter for Einstein to sign, which was addressed to Alexander Sachs but intended for him to convey to the president. It warned of all the work on uranium they heard was being done in Berlin. Given the progress being made in producing chain reactions with huge explosive potential, the letter urged the president to consider whether the American work was proceeding quickly enough.[16]

Roosevelt reacted by calling for a conference designed to spur greater urgency, and he told officials to make sure that Einstein could attend. But Einstein had no desire to be more involved. He replied by saying he had a cold—somewhat of a convenient excuse—and did not need to be at the meeting. But he did urge the group to get moving: "I am convinced of the wisdom and urgency of creating the conditions under which work can be carried out with greater speed and on a larger scale."[17]

Even if Einstein had wanted to take part in the meetings, which led to the Manhattan Project that developed the atom bomb, he may not have been welcome. Amazingly, the man who had helped get the project launched was considered, by some, to be too great a potential security risk to be permitted to know about the work.

Brigadier General Sherman Miles, the acting Army chief of staff who was organizing the new committee, sent a letter in July 1940 to J. Edgar Hoover, who had already been the director of the FBI for sixteen years and would remain so for another thirty-two. By addressing him by his national guard rank as "Colonel Hoover," the general was subtly pulling rank when it came to controlling intelligence decisions. But Hoover was assertive when Miles asked for a summary of information the Bureau had on Einstein.[18]

Hoover began by providing General Miles with the letter from Mrs. Frothingham's Woman Patriot Corporation, which had argued in

1932 that Einstein should be denied a visa and raised alarms about various pacifist and political groups he had supported.[19] The Bureau made no attempt to verify or assess any of the charges.

Hoover went on to say that Einstein had been involved in the World Antiwar Congress in Amsterdam in 1932, which had some European communists on its committee. This was the conference that Einstein, as noted earlier, had specifically and publicly declined to attend or even support; as he wrote the organizer, "Because of the glorification of Soviet Russia it includes, I cannot bring myself to sign it." Einstein had gone on in that letter to denounce Russia, where "there seems to be complete suppression of the individual and of freedom of speech." Nevertheless, Hoover implied that Einstein had supported the conference and was thus pro-Soviet.[20]

Hoover's letter had six more paragraphs making similar allegations about a variety of alleged Einstein associations, ranging from pacifist groups to those supporting Spain's loyalists. Appended was a biographical sketch filled with trivial misinformation ("has one child") and wild allegations. It called him "an extreme radical," which he certainly was not, and said he "has contributed to communist magazines," which he hadn't. General Miles was so taken aback by the memo that he wrote a note in the margin, warning, "There is some possibility of flameback" if it ever leaked.[21]

The conclusion of the unsigned biographical sketch was stark: "In view of this radical background, this office would not recommend the employment of Dr. Einstein on matters of a secret nature, without a very careful investigation, as it seems unlikely that a man of his background could, in such a short time, become a loyal American citizen." In a memo the following year, it was reported that the Navy had assented to giving Einstein a security clearance, but "the Army could not clear."[22]

Citizen Einstein

Just as the Army's decision was being made, Einstein was in fact eagerly doing something the likes of which he had not done for forty years, ever since he had saved up his money so that he could become a

Swiss citizen after leaving Germany. He was voluntarily and proudly becoming a citizen of the United States, a process that had begun five years earlier when he sailed to Bermuda so that he could return on an immigration visa. He still had his Swiss citizenship and passport, so he did not need to do this. But he wanted to.

He took his citizenship test on June 22, 1940, in front of a federal judge in Trenton. To celebrate the process, he agreed to give a radio interview as part of the immigration service's *I Am an American* series. The judge served lunch and had the radio folks set up in his chambers to make the process easier for Einstein.[23]

It was an inspiring day, partly because Einstein showed just what type of free-speaking citizen he would be. In his radio talk, he argued that, to prevent wars in the future, nations would have to give up some of their sovereignty to an armed international federation of nations. "A worldwide organization cannot insure peace effectively unless it has control over the entire military power of its members," he said.[24]

Einstein passed his test and he was sworn in—along with his stepdaughter Margot, his assistant Helen Dukas, and eighty-six other new citizens—on October 1. Afterward, he praised America to the reporters covering his naturalization. The nation, he said, would prove that democracy is not just a form of government but "a way of life tied to a great tradition, the tradition of moral strength." Asked if he would renounce other loyalties, he joyously declared that he "would even renounce my cherished sailboat" if that were necessary.[25] It was not, however, necessary for him to renounce his Swiss citizenship, and he did not.

When he first arrived in Princeton, Einstein had been impressed that America was, or could be, a land free of the rigid class hierarchies and servility in Europe. But what grew to impress him more—and what made him fundamentally such a good American but also a controversial one—was the country's tolerance of free thought, free speech, and nonconformist beliefs. That had been a touchstone of his science, and now it was a touchstone of his citizenship.

He had forsaken Nazi Germany with the public pronouncement that he would not live in a country where people were denied the freedom to hold and express their own thoughts. "At that time, I did not understand how right I was in my choice of America as such a place," he wrote in an unpublished essay just after becoming a citizen. "On every side I hear men and women expressing their opinion on candidates for office and the issues of the day without fear of consequences."

The beauty of America, he said, was that this tolerance of each person's ideas existed without the "brute force and fear" that had arisen in Europe. "From what I have seen of Americans, I think that life would not be worth living to them without this freedom of self expression."[26] The depth of his appreciation for America's core value would help explain Einstein's cold public anger and dissent when, during the McCarthy era a few years later, the nation lapsed into a period marked by the intimidation of those with unpopular views.

More than two years after Einstein and his colleagues had urged attention to the possibility of building atomic weapons, the United States launched the supersecret Manhattan Project. It happened on December 6, 1941, which turned out to be, fittingly enough, the day before Japan launched its attack on Pearl Harbor that brought the nation into the war.

Because so many fellow physicists, such as Wigner, Szilárd, Oppenheimer, and Teller, had disappeared to obscure towns, Einstein was able to surmise that the bomb-making work he had recommended was now proceeding with greater urgency. But he was not asked to join the Manhattan Project, nor was he officially told about it.

There were many reasons he was not secretly summoned to places like Los Alamos or Oak Ridge. He was not a nuclear physicist or a practicing expert in the scientific issues at hand. He was, as noted, considered by some a security risk. And even though he had put aside his pacifist sentiments, he never expressed any desire or made any requests to enlist in the endeavor.

He was, however, offered a bit part that December. Vannevar Bush, the director of the Office of Scientific Research and Development, which oversaw the Manhattan Project, contacted Einstein through the man who had succeeded Flexner as the head of the Institute for Advanced Study in Princeton, Frank Aydelotte, and asked for his help on a problem involving the separation of isotopes that shared chemical traits. Einstein was happy to comply. Drawing on his old expertise in osmosis and diffusion, he worked on a process of gaseous diffusion in which uranium was converted into a gas and forced through filters. To preserve secrecy, he was not even allowed to have Helen Dukas or anyone else type up his work, so he sent it back in his careful handwriting.

"Einstein was very much interested in your problem, has worked on it for a couple of days and produced the solution, which I enclose," Aydelotte wrote Bush. "Einstein asks me to say that if there are other angles of the problem that you want him to develop or if you wish any parts of this amplified, you need only let him know and he will do anything in his power. I very much hope that you will make use of him in any way that occurs to you, because I know how deep is his satisfaction at doing anything which might be useful in the national effort." As an afterthought, Aydelotte added, "I hope you can read his handwriting."[27]

The scientists who received Einstein's paper were impressed, and they discussed it with Vannevar Bush. But in order for Einstein to be more useful, they said, he should be given more information about how the isotope separation fit in with other parts of the bomb-making challenge.

Bush refused. He knew that Einstein would have trouble getting a security clearance. "I do not feel that I ought to take him into confidence on the subject to the extent of showing just where this thing fits into the defense picture," Bush wrote Aydelotte. "I wish very much that I could place the whole thing before him and take him fully into confidence, but this is utterly impossible in view of the attitude of people here in Washington who have studied his whole history."[28]

Later, during the war, Einstein helped with less secret matters. A Navy lieutenant came to visit him at the Institute to enlist him in analyzing ordnance capabilities. He was enthusiastic. As Aydelotte noted, he had felt neglected since his brief flurry of work on uranium isotopes. Among the issues Einstein explored, as part of a $25-per-day consulting arrangement, were ways to shape the placement of sea mines in Japanese harbors, and his friend the physicist George Gamow got to come pick his brain on a variety of topics. "I am in the Navy, but not required to get a Navy haircut," Einstein joked to colleagues, who probably had trouble picturing him with a crew cut.[29]

Einstein also helped the war effort by donating a manuscript of his special relativity paper to be auctioned off for a War Bond drive. It was not the original version; he had thrown that away back when it was published in 1905, not knowing it would ever be worth millions. To re-create the manuscript, he had Helen Dukas read the paper to him aloud as he copied down the words. "Did I really say it that way?" he griped at one point. When Dukas assured him that he had, Einstein lamented, "I could have put it much more simply." When he heard that the manuscript, along with one other, had sold for $11.5 million, he declared that "economists will have to revise their theories of value."[30]

Atomic Fears

The physicist Otto Stern, who had been one of Einstein's friends since their days together in Prague, had been secretly working on the Manhattan Project, mainly in Chicago, and had a good sense by the end of 1944 that it would be successful. That December, he made a visit to Princeton. What Einstein heard upset him. Whether or not the bomb was used in the war, it would change the nature of both war and peace forever. The policymakers weren't thinking about that, he and Stern agreed, and they must be encouraged to do so before it was too late.

So Einstein decided to write to Niels Bohr. They had sparred over quantum mechanics, but Einstein trusted his judgment on more earthly issues. Einstein was one of the few people to know that Bohr, who was half Jewish, was secretly in the United States. When the Nazis overran Denmark, he had made a daring escape by sailing with his son in a small boat to Sweden. From there he had been flown to Britain, given a fake passport with the name Nicholas Baker, then sent to America to join the Manhattan Project at Los Alamos.

Einstein wrote to Bohr, using his real name, in care of Denmark's embassy in Washington, and somehow the letter got to him. In it Einstein described his worrisome talk with Stern about the dearth of thinking about how to control atomic weapons in the future. "The politicians do not appreciate the possibilities and consequently do not know the extent of the menace," Einstein wrote. Once again, he made his argument that it would take an empowered world government to prevent an arms race once the age of atomic weaponry arrived. "Scientists who know how to get a hearing with political leaders," Einstein urged, "should bring pressure on the political leaders in their countries in order to bring about an internationalization of military power."[31]

Thus began what would be the political mission that would dominate the remaining decade of Einstein's life. Since his days as a teenager in Germany, he had been repulsed by nationalism, and he had long argued that the best way to prevent wars was to create a world authority that had the right to resolve disputes and the military power to impose its resolutions. Now, with the impending advent of a weapon so awesome that it could transform both war and peace, Einstein viewed this approach as no longer an ideal but a necessity.

Bohr was unnerved by Einstein's letter, but not for the reason Einstein would have hoped. The Dane shared his desire for the internationalization of atomic weaponry, and he had advocated that approach in meetings with Churchill, and then with Roosevelt, earlier in the year. But instead of persuading them, he had prompted the two leaders to issue a joint order to their intelligence agencies saying that "enquiries should be made regarding the activities of Professor Bohr and steps taken to ensure that he is responsible for no leakage of information, particularly to the Russians."[32]

So upon receiving Einstein's letter, Bohr hurried to Princeton. He wanted to protect his friend by warning him to be circumspect, and he also hoped to repair his own reputation by reporting to government officials on what Einstein said.

During their private talk at the Mercer Street house, Bohr told Einstein that there would be "the most deplorable consequences" if anyone who knew about the development of the bomb shared that information. Responsible statesmen in Washington and London, Bohr assured him, were aware of the threat caused by the bomb as well as

"the unique opportunity for furthering a harmonious relationship between nations."

Einstein was persuaded. He promised that he would refrain from
sharing any information he had surmised and would urge his friends
not do anything to complicate American or British foreign policy. And
he immediately set out to make good on his word by writing a letter to
Stern that was, for Einstein, remarkable in its circumspection. "I have
the impression that one must strive seriously to be responsible, that one
does best not to speak about the matter for the time being, and that it
would in no way help, at the present moment, to bring it to public notice," he said. He was careful not to reveal anything, even that he had
met with Bohr. "It is difficult for me to speak in such a nebulous way,
but for the moment I cannot do anything else."[33]

Einstein's only intervention before the end of the war was
prompted again by Szilárd, who came to visit in March 1945 and expressed anxiety about how the bomb might be used. It was clear that
Germany, now weeks away from defeat, was not making a bomb. So
why should the Americans rush to complete one? And shouldn't policymakers think twice about using it against Japan when it might not be
needed to secure victory?

Einstein agreed to write another letter to President Roosevelt urging him to meet with Szilárd and other concerned scientists, but he
went out of his way to feign ignorance. "I do not know the substance of
the considerations and recommendations which Dr. Szilárd proposes
to submit to you," Einstein wrote. "The terms of secrecy under which
Dr. Szilárd is working at present do not permit him to give me information about his work; however, I understand that he now is greatly
concerned about the lack of adequate contact between scientists who
are doing this work and those members of your Cabinet who are responsible for formulating policy."[34]

Roosevelt never read the letter. It was found in his office after he
died on April 12 and was passed on to Harry Truman, who in turn gave
it to his designated secretary of state, James Byrnes. The result was a
meeting between Szilárd and Byrnes in South Carolina, but Byrnes
was neither moved nor impressed.

The atom bomb was dropped, with little high-level debate, on August 6, 1945, on the city of Hiroshima. Einstein was at the cottage he rented that summer on Saranac Lake in the Adirondacks, taking an afternoon nap. Helen Dukas informed him when he came down for tea. "Oh, my God," is all he said.[35]

Three days later, the bomb was used again, this time on Nagasaki. The following day, officials in Washington released a long history, compiled by Princeton physics professor Henry DeWolf Smyth, of the secret endeavor to build the weapon. The Smyth report, much to Einstein's lasting discomfort, assigned great historic weight for the launch of the project to the 1939 letter he had written to Roosevelt.

Between the influence imputed to that letter and the underlying relationship between energy and mass that he had formulated forty years earlier, Einstein became associated in the popular imagination with the making of the atom bomb, even though his involvement was marginal. *Time* put him on its cover, with a portrait showing a mushroom cloud erupting behind him with $E=mc^2$ emblazoned on it. In a story that was overseen by an editor named Whittaker Chambers, the magazine noted with its typical prose flair from the period:

> Through the incomparable blast and flame that will follow, there will be dimly discernible, to those who are interested in cause & effect in history, the features of a shy, almost saintly, childlike little man with the soft brown eyes, the drooping facial lines of a world-weary hound, and hair like an aurora borealis . . . Albert Einstein did not work directly on the atom bomb. But Einstein was the father of the bomb in two important ways: 1) it was his initiative which started U.S. bomb research; 2) it was his equation ($E = mc^2$) which made the atomic bomb theoretically possible.[36]

It was a perception that plagued him. When *Newsweek* did a cover on him, with the headline "The Man Who Started It All," Einstein offered a memorable lament. "Had I known that the Germans would not succeed in producing an atomic bomb," he said, "I never would have lifted a finger."[37]

Of course, neither he nor Szilárd nor any of their friends involved with the bomb-building effort, many of them refugees from Hitler's horrors, could know that the brilliant scientists they had left behind in Berlin, such as Heisenberg, would fail to unlock the secrets. "Perhaps I can be forgiven," Einstein said a few months before his death in a conversation with Linus Pauling, "because we all felt that there was a high probability that the Germans were working on this problem and they might succeed and use the atomic bomb and become the master race."[38]

ONE-WORLDER

1945–1948

Portrait by Philippe Halsman, 1947

Arms Control

For a few weeks after the dropping of the atom bomb, Einstein was uncharacteristically reticent. He fended off reporters who were knocking at his door in Saranac Lake, and he even declined to give a quote to his summer neighbor Arthur Hays Sulzberger, publisher of the *New York Times,* when he called.[1]

It was only as he was about to leave his summer rental in mid-September, more than a month after the bombs had been dropped, that Einstein agreed to discuss the issue with a wire service reporter who came calling. The point he stressed was that the bomb reinforced his longtime support for world federalism. "The only salvation for civilization and the human race lies in the creation of world government," he said. "As long as sovereign states continue to have armaments and armaments secrets, new world wars will be inevitable."[2]

As in science, so it was in world politics for Einstein: he sought a unified set of principles that could create order out of anarchy. A system based on sovereign nations with their own military forces, competing ideologies, and conflicting national interests would inevitably produce more wars. So he regarded a world authority as realistic rather than idealistic, as practical rather than naïve.

He had been circumspect during the war years. He was a refugee in a nation that was using its military might for noble rather than nationalistic goals. But the end of the war changed things. So did the dropping of the atom bombs. The increase in the destructive power of offensive weaponry led to a commensurate increase in the need to find a world structure for security. It was time for him to become politically outspoken again.

For the remaining ten years of his life, his passion for advocating a unified governing structure for the globe would rival that for finding a unified field theory that could govern all the forces of nature. Although distinct in most ways, both quests reflected his instincts for transcendent order. In addition, both would display Einstein's willingness to be a nonconformist, to be serenely secure in challenging prevailing attitudes.

The month after the bombs were dropped, a group of scientists signed a statement urging that a council of nations be created to control atomic weaponry. Einstein responded with a letter to J. Robert Oppenheimer, who had so successfully led the scientific efforts at Los Alamos. He was pleased with the sentiments behind the statement, Einstein said, but he criticized the political recommendations as "obviously inadequate" because they retained sovereign nations as the ultimate powers. "It is unthinkable that we can have peace without a real governmental organization to create and enforce law on individuals in their international relations."

Oppenheimer politely pointed out that "the statements you attributed to me are not mine." They had been written by another group of scientists. He did, nevertheless, challenge Einstein's argument for a full-fledged world government: "The history of this nation up through the Civil War shows how difficult the establishment of a federal authority can be when there are profound differences in the values of the societies it attempts to integrate."³ Oppenheimer thus became the first of many postwar realists to disparage Einstein for being allegedly too idealistic. Of course, one could flip his argument by noting that the Civil War showed in gruesome terms the danger of *not* having a secure federal authority instead of state military sovereignty when there are differences of values among member states.

What Einstein envisioned was a world "government" or "authority" that had a monopoly on military power. He called it a "supranational" entity, rather than an "international" one, because it would exist *above* its member nations rather than as a mediator among sovereign nations.⁴ The United Nations, which was founded in October 1945, did not come close to meeting these criteria, Einstein felt.

Over the next few months, Einstein fleshed out his proposals in a series of essays and interviews. The most important arose from an exchange of fan letters he had with Raymond Gram Swing, a commentator on ABC radio. Einstein invited Swing to visit him in Princeton, and the result was an article by Einstein, as told to Swing, in the November 1945 issue of the *Atlantic* called "Atomic War or Peace."⁵

The three great powers—the United States, Britain, and Russia—should jointly establish the new world government, Einstein said in the article, and then invite other nations to join. Using a somewhat misleading phrase that was part of the popular debate of the time, he said that "the secret of the bomb" should be given to this new organization by Washington.⁶ The only truly effective way to control atomic arms, he believed, was by ceding the monopoly on military power to a world government.

By then, in late 1945, the cold war was under way. America and Britain had begun to clash with Russia for imposing communist regimes in Poland and other eastern European areas occupied by the Red Army. For its part, Russia zealously sought a security perimeter and was neuralgic about any perceived attempt to interfere in its domestic affairs, which made its leaders resist surrendering any sovereignty to a world authority.

So Einstein sought to make it clear that the world government he envisioned would not try to impose a Western-style liberal democracy everywhere. He advocated a world legislature that would be elected directly by the people of each member country, in secret ballot, rather than appointed by the nation's rulers. However, "it should not be necessary to change the internal structure of the three great powers," he added as a reassurance to Russia. "Membership in a supranational security system should not be based on any arbitrary democratic standards."

One issue that Einstein could not resolve neatly was what right this world government would have to intervene in the internal affairs of nations. It must be able "to interfere in countries where a minority is oppressing a majority," he said, citing Spain as an example. Yet that caused him contortions about whether this standard applied to Russia. "One must bear in mind that the people in Russia have not had a long tradition of political education," he rationalized. "Changes to improve conditions in Russia had to be effected by a minority because there was no majority capable of doing so."

Einstein's efforts to prevent future wars were motivated not only by his old pacifist instincts but also, he admitted, by his guilty feelings about the role he had played in encouraging the atom bomb project. At a Manhattan dinner given by the Nobel Prize committee in December, he noted that Alfred Nobel, the inventor of dynamite, had created the award "to atone for having invented the most powerful explosives ever known up to his time." He was in a similar situation. "Today, the physicists who participated in forging the most formidable and dangerous weapon of all times are harassed by an equal feeling of responsibility, not to say guilt," he said.[7]

These sentiments prompted Einstein, in May 1946, to take on the most prominent public policy role in his career. He became chairman of the newly formed Emergency Committee of Atomic Scientists, which was dedicated to nuclear arms control and world government. "The unleashed power of the atom has changed everything save our modes of thinking," Einstein wrote in a fund-raising telegram that month, "and thus we drift towards unparalleled catastrophe."[8]

Leó Szilárd served as the executive director and did most of the organizational work. But Einstein, who served until the end of 1948, gave speeches, chaired meetings, and took his role seriously. "Our generation has brought into the world the most revolutionary force since prehistoric man's discovery of fire," he said. "This basic power of the universe cannot be fitted into the outmoded concept of narrow nationalisms."[9]

The Truman administration proposed a variety of plans for the international control of atomic power, but none were able, intentionally or not, to win the support of Moscow. As a result, the battle over the best approach quickly created a political divide.

On one side were those who celebrated the success of America and Britain in winning the race to develop such weapons. They saw the bomb as a guarantor of the freedoms of the West, and they wanted to guard what they saw as "the secret." On the other side were arms control advocates like Einstein. "The secret of the atomic bomb is to America what the Maginot Line was to France before 1939," he told *Newsweek*. "It gives us imaginary security, and in this respect it is a great danger."[10]

Einstein and his friends realized that the battle for public sentiment needed to be fought not only in Washington but also in the realm of popular culture. This led to an amusing—and historically illustrative—tangle in 1946 pitting them against Louis B. Mayer and a coterie of earnest Hollywood moviemakers.

It began when a Metro-Goldwyn-Mayer scriptwriter named Sam Marx asked if he could come to Princeton to get Einstein's cooperation on a docudrama about the making of the bomb. Einstein sent back word that he had no desire to help. A few weeks later Einstein got an anxious letter from an official with the Association of Manhattan Project Scientists saying that the movie seemed to be taking a very pro-military slant, celebrating the creation of the bomb and the security it gave to America. "I know that you will not want to lend your name to a picture which misrepresents the military and political implications of the bomb," the letter said. "I hope that you will see fit to make the use of your name conditional on your personal approval of the script."[11]

The following week Szilárd came to see Einstein about the issue, and soon a bevy of peace-loving physicists was bombarding him with concerns. So Einstein read the script and agreed to join the campaign to stop the movie. "The presentation of facts was so utterly misleading that I declined any cooperation or permission of the use of my name," he said.

He also sent a spiky letter to the famed mogul that attacked the proposed movie and also, for good measure, the tone of previous ones that Mayer had made. "Although I am not much of a moviegoer, I do know from the tenor of earlier films that have come out of your studio that you will understand my reasons," he wrote. "I find that the whole film is written too much from the point of view of the Army and the Army leader of the project, whose influence was not always in the direction which one would desire from the point of view of humanity."[12]

Mayer turned Einstein's letter over to the film's chief editor, who responded with a memo that Mayer sent back to Einstein. President Truman, it said, "was most anxious to have the picture made" and had personally read and approved the script, an argument not likely to reassure Einstein. "As American citizens we are bound to respect the viewpoint of our government." That, too, was not the best argument to use on Einstein. There followed an even less persuasive argument: "It must be realized that dramatic truth is just as compelling a requirement to us as veritable truth is to a scientist."

The memo concluded by promising that the moral issues raised by the scientists would be given a proper airing through the character of a fictional young scientist played by an actor named Tom Drake. "We selected among our young male players the one who best typifies earnestness and a spiritual quality," it said reassuringly. "You need only recall his performance in 'The Green Years.' "[13]

Not surprisingly, this did not turn Einstein around. When Sam Marx, the scriptwriter, wrote beseeching him to change his mind and allow himself to be portrayed, Einstein replied curtly: "I have explained my point of view in a letter to Mr. Louis Mayer." Marx was persistent. "When the picture is complete," he wrote back, "the audience will feel in greatest sympathy with the young scientist." And from later the same day: "Here is a new and revised script."[14]

The ending was not that hard to predict. The new script was more pleasing to the scientists, and they were not immune to the lure of being glorified on the big screen. Szilárd sent Einstein a telegram saying, "Have received new script from MGM and am writing that I have no objection to use of my name in it." Einstein relented. "Agree with use of my name on basis of the new script," he scribbled in English on the back of the telegram. The only change he requested was in the scene of Szilárd's 1939 visit to him on Long Island. The script said that he had not met Roosevelt before then, but he had.[15]

The Beginning or the End, which was the name of the movie, opened to good reviews in February 1947. "A sober, intelligent account of the development and deployment of the Atom Bomb," Bosley Crowther declared in the *New York Times,* "refreshingly free of propagandizing." Einstein was played by a character actor named Ludwig Stossel, who had a small part in *Casablanca* as a German Jew trying to get to America and would later have a flicker of fame in Swiss Colony wine commercials in the 1960s in which he spoke the tagline "That little old winemaker, me."[16]

Einstein's efforts on behalf of arms control and his advocacy of world government in the late 1940s got him tagged as woolly-headed and naïve. Woolly-headed he may have been, at least in appearance, but was it right to dismiss him as naïve?

Most Truman administration officials, even those working on behalf of arms control, thought so. William Golden was an example. An Atomic Energy Commission staffer who was preparing a report for Secretary of State George Marshall, he went to Princeton to consult with Einstein. Washington needed to try harder to enlist Moscow in an arms control plan, Einstein argued. Golden felt he was speaking "with almost childlike hope for salvation and without appearing to have thought through the details of his solution." He reported back to Marshall, "It was surprising, though perhaps it should not have been, that, out of his métier of mathematics, he seemed naïve in the field of international politics. The man who popularized the concept of a fourth dimension could think in only two of them in considerations of World Government."[17]

To the extent that Einstein was naïve, it was not because he had a benign view of human nature. Having lived in Germany in the first half of the twentieth century, there was little chance of that. When the famed photographer Philippe Halsman, who had escaped the Nazis with Einstein's help, asked whether he thought there would ever be lasting peace, Einstein answered, "No, as long as there will be man, there will be war." At that moment Halsman clicked his shutter and captured Einstein's sadly knowing eyes for what became a famous portrait (reproduced on page 487).[18]

Einstein's advocacy of an empowered world authority was based not on gooey sentiments but on this hardnosed assessment of human nature. "If the idea of world government is not realistic," he said in 1948, "then there is only one realistic view of our future: wholesale destruction of man by man."[19]

Like some of his scientific breakthroughs, Einstein's approach involved abandoning entrenched suppositions that others considered verities. National sovereignty and military autonomy had been an underpinning of the world order for centuries, just as absolute time and absolute space had been the underpinning of the cosmic order. To advocate transcending that approach was a radical idea, the product of a nonconformist thinker. But like many of Einstein's ideas that at first seemed so radical, it may have looked less so had it come to be accepted.

The world federalism that Einstein—and indeed many sober and established political leaders—advocated during the early years of America's atomic monopoly was not unthinkable. To the extent that he was naïve, it was because he put forth his idea in a simple fashion and did not consider complex compromises. Physicists are not used to trimming or compromising their equations in order to get them accepted. Which is why they do not make good politicians.

At the end of the 1940s, when it was becoming clear to him that the effort to control nuclear weaponry would fail, Einstein was asked what the next war would look like. "I do not know how the Third World War will be fought," he answered, "but I can tell you what they will use in the Fourth—rocks."[20]

Russia

Those who wanted international control of the bomb had one big issue to confront: how to deal with Russia. A growing number of

Americans, along with their elected leaders, came to view Moscow's communists as dangerously expansionist and deceitful. The Russians, for their part, did not seem all that eager for arms control or world governance either. They had deeply ingrained fears about their security, a desire for a bomb of their own, and leaders who recoiled at any hint of outside meddling in their nation's internal affairs.

There was a typical nonconformity in Einstein's attitudes toward Russia. He did not swing as far as many others did toward glorifying the Russians when they became allies during the war, nor did he swing as far toward demonizing them when the cold war began. But by the late 1940s, this put him increasingly outside mainstream American sentiments.

He disliked communist authoritarianism, but he did not see it as an imminent danger to American liberty. The greater danger, he felt, was rising hysteria about the supposed Red menace. When Norman Cousins, editor of the *Saturday Review* and the journalistic patron of America's internationalist intelligentsia, wrote a piece calling for international arms control, Einstein responded with a fan letter but added a caveat. "What I object to in your article is that you not only fail to oppose the widespread hysterical fear in our country of Russian aggression but actually encourage it," he said. "All of us should ask ourselves which of the two countries is objectively more justified in fearing the aggressive intentions of the other." [21]

As for the repression inside Russia, Einstein tended to offer only mild condemnations diluted by excuses. "It is undeniable that a policy of severe coercion exists in the political sphere," he said in one talk. "This may, in part, be due to the need to break the power of the former ruling class and to convert a politically inexperienced, culturally backward people into a nation well organized for productive work. I do not presume to pass judgment in these difficult matters." [22]

Einstein consequently became the target of critics who saw him as a Soviet sympathizer. Mississippi Congressman John Rankin said that Einstein's world government plan was "simply carrying out the Communist line." Speaking on the House floor, Rankin also denounced Einstein's science: "Ever since he published his book on relativity to try to convince the world that light had weight, he has capitalized on his reputation as a scientist . . . and has been engaged in communistic activities."[23]

Einstein continued his long-running exchanges on Russia with Sidney Hook, the social philosopher who had once been a communist and then become strongly anticommunist. These were not as exalted as his exchanges with Bohr, on either side, but they got as intense. "I am not blind to the serious weakness of the Russian system of government," Einstein replied to one of Hook's missives. "But it has, on the other side, great merits and it is difficult to decide whether it would have been possible for the Russians to survive by following softer methods."[24]

Hook took it upon himself to convince Einstein of the error of his ways and sent him long and rather frequent letters, most of which Einstein ignored. On the occasions he did answer, Einstein generally agreed that Russia's oppression was wrong, but he tended to balance such judgments by adding that it was also somewhat understandable. As he juggled it in one 1950 response:

> I do not approve of the interference by the Soviet government in intellectual and artistic matters. Such interference seems to me objectionable, harmful, and even ridiculous. Regarding the centralization of political power and the limitations of the freedom of action for the individual, I think that these restrictions should not exceed the limit demanded by security, stability, and the necessities resulting from a planned economy. An outsider is hardly able to judge the facts and possibilities. In any case it cannot be doubted that the achievements of the Soviet regime are considerable in the fields of education, public health, social welfare, and economics, and that the people as a whole have greatly gained by these achievements.[25]

Despite these qualified excuses for some of Moscow's behavior, Einstein was not the Soviet supporter that some tried to paint him. He had always rejected invitations to Moscow and rebuffed attempts by friends on the left to embrace him as a comrade. He denounced Moscow's repeated use of the veto at the United Nations and its resistance to the idea of world government, and he became even more critical when the Soviets made it clear that they had no appetite for arms control.

This was evident when an official group of Russian scientists attacked Einstein in a 1947 Moscow newspaper article, "Dr. Einstein's Mistaken Notions." His vision for a world government, they declared, was a plot by capitalists. "The proponents of a world super-state are asking us voluntarily to surrender independence for the sake of world government, which is nothing but a flamboyant signboard for the supremacy of capitalist monopolies," they wrote. They denounced Einstein for recommending a directly elected supranational parliament. "He has gone so far as to declare that if the Soviet Union refuses to join this new-fangled organization, other countries would have every right to go ahead without it. Einstein is supporting a political fad which plays into the hands of the sworn enemies of sincere international cooperation and enduring peace."[26]

Soviet sympathizers at the time were willing to follow almost any party line that Moscow dictated. Such conformity was not in Einstein's nature. When he disagreed with someone, he merrily said so. He was happy to take on the Russian scientists.

Although he reiterated his support for democratic socialist ideals, he rebutted the Russians' faith in communist dogma. "We should not make the mistake of blaming capitalism for all existing social and political evils, nor of assuming that the very establishment of socialism would be sufficient to cure the social and political ills of humanity," he wrote. Such thinking led to the "fanatical intolerance" that infected the Communist Party faithful, and it opened the way to tyranny.

Despite his criticisms of untrammeled capitalism, what repelled him more—and had repelled him his entire life—was repression of free thought and individuality. "Any government is evil if it carries within it the tendency to deteriorate into tyranny," he warned the Russian scientists. "The danger of such deterioration is more acute in a country in which the government has authority not only over the armed forces but also over every channel of education and information as well as over the existence of every single citizen."[27]

Just as his dispute with the Russian scientists was breaking, Einstein was working with Raymond Gram Swing to update the article in the *Atlantic* that they had done two years earlier. This time Einstein attacked Russia's rulers. Their reasons for not supporting a world government, he said, "quite obviously are pretexts." Their real fear was that their repressive communist command system might not survive in such an environment. "The Russians may be partly right about the difficulty of retaining their present social structure in a supranational regime, though in time they may be brought to see that this is a far lesser loss than remaining isolated from a world of law."[28]

The West should proceed with creating a world government without Russia, he said. They would eventually come around, he thought: "I believe that if this were done intelligently (rather than in clumsy Truman style!) Russia would cooperate once she realized that she was no longer able to prevent world government anyhow."[29]

From then on, Einstein seemed to take a perverse pride in disputing those who blamed the Russians for everything, and those who blamed them for nothing. When a left-leaning pacifist he knew sent him a book he had written on arms control, expecting Einstein's endorsement, he got instead a rebuff. "You have presented the whole problem as an advocate of the Soviet point of view," Einstein wrote, "but you have kept silent about everything which is not favorable for the Soviets (and this is not little)."[30]

Even his longtime pacifism developed a hard, realistic edge when it came to dealing with Russia, just as it had after the Nazis rose to power in Germany. Pacifists liked to think that Einstein's break with their philosophy in the 1930s was an aberration caused by the unique threat posed by the Nazis, and some biographers likewise treat it as a temporary anomaly.[31] But that minimizes the shift in Einstein's thinking. He was never again a pure pacifist.

When he was asked, for example, to join a campaign to persuade American scientists to refuse to work on atomic weapons, he not only declined but berated the organizers for advocating unilateral disarmament. "Disarmament cannot be effective unless all countries participate," he lectured. "If even one nation continues to arm, openly or secretly, the disarmament of the others will involve disastrous consequences."

Pacifists like himself had made a mistake in the 1920s by encouraging Germany's neighbors not to rearm, he explained. "This merely served to encourage the arrogance of the Germans." There were parallels now with Russia. "Similarly, your proposition would, if effective, surely lead to a serious weakening of the democracies," he wrote those pushing the antimilitary petition. "For we must realize that we are probably not able to exert any significant influence on the attitude of our Russian colleagues."[32]

He took a similar stance when his former colleagues in the War Resisters' League asked him to rejoin in 1948. They flattered him by quoting one of his old pacifist proclamations, but Einstein rebuffed them. "That statement accurately expresses the views I held on war resistance in the period from 1918 to the early thirties," he replied. "Now, however, I feel that policy, which involves the refusal of individuals to participate in military activities, is too primitive."

Simplistic pacifism could be dangerous, he warned, especially given the internal policies and external attitude of Russia. "The war resistance movement actually serves to weaken the nations with a more liberal type of government and, indirectly, to support the policies of the existing tyrannical governments," he argued. "Antimilitaristic activities, through refusal of military service, are wise only if they are feasible everywhere throughout the world. Individual antimilitarism is impossible in Russia."[33]

Some pacifists argued that world socialism, rather than world government, would be the best foundation for lasting peace. Einstein disagreed. "You say that socialism by its very nature rejects the remedy of war," Einstein replied to one such advocate. "I do not believe that. I can easily imagine that two socialist states might fight a war against each other."[34]

One of the early flashpoints of the cold war was Poland, where the occupying Red Army had installed a pro-Soviet regime without the open elections that Moscow had promised. When that new Polish government invited Einstein to a conference, they got a taste of his independence from party dogma. He politely explained that he no longer traveled overseas, and he sent a careful message that offered encouragement but also stressed his call for a world government.

The Poles decided to delete the parts about world government, which Moscow opposed. Einstein was furious, and he released his undelivered full message to the *New York Times*. "Mankind can gain protection against the danger of unimaginable destruction and wanton annihilation only if a supranational organization has alone the authority to produce or possess these weapons," it said. He also complained to the British pacifist who presided over the meeting that the communists were trying to enforce conformity to a party line: "I am convinced that our colleagues on the other side of the fence are completely unable to express their real opinions."[35]

The FBI Files

He had criticized the Soviet Union, refused to visit there, and opposed the sharing of atomic secrets unless a world government could be created. He had never worked on the bomb-making project and knew no classified information about its technology. Nevertheless, Einstein was unwittingly caught up in a chain of events that showed how suspicious, intrusive, and inept the FBI could be back then when pursuing the specter of Soviet communism.

The Red Scares and investigations into communist subversion originally had some legitimate justifications, but eventually they included bumbling inquisitions that resembled witch hunts. They began in earnest at the start of 1950, after America was stunned by news that the Soviets had developed their own bomb. During the first few weeks of that year, President Truman launched a program to build a hydrogen bomb, a refugee German physicist working in Los Alamos named Klaus Fuchs was arrested as a Soviet spy, and Senator Joseph McCarthy gave his famous speech, claiming that he had a list of card-carrying communists in the State Department.

As the head of the Emergency Committee of Atomic Scientists, Einstein had dismayed Edward Teller by not supporting the building of the hydrogen bomb. But Einstein also had not opposed it outright. When A. J. Muste, a prominent pacifist and socialist activist, asked him to join an appeal to delay construction of the new weapon, Einstein declined. "Your new proposal seems quite impractical to me," he said. "As long as competitive armament prevails, it will not be possible to halt the process in one country."[36] It was more sensible, he felt, to push for a global solution that included a world government.

The day after Einstein wrote that letter, Truman made his announcement of a full-scale effort to produce the H-bomb. From his Princeton home, Einstein taped a three-minute appearance for the premiere of a Sunday evening NBC show called *Today with Mrs. Roosevelt*. The former first lady had become a voice of progressivism after the death of her husband. "Each step appears as the inevitable consequence of the one that went before," he said of the arms race. "And at the end, looming ever clearer, lies general annihilation." The headline in the *New York Post* the next day was, "Einstein Warns World: Outlaw H-Bomb or Perish."[37]

Einstein made another point in his televised talk. He expressed his growing concern over the U.S. government's increased security measures and willingness to compromise the liberties of its citizens. "The loyalty of citizens, particularly civil servants, is carefully supervised by a police force growing more powerful every day," he warned. "People of independent thought are harassed."

As if to prove him right, J. Edgar Hoover, who hated communists and Eleanor Roosevelt with almost equal passion, the very next day called in the FBI's chief of domestic intelligence and ordered a report on Einstein's loyalty and possible communist connections.

The resulting fifteen-page document, produced two days later, listed thirty-four organizations, some purportedly communist fronts, that Einstein had been affiliated with or lent his name to, including the Emergency Committee of Atomic Scientists. "He is principally a pacifist and could be considered a liberal thinker," the memo concluded somewhat benignly, and it did not charge him with being either a communist or someone who gave information to subversives.[38]

Indeed, there was nothing that linked Einstein to any security threat. A reading of the dossier, however, makes the FBI agents look like Keystone Kops. They bumbled around, unable to answer questions such as whether Elsa Einstein was his first wife, whether Helen Dukas was a Soviet spy while in Germany, and whether Einstein had been responsible for bringing Klaus Fuchs into the United States. (In all three cases, the correct answer was no.)

The agents also tried to pin down a tip that Elsa had told a friend in California that they had a son by the name of Albert Einstein Jr. who was being held in Russia. In fact, Hans Albert Einstein was by then an engineering professor at Berkeley. Neither he nor Eduard, still in a Swiss sanatorium, had ever been to Russia. (If there was any basis to the rumor, it was that Elsa's daughter Margot had married a Russian, who returned there after they divorced, though the FBI never found that out.)

The FBI had been gathering rumors about Einstein ever since the 1932 screed from Mrs. Frothingham and her women patriots. Now it began systematically keeping track of that material in one growing dossier. It included such tips as one from a Berlin woman who sent him a mathematical scheme for winning the Berlin lottery and had concluded he was a communist when he did not respond to her.[39] By the time he died, the Bureau would amass 1,427 pages stored in fourteen boxes, all stamped *Confidential* but containing nothing incriminating.[40]

What is most notable, in retrospect, about Einstein's FBI file is not all the odd tips it contained, but the one relevant piece of information that was completely missing. Einstein did in fact consort with a Soviet spy, unwittingly. But the FBI remained clueless about it.

The spy was Margarita Konenkova, who lived in Greenwich Village with her husband, the Russian realist sculptor Sergei Konenkov, mentioned earlier. A former lawyer who spoke five languages and had an engaging way with men, so to speak, her job as a Russian secret agent was to try to influence American scientists. She had been introduced to Einstein by Margot, and she became a frequent visitor to Princeton during the war.

Out of duty or desire, she embarked on an affair with the widowed Einstein. One weekend during the summer of 1941, she and some friends invited him to a cottage on Long Island, and to everyone's surprise he accepted. They packed a lunch of boiled chicken, took the train from Penn Station, and spent a pleasant weekend during which Einstein sailed on the Sound and scribbled equations on the porch. At one point they went to a secluded beach to watch the sunset and almost got arrested by a local policeman who had no idea who Einstein was. "Can't you read," the officer said, pointing to a no-trespassing sign. He and Konenkova remained lovers until she returned to Moscow in 1945 at age 51.[41]

She succeeded in introducing him to the Soviet vice consul in New York, who was also a spy. But Einstein had no secrets to share, nor is there any evidence that he had any inclination at all to help the Soviets in any way, and he rebuffed her attempts to get him to visit Moscow.

The affair and potential security issue came to light not because of any FBI sleuthing but because a collection of nine amorous letters written by Einstein to Konenkova in the 1940s became public in 1998. In addition, a former Soviet spy, Pavel Sudoplatov, published a rather explosive but not totally reliable memoir in which he revealed that she was an agent code-named "Lukas."[42]

Einstein's letters to Konenkova were written the year after she left America. Neither she nor Sudoplatov, nor anyone else, ever claimed that Einstein passed along any secrets, wittingly or unwittingly. However, the letters do make clear that, at age 66, he was still able to be amorous in prose and probably in person. "I recently washed my hair myself, but not with great success," he said in one. "I am not as careful as you are."

Even with his Russian lover, however, Einstein made clear that he was not an unalloyed lover of Russia. In one letter he denigrated Moscow's militaristic May Day celebration, saying, "I watch these exaggerated patriotic exhibits with concern."[43] Any expressions of excess nationalism and militarism had always made him uncomfortable, ever since he had watched German soldiers march by when he was a boy, and Russia's were no different.

Einstein's Politics

Despite Hoover's suspicions, Einstein was a solid American citizen, and he considered his opposition to the wave of security and loyalty investigations to be a defense of the nation's true values. Tolerance of free expression and independence of thought, he repeatedly argued, were the core values that Americans, to his delight, most cherished.

His first two presidential votes had been cast for Franklin Roosevelt, whom he publicly and enthusiastically endorsed. In 1948, dismayed by Harry Truman's cold war policies, Einstein voted for the
Progressive Party candidate Henry Wallace, who advocated greater cooperation with Russia and increased social welfare spending.

Throughout his life, Einstein was consistent in the fundamental
premises of his politics. Ever since his student days in Switzerland, he
had supported socialist economic policies tempered by a strong instinct for individual freedom, personal autonomy, democratic institutions, and protection of liberties. He befriended many of the
democratic socialist leaders in Britain and America, such as Bertrand
Russell and Norman Thomas, and in 1949 he wrote an influential essay
for the inaugural issue of the *Monthly Review* titled "Why Socialism?"

In it he argued that unrestrained capitalism produced great disparities of wealth, cycles of boom and depression, and festering levels of
unemployment. The system encouraged selfishness instead of cooperation, and acquiring wealth rather than serving others. People were educated for careers rather than for a love of work and creativity. And
political parties became corrupted by political contributions from
owners of great capital.

These problems could be avoided, Einstein argued in his article,
through a socialist economy, if it guarded against tyranny and centralization of power. "A planned economy, which adjusts production to the
needs of the community, would distribute the work to be done among
all those able to work and would guarantee a livelihood to every man,
woman, and child," he wrote. "The education of the individual, in addition to promoting his own innate abilities, would attempt to develop
in him a sense of responsibility for his fellow-men in place of the glorification of power and success in our present society."

He added, however, that planned economies faced the danger of
becoming oppressive, bureaucratic, and tyrannical, as had happened in
communist countries such as Russia. "A planned economy may be accompanied by the complete enslavement of the individual," he warned.
It was therefore important for social democrats who believed in individual liberty to face two critical questions: "How is it possible, in view
of the far-reaching centralization of political and economic power, to

prevent bureaucracy from becoming all-powerful and overweening? How can the rights of the individual be protected?"[44]

That imperative—to protect the rights of the individual—was Einstein's most fundamental political tenet. Individualism and freedom were necessary for creative art and science to flourish. Personally, politically, and professionally, he was repulsed by any restraints.

That is why he remained outspoken about racial discrimination in America. In Princeton during the 1940s, movie theaters were still segregated, blacks were not allowed to try on shoes or clothes at department stores, and the student newspaper declared that equal access for blacks to the university was "a noble sentiment but the time had not yet come."[45]

As a Jew who had grown up in Germany, Einstein was acutely sensitive to such discrimination. "The more I feel an American, the more this situation pains me," he wrote in an essay called "The Negro Question" for *Pageant* magazine. "I can escape the feeling of complicity in it only by speaking out."[46]

Although he rarely accepted in person the many honorary degrees offered to him, Einstein made an exception when he was invited to Lincoln University, a black institution in Pennsylvania. Wearing his tattered gray herringbone jacket, he stood at a blackboard and went over his relativity equations for students, and then he gave a graduation address in which he denounced segregation as "an American tradition which is uncritically handed down from one generation to the next."[47] As if to break the pattern, he met with the 6-year-old son of Horace Bond, the university's president. That son, Julian, went on to become a Georgia state senator, one of the leaders of the civil rights movement, and chairman of the NAACP.

There was, however, one group for which Einstein could feel little tolerance after the war. "The Germans, as a whole nation, are responsible for these mass killings and should be punished as a people," he declared.[48] When a German friend, James Franck, asked him at the end of 1945 to join an appeal calling for a lenient treatment of the German economy, Einstein angrily refused. "It is absolutely necessary to prevent the restoration of German industrial policy for many years," he said. "Should your appeal be circulated, I shall do whatever I can to oppose it." When Franck persisted, Einstein became even more adamant. "The Germans butchered millions of civilians according to a well-prepared plan," he wrote. "They would do it again if only they were able to. Not a trace of guilt or remorse is to be found among them."[49]

Einstein would not even permit his books to be sold in Germany again, nor would he allow his name to be placed back on the rolls of any German scientific society. "The crimes of the Germans are really the most abominable ever to be recorded in the history of the so-called civilized nations," he wrote the physicist Otto Hahn. "The conduct of the German intellectuals—viewed as a class—was no better than that of the mob."[50]

Like many Jewish refugees, his feelings had a personal basis. Among those who suffered under the Nazis was his first cousin Roberto, son of Uncle Jakob. When German troops were retreating from Italy near the end of the war, they wantonly killed his wife and two daughters, then burned his home while he hid in the woods. Roberto wrote to Einstein, giving the horrible details, and committed suicide a year later.[51]

The result was that Einstein's national and tribal kinship became even more clear in his own mind. "I am not a German but a Jew by nationality," he declared as the war ended.[52]

Yet in ways that were subtle yet real, he had become an American as well. After settling in Princeton in 1933, he never once in the remaining twenty-two years of his life left the United States, except for the brief cruise to Bermuda that was necessary to launch his immigration process.

Admittedly, he was a somewhat contrarian citizen. But in that regard he was in the tradition of some venerable strands in the fabric of American character: fiercely protective of individual liberties, often cranky about government interference, distrustful of great concentrations of wealth, and a believer in the idealistic internationalism that gained favor among American intellectuals after both of the great wars of the twentieth century.

His penchant for dissent and nonconformity did not make him a worse American, he felt, but a better one. On the day in 1940 when he was naturalized as a citizen, Einstein had touched on these values in a radio talk. After the war ended, Truman proclaimed a day in honor of all new citizens, and the judge who had naturalized Einstein sent out thousands of form letters inviting anyone he had sworn in to come to a park in Trenton to celebrate. To the judge's amazement, ten thousand people showed up. Even more amazing, Einstein and his household decided to come down for the festivities. During the ceremony, he sat smiling and waving, with a young girl sitting on his lap, happy to be a small part of "I Am an American" Day.[53]

LANDMARK

1948–1953

With Israeli Prime Minister David Ben-Gurion in Princeton, 1951

The Endless Quest

The problems of the world were important to Einstein, but the problems of the cosmos helped him to keep earthly matters in perspective. Even though he was producing little of scientific significance, physics rather than politics would remain his defining endeavor until the day he died. One morning when walking to work with his scientific assistant and fellow arms control advocate Ernst Straus, Einstein mused at their ability to divide their time between the two realms. "But our equations are much more important to me," Einstein added. "Politics is for the present, while our equations are for eternity."[1]

Einstein had officially retired from the Institute for Advanced Study at the end of the war, when he turned 66. But he continued to work in a small office there every day, and he was still able to enlist the aid of loyal assistants willing to pursue what had come to be considered his quaint quest for a unified field theory.

Each weekday, he would wake at a civilized hour, eat breakfast and read the papers, and then around ten walk slowly up Mercer Street to the Institute, trailing stories both real and apocryphal. His colleague Abraham Pais recalled "one occasion when a car hit a tree after the driver suddenly recognized the face of the beautiful old man walking along the street, the black woolen knit cap firmly planted on his long white hair."[2]

Soon after the war ended, J. Robert Oppenheimer came from Los Alamos to take over as director of the Institute. A brilliant, chain-smoking theoretical physicist, he proved charismatic and competent enough to be an inspiring leader for the scientists who built the atomic bomb. With his charm and biting wit, he tended to produce either acolytes or enemies, but Einstein fell into neither category. He and Oppenheimer viewed each other with a mixture of amusement and respect, which allowed them to develop a cordial though not close relationship.[3]

When Oppenheimer first visited the Institute in 1935, he called it a "madhouse" with "solipsistic luminaries shining in separate and hapless desolation." As for the greatest of these luminaries, Oppenheimer declared, "Einstein is completely cuckoo," though he seemed to mean it in an affectionate way.[4]

Once they became colleagues, Oppenheimer became more adroit at dealing with his luminous charges and his jabs became more subtle. Einstein, he declared, was "a landmark but not a beacon," meaning he was admired for his great triumphs but attracted few apostles in his current endeavors, which was true. Years later, he provided another telling description of Einstein: "There was always in him a powerful purity at once childlike and profoundly stubborn."[5]

Einstein became a closer friend, and a walking partner, of another iconic figure at the Institute, the intensely introverted Kurt Gödel, a German-speaking mathematical logician from Brno and Vienna. Gödel was famous for his "incompleteness theory," a pair of logical proofs that purport to show that any useful mathematical system will have some propositions that cannot be proven true or false based on the postulates of that system.

Out of the supercharged German-speaking intellectual world, in which physics and mathematics and philosophy intertwined, three jarring theories of the twentieth century emerged: Einstein's relativity, Heisenberg's uncertainty, and Gödel's incompleteness. The surface similarity of the three words, all of which conjure up a cosmos that is tentative and subjective, oversimplifies the theories and the connections between them. Nevertheless, they all seemed to have philosophical resonance, and this became the topic of discussion when Gödel and Einstein walked to work together.[6]

They were very different personalities. Einstein was filled with good humor and sagacity, both qualities lacking in Gödel, whose intense logic sometimes overwhelmed common sense. This was on glorious display when Gödel decided to become a U.S. citizen in 1947. He took his preparation for the exam very seriously, studied the Constitution carefully, and (as might be expected by the formulator of the incompleteness theory) found what he believed was a logical flaw. There was an internal inconsistency, he insisted, that could allow the entire government to degenerate into tyranny.

Concerned, Einstein decided to accompany—or chaperone—Gödel on his visit to Trenton to take the citizenship test, which was to be administered by the same judge who had done so for Einstein. On the drive, he and a third friend tried to distract Gödel and dissuade him from mentioning this perceived flaw, but to no avail. When the judge asked him about the Constitution, Gödel launched into his proof that its internal inconsistency made a dictatorship possible. Fortunately, the judge, who by now cherished his connection to Einstein, cut Gödel off. "You needn't go into all that," he said, and Gödel's citizenship was saved.[7]

During their walks, Gödel explored some of the implications of relativity theory, and he came up with an analysis that called into question whether time, rather than merely being relative, could be said to exist at all. Einstein's equations, he figured, could describe a universe that was rotating rather than (or in addition to) expanding. In such a case, the relationship between space and time could become, mathematically, mixed up. "The existence of an objective lapse of time," he wrote, "means that reality consists of an infinity of layers of 'now' which come into existence successively. But if simultaneity is something relative, each observer has his own set of 'nows,' and none of these various layers can claim the prerogative of representing the objective lapse of time."[8]

As a result, Gödel argued, time travel would be possible. "By making a round trip on a rocket ship in a sufficiently wide curve, it is possible in these worlds to travel into any region of the past, present and future, and back again." That would be absurd, he noted, because then we could go back and chat with a younger version of ourselves (or, even more discomforting, our older version could come back and chat with us). "Gödel had achieved an amazing demonstration that time travel, strictly understood, was consistent with the theory of relativity," writes Boston University philosophy professor Palle Yourgrau in his book on Gödel's relationship with Einstein, *World Without Time*. "The primary result was a powerful argument that if time travel is possible, time itself is not."[9]

Einstein responded to Gödel's essay along with a variety of others that had been collected in a book, and he seemed to be mildly impressed but also not totally engaged by the argument. In his brief assessment, Einstein called Gödel's "an important contribution" but noted that he had thought of the issue long ago and "the problem here involved disturbed me already." He implied that although time travel may be true as a mathematical conceivability, it might not be possible in reality. "It will be interesting to weigh whether these are not to be excluded on physical grounds," Einstein concluded.[10]

For his part, Einstein remained focused on his own white whale, which he pursued not with the demonic drive of Ahab but the dutiful serenity of Ishmael. In his quest for a unified field theory, he still had no compelling physical insight—such as the equivalence of gravity and acceleration, or the relativity of simultaneity—to guide his way, so his endeavors remained a groping through clouds of abstract mathematical equations with no ground lights to orient him. "It's like being in an airship in which one can cruise around in the clouds but cannot see clearly how one can return to reality, i.e., earth," he lamented to a friend.[11]

His goal, as it had been for decades, was to come up with a theory that encompassed both the electromagnetic and the gravitational fields, but he had no compelling reason to believe that they in fact *had* to be part of the same unified structure, other than his intuition that nature liked the beauty of simplicity.

Likewise, he was still hoping to explain the existence of particles in terms of a field theory by finding permissible pointlike solutions to his field equations. "He argued that if one believed wholeheartedly in the basic idea of a field theory, matter should enter not as an interloper but as an honest part of the field itself," recalled one of his Princeton collaborators, Banesh Hoffmann. "Indeed, one might say that he wanted to build matter out of nothing but convolutions of spacetime." In the process he used all sorts of mathematical devices, but constantly searched for others. "I need more mathematics," he lamented at one point to Hoffmann.[12]

Why did he persist? Deep inside, such disjunctures and dualities—different field theories for gravity and electromagnetism, distinctions between particles and fields—had always discomforted him. Simplicity and unity, he intuitively believed, were hallmarks of the Old One's handiwork. "A theory is more impressive the greater the simplicity of its premises, the more different things it relates, and the more expanded its area of applicability," he wrote.[13]

In the early 1940s, Einstein returned for a while to the five-dimensional mathematical approach that he had adopted from Theodor Kaluza two decades earlier. He even worked on it with Wolfgang Pauli, the quantum mechanics pioneer, who had spent some of the war years in Princeton. But he could not get his equations to describe particles.[14]

So he moved on to a strategy dubbed "bivector fields." Einstein seemed to be getting a little desperate. This new approach, he admitted, might require surrendering the principle of locality that he had sanctified in some of his thought-experiments assaulting quantum mechanics.[15] In any event, it was soon abandoned as well.

Einstein's final strategy, which he pursued for the final decade of his life, was a resurrection of one he had tried during the 1920s. It used a Riemannian metric that was not assumed to be symmetric, which opened the way for sixteen quantities. Ten combinations of them were used for gravity, and the remaining ones for electromagnetism.

Einstein sent early versions of this work to his old comrade Schrödinger. "I am sending them to nobody else, because you are the only person known to me who is not wearing blinders in regard to the fundamental questions in our science," Einstein wrote. "The attempt depends on an idea that at first seems antiquated and unprofitable, the introduction of a non-symmetrical tensor . . . Pauli stuck his tongue out at me when I told him about it." [16]

Schrödinger spent three days poring over Einstein's work and wrote back to say how impressed he was. "You are after big game," he said.

Einstein was thrilled with such support. "This correspondence gives me great joy," he replied, "because you are my closest brother and your brain runs so similarly to mine." But he soon began to realize that the gossamer theories he was spinning were mathematically elegant but never seemed to relate to anything physical. "Inwardly I am not so certain as I previously asserted," he confessed to Schrödinger a few months later. "We have squandered a lot of time on this, and the result looks like a gift from the devil's grandmother." [17]

And yet he soldiered on, churning out papers and producing the occasional headline. When a new edition of his book, *The Meaning of Relativity,* was being prepared in 1949, he added the latest version of the paper he had shown Schrödinger as an appendix. The *New York Times* reprinted an entire page of complex equations from the manuscript, along with a front-page story headlined "New Einstein Theory Gives a Master Key to Universe: Scientist, after 30 Years' Work, Evolves Concept That Promises to Bridge Gap between the Star and the Atom." [18]

But Einstein soon realized that it still wasn't right. During the six weeks between when he submitted the chapter and when it went to the printers, he had second thoughts and revised it yet again.

In fact, he continued to revise the theory repeatedly, but to no avail. His growing pessimism was visible in the lamentations he sent to his old friend from the Olympia Academy days, Maurice Solovine, then

Einstein's publisher in Paris. "I shall never ever solve it," he wrote in 1948. "It will be forgotten and must later be rediscovered again." Then, the following year: "I am uncertain as to whether I was even on the right track. The current generation sees in me both a heretic and a reactionary who has, so to speak, outlived himself." And, with some resignation, in 1951: "The unified field theory has been put into retirement. It is so difficult to employ mathematically that I have not been able to verify it. This state of affairs will last for many more years, mainly because physicists have no understanding of logical and philosophical arguments."[19]

Einstein's quest for a unified theory was destined to produce no tangible results that added to the framework of physics. He was able to come up with no great insights or thought experiments, no intuitions about underlying principles, to help him visualize his goal. "No pictures came to our aid," his collaborator Hoffmann lamented. "It is intensely mathematical, and over the years, with helpers and alone, Einstein surmounted difficulty after difficulty, only to find new ones awaiting him."[20]

Perhaps the search was futile. And if it turns out a century from now that there is indeed no unified theory to be found, it will also look misconceived. But Einstein never regretted his dedication to it. When a colleague asked him one day why he was spending—perhaps squandering—his time in this lonely endeavor, he replied that even if the chance of finding a unified theory was small, the attempt was worthy. He had already made his name, he noted. His position was secure, and he could afford to take the risk and expend the time. A younger theorist, however, could not take such a risk, for he might thus sacrifice a promising career. So, Einstein said, it was his duty to do it.[21]

Einstein's repeated failures in seeking a unified theory did not soften his skepticism about quantum mechanics. Niels Bohr, his frequent sparring partner, came to the Institute for a stay in 1948 and spent part of his time writing an essay on their debates at the Solvay Conferences before the war.[22] Struggling with the article in his office one floor above Einstein's, he developed writer's block and called in Abraham Pais to help him. As Bohr paced furiously around an oblong table, Pais coaxed him and took notes.

When he got frustrated, Bohr sometimes would simply sputter the same word over and over. Soon he was doing so with Einstein's name. He walked to the window and kept muttering, over and over, "Einstein . . . Einstein . . ."

At one such moment, Einstein softly opened the door, tiptoed in, and signaled to Pais not to say anything. He had come to steal a bit of tobacco, which his doctor had ordered him not to buy. Bohr kept muttering, finally spurting out one last loud "Einstein" and then turning around to find himself staring at the cause of his anxieties. "It is an understatement to say that for a moment Bohr was speechless," Pais recalled. Then, after an instant, they all burst into laughter.[23]

Another colleague who tried and failed to convert Einstein was John Wheeler, Princeton University's renowned theoretical physicist. One afternoon he came by Mercer Street to explain a new approach to quantum theory (known as the sum-over-histories approach) that he was developing with his graduate student, Richard Feynman. "I had gone to Einstein with the hope to persuade him of the naturalness of the quantum theory when seen in this new light," Wheeler recalled. Einstein listened patiently for twenty minutes, but when it was over repeated his very familiar refrain: "I still cannot believe that the good Lord plays dice."

Wheeler showed his disappointment, and Einstein softened his pronouncement slightly. "Of course, I may be wrong," he said in a slow and humorous cadence. Pause. "But perhaps I have earned the right to make my mistakes." Einstein later confided to a woman friend, "I don't think I'll live to find out who is correct."

Wheeler kept coming back, sometimes bringing his students, and Einstein admitted that he found many of his arguments "sensible." But he was never converted. Near the end of his life, Einstein regaled a small group of Wheeler's students. When the talk turned to quantum mechanics, he once again tried to poke holes in the idea that our observations can affect and determine realities. "When a mouse observes," Einstein asked them, "does that change the state of the universe?"[24]

The Lion in Winter

Mileva Marić, her health deteriorating due to a succession of minor strokes, was still living in Zurich and trying to take care of their institutionalized son, Eduard, whose behavior had become increasingly erratic and violent. Financial problems again plagued her and revived the tension with her former husband. The portion of the money that he had put into trust for her in America from the Nobel Prize had slipped away during the Depression, and two of her three apartment houses had been sold to help pay for Eduard's care. By late 1946, Einstein was pushing to sell the remaining house and give control of the money to a legal guardian who would be appointed for Eduard. But Marić had the usufruct of the house and its proceeds, as well as power of attorney over it, and she was terrified of surrendering any control.[25]

One cold day later that winter, she slipped on the ice on the way to see Eduard and ended up lying unconscious until strangers found her. She knew she was going to die soon, and she had recurring nightmares about struggling through the snow, unable to reach Eduard. She was panicked about what would happen to him, and wrote heartwrenching letters to Hans Albert.[26]

Einstein succeeded in selling her house by early 1948, but with her power of attorney she blocked the proceeds from being sent to him. He wrote to Hans Albert, giving him all the details and promising him that, whatever happened, he would take care of Eduard "even if it costs me all my savings."[27] That May, Marić had a stroke and lapsed into a trance in which she repeatedly muttered only "No, no!" until she died three months later. The money from the sale of her apartment, 85,000 Swiss francs, was found under her mattress.

Eduard lapsed into a daze and never spoke of his mother again. Carl Seelig, a friend of Einstein's who lived nearby, visited him frequently and sent back regular reports to Einstein. Seelig hoped to get him to make contact with his son, but he never did. "There is something blocking me that I am unable to analyze fully," Einstein told Seelig. "I believe I would be arousing painful feelings of various kinds in him if I made an appearance in whatever form."[28]

Einstein's own health began to decline in 1948 as well. For years he had been plagued by stomach ailments and anemia, and late that year, after an attack of sharp pains and vomiting, he checked into the Jewish Hospital in Brooklyn. Exploratory surgery revealed an aneurysm in the abdominal aorta,* but doctors decided there was not much they could do about it. It was assumed, correctly, that it was likely to kill him one day, but in the meantime he could live on borrowed time and a healthy diet.[29]

To recuperate, he went on the longest trip he would make during his twenty-two years as a Princeton resident: down to Sarasota, Florida. For once, he successfully avoided publicity. "Einstein Elusive Sarasota Visitor," the local paper lamented.

Helen Dukas accompanied him. After Elsa's death, she had become even more of a loyal guardian, and she even shielded Einstein from letters written by Hans Albert's daughter, Evelyn. Hans Albert suspected that Dukas may have had an affair with his father, and said so to others. "On many occasions, Hans Albert told me of his long-held suspicion," family friend Peter Bucky later recalled. But others who knew Dukas found the suggestion to be implausible.[30]

By then, Einstein had become much friendlier with his son, now a respected engineering professor at Berkeley. "Whenever we met," Hans Albert later recalled of his trips east to see his father, "we mutually reported on all the interesting developments in our field and in our work." Einstein particularly loved learning about new inventions and solutions to puzzles. "Maybe both, inventions and puzzles, reminded him of the happy, carefree, and successful days at the patent office in Bern," said Hans Albert.[31]

Einstein's beloved sister, Maja, the closest intimate of his life, was also in declining health. She had come to Princeton when Mussolini enacted anti-Jewish laws, but her husband, Paul Winteler, from whom she had been drifting apart for many years,[32] moved to Switzerland to be with his own sister and her husband, Michele Besso. They corresponded often, but never rejoined one another.

---

* An aneurysm is the ballooning or dilation of a blood vessel, as if it were blistering. The abdominal aorta is one of the large arteries from the heart, in the region between the diaphragm and the abdomen.

Maja began, as Elsa had, to look more like Einstein, with radiating silver hair and a devilish smile. The inflection of her voice and the slightly skeptical wry tone she used when asking questions were similar to his. Although she was a vegetarian, she loved hot dogs, so Einstein decreed that they were a vegetable, and that satisfied her.[33]

Maja had suffered a stroke and, by 1948, was confined to bed most of the time. Einstein doted on her as he did no other person. Every evening he read aloud to her. Sometimes the fare was heavy, such as the arguments of Ptolemy against Aristarchus's opinion that the world rotates around the sun. "I could not help thinking of certain arguments of present-day physicists: learned and subtle, but without insight," he wrote Solovine about that evening. Other times, the readings were lighter but perhaps just as revealing, such as the evenings he read from *Don Quixote;* he sometimes compared his own quixotic parries against the prevailing windmills of science with that of the old knight with a ready lance.[34]

When Maja died in June 1951, Einstein was grief-stricken. "I miss her more than can be imagined," he wrote a friend. He sat on the back porch of his Mercer Street home for hours, pale and tense, staring into space. When his stepdaughter Margot came to console him, he pointed to the sky and said, as if reassuring himself, "Look into nature, and then you will understand it better."[35]

Margot had likewise left her husband, who responded by writing, as he had long wanted to, an unauthorized biography of Einstein. She worshipped Einstein, and each year they grew closer. He found her presence charming. "When Margot speaks," he said, "you see flowers growing."[36]

His ability to engender and feel such affection belied his reputation for being emotionally distant. Both Maja and Margot preferred living with him to living with their own husbands as they got older. He had been a difficult husband and father because he did not take well to any constricting bonds, but he could also be intense and passionate, both with family and friends, when he found himself engaged rather than confined.

Einstein was human, and thus both good and flawed, and the greatest of his failings came in the realm of the personal. He had lifelong friends who were devoted to him, and he had family members who doted on him, but there were also those few—Mileva and Eduard foremost among them—whom he simply walled out when the relationship became too painful.

As for his colleagues, they saw his kindly side. He was gentle and generous with partners and subordinates, both those who agreed with him and those who didn't. He had deep friendships lasting for decades. He was unfailingly benevolent to his assistants. His warmth, sometimes missing at home, radiated on the rest of humanity. So as he grew old, he was not only respected and revered by his colleagues, he was loved.

They honored him, with the blend of scientific and personal camaraderie he had enjoyed since his student days, at a seventieth birthday convocation upon his return from his Florida recuperation. Although the talks were supposed to focus on Einstein's science, most dwelled on his sweetness and humanity. When he walked in, there was a hush, then thunderous applause. "Einstein just had no sense at all about what absolute reverence there was for him," one of his assistants recalled.[37]

His closest friends at the Institute bought him a present, an advanced AM-FM radio and high-fidelity record player, which they installed in his home secretly when he was at work one day. Einstein was thrilled and used it not only for music but for news. In particular, he liked to catch Howard K. Smith's commentaries.

He had pretty much given up the violin by then. It was too hard on his aging fingers. Instead, he focused on the piano, which he was not quite as good at playing. Once, after repeatedly stumbling on a passage, he turned to Margot and smiled. "Mozart wrote such nonsense here," he said.[38]

He came to look even more like a prophet, with his hair getting longer, his eyes a bit sadder and more weary. His face grew more deeply etched yet somehow more delicate. It showed wisdom and wear but still a vitality. He was dreamy, as he was when a child, but also now serene.

"I am generally regarded as sort of a petrified object," he noted to Max Born, then a professor in Edinburgh, one of those friends whose affection had lasted so long. "I find this role not too distasteful, as it corresponds very well with my temperament . . . I simply enjoy giving more than receiving in every respect, do not take myself nor the doings of the masses seriously, am not ashamed of my weaknesses and vices, and naturally take things as they come with equanimity and humor."[39]

Israel's Presidency

Before the Second World War, Einstein had stated his opposition to a Jewish state when speaking to three thousand celebrants at a Manhattan hotel seder. "My awareness of the essential nature of Judaism resists the idea of a Jewish state with borders, an army, and a measure of temporal power," he said. "I am afraid of the inner damage Judaism will sustain—especially from the development of a narrow nationalism within our ranks. We are no longer the Jews of the Maccabee period."[40]

After the war, he took the same stance. When he testified in Washington in 1946 to an international committee looking into the situation in Palestine, he denounced the British for pitting Jews against Arabs, called for more Jewish immigration, but rejected the idea that the Jews should be nationalistic. "The State idea is not in my heart," he said in a quiet whisper that reverberated through the shocked audience of ardent Zionists. "I cannot understand why it is needed."[41] Rabbi Stephen Wise was flabbergasted that Einstein would break ranks with true Zionists at such a public hearing, and he got him to sign a clarifying statement that was, in fact, not clarifying at all.

Einstein was especially dismayed by the militaristic methods used by Menachem Begin and other Jewish militia leaders, and he joined with his occasional antagonist Sidney Hook to sign a petition in the New York Times denouncing Begin as a "terrorist" and "closely akin" to the fascists.[42] The violence was contrary to Jewish heritage. "We imitate the stupid nationalism and racial nonsense of the goyim," he wrote a friend in 1947.

But when the State of Israel was declared in 1948, Einstein wrote the same friend to say that his attitude had changed. "I have never considered the idea of a state a good one, for economic, political and military reasons," he conceded. "But now, there is no going back, and one has to fight it out."[43]

The creation of Israel caused him, yet again, to back away from the pure pacifism he had once embraced. "We may regret that we have to use methods that are repulsive and stupid to us," he wrote to a Jewish group in Uruguay, "but to bring about better conditions in the international sphere, we must first of all maintain our experience by all means at our disposal."[44]

Chaim Weizmann, the indefatigable Zionist who brought Einstein to America in 1921, had become Israel's first president, a prestigious but generally ceremonial post in a system that vested most power in the prime minister and cabinet. When he died in November 1952, a Jerusalem newspaper began urging that Einstein be tapped to replace him. Prime Minister David Ben-Gurion bowed to the pressure, and word quickly spread that Einstein would be asked.

It was an idea that was at once both astonishing and obvious—and also impractical. Einstein first learned of it from a small article in the *New York Times* a week after Weizmann's death. At first he and the women in his house laughed it off, but then reporters started to call. "This is very awkward, very awkward," he told a visitor. A few hours later, a telegram arrived from Israel's ambassador in Washington, Abba Eban. Could the embassy, it asked, send someone the next day to see him officially?

"Why should that man come all that way," Einstein lamented, "when I only will have to say no?"

Helen Dukas came up with the idea of simply giving Ambassador Eban a phone call. In those days, impromptu long-distance calls were somewhat novel. To her surprise, she was able to track Eban down in Washington and put him on the line with Einstein.

"I am not the person for that and I cannot possibly do it," Einstein said.

"I cannot tell my government that you phoned me and said no," Eban replied. "I have to go through the motions and present the offer officially."

Eban ended up sending a deputy, who handed Einstein a formal letter asking if he would take on the presidency. "Acceptance would entail moving to Israel and taking its citizenship," Eban's letter noted (presumably in case Einstein harbored any fantasy that he could pre-side over Israel from Princeton). Eban hastened to reassure Einstein, however: "Freedom to pursue your great scientific work would be afforded by a government and people who are fully conscious of the supreme significance of your labors." In other words, it was a job that would require his presence, but not much else.

Even though the offer seemed somewhat strange, it was a powerful testament to Einstein's unsurpassed standing as a hero of world Jewry. It "embodies the deepest respect which the Jewish people can repose in any of its sons," Eban said.

Einstein had already prepared his note of rejection, which he handed to Eban's envoy as soon as he arrived. "I have been a lawyer all my life," the visitor joked, "and I have never gotten a rebuttal before I have stated my case."

He was "deeply moved" by the offer, Einstein said in his prepared response, and "at once saddened and ashamed" that he would not accept it. "All my life I have dealt with objective matters, hence I lack both the natural aptitude and the experience to deal properly with people and to exercise official function," he explained. "I am the more distressed over these circumstances because my relationship with the Jewish people became my strongest human tie once I achieved complete clarity about our precarious position among the nations of the world."[45]

Offering Einstein the presidency of Israel was a clever idea, but Einstein was right to realize that sometimes a brilliant idea is also a very bad one. As he noted with his usual wry self-awareness, he did not have the natural aptitude to deal with people in the way the role would require, nor did he have the temperament to be an official functionary. He was not cut out to be either a statesman or a figurehead.

He liked to speak his mind, and he had no patience for the compromises necessary to manage, or even symbolically lead, complex organizations. Back when he was involved as a figurehead leader in the establishing of Hebrew University, he had not possessed the talent to handle, nor the temperament to ignore, all of the maneuverings involved. Likewise, he had more recently had the same unpleasant experiences with a group creating Brandeis University near Boston, which caused him to resign from that endeavor.[46]

In addition, he had never displayed a discernible ability to run anything. The only formal administrative duty he had ever undertaken was to head a new physics institute at the University of Berlin. He did little other than hire his stepdaughter to handle some clerical tasks and give a job to the astronomer trying to confirm his theories.

Einstein's brilliance sprang from being a rebel and nonconformist who recoiled at any attempt to restrain his free expression. Are there any worse traits for someone who is supposed to be a political conciliator? As he explained in a polite letter to the Jerusalem newspaper that had been campaigning for him, he did not want to face the chance that he would have to go along with a government decision that "might create a conflict with my conscience."

In society as in science, he was better off remaining a nonconformist. "It is true that many a rebel has in the end become a figure of responsibility," Einstein conceded to a friend that week, "but I cannot bring myself to do so."[47]

Ben-Gurion was secretly relieved. He had begun to realize that the idea was a bad one. "Tell me what to do if he says yes!" he joked to his assistant. "I've had to offer the post to him because it's impossible not to. But if he accepts, we are in for trouble." Two days later, when Ambassador Eban ran into Einstein at a black-tie reception in New York, he was happy that the issue was behind them. Einstein was not wearing socks.[48]

RED SCARE

1951–1954

With J. Robert Oppenheimer, 1947

The Rosenbergs

The rush to build the H-Bomb, rising anticommunist fervor, and Senator Joseph McCarthy's increasingly untethered security investigations unnerved Einstein. The atmosphere reminded him of the rising Nazism and anti-Semitism of the 1930s. "The German calamity of years ago repeats itself," he lamented to the queen mother of Belgium in early 1951. "People acquiesce without resistance and align themselves with the forces for evil."[1]

He tried to maintain a middle ground between those who were reflexively anti-American and those who were reflexively anti-Soviet. On the one hand, he rebuked his collaborator Leopold Infeld, who wanted him to support statements by the World Peace Council, which Einstein rightly suspected was Soviet-influenced. "In my view they are more or less propaganda," he said. He did the same to a group of Russian students who pressed him to join a protest against what they alleged was America's use of biological weapons during the Korean War. "You cannot expect me to protest against incidents which possibly, and very probably, have never taken place," he replied.[2]

On the other hand, Einstein refrained from signing a petition circulated by Sidney Hook denouncing the perfidy of those who made such charges against America. He was enamored of neither extreme. As he put it, "Every reasonable person must strive to promote moderation and a more objective judgment."[3]

In what he presumed would be a quiet effort at promoting such moderation, Einstein wrote a private letter asking that Julius and Ethel Rosenberg, who had been convicted of turning over atomic secrets to the Soviets, be spared the death penalty. He had avoided making any statements about the case, which had divided the nation with a frenzy seldom seen before the advent of the cable-TV age. Instead, he sent the letter to the judge, Irving Kaufman, with a promise not to publicize it. Einstein did not contend that the Rosenbergs were innocent. He merely argued that a death penalty was too harsh in a case where the facts were murky and the outcome was driven more by popular hysteria than objectivity.[4]

In a reflection of the tenor of the time, Judge Kaufman took the private letter and turned it over to the FBI. Not only was it put into Einstein's file, but it was investigated to see if it could be construed as disloyalty. After three months, a report was sent to Hoover saying no further incriminating evidence had been found, but the letter remained in the file.[5]

When Judge Kaufman went ahead and imposed a death penalty, Einstein wrote to President Harry Truman, who was about to leave office, to ask him to commute the sentence. He drafted the letter first in German and then in English on the back of a piece of scrap paper that he had filled with a variety of equations that apparently, given how they trail off, led to nothing.[6] Truman bucked the decision to incoming President Eisenhower, who allowed the executions to proceed.

Einstein's letter to Truman was released publicly, and the *New York Times* ran a front-page story headlined "Einstein Supports Rosenberg

Appeal."[7] More than a hundred angry letters swept in from across the nation. "You need some common sense plus some appreciation for what America has given you," wrote Marian Rawles of Portsmouth, Virginia. "You place the Jew first and the United States second," said Charles Williams of White Plains, New York. From Corporal Homer Greene, serving in Korea: "You evidently like to see our GI's killed. Go to Russia or back where you came from, because I don't like Americans like you living off this country and making un-American statements."[8]

There were not as many positive letters, but Einstein did have a pleasant exchange with the liberal Supreme Court Justice William O. Douglas, who had unsuccessfully tried to stop the executions. "You have struggled so devotedly for the creation of a healthy public opinion in our troubled time," Einstein wrote in a note of appreciation. Douglas sent back a handwritten reply: "You have paid me a tribute which brightens the burdens of this dark hour—a tribute I will always cherish."[9]

Many of the critical letters asked Einstein why he was willing to speak out for the Rosenbergs but not for the nine Jewish doctors whom Stalin had put on trial as part of an alleged Zionist conspiracy to murder Russian leaders. Among those who publicly challenged what they saw as Einstein's double standard were the publisher of the *New York Post* and the editor of the *New Leader*.[10]

Einstein agreed that the Russian actions should be denounced. "The perversion of justice which manifests itself in all the official trials staged by the Russian government deserves unconditional condemnation," he wrote. He added that individual appeals to Stalin would probably not do much, but perhaps a joint declaration from a group of scholars would help. So he got together with the chemistry Nobel laureate Harold Urey and others to issue one. "Einstein and Urey Hit Reds' Anti-Semitism," the *New York Times* reported.[11] (After Stalin died a few weeks later, the doctors were freed.)

On the other hand, he stressed in scores of letters and statements that Americans should not let the fear of communism cause them to surrender the civil liberties and freedom of thought that they cherished. There were a lot of domestic communists in England, but the people there did not get themselves whipped into a frenzy by internal security investigations, he pointed out. Americans need not either.

William Frauenglass

Every year, Lord & Taylor department stores gave an award that, especially in the early 1950s, might have seemed unusual. It honored independent thinking, and Einstein, fittingly, won it in 1953 for his "nonconformity" in scientific matters.

Einstein took pride in that trait, which he knew had served him well over the years. "It gives me great pleasure to see the stubbornness of an incorrigible nonconformist warmly acclaimed," he said in his radio talk accepting the award.

Even though he was being honored for his nonconformity in the field of science, Einstein used the occasion to turn attention to the McCarthy-style investigations. For him, freedom in the realm of thought was linked to freedom in the realm of politics. "To be sure, we are concerned here with nonconformism in a remote field of endeavor," he said, meaning physics. "No Senatorial committee has as yet felt compelled to tackle the task of combating in this field the dangers that threaten the inner security of the uncritical or intimidated citizen." [12]

Listening to his talk was a Brooklyn schoolteacher, William Frauenglass, who had a month earlier been called to testify in Washington before a Senate Internal Security Subcommittee looking into communist influence in high schools. He had refused to talk, and now he wanted Einstein to say whether he had been right.

Einstein crafted a reply and told Frauenglass he could make it public. "The reactionary politicians have managed to instill suspicions of all intellectual efforts," he wrote. "They are now proceeding to suppress the freedom of teaching." What should intellectuals do against this evil? "Frankly, I can only see the revolutionary way of non-cooperation in the sense of Gandhi's," Einstein declared. "Every intellectual who is called before one of the committees ought to refuse to testify." [13]

Einstein's lifelong comfort in resisting prevailing winds made him serenely stubborn during the McCarthy era. At a time when citizens were asked to name names and testify at inquiries into their loyalty and that of their colleagues, he took a simple approach. He told people not to cooperate.

He felt, as he told Frauenglass, that this should be done based on the free speech guarantees of the First Amendment, rather than the "subterfuge" of invoking the Fifth Amendment's protection against possible self-incrimination. Standing up for the First Amendment was particularly a duty of intellectuals, he said, because they had a special role in society as preservers of free thought. He was still horrified that most intellectuals in Germany had not risen in resistance when the Nazis came to power.

When his letter to Frauenglass was published, there was an even greater public uproar than had been provoked by his Rosenberg appeal. Editorial writers across the nation pulled out all the stops for their denunciatory chords.

> The *New York Times:* "To employ unnatural and illegal forces of civil disobedience, as Professor Einstein advises, is in this case to attack one evil with another. The situation which Professor Einstein rebels against certainly needs correction, but the answer does not lie in defying the law."
> The *Washington Post:* "He has put himself in the extremist category by his irresponsible suggestion. He has proved once more that genius in science is no guarantee of sagacity in political affairs."
> The *Philadelphia Inquirer:* "It is particularly regrettable when a scholar of his attainments, full of honors, should permit himself to be used as an instrument of propaganda by the enemies of the country that has given him such a secure refuge . . . Dr. Einstein has come down from the stars to dabble in ideological politics, with lamentable results."
> The *Chicago Daily Tribune:* "It is always astonishing to find that a man of great intellectual power in some directions is a simpleton or even a jackass in others."
> The *Pueblo* (Colorado) *Star-Journal:* "He, of all people, should know better. This country protected him from Hitler." [14]

Ordinary citizens wrote as well. "Look in the mirror and see how disgraceful you look without a haircut like a wild man and wear a Russian wool cap like a Bolshevik," said Sam Epkin of Cleveland. The anticommunist columnist Victor Lasky sent a handwritten screed: "Your most recent blast against the institutions of this great nation finally convinces me that, despite your great scientific knowledge, you are an idiot, a menace to this country." And George Stringfellow of East Orange, New Jersey, noted incorrectly, "Don't forget that you left a communist country to come here where you could have freedom. Don't abuse that freedom sir."[15]

Senator McCarthy also issued a denunciation, though it seemed slightly muted due to Einstein's stature. "Anyone who advises Americans to keep secret information which they have about spies and saboteurs is himself an enemy of America," he said, not quite aiming directly at Einstein or what he had written.[16]

This time, however, there were actually more letters in support of Einstein. Among the more amusing ripostes came from his friend Bertrand Russell. "You seem to think that one should always obey the law, however bad," the philosopher wrote to the *New York Times*. "I am compelled to suppose that you condemn George Washington and hold that your country ought to return to allegiance to Her Gracious Majesty, Queen Elisabeth II. As a loyal Briton, I of course applaud this view; but I fear it may not win much support in your country." Einstein wrote Russell a thank-you letter, lamenting, "All the intellectuals in this country, down to the youngest student, have become completely intimidated."[17]

Abraham Flexner, now retired from the Institute for Advanced Study and living on Fifth Avenue, took the opportunity to restore his relationship with Einstein. "I am grateful to you as a native American for your fine letter to Mr. Frauenglass," he wrote. "American citizens in general will occupy a more dignified position if they absolutely refuse to say a word if questioned about their personal opinions and beliefs."[18]

Among the most poignant notes was from Frauenglass's teenage son, Richard. "In these troubled times, your statement is one that might alter the course of this nation," he said, which had a bit of truth to it. He noted that he would cherish Einstein's letter for the rest of his life, then added a P.S.: "My favorite subjects are your favorite too—math and physics. Now I am taking trigonometry."[19]

Passive Resistance

Dozens of dissenters subsequently begged Einstein to intervene on their behalf, but he declined. He had made his point and did not see the need to keep thrusting himself into the fray.

But one person did get through: Albert Shadowitz, a physics professor who had worked as an engineer during the war and helped form a union that was eventually expelled from the labor movement for having communists on its board. Senator McCarthy wanted to show that the union had ties to Moscow and had endangered the defense industry. Shadowitz, who had been a member of the Communist Party, decided to invoke the protections of the First, not the Fifth, Amendment, as Einstein had recommended to Frauenglass.[20]

Shadowitz was so worried about his plight that he decided to call Einstein for support. But Einstein's number was unlisted. So he got into his car in northern New Jersey, drove to Princeton, and showed up at Einstein's house, where he was met by the zealous guardian Dukas. "Do you have an appointment?" she demanded. He admitted he didn't. "Well, you can't just come in and speak to Professor Einstein," she declared. But when he explained his story, she stared at him for a while, then waved him in.

Einstein was wearing his usual attire: a baggy sweatshirt and corduroy trousers. He took Shadowitz upstairs to his study and assured him that his actions were right. He was an intellectual, and it was the special duty of intellectuals to stand up in such cases. "If you take this path then feel free to use my name in any way that you wish," Einstein generously offered.

Shadowitz was surprised by the blank check, but happy to use it. McCarthy's chief counsel, Roy Cohn, did the questioning as McCarthy listened during the initial closed hearing. Was he a communist? Shadowitz replied: "I refuse to answer that and I am following the advice of Professor Einstein." McCarthy suddenly took over the questioning. Did he know Einstein? Not really, Shadowitz answered, but I've met him. When the script was replayed in an open hearing, it made the same type of headlines, and provoked the same spurt of mail, as the Frauenglass case had.

Einstein believed he was being a good, rather than a disloyal, citizen. He had read the First Amendment and felt that upholding its spirit was at the core of America's cherished freedom. One angry critic sent him a copy of a card that contained what he called "The American Creed." It read, in part, "It is my duty to my country to love it; to support its Constitution; to obey its laws." Einstein wrote on the edge, "This is precisely what I have done."[21]

When the great black scholar W.E.B. Du Bois was indicted on charges stemming from helping to circulate a petition initiated by the World Peace Council, Einstein volunteered to testify as a character witness on his behalf. It represented a union of Einstein's sentiments on behalf of civil rights and of free speech. When Du Bois's lawyer informed the court that Einstein would appear, the judge rather quickly decided to dismiss the case.[22]

Another case hit closer to home: that of J. Robert Oppenheimer. After leading the scientists who developed the atom bomb and then becoming head of the Institute where Einstein still puttered in to work, Oppenheimer remained an adviser to the Atomic Energy Commission and kept his security clearance. By initially opposing the development of the hydrogen bomb, he had turned Edward Teller into an adversary, and he also alienated AEC commissioner Lewis Strauss. Oppenheimer's wife, Kitty, and his brother, Frank, had been members of the Communist Party before the war, and Oppenheimer himself had associated freely with party members and with scientists whose loyalty came under question.[23]

All of this prompted an effort in 1953 to strip Oppenheimer of his security clearance. It would have expired soon anyway, and everyone could have allowed the matter to be resolved quietly, but in the heated atmosphere neither Oppenheimer nor his adversaries wanted to back away from what they saw as a matter of principle. So a secret hearing was scheduled in Washington.

One day at the Institute, Einstein ran into Oppenheimer, who was preparing for the hearings. They chatted for a few minutes, and when Oppenheimer got to his car he recounted the conversation to a friend. "Einstein thinks that the attack on me is so outrageous that I should just resign," he said. Einstein considered Oppenheimer "a fool" for even answering the charges. Having served his country admirably, he had no obligations to subject himself to a "witch hunt."[24]

A few days after the secret hearings finally began—in April 1954, just as CBS journalist Edward R. Murrow was taking on Joseph McCarthy and the controversy over security investigations was at its height—they became public through a page-1 exclusive by James Reston of the *New York Times*.[25] The issue of the government's investigation of Oppenheimer's loyalty instantly became another polarizing public debate.

Warned that the story was about to break, Abraham Pais went to Mercer Street to make sure that Einstein was prepared for the inevitable press calls. He was bitterly amused when Pais told him that Oppenheimer continued to insist on a hearing rather than simply cutting his ties with the government. "The trouble with Oppenheimer is that he loves a woman who doesn't love him—the United States government," Einstein said. All Oppenheimer had to do, Einstein told Pais, was "go to Washington, tell the officials that they were fools, and then go home."[26]

Oppenheimer lost. The AEC voted that he was a loyal American but also a security risk and—one day before it would have expired anyway—revoked his clearance. Einstein visited him at the Institute the next day and found him depressed. That evening he told a friend that he did not "understand why Oppenheimer takes the business so seriously."

When a group of Institute faculty members circulated a petition affirming support for their director, Einstein immediately signed up. Others initially declined, some partly out of fear. This galvanized Einstein. He "put his 'revolutionary talents' into action to garner support," a friend recalled. After a few more meetings, Einstein had helped to convince or shame everyone into signing the statement.[27]

Lewis Strauss, Oppenheimer's AEC antagonist, was on the board of the Institute, which worried the faculty. Would he try to get Oppenheimer fired? Einstein wrote his friend Senator Herbert Lehman of New York, another trustee, calling Oppenheimer "by far the most capable Director the Institute has ever had." Dismissing him, he said, "would arouse the justified indignation of all men of learning."[28] The trustees voted to keep him.

Soon after the Oppenheimer affair, Einstein was visited in Princeton by Adlai Stevenson, the once and future Democratic nominee for president, who was a particular darling among intellectuals. Einstein expressed concern at the way politicians were whipping up fear of communism. Stevenson replied somewhat circumspectly. The Russians were, in fact, a danger. After some more gentle back and forth, Stevenson thanked Einstein for endorsing him in 1952. There was no need for thanks, Einstein replied, as he had done so only because he trusted Eisenhower even less. Stevenson said he found such honesty refreshing, and Einstein decided that he was not quite as pompous as he had originally seemed.[29]

Einstein's opposition to McCarthyism arose partly out of his fear of fascism. America's most dangerous internal threat, he felt, came not from communist subversives but from those who used the fear of communists to trample civil liberties. "America is incomparably less endangered by its own Communists than by the hysterical hunt for the few Communists that are here," he told the socialist leader Norman Thomas.

Even to people he did not know, Einstein expressed his disgust in unvarnished terms. "We have come a long way toward the establishment of a Fascist regime," he replied to an eleven-page letter sent to him by a New Yorker he had never met. "The similarity of general conditions here to those in the Germany of 1932 is quite obvious."[30]

Some colleagues worried that Einstein's vocal opinions would cause controversy for the Institute. Such concerns, he joked, made his hair turn gray. Indeed, he took a boyish American glee at his freedom to say whatever he felt. "I have become a kind of *enfant terrible* in my new homeland due to my inability to keep silent and to swallow everything that happens," he wrote Queen Mother Elisabeth. "Besides, I believe that older people who have scarcely anything to lose ought to be willing to speak out in behalf of those who are young and are subject to much greater restraint."[31]

He even announced, in tones both grave and a bit playful, that he would not have become a professor given the political intimidation that now existed. "If I were a young man again and had to decide how to make a living, I would not try to become a scientist or scholar or teacher," he intoned to Theodore White of the *Reporter* magazine. "I would rather choose to be a plumber or a peddler, in the hope of finding that modest degree of independence still available."[32]

That earned him an honorary membership card from a plumbers' union, and it sparked a national debate on academic freedom. Even slightly frivolous remarks made by Einstein carried a lot of momentum.

Einstein was right that academic freedom was under assault, and the damage done to careers was real. For example, David Bohm, a great theoretical physicist who worked with Oppenheimer and Einstein in Princeton and refined certain aspects of quantum mechanics, was called before the House Un-American Activities Committee, pleaded the Fifth Amendment, lost his job, and ended up moving to Brazil.

Nevertheless, Einstein's remark—and his litany of doom—turned out to be overstated. Despite his impolitic utterances, there was no serious attempt to muzzle him or threaten his job. Even the slapstick FBI efforts to compile a dossier on him did not curtail his free speech. At the end of the Oppenheimer investigation, both he and Einstein were still harbored safely in their haven in Princeton, free to think and speak as they chose. The fact that both men had their loyalty questioned and, at times, their security clearances denied was shameful. But it was not like Nazi Germany, not anything close, despite what Einstein sometimes said.

Einstein and some other refugees tended, understandably, to view McCarthyism as a descent into the black hole of fascism, rather than as one of those ebbs and flows of excess that happen in a democracy. As it turned out, American democracy righted itself, as it always has. McCarthy was relegated to his own disgrace in 1954 by Army lawyers, his Senate colleagues, President Eisenhower, and journalists such as Drew Pearson and Edward R. Murrow. When the transcript of the Oppenheimer case was published, it ended up hurting the reputation of Lewis Strauss and Edward Teller, at least within the academic and scientific establishment, as much as that of Oppenheimer.

Einstein was not used to self-righting political systems. Nor did he fully appreciate how resilient America's democracy and its nurturing of individual liberty could be. So for a while his disdain deepened. But he was saved from serious despair by his wry detachment and his sense of humor. He was not destined to die a bitter man.

THE END

1955

Intimations of Mortality

For his seventy-fifth birthday in March 1954, Einstein received from a medical center, unsolicited, a pet parrot that was delivered in a box to his doorstep. It had been a difficult journey, and the parrot seemed traumatized. At the time, Einstein was seeing a woman who worked in one of Princeton University's libraries named Johanna Fantova, whom he had met back in Germany in the 1920s. "The pet parrot is depressed after his traumatic delivery and Einstein is trying to cheer him up with his jokes, which the bird doesn't seem to appreciate," she wrote in the wonderful journal she kept of their dates and conversations.[1]

The parrot rebounded psychologically and was soon eating out of Einstein's hand, but it developed an infection. That necessitated a series of injections, and Einstein worried that the bird would not survive. But it was a tough bird, and after only two injections he bounced back.

Einstein likewise had repeatedly bounced back from bouts of anemia and stomach ailments. But he knew that the aneurysm on his abdominal aorta should soon prove fatal, and he began to display a peaceful sense of his own mortality. When he stood at the graveside and eulogized the physicist Rudolf Ladenberg, who had been his colleague in Berlin and then Princeton, the words seemed to be ones he felt personally. "Brief is this existence, as a fleeting visit in a strange house," he said. "The path to be pursued is poorly lit by a flickering consciousness."[2]

He seemed to sense that this final transition he was going through was at once natural and somewhat spiritual. "The strange thing about growing old is that the intimate identification with the here and now is slowly lost," he wrote his friend the queen mother of Belgium. "One feels transposed into infinity, more or less alone."[3]

After his colleagues updated, as a seventy-fifth birthday gift, the music system they had given him five years earlier, Einstein began repeatedly to play an RCA Victor recording of Beethoven's *Missa Solemnis*. It was an unusual choice for two reasons. He tended to regard Beethoven, who was not his favorite composer, as "too personal, almost naked."[4] Also, his religious instincts did not usually include these sorts of trappings. "I am a deeply religious nonbeliever," he noted to a friend who had sent him birthday greetings. "This is a somewhat new kind of religion."[5]

It was time for reminiscing. When his old friends Conrad Habicht and Maurice Solovine wrote a postcard from Paris recalling their time together in Bern, more than a half century earlier, as members of their self-proclaimed Olympia Academy, Einstein replied with a paean addressed to that bygone institution: "Though somewhat decrepit, we still follow the solitary path of our life by your pure and inspiring light." As he later lamented in another letter to Solovine, "The devil counts out the years conscientiously."[6]

Despite his stomach problems, he still loved to walk. Sometimes it was with Gödel to and from the Institute, at other times it was in the woods near Princeton with his stepdaughter Margot. Their relationship had become even closer, but their walks were usually enjoyed in silence. She noticed that he was becoming mellower, both personally and politically. His judgments were mild, even sweet, rather than harsh.[7]

He had, in particular, made his peace with Hans Albert. Shortly after he celebrated his seventy-fifth birthday, his son turned 50. Einstein, thanks to a reminder from his son's wife, wrote him a letter that was slightly formal, as if created for a special occasion. But it contained a nice tribute both to his son and to the value of a life in science: "It is a joy for me to have a son who has inherited the main traits of my personality: the ability to rise above mere existence by sacrificing one's self through the years for an impersonal goal."[8] That fall, Hans Albert came east for a visit.

By then Einstein had finally discovered what was fundamental about America: it can be swept by waves of what may seem, to outsiders, to be dangerous political passions but are, instead, passing sentiments that are absorbed by its democracy and righted by its constitutional gyroscope. McCarthyism had died down, and Eisenhower had proved a calming influence. "God's own country becomes stranger and stranger," Einstein wrote Hans Albert that Christmas, "but somehow they manage to return to normality. Everything—even lunacy—is mass produced here. But everything goes out of fashion very quickly."[9]

Almost every day he continued to amble to the Institute to wrestle with his equations and try to push them a little closer toward the horizon of a unified field theory. He would come in with his new ideas, often clutching equations on scraps of paper he had scribbled the night before, and go over them with his assistant of that final year, Bruria Kaufman, a physicist from Israel.

She would write the new equations on a blackboard so they could ponder them together, and point out problems. Einstein would then try to counter them. "He had certain criteria by which to judge whether this is relevant to physical reality or not," she recounted. Even when they were defeated by the obstacles to a new approach, as they invariably were, Einstein remained optimistic. "Well, we've learned something," he would say as the clock ticked down.[10]

In the evening, he would often explain his last-ditch efforts to his companion, Johanna Fantova, and she would record them in her journal. The entries for 1954 were littered with hopes raised and dashed. February 20: "Thinks he found a new angle to his theory, something very important that would simplify it. Hopes he won't find any errors." February 21: "Didn't find any errors, but the new work isn't as exciting as he had thought the day before." August 25: "Einstein's equations are looking good—maybe something will come of them—but it's damned hard work." September 21: "He's making some progress with what was at first only a theory but is now looking good." October 14: "Found an error in his work today, which is a setback." October 24: "He calculated like crazy today but accomplished nothing."[11]

That year Wolfgang Pauli, the quantum mechanics pioneer, came to visit. Again the old debate over whether God would play dice was reengaged, as it had been a quarter-century earlier at the Solvay Conferences. Einstein told Pauli that he still objected to the fundamental tenet in quantum mechanics that a system can be defined only by specifying the experimental method of observing it. There was a reality, he insisted, that was independent of how we observed it. "Einstein has the philosophical prejudice that a state, termed 'real,' can be defined objectively under any circumstances, that is, without specification of the experimental arrangement used to examine the system," Pauli marveled in a letter to Max Born.[12]

He also clung to his belief that physics should be based, as he told his old friend Besso, "on the field concept, i.e., on continuous structures." Seventy years earlier, his awe at contemplating a compass caused him to marvel at the concept of fields, and they had guided his theories ever since. But what would happen, he worried to Besso, if field theory turned out to be unable to account for particles and quantum mechanics? "In that case *nothing* remains of my entire castle in the air, gravitation theory included."[13]

So even as Einstein apologized for his stubbornness, he proudly refused to abandon it. "I must seem like an ostrich who forever buries its head in the relativistic sand in order not to face the evil quanta," he wrote Louis de Broglie, another of his colleagues in the long struggle. He had found his gravitational theories by trusting an underlying principle, and that made him a "fanatic believer" that comparable methods would eventually lead to a unified field theory. "This should explain the ostrich policy," he wryly told de Broglie.[14]

He expressed this more formally in the concluding paragraph of his final updated appendix to his popular book, *Relativity: The Special and General Theory*. "The conviction prevails that the experimentally assured duality (corpuscular and wave structure) can be realized only by such a weakening of the concept of reality," he wrote. "I think that such a far-reaching theoretical renunciation is not for the present justified by our actual knowledge, and that one should not desist from pursuing to the end the path of the relativistic field theory."[15]

Bertrand Russell encouraged him to continue, in addition, the search for a structure that would ensure peace in the atomic age. They had both opposed the First World War, Russell recalled, and supported the Second. Now it was imperative to prevent a third. "I think that eminent men of science ought to do something dramatic to bring home to the governments the disasters that may occur," Russell wrote. Einstein replied by proposing a "public declaration" that they and perhaps a few other eminent scientists and thinkers could sign.[16]

Einstein set to work enlisting his old friend and sparring partner, Niels Bohr. "Don't frown like that!" Einstein joked, as if he were face-to-face with Bohr rather than writing to him in Copenhagen. "This has nothing to do with our old controversy on physics, but rather concerns a matter on which we are in complete agreement." Einstein admitted that his own name might carry some influence abroad, but not in America, "where I am known as a black sheep (and not merely in scientific matters)."[17]

Alas, Bohr declined, but nine other scientists, including Max Born, agreed to join the effort. Russell concluded the proposed document with a simple plea: "In view of the fact that in any future world war nuclear weapons will certainly be employed, and that such weapons threaten the continued existence of mankind, we urge the governments of the world to realize, and to acknowledge publicly, that their purpose cannot be furthered by a world war, and we urge them, consequently, to find peaceful means for the settlement of all matters of dispute between them."[18]

Einstein made it to his seventy-sixth birthday, but he was not well enough to come outside to wave to the reporters and photographers gathered in front of 112 Mercer Street. The mailman delivered presents, Oppenheimer came by with papers, the Bucky family brought some puzzles, and Johanna Fantova was there to record the events.

Among the presents was a tie sent by the fifth grade of the Farmingdale Elementary School in New York, which presumably had seen pictures of him and thought he could use one. "Neckties exist for me only as remote memories," he admitted politely in his letter of thanks.[19]

A few days later, he learned of the death of Michele Besso, the personal confessor and scientific sounding board he had met six decades earlier upon arriving as a student in Zurich. As if he knew that he had only a few more weeks, Einstein ruminated on the nature of death and time in the condolence letter he wrote to Besso's family. "He has departed from this strange world a little ahead of me. That means nothing. For us believing physicists, the distinction between past, present and future is only a stubborn illusion."

Einstein had introduced Besso to his wife, Anna Winteler, and he marveled as his friend made the marriage survive despite some difficult patches. Besso's most admirable personal trait, Einstein said, was to live in harmony with a woman, "an undertaking in which I twice failed rather miserably."[20]

One Sunday in April, the Harvard historian of science I. Bernard Cohen went to see Einstein. His face, deeply lined, struck Cohen as tragic, yet his sparkling eyes made him seem ageless. He spoke softly yet laughed loudly. "Every time he made a point that he liked," Cohen recalled, "he would burst into booming laughter."

Einstein was particularly amused by a scientific gadget, designed to show the equivalence principle, that he had recently been given. It was a version of the old-fashioned toy in which a ball that hangs by a string from the end of a stick has to be swung up so that it lands in a cup atop the stick. This one was more complex; the string tied to the ball went through the bottom of the cup and was attached to a loose spring inside the handle of the contraption. Random shaking would get the ball in the cup every now and then. The challenge: Was there a method that would get the ball in the cup every time?

As Cohen was leaving, a big grin came over Einstein's face as he said he would explain the answer to the gadget. "Now the equivalence principle!" he announced. He poked the stick upward until it almost touched the ceiling. Then he let it drop straight down. The ball, while in free fall, behaved as if it was weightless. The spring inside the contraption instantly pulled it into the cup.[21]

Einstein was now entering the last week of his life, and it is fitting that he focused on the matters most important to him. On April 11, he signed the Einstein-Russell manifesto. As Russell later declared, "He remained sane in a mad world."[22] Out of that document grew the Pugwash Conferences, in which scientists and thinkers gathered annually to discuss how to control nuclear weapons.

Later that same afternoon, Israeli Ambassador Abba Eban arrived at Mercer Street to discuss a radio address Einstein was scheduled to give to commemorate the seventh anniversary of the Jewish state. He would be heard, Eban told him, by as many as 60 million listeners. Einstein was amused. "So, I shall now have a chance to become world famous," he smiled.

After rattling around in the kitchen to make Eban a cup of coffee, Einstein told him that he saw the birth of Israel as one of the few political acts in his lifetime that had a moral quality. But he was concerned that the Jews were having trouble learning to live with the Arabs. "The attitude we adopt toward the Arab minority will provide the real test of our moral standards as a people," he had told a friend a few weeks earlier. He wanted to broaden his speech, which he was scribbling in German in a very tight and neat handwriting, to urge the creation of a world government to preserve peace.[23]

Einstein went in to work at the Institute the next day, but he had a pain in his groin and it showed on his face. Is everything all right? his assistant asked. Everything is all right, he replied, but I am not.

He stayed at home the following day, partly because the Israeli consul was coming and partly because he was still not feeling well. After the visitors left, he lay down for a nap. But Dukas heard him rush to the bathroom in the middle of the afternoon, where he collapsed. The doctors gave him morphine, which helped him sleep, and Dukas set up her bed right next to his so that she could put ice on his dehydrated lips throughout the night. His aneurysm had started to break.[24]

A group of doctors convened at his home the next day, and after some consultation they recommended a surgeon who might be able, though it was thought unlikely, to repair the aorta. Einstein refused. "It is tasteless to prolong life artificially," he told Dukas. "I have done my share, it is time to go. I will do it elegantly."

He did ask, however, whether he would suffer "a horrible death." The answer, the doctors said, was unclear. The pain of an internal hemorrhage could be excruciating. But it may take only a minute, or maybe an hour. To Dukas, who became overwrought, he smiled and said, "You're really hysterical—I have to pass on sometime, and it doesn't really matter when."[25]

Dukas found him the next morning in agony, unable to lift his head. She rushed to the telephone, and the doctor ordered him to the hospital. At first he refused, but he was told he was putting too much of a burden on Dukas, so he relented. The volunteer medic in the ambulance was a political economist at Princeton, and Einstein was able to carry on a lively conversation with him. Margot called Hans Albert, who caught a plane from San Francisco and was soon by his father's bedside. The economist Otto Nathan, a fellow German refugee who had become his close friend, arrived from New York.

But Einstein was not quite ready to die. On Sunday, April 17, he woke up feeling better. He asked Dukas to get him his glasses, papers, and pencils, and he proceeded to jot down a few calculations. He talked to Hans Albert about some scientific ideas, then to Nathan about the dangers of allowing Germany to rearm. Pointing to his equations, he lamented, half jokingly, to his son, "If only I had more mathematics."[26] For a half century he had been bemoaning both German nationalism and the limits of his mathematical toolbox, so it was fitting that these should be among his final utterances.

He worked as long as he could, and when the pain got too great he went to sleep. Shortly after one a.m. on Monday, April 18, 1955, the nurse heard him blurt out a few words in German that she could not understand. The aneurysm, like a big blister, had burst, and Einstein died at age 76.

At his bedside lay the draft of his undelivered speech for Israel Independence Day. "I speak to you today not as an American citizen and not as a Jew, but as a human being," it began.[27]

Also by his bed were twelve pages of tightly written equations, littered with cross-outs and corrections.[28] To the very end, he struggled to find his elusive unified field theory. And the final thing he wrote, before he went to sleep for the last time, was one more line of symbols and numbers that he hoped might get him, and the rest of us, just a little step closer to the spirit manifest in the laws of the universe.

$$U_i{}^n{}_k \, U_g{}^q{}_k \left(-\frac{16}{9} + \frac{2}{9} - \frac{4}{9} + \frac{2}{9} + \frac{2}{9} + \frac{2}{9} \right) + U_k{}^n{}_n \, U_g{}^q{}_i \left(\frac{4}{9} + \frac{2}{9} - \frac{1}{9} + \frac{3}{9} + \frac{1}{9} \neq \frac{1}{9} \right)$$

EINSTEIN'S BRAIN
AND EINSTEIN'S MIND

Einstein's study, as he left it

When Sir Isaac Newton died, his body lay in state in the Jerusalem chamber of Westminster Abbey, and his pallbearers included the lord high chancellor, two dukes, and three earls. Einstein could have had a similar funeral, glittering with dignitaries from around the world. Instead, in accordance with his wishes, he was cremated in Trenton on the afternoon that he died, before most of the world had heard the news. There were only twelve people at the crematorium, including Hans Albert Einstein, Helen Dukas, Otto Nathan, and four members of the Bucky family. Nathan recited a few lines from Goethe, and then took Einstein's ashes to the nearby Delaware River, where they were scattered.[1]

"No other man contributed so much to the vast expansion of 20th century knowledge," President Eisenhower declared. "Yet no other man was more modest in the possession of the power that is knowledge, more sure that power without wisdom is deadly." The *New York Times* ran nine stories plus an editorial about his death the next day: "Man stands on this diminutive earth, gazes at the myriad stars and upon billowing oceans and tossing trees—and wonders. What does it all mean? How did it come about? The most thoughtful wonderer who appeared among us in three centuries has passed on in the person of Albert Einstein."[2]

Einstein had insisted that his ashes be scattered so that his final resting place would not become the subject of morbid veneration. But there was one part of his body that was not cremated. In a drama that would seem farcical were it not so macabre, Einstein's brain ended up being, for more than four decades, a wandering relic.[3]

Hours after Einstein's death, what was supposed to be a routine autopsy was performed by the pathologist at Princeton Hospital, Thomas Harvey, a small-town Quaker with a sweet disposition and rather dreamy approach to life and death. As a distraught Otto Nathan watched silently, Harvey removed and inspected each of Einstein's major organs, ending by using an electric saw to cut through his skull and remove his brain. When he stitched the body back up, he decided, without asking permission, to embalm Einstein's brain and keep it.

The next morning, in a fifth-grade class at a Princeton school, the teacher asked her students what news they had heard. "Einstein died," said one girl, eager to be the first to come up with that piece of information. But she quickly found herself topped by a usually quiet boy who sat in the back of the class. "My dad's got his brain," he said.[4]

Nathan was horrified when he found out, as was Einstein's family. Hans Albert called the hospital to complain, but Harvey insisted that there may be scientific value to studying the brain. Einstein would have wanted that, he said. The son, unsure what legal and practical rights he now had in this matter, reluctantly went along.[5]

Soon Harvey was besieged by those who wanted Einstein's brain or a piece of it. He was summoned to Washington to meet with officials of the U.S. Army's pathology unit, but despite their requests he refused to show them his prized possession. Guarding it had become a mission. He finally decided to have friends at the University of Pennsylvania turn part of it into microscopic slides, and so he put Einstein's brain, now chopped into pieces, into two glass cookie jars and drove it there in the back of his Ford.

Over the years, in a process that was at once guileless as well as bizarre, Harvey would send off slides or chunks of the remaining brain to random researchers who struck his fancy. He demanded no rigorous studies, and for years none were published. In the meantime, he quit Princeton Hospital, left his wife, remarried a couple of times, and moved around from New Jersey to Missouri to Kansas, often leaving no forwarding address, the remaining fragments of Einstein's brain always with him.

Every now and then, a reporter would stumble across the story and track Harvey down, causing a minor media flurry. Steven Levy, then of *New Jersey Monthly* and later of *Newsweek,* found him in 1978 in Wichita, where he pulled a Mason jar of Einstein's brain chunks from a box labeled "Costa Cider" in the corner of his office behind a red plastic picnic cooler.[6] Twenty years later, Harvey was tracked down again, by Michael Paterniti, a free-spirited and soulful writer for *Harper's,* who turned his road trip in a rented Buick across America with Harvey and the brain into an award-winning article and best-selling book, *Driving Mr. Albert.*

Their destination was California, where they paid a call on Einstein's granddaughter, Evelyn Einstein. She was divorced, marginally employed, and struggling with poverty. Harvey's perambulations with the brain struck her as creepy, but she had a particular interest in one secret it might hold. She was the adopted daughter of Hans Albert and his wife Frieda, but the timing and circumstances of her birth were murky. She had heard rumors that made her suspect that possibly, just possibly, she might actually be Einstein's own daughter. She had been born after Elsa's death, when Einstein was spending time with a variety of women. Perhaps she had been the result of one of those liaisons, and he had arranged for her to be adopted by Hans Albert. Working with Robert Schulmann, an early editor of the Einstein papers, she hoped to see what could be learned by studying the DNA from Einstein's brain.

Unfortunately, it turned out that the way Harvey had embalmed the brain made it impossible to extract usable DNA. And so her questions were never answered.[7]

In 1998, after forty-three years as the wandering guardian of Einstein's brain, Thomas Harvey, by then 86, decided it was time to pass on the responsibility. So he called the person who currently held his old job as pathologist at Princeton Hospital and went by to drop it off.[8]

Of the dozens of people to whom Harvey doled out pieces of Einstein's brain over the years, only three published significant scientific studies. The first was by a Berkeley team led by Marian Diamond.[9] It reported that one area of Einstein's brain, part of the parietal cortex, had a higher ratio of what are known as glial cells to neurons. This could, the authors said, indicate that the neurons used and needed more energy.

One problem with this study was that his 76-year-old brain was compared to eleven others from men who had died at an average age of 64. There were no other geniuses in the sample to help determine if the findings fit a pattern. There was also a more fundamental problem: with no ability to trace the development of the brain over a lifetime, it was unclear which physical attributes might be the *cause* of greater intelligence and which might instead be the *effect* of years spent using and exercising certain parts of the brain.

A second paper, published in 1996, suggested that Einstein's cerebral cortex was thinner than in five other sample brains, and the density of his neurons was greater. Once again, the sample was small and evidence of any pattern was sketchy.

The most cited paper was done in 1999 by Professor Sandra Witelson and a team at McMaster University in Ontario. Harvey had sent her a fax, unprompted, offering samples for study. He was in his eighties, but he personally drove up to Canada by himself, transporting a hunk that amounted to about one-fifth of Einstein's brain, including the parietal lobe.

When compared to brains of thirty-five other men, Einstein's had a much shorter groove in one area of his inferior parietal lobe, which is thought to be key to mathematical and spatial thinking. His brain was also 15 percent wider in this region. The paper speculated that these traits may have produced richer and more integrated brain circuits in this region.[10]

But any true understanding of Einstein's imagination and intuition will not come from poking around at his patterns of glia and grooves. The relevant question was how his *mind* worked, not his brain.

The explanation that Einstein himself most often gave for his mental accomplishments was his curiosity. As he put it near the end of his life, "I have no special talents, I am only passionately curious."[11]

That trait is perhaps the best place to begin when sifting through the elements of his genius. There he is, as a young boy sick in bed, trying to figure out why the compass needle points north. Most of us can recall seeing such needles swing into place, but few of us pursue with passion the question of how a magnetic field might work, how fast it might propagate, how it could possibly interact with matter.

What would it be like to race alongside a light beam? If we are moving through curved space the way a beetle moves across a curved leaf, how would we notice it? What does it mean to say that two events are simultaneous? Curiosity, in Einstein's case, came not just from a desire to question the mysterious. More important, it came from a childlike sense of marvel that propelled him to question the familiar, those concepts that, as he once said, "the ordinary adult never bothers his head about."[12]

He could look at well-known facts and pluck out insights that had escaped the notice of others. Ever since Newton, for example, scientists had known that inertial mass was equivalent to gravitational mass. But Einstein saw that this meant that there was an equivalence between gravity and acceleration that would unlock an explanation of the universe.[13]

A tenet of Einstein's faith was that nature was not cluttered with extraneous attributes. Thus, there must be a purpose to curiosity. For Einstein, it existed because it created minds that question, which produced an appreciation for the universe that he equated with religious feelings. "Curiosity has its own reason for existing," he once explained. "One cannot help but be in awe when one contemplates the mysteries of eternity, of life, of the marvelous structure of reality."[14]

From his earliest days, Einstein's curiosity and imagination were expressed mainly through visual thinking—mental pictures and thought experiments—rather than verbally. This included the ability to visualize the physical reality that was painted by the brush strokes of mathematics. "Behind a formula he immediately saw the physical content, while for us it only remained an abstract formula," said one of his first students.[15] Planck came up with the concept of the quanta, which he viewed as mainly a mathematical contrivance, but it took Einstein to understand their physical reality. Lorentz came up with mathematical transformations that described bodies in motion, but it took Einstein to create a new theory of relativity based on them.

One day during the 1930s, Einstein invited Saint-John Perse to Princeton to find out how the poet worked. "How does the idea of a poem come?" Einstein asked. The poet spoke of the role played by intuition and imagination. "It's the same for a man of science," Einstein responded with delight. "It is a sudden illumination, almost a rapture. Later, to be sure, intelligence analyzes and experiments confirm or invalidate the intuition. But initially there is a great forward leap of the imagination."[16]

There was an aesthetic to Einstein's thinking, a sense of beauty. And one component to beauty, he felt, was simplicity. He had echoed Newton's dictum "Nature is pleased with simplicity" in the creed he declared at Oxford the year he left Europe for America: "Nature is the realization of the simplest conceivable mathematical ideas."[17]

Despite Occam's razor and other philosophical maxims along these lines, there is no self-evident reason this has to be true. Just as it is possible that God might actually play dice, so too it is possible that he might delight in Byzantine complexities. But Einstein didn't think so. "In building a theory, his approach had something in common with that of an artist," said Nathan Rosen, his assistant in the 1930s. "He would aim for simplicity and beauty, and beauty for him was, after all, essentially simplicity."[18]

He became like a gardener weeding a flower bed. "I believe what allowed Einstein to achieve so much was primarily a moral quality," said physicist Lee Smolin. "He simply cared far more than most of his colleagues that the laws of physics have to explain everything in nature coherently and consistently."[19]

Einstein's instinct for unification was ingrained in his personality and reflected in his politics. Just as he sought a unified theory in science that could govern the cosmos, so he sought one in politics that could govern the planet, one that would overcome the anarchy of unfettered nationalism through a world federalism based on universal principles.

Perhaps the most important aspect of his personality was his willingness to be a nonconformist. It was an attitude that he celebrated in a foreword he wrote near the end of his life to a new edition of Galileo. "The theme that I recognize in Galileo's work," he said, "is the passionate fight against any kind of dogma based on authority."[20]

Planck and Poincaré and Lorentz all came close to some of the breakthroughs Einstein made in 1905. But they were a little too confined by dogma based on authority. Einstein alone among them was rebellious enough to throw out conventional thinking that had defined science for centuries.

This joyous nonconformity made him recoil from the sight of Prussian soldiers marching in lockstep. It was a personal outlook that became a political one as well. He bristled at all forms of tyranny over free minds, from Nazism to Stalinism to McCarthyism.

Einstein's fundamental creed was that freedom was the lifeblood of creativity. "The development of science and of the creative activities of the spirit," he said, "requires a freedom that consists in the independence of thought from the restrictions of authoritarian and social prejudice." Nurturing that should be the fundamental role of government, he felt, and the mission of education.[21]

There was a simple set of formulas that defined Einstein's outlook. Creativity required being willing not to conform. That required nurturing free minds and free spirits, which in turn required "a spirit of tolerance." And the underpinning of tolerance was humility—the belief that no one had the right to impose ideas and beliefs on others.

The world has seen a lot of impudent geniuses. What made Einstein special was that his mind and soul were tempered by this humility. He could be serenely self-confident in his lonely course yet also humbly awed by the beauty of nature's handiwork. "A spirit is manifest in the laws of the universe—a spirit vastly superior to that of man, and one in the face of which we with our modest powers must feel humble," he wrote. "In this way the pursuit of science leads to a religious feeling of a special sort." [22]

For some people, miracles serve as evidence of God's existence. For Einstein it was the absence of miracles that reflected divine providence. The fact that the cosmos is comprehensible, that it follows laws, is worthy of awe. This is the defining quality of a "God who reveals himself in the harmony of all that exists." [23]

Einstein considered this feeling of reverence, this cosmic religion, to be the wellspring of all true art and science. It was what guided him. "When I am judging a theory," he said, "I ask myself whether, if I were God, I would have arranged the world in such a way." [24] It is also what graced him with his beautiful mix of confidence and awe.

He was a loner with an intimate bond to humanity, a rebel who was suffused with reverence. And thus it was that an imaginative, impertinent patent clerk became the mind reader of the creator of the cosmos, the locksmith of the mysteries of the atom and the universe.

SOURCES

EINSTEIN'S CORRESPONDENCE AND WRITINGS

The Collected Papers of Albert Einstein, vols. 1–10. 1987–2006. Princeton: Princeton University Press. (Abbreviated CPAE)

    The founding editor was John Stachel. The current general editor is Diana Kormos Buchwald. Other editors over the years include David Cassidy, Robert Schulmann, Jürgen Renn, Martin Klein, A. J. Knox, Michel Janssen, Jósef Illy, Christoph Lehner, Daniel Kennefick, Tilman Sauer, Ze'ev Rosenkranz, and Virginia Iris Holmes.

    These volumes cover the years 1879–1920. Each volume comes in a German version and an English translation. The page numbers in each differ, but the document numbers are the same. In cases where I cite some information that is in one version but not the other (such as an editor's essay or footnote), I designate the volume and language version and cite the page number.

Albert Einstein Archives. (Abbreviated AEA)

    These archives are now at Hebrew University in Jerusalem with copies at the Einstein Papers Project at Caltech and in the Princeton University library. Documents from the archives are cited both by date and by the AEA folder (reel) and document number. In the case of most of the untranslated German documents, I have relied on translations made for me by James Hoppes and Natasha Hoffmeyer.

FREQUENTLY CITED WORKS

Abraham, Carolyn. 2001. *Possessing Genius*. New York: St. Martin's Press.

Aczel, Amir. 1999. *God's Equation: Einstein, Relativity, and the Expanding Universe*. New York: Random House.

———. 2002. *Entanglement: The Unlikely Story of How Scientists, Mathematicians, and Philosophers Proved Einstein's Spookiest Theory*. New York: Plume.

Baierlein, Ralph. 2001. *Newton to Einstein: The Trail of Light, an Excursion to the Wave-Particle Duality and the Special Theory of Relativity*. New York: Cambridge University Press.

Barbour, Julian, and Herbert Pfister, eds. 1995. *Mach's Principle: From Newton's Bucket to Quantum Gravity*. Boston: Birkhäuser.

Bartusiak, Marcia. 2000. *Einstein's Unfinished Symphony*. New York: Berkley.

Batterson, Steve. 2006. *Pursuit of Genius*. Wellesley, Mass.: A. K. Peters.

Beller, Mara, et al., eds. 1993. *Einstein in Context*. Cambridge, England: Cambridge University Press.

Bernstein, Jeremy. 1973. *Einstein*. Modern Masters Series. New York: Viking.

———. 1991. *Quantum Profiles*. Princeton: Princeton University Press.

———. 1996a. *Albert Einstein and the Frontiers of Physics*. New York: Oxford University Press.

———. 1996b. *A Theory for Everything*. New York: Springer-Verlag.

———. 2001. *The Merely Personal*. Chicago: Ivan Dee.

———. 2006. *Secrets of the Old One: Einstein, 1905*. New York: Copernicus.

Besso, Michele. 1972. *Correspondence 1903–1955*. In German with parallel French translation by Pierre Speziali. Paris: Hermann.

Bird, Kai, and Martin J. Sherwin. 2005. *American Prometheus: The Triumph and Tragedy of J. Robert Oppenheimer*. New York: Knopf.

Bodanis, David. 2000. *E=mc²: A Biography of the World's Most Famous Equation*. New York: Walker.

Bolles, Edmund Blair. 2004. *Einstein Defiant: Genius versus Genius in the Quantum Revolution*. Washington, D.C.: Joseph Henry.

Born, Max. 1978. *My Life: Recollections of a Nobel Laureate*. New York: Scribner's.

———. 2005. *Born-Einstein Letters*. New York: Walker Publishing. (Originally published in 1971, with new material for the 2005 edition)

Brian, Denis. 1996. *Einstein: A Life*. Hoboken, N.J.: Wiley.

———. 2005. *The Unexpected Einstein*. Hoboken, N.J.: Wiley.

Brockman, John, ed. 2006. *My Einstein*. New York: Pantheon.

Bucky, Peter. 1992. *The Private Albert Einstein*. Kansas City, Mo.: Andrews and McMeel.

Cahan, David. 2000. "The Young Einstein's Physics Education." In Howard and Stachel 2000.

Calaprice, Alice, ed. 2005. *The New Expanded Quotable Einstein*. Princeton: Princeton University Press.

Calder, Nigel. 1979. *Einstein's Universe: A Guide to the Theory of Relativity*. New York: Viking Press. (Reissued by Penguin Press in 2005)

Carroll, Sean M. 2003. *Spacetime and Geometry: An Introduction to General Relativity*. Boston: Addison-Wesley.

Cassidy, David C. 2004. *Einstein and Our World*. Amherst, N.Y.: Humanity Books.

Clark, Ronald. 1971. *Einstein: The Life and Times*. New York: HarperCollins.

Corry, Leo, Jürgen Renn, and John Stachel. 1997. "Belated Decision in the Hilbert-Einstein Priority Dispute." *Science* 278: 1270–1273.

Crelinsten, Jeffrey. 2006. *Einstein's Jury: The Race to Test Relativity*. Princeton: Princeton University Press.

Damour, Thibault. 2006. *Once upon Einstein*. Wellesley, Mass.: A. K. Peters.

Douglas, Vibert. 1956. *The Life of Arthur Stanley Eddington*. London: Thomas Nelson.

Dukas, Helen, and Banesh Hoffmann, eds. 1979. *Albert Einstein: The Human Side. New Glimpses from His Archives*. Princeton: Princeton University Press.

Dyson, Freeman. 2003. "Clockwork Science." (Review of Galison). *New York Review of Books*, Nov. 6.

Earman, John. 1978. *World Enough and Space-Time*. Cambridge, Mass.: MIT Press.

Earman, John, Clark Glymour, and Robert Rynasiewicz. 1982. "On Writing the History of Special Relativity." *Philosophy of Science Association Journal* 2: 403–416.

Earman, John, et al., eds. 1993. *The Attraction of Gravitation: New Studies in the History of General Relativity*. Boston: Birkhäuser.

Einstein, Albert. 1916. *Relativity: The Special and the General Theory*. (Written as a popular account, this book was published in German in December 1916. An authorized English translation was first published in 1920 by Methuen in London and Henry Holt in New York. It went through fifteen English-language editions in his lifetime, and he added appendixes up until 1952. It is available now from multiple publishers. The version I cite is the 1995 Random House edition. The book can be found at www.bartleby.com/173/ and at www.gutenberg.org/etext/5001.)

———. 1922a. *The Meaning of Relativity*. Princeton: Princeton University Press. (A technical exposition based on his 1921 lectures at Princeton. The fifth edition, published in 1954, contains an appendix revising his attempt at a unified field theory. The 2005 edition from Princeton University Press contains an introduction by Brian Greene.)

———. 1922b. *Sidelights on Relativity*. New York: Dutton.

———. 1922c. "How I Created the Theory of Relativity." Talk in Kyoto, Japan, Dec. 14. (I have used a new, corrected, and heretofore unpublished translation. Einstein's Kyoto talk was published in Japanese in 1923 by theoretical physicist Jun Ishiwara, who was present and took notes. His version was translated into English by Yoshimasa A. Ono and published in *Physics Today* in August 1982. This translation, which has been used by most previous writers on Einstein, is flawed, especially in the parts where Einstein refers to the Michelson-Morley experiments; see Ryoichi Itagaki, "Einstein's Kyoto Lecture," *Science* magazine, vol. 283, March 5, 1999. A proper and corrected translation by Prof. Itagaki will appear in a forthcoming volume of CPAE. I am grateful to Gerald Holton for providing me with a copy of this translation. See also Seiya Abiko, "Einstein's Kyoto Address," *Historical Studies in the Physical and Biological Sciences* 31 (2000): 1–35.)

———. 1934. *Essays in Science*. New York: Philosophical Library.

———. 1949a. *The World As I See It*. New York: Philosophical Library. (Based on *Mein Weltbild*, edited by Carl Seelig.)

———. 1949b. "Autobiographical Notes." In Schilpp 1949, 3–94.

———. 1950a. *Out of My Later Years*. New York: Philosophical Library.

———. 1950b. *Einstein on Humanism*. New York: Philosophical Library.

———. 1954. *Ideas and Opinions.* New York: Random House.

———. 1956. "Autobiographische Skizze." In Seelig 1956b.

Einstein, Albert, and Leopold Infeld. 1938. *The Evolution of Physics: The Growth of Ideas from Early Concepts to Relativity and Quanta.* New York: Simon & Schuster.

Einstein, Elizabeth Roboz. 1991. *Hans Albert Einstein: Reminiscences of Our Life Together.* Iowa City: University of Iowa Press.

Einstein, Maja. 1923. "Albert Einstein—A Biographical Sketch." CPAE 1: xv. (This sketch was originally written in 1923 as the start of a book she hoped to write, but it was never published by her. It tracks her brother's life only until 1905. See lorentz.phl.jhu.edu/AnnusMirabilis/AeReserveArticles/maja.pdf.)

Eisenstaedt, Jean, and A. J. Kox, eds. 1992. *Studies in the History of General Relativity.* Boston: Birkhäuser.

Elon, Amos. 2002. *The Pity of It All: A History of the Jews in Germany, 1743–1933.* New York: Henry Holt.

Elzinga, Aant. 2006. *Einstein's Nobel Prize.* Sagamore Beach, Mass.: Science History Publications.

Fantova, Johanna. "Journal of Conversations with Einstein, 1953–55." In Princeton University Einstein Papers archives and published as an appendix in Calaprice 2005. (For clarity and because the page numbers vary in different editions of Calaprice, I identify Fantova's entries by date.)

Federal Bureau of Investigation, Files on Einstein. Available through the Freedom of Information Act website, foia.fbi.gov/foiaindex/einstein.htm.

Feynman, Richard. 1997. *Six Not-So-Easy Pieces: Einstein's Relativity, Symmetry, and Space-Time.* Boston: Addison-Wesley.

———. 1999. *The Pleasure of Finding Things Out.* Cambridge, England: Perseus.

———. 2002. *The Feynman Lectures on Gravitation.* Boulder, Colo.: Westview Press.

Fine, Arthur. 1996. *The Shaky Game: Einstein, Realism, and the Quantum Theory.* Chicago: University of Chicago Press. (Revised edition of original 1986 publication.)

Flexner, Abraham. 1960. *An Autobiography.* New York: Simon & Schuster.

Flückiger, Max. 1974. *Albert Einstein in Bern.* Bern: Haupt.

Fölsing, Albrecht. 1997. *Albert Einstein: A Biography.* Translated and abridged by Ewald Osers. New York: Viking. (Original unabridged edition in German published in 1993.)

Frank, Philipp. 1947. *Einstein: His Life and Times.* Translated by George Rosen. New York: Da Capo Press. (Reprinted in 2002.)

———. 1957. *Philosophy of Science.* Saddle River, N.J.: Prentice-Hall.

French, A. P., ed. 1979. *Einstein: A Centenary Volume.* Cambridge, Mass.: Harvard University Press.

Friedman, Alan J., and Carol C. Donley. 1985. *Einstein as Myth and Muse.* Cambridge, England: Cambridge University Press.

Friedman, Robert Marc. 2005. "Einstein and the Nobel Committee." *Europhysics News,* July/Aug.

Galileo Galilei. 1632. *Dialogue Concerning the Two Chief World Systems: Ptolemaic and Copernican*. (I use the 2001 Modern Library edition translated by Stillman Drake, foreword by Albert Einstein, introduction by John Heilbron.)

Galison, Peter. 2003. *Einstein's Clocks, Poincaré's Maps*. New York: Norton.

Gamow, George. 1966. *Thirty Years That Shook Physics: The Story of Quantum Theory*. New York: Dover.

————. 1970. *My World Line*. New York: Viking.

————. 1993. *Mr. Tompkins in Paperback*. New York: Cambridge University Press.

Gardner, Martin. 1976. *The Relativity Explosion*. New York: Vintage.

Gell-Mann, Murray. 1994. *The Quark and the Jaguar*. New York: Henry Holt.

Goenner, Hubert. 2004. "On the History of Unified Field Theories." Living Reviews in Relativity website, relativity.livingreviews.org/.

————. 2005. *Einstein in Berlin*. Munich: Beck Verlag.

Goenner, Hubert, et al., eds. 1999. *The Expanding Worlds of General Relativity*. Boston: Birkhäuser.

Goldberg, Stanley. 1984. *Understanding Relativity: Origin and Impact of a Scientific Revolution*. Boston: Birkhäuser.

Goldsmith, Maurice, et al. 1980. *Einstein: The First Hundred Years*. New York: Pergamon Press.

Goldstein, Rebecca. 2005. *Incompleteness: The Proof and Paradox of Kurt Gödel*. New York: Atlas/Norton.

Greene, Brian. 1999. *The Elegant Universe: Superstrings, Hidden Dimensions, and the Quest for the Ultimate Theory*. New York: Norton.

————. 2004. *The Fabric of the Cosmos: Space, Time, and the Texture of Reality*. New York: Knopf.

Gribbin, John, and Mary Gribbin. 2005. *Annus Mirabilis: 1905, Albert Einstein, and the Theory of Relativity*. New York: Chamberlain Brothers.

Haldane, Richard. 1921. *The Reign of Relativity*. London: Murray. (Reprinted in 2003 by the University Press of the Pacific in Honolulu.)

Hartle, James. 2002. *Gravity: An Introduction to Einstein's General Relativity*. Boston: Addison-Wesley.

Hawking, Stephen. 1999. "A Brief History of Relativity." *Time*, Dec. 31.

————. 2001. *The Universe in a Nutshell*. New York: Bantam.

————. 2005. "Does God Play Dice?" Available at www.hawking.org.uk/lectures/lindex.html.

Hawking, Stephen, and Roger Penrose. 1996. *The Nature of Space and Time*. Princeton: Princeton University Press.

Heilbron, John. 2000. *The Dilemmas of an Upright Man: Max Planck and the Fortunes of German Science*. Cambridge, Mass.: Harvard University Press. (Revised edition of 1986 book.)

Heisenberg, Werner. 1958. *Physics and Philosophy*. New York: Harper.

————. 1971. *Physics and Beyond: Encounters and Conversations*. New York: Harper & Row.

————. 1989. *Encounters with Einstein*. Princeton: Princeton University Press.

Highfield, Roger, and Paul Carter. 1994. *The Private Lives of Albert Einstein.* New York: St. Martin's Press.

Hoffmann, Banesh, with the collaboration of Helen Dukas. 1972. *Albert Einstein: Creator and Rebel.* New York: Viking.

Hoffmann, Banesh. 1983. *Relativity and Its Roots.* New York: Scientific American Books.

Holmes, Frederick L., Jürgen Renn, and Hans-Jörg Rheinberger, eds. 2003. *Reworking the Bench: Research Notebooks in the History of Science.* Dordrecht: Kluwer.

Holton, Gerald. 1973. *Thematic Origins of Scientific Thought: Kepler to Einstein.* Cambridge, Mass.: Harvard University Press.

———. 2000. *Einstein, History, and Other Passions: The Rebellion against Science at the End of the Twentieth Century.* Cambridge, Mass.: Harvard University Press.

———. 2003. "Einstein's Third Paradise." *Daedalus* 132, no. 4 (fall): 26–34. Available at www.physics.harvard.edu/holton/3rdParadise.pdf.

Holton, Gerald, and Stephen Brush. 2004. *Physics, the Human Adventure.* New Brunswick, N.J.: Rutgers University Press.

Holton, Gerald, and Yehuda Elkana, eds. 1997. *Albert Einstein: Historical and Cultural Perspectives.* The Centennial Symposium in Jerusalem. Mineola, N.Y.: Dover Publications.

Howard, Don. 1985. "Einstein on Locality and Separability." *Studies in History and Philosophy of Science* 16: 171–201.

———. 1990a. "Einstein and Duhem." *Synthese* 83: 363–384.

———. 1990b. " 'Nicht sein kann was nicht sein darf,' or The Prehistory of EPR, 1909–1935. Einstein's Early Worries about the Quantum Mechanics of Composite Systems." In Arthur Miller, ed., *Sixty-two Years of Uncertainty: Historical, Philosophical, and Physical Inquiries into the Foundations of Quantum Mechanics.* New York: Plenum, 61–111.

———. 1993. "Was Einstein Really a Realist?" *Perspectives on Science* 1: 204–251.

———. 1997. "A Peek behind the Veil of Maya: Einstein, Schopenhauer, and the Historical Background of the Conception of Space as a Ground for the Individuation of Physical Systems." In John Earman and John D. Norton, eds., *The Cosmos of Science: Essays of Exploration.* Pittsburgh: University of Pittsburgh Press, 87–150.

———. 2004. "Albert Einstein, Philosophy of Science." *Stanford Encyclopedia of Philosophy.* Available at plato.stanford.edu/entries/einstein-philscience/.

———. 2005. "Albert Einstein as a Philosopher of Science." *Physics Today,* Dec., 34.

Howard, Don, and John Norton. 1993. "Out of the Labyrinth? Einstein, Hertz, and the Göttingen Answer to the Hole Argument." In Earman et al. 1993.

Howard, Don, and John Stachel, eds. 1989. *Einstein and the History of General Relativity.* Boston: Birkhäuser.

———, eds. 2000. *Einstein: The Formative Years, 1879–1909.* Boston: Birkhäuser.

Illy, József, ed. 2005, February. "Einstein Due Today." Manuscript. (Courtesy of the Einstein Papers Project, Pasadena. Includes newspaper clippings about Einstein's 1921 visit. Forthcoming publication planned as *Albert Meets America.* Baltimore: Johns Hopkins University Press.)

Infeld, Leopold. 1950. *Albert Einstein: His Work and Its Influence on Our World*. New York: Scribner's.

Jammer, Max. 1989. *The Conceptual Development of Quantum Mechanics*. Los Angeles: American Institute of Physics.

——. 1999. *Einstein and Religion: Physics and Theology*. Princeton: Princeton University Press.

Janssen, Michel. 1998. "Rotation as the Nemesis of Einstein's Entwurf Theory." In Goenner et al. 1999.

——. 2002. "The Einstein-Besso Manuscript: A Glimpse behind the Curtain of the Wizard." Available at www.tc.umn.edu/~janss011/.

——. 2004. "Einstein's First Systematic Exposition of General Relativity." Available at philsci-archive.pitt.edu/archive/00002123/01/annalen.pdf.

——. 2005. "Of Pots and Holes: Einstein's Bumpy Road to General Relativity." *Annalen der Physik* 14 (Supplement): 58–85.

——. 2006. "What Did Einstein Know and When Did He Know It? A Besso Memo Dated August 1913." Available at www.tc.umn.edu/~janss011/.

Janssen, Michel, and Jürgen Renn. 2004. "Untying the Knot: How Einstein Found His Way Back to Field Equations Discarded in the Zurich Notebook." Available at www.tc.umn.edu/~janss011/pdf%20files/knot.pdf.

Jerome, Fred. 2002. *The Einstein File: J. Edgar Hoover's Secret War against the World's Most Famous Scientist*. New York: St. Martin's Press.

Jerome, Fred, and Rodger Taylor. 2005. *Einstein on Race and Racism*. New Brunswick, N.J.: Rutgers University Press.

Kaku, Michio. 2004. *Einstein's Cosmos: How Albert Einstein's Vision Transformed Our Understanding of Space and Time*. New York: Atlas Books.

Kessler, Harry. 1999. *Berlin in Lights: The Diaries of Count Harry Kessler (1918–1937)*. Translated and edited by Charles Kessler. New York: Grove Press.

Klein, Martin J. 1970a. *Paul Ehrenfest: The Making of a Theoretical Physicist*. New York: American Elsevier.

——. 1970b. "The First Phase of the Bohr-Einstein Dialogue." *Historical Studies in the Physical Sciences* 2: 1–39.

Kox, A. J., and Jean Eisenstaedt, eds. 2005. *The Universe of General Relativity. Vol. II of Einstein Studies*. Boston: Birkhäuser.

Krauss, Lawrence. 2005. *Hiding in the Mirror*. New York: Viking.

Levenson, Thomas. 2003. *Einstein in Berlin*. New York: Bantam Books.

Levy, Steven. 1978. "My Search for Einstein's Brain." *New Jersey Monthly*, Aug.

Lightman, Alan. 1993. *Einstein's Dreams*. New York: Pantheon Books.

——. 1999. "A New Cataclysm of Thought." *Atlantic Monthly*, Jan.

——. 2005. *The Discoveries*. New York: Pantheon.

Lightman, Alan, et al. 1975. *Problem Book in Relativity and Gravitation*. Princeton: Princeton University Press.

Marianoff, Dimitri. 1944. *Einstein: An Intimate Study of a Great Man*. New York: Doubleday. (Marianoff married and then divorced Margot Einstein, a daughter of Einstein's second wife Elsa, and Einstein denounced this book.)

Mehra, Jagdish. 1975. *The Solvay Conferences on Physics: Aspects of the Development of Physics Since 1911.* Dordrecht: D. Reidel.

Mermin, N. David. 2005. *It's about Time: Understanding Einstein's Relativity.* Princeton: Princeton University Press.

Michelmore, Peter. 1962. *Einstein: Profile of the Man.* New York: Dodd, Mead.

Miller, Arthur I. 1981. *Albert Einstein's Special Theory of Relativity: Emergence (1905) and Early Interpretation (1905–1911).* Boston: Addison-Wesley.

————. 1984. *Imagery in Scientific Thought.* Boston: Birkhäuser.

————. 1992. "Albert Einstein's 1907 Jahrbuch Paper: The First Step from SRT to GRT." In Eisenstaedt and Kox 1992, 319–335.

————. 1999. *Insights of Genius.* New York: Springer-Verlag.

————. 2001. *Einstein, Picasso: Space, Time and the Beauty That Causes Havoc.* New York: Basic Books.

————. 2005. *Empire of the Stars.* New York: Houghton Mifflin.

Misner, Charles, Kip Thorne, and John Archibald Wheeler. 1973. *Gravitation.* San Francisco: Freeman.

Moore, Ruth. 1966. *Niels Bohr: The Man, His Science, and the World They Changed.* New York: Knopf.

Moszkowski, Alexander. 1921. *Einstein the Searcher: His Work Explained from Dialogues with Einstein.* New York: Dutton.

Nathan, Otto, and Heinz Norden, eds. 1960. *Einstein on Peace.* New York: Simon & Schuster.

Neffe, Jürgen. 2005. *Einstein: Eine Biographie.* Hamburg: Rowohlt.

Norton, John D. 1984. "How Einstein Found His Field Equations." *Historical Studies in the Physical Sciences.* Reprinted in Howard and Stachel 1989, 101–159.

————. 1985. "What Was Einstein's Principle of Equivalence?" *Studies in History and Philosophy of Science* 16: 203–246. Reprinted in Howard and Stachel 1989, 5–47.

————. 1991. "Thought Experiments in Einstein's Work." In Tamara Horowitz and Gerald Massey, eds., *Thought Experiments in Science and Philosophy.* Savage, Md.: Rowman and Littlefield, 129–148.

————. 1993. "General Covariance and the Foundations of General Relativity: Eight Decades of Dispute." *Reports on Progress in Physics* 56: 791–858.

————. 1995a. "Eliminative Induction as a Method of Discovery: Einstein's Discovery of General Relativity." In Jarrett Leplin, ed., *The Creation of Ideas in Physics: Studies for a Methodology of Theory Construction.* Dordrecht: Kluwer, 29–69.

————. 1995b. "Did Einstein Stumble? The Debate over General Covariance." *Erkenntnis* 42: 223–245.

————. 1995c. "Mach's Principle before Einstein." Available at www.pitt.edu/~jdnorton/papers/MachPrinciple.pdf.

————. 2000. "Nature Is the Realization of the Simplest Conceivable Mathematical Ideas: Einstein and the Canon of Mathematical Simplicity." *Studies in the History and Philosophy of Modern Physics* 31: 135–170.

————. 2002. "Einstein's Triumph Over the Spacetime Coordinate System." *Dialogos* 79: 253–262.

————. 2004. "Einstein's Investigations of Galilean Covariant Electrodynamics prior to 1905." *Archive for History of Exact Sciences* 59: 45–105.

————. 2005a. "How Hume and Mach Helped Einstein Find Special Relativity." Available at www.pitt.edu/~jdnorton.

————. 2005b. "A Conjecture on Einstein, the Independent Reality of Spacetime Coordinate Systems and the Disaster of 1913." In Kox and Eisenstaedt 2005.

————. 2006a. "Einstein's Special Theory of Relativity and the Problems in the Electrodynamics of Moving Bodies That Led Him to It." Available at www.pitt.edu/~jdnorton/homepage/cv.html.

————. 2006b. "What Was Einstein's 'Fateful Prejudice'?" In Jürgen Renn, *The Genesis of General Relativity*, vol. 2. Dordrecht: Kluwer.

————. 2006c. "Atoms, Entropy, Quanta: Einstein's Miraculous Argument of 1905." Available at www.pitt.edu/~jdnorton.

Overbye, Dennis. 2000. *Einstein in Love: A Scientific Romance*. New York: Viking.

Pais, Abraham. 1982. *Subtle Is the Lord: The Science and Life of Albert Einstein*. New York: Oxford University Press.

————. 1991. *Niels Bohr's Times in Physics, Philosophy, and Polity*. Oxford: Clarendon Press.

————. 1994. *Einstein Lived Here: Essays for the Layman*. New York: Oxford University Press.

Panek, Richard. 2004. *The Invisible Century: Einstein, Freud, and the Search for Hidden Universes*. New York: Viking.

Parzen, Herbert. 1974. *The Hebrew University: 1925–1935*. New York: KTAV.

Paterniti, Michael. 2000. *Driving Mr. Albert*. New York: Dial.

Pauli, Wolfgang. 1994. *Writings on Physics and Philosophy*. Berlin: Springer-Verlag.

Penrose, Roger. 2005. *The Road to Reality*. New York: Knopf.

Poincaré, Henri. 1902. *Science and Hypothesis*. Available at spartan.ac.brocku.ca/~lward/Poincare/Poincare_1905_toc.html.

Popović, Milan. 2003. *In Albert's Shadow: The Life and Letters of Mileva Marić*. Baltimore: Johns Hopkins University Press.

Powell, Corey. 2002. *God in the Equation*. New York: Free Press.

Pyenson, Lewis. 1985. *The Young Einstein*. Boston: Adam Hilger.

Regis, Ed. 1988. *Who Got Einstein's Office?* New York: Addison-Wesley.

Reid, Constance. 1986. *Hilbert-Courant*. New York: Springer-Verlag.

Reiser, Anton. 1930. *Albert Einstein: A Biographical Portrait*. New York: Boni. (Reiser was the pseudonym of Rudoph Kayser, who married Ilse Einstein, the daughter of Einstein's second wife Elsa.)

Renn, Jürgen. 1994. "The Third Way to General Relativity." Max Planck Institute, www.mpiwg-berlin.mpg.de/Preprints/P9.pdf.

————. 2005a. "Einstein's Controversy with Drude and the Origin of Statistical Mechanics." In Howard and Stachel 2000.

————. 2005b. "Standing on the Shoulders of a Dwarf." In Kox and Eisenstaedt 2005.

————. 2005c. "Before the Riemann Tensor: The Emergence of Einstein's Double Strategy." In Kox and Eisenstaedt 2005.

———. 2005d. *Albert Einstein: Chief Engineer of the Universe. One Hundred Authors for Einstein.* Hoboken, N. J.: Wiley.

———. 2006. *Albert Einstein: Chief Engineer of the Universe. Einstein's Life and Work in Context and Documents of a Life's Pathway.* Hoboken, N. J.: Wiley.

Renn, Jürgen, and Tilman Sauer. 1997. "The Rediscovery of General Relativity in Berlin." Max Planck Institute, www.mpiwg-berlin.mpg.de/en/forschung/ Preprints/P63.pdf.

———. 2003. "Errors and Insights: Reconstructing the Genesis of General Relativity from Einstein's Zurich Notebook." In Holmes et al. 2003, 253–268.

———. 2006. "Pathways out of Classical Physics: Einstein's Double Strategy in Searching for the Gravitational Field Equation." Available at www.hss .caltech.edu/~tilman/.

Renn, Jürgen, and Robert Schulmann, eds. 1992. *Albert Einstein and Mileva Marić: The Love Letters.* Princeton: Princeton University Press.

Rhodes, Richard. 1987. *The Making of the Atom Bomb.* New York: Simon & Schuster.

Rigden, John. 2005. *Einstein 1905: The Standard of Greatness.* Cambridge, England: Cambridge University Press.

Robinson, Andrew. 2005. *Einstein: A Hundred Years of Relativity.* New York: Abrams.

Rosenkranz, Ze'ev. 1998. *Albert through the Looking Glass: The Personal Papers of Albert Einstein.* Jerusalem: Hebrew University Press.

———. 2002. *The Einstein Scrapbook.* Baltimore: Johns Hopkins University Press.

Rowe, David E., and Robert Schulmann, eds. 2007. *Einstein's Political World.* Princeton: Princeton University Press.

Rozental, Stefan, ed. 1967. *Niels Bohr: His Life and Work As Seen by His Friends and Colleagues.* Hoboken, N. J.: Wiley.

Ryan, Dennis P., ed. 1987. *Einstein and the Humanities.* New York: Greenwood Press.

Ryckman, Thomas. 2005. *The Reign of Relativity.* Oxford: Oxford University Press.

Rynasiewicz, Robert. 1988. "Lorentz's Local Time and the Theorem of Corresponding States." *Philosophy of Science Association Journal* 1: 67–74.

———. 2000. "The Construction of the Special Theory: Some Queries and Considerations." In Howard and Stachel 2000.

Rynasiewicz, Robert, and Jürgen Renn. 2006. "The Turning Point for Einstein's Annus Mirabilis." *Studies in the History and Philosophy of Modern Physics* 37, Mar.

Sartori, Leo. 1996. *Understanding Relativity.* Berkeley: Univ. of California Press.

Sauer, Tilman. 1999. "The Relativity of Discovery: Hilbert's First Note on the Foundations of Physics." *Archive for History of Exact Sciences* 53: 529–575.

———. 2005. "Einstein Equations and Hilbert Action: What Is Missing on Page 8 of the Proofs for Hilbert's First Communication on the Foundations of Physics?" *Archive for History of Exact Sciences* 59: 577.

Sayen, Jamie. 1985. *Einstein in America: The Scientist's Conscience in the Age of Hitler and Hiroshima.* New York: Crown.

Schilpp, Paul Arthur, ed. 1949. *Albert Einstein: Philosopher–Scientist*. La Salle, Ill.: Open Court Press.

Seelig, Carl. 1956a. *Albert Einstein: A Documentary Biography*. Translated by Mervyn Savill. London: Staples Press. (Translation of *Albert Einstein: Eine Dokumentarische Biographie*, a revision of *Albert Einstein und die Schweiz*. Zürich: Europa-Verlag, 1952.)

———, ed. 1956b. *Helle Zeit, Dunkle Zeit: In Memoriam Albert Einstein*. Zürich: Europa-Verlag.

Singh, Simon. 2004. *Big Bang: The Origin of the Universe*. New York: Harper-Collins.

Solovine, Maurice. 1987. *Albert Einstein: Letters to Solovine*. New York: Philosophical Library.

Sonnert, Gerhard. 2005. *Einstein and Culture*. Amherst, N.Y.: Humanity Books.

Speziali, Maurice, ed. 1956. *Albert Einstein–Michele Besso, Correspondence 1903–1955*. Paris: Hermann.

Stachel, John. 1980. "Einstein and the Rigidly Rotating Disk." In A. Held, ed., *General Relativity and Gravitation: A Hundred Years after the Birth of Einstein*. New York: Plenum, 1–15.

———. 1987. "How Einstein Discovered General Relativity." In M. A. H. Mac-Callum, ed., *General Relativity and Gravitation: Proceedings of the 11th International Conference on General Relativity and Gravitation*. Cambridge, England: Cambridge University Press, 200–208.

———. 1989a. "The Rigidly Rotating Disk as the Missing Link in the History of General Relativity." In Howard and Stachel 1989.

———. 1989b. "Einstein's Search for General Covariance, 1912–1915." In Howard and Stachel 1989.

———. 1998. *Einstein's Miraculous Year: Five Papers That Changed the Face of Physics*. Princeton: Princeton University Press.

———. 2002a. *Einstein from "B" to "Z."* Boston: Birkhäuser.

———. 2002b. "What Song the Syrens Sang: How Did Einstein Discover Special Relativity?" In Stachel 2002a.

———. 2002c. "Einstein and Ether Drift Experiments." In Stachel 2002a.

Stern, Fritz. 1999. *Einstein's German World*. Princeton: Princeton University Press.

Talmey, Max. 1932. *The Relativity Theory Simplified, and the Formative Period of Its Inventor*. New York: Falcon Press.

Taylor, Edwin, and J. Archibald Wheeler. 1992. *Spacetime Physics: Introduction to Special Relativity*. New York: W. H. Freeman.

———. 2000. *Exploring Black Holes*. New York: Benjamin/Cummings.

Thorne, Kip. 1995. *Black Holes and Time Warps: Einstein's Outrageous Legacy*. New York: Norton.

Trbuhovic-Gjuric, Desanka. 1993. *In the Shadow of Albert Einstein*. Bern: Verlag Paul Haupt.

Vallentin, Antonina. 1954. *The Drama of Albert Einstein*. New York: Doubleday.

van Dongen, Jeroen. 2002. "Einstein's Unification: General Relativity and the Quest for Mathematical Naturalness." Ph.D. dissertation, Univ. of Amsterdam.

Viereck, George Sylvester. 1930. *Glimpses of the Great*. New York: Macauley. (Einstein profile first published as "What Life Means to Einstein," *Saturday Evening Post*, Oct. 26, 1929.)

Walter, Scott. 1998. "Minkowski, Mathematicians, and the Mathematical Theory of Relativity." In Goenner et al. 1999.

Weart, Spencer, and Gertrud Weiss Szilard, eds. 1978. *Leo Szilard: His Version of the Facts*. Cambridge, Mass.: MIT Press.

Weizmann, Chaim. 1949. *Trail and Error*. New York: Harper.

Wertheimer, Max. 1959. *Productive Thinking*. New York: Harper.

Whitaker, Andrew. 1996. *Einstein, Bohr and the Quantum Dilemma*. Cambridge, England: Cambridge University Press.

White, Michael, and John Gribbin.1994. *Einstein: A Life in Science*. New York: Dutton.

Whitrow, Gerald J. 1967. *Einstein: The Man and His Achievement*. London: BBC.

Wolfson, Richard. 2003. *Simply Einstein*. New York: Norton.

Yourgrau, Palle. 1999. *Gödel Meets Einstein*. La Salle, Ill.: Open Court Press.

———. 2005. *A World without Time: The Forgotten Legacy of Gödel and Einstein*. New York: Basic Books.

Zackheim, Michele. 1999. *Einstein's Daughter*. New York: Riverhead.

NOTES

Einstein's letters and writings through 1920 have been published in *The Collected Papers of Albert Einstein* series, and they are identified by the dates used in those volumes. Unpublished material that is in the Albert Einstein Archives (AEA) is identified using the folder (reel)-document numbering format of the archives. For some of the material, especially that previously unpublished, I have used translations made for me by James Hoppes and Natasha Hoffmeyer.

EPIGRAPH

1. Einstein to Eduard Einstein, Feb. 5, 1930. Eduard was suffering from deepening mental illness at the time. The exact quote is: "Beim Menschen ist es wie beim Velo. Nur wenn er faehrt, kann er bequem die Balance halten." A more literal translation is: "It is the same with people as it is with riding a bike. Only when moving can one comfortably maintain one's balance." Courtesy of Barbara Wolff, Einstein archives, Hebrew University, Jerusalem.

CHAPTER ONE: THE LIGHT-BEAM RIDER

1. Einstein to Conrad Habicht, May 18 or 25, 1905.
2. These ideas are drawn from essays I wrote in *Time*, Dec. 31, 1999, and *Discover*, Sept. 2004.
3. Dudley Herschbach, "Einstein as a Student," Mar. 2005, unpublished paper provided to the author. Herschbach says, "Efforts to improve science education and literacy face a root problem: science and mathematics are regarded not as part of the general culture, but rather as the province of priest-like experts. Einstein is seen as a towering icon, the exemplar par excellence of lonely genius. That fosters an utterly distorted view of science."
4. Frank 1957, xiv; Bernstein 1996b, 18.
5. Vivienne Anderson to Einstein, Apr. 27, 1953, AEA 60-714; Einstein to Vivienne Anderson, May 12, 1953, AEA 60-716.
6. Viereck, 377. See also Thomas Friedman, "Learning to Keep Learning," *New York Times*, Dec. 13, 2006.

7. Einstein to Mileva Marić, Dec. 12, 1901; Hoffmann and Dukas, 24. Hoffmann was Einstein's friend in the late 1930s in Princeton. He notes, "His early suspicion of authority, which never wholly left him, was to prove of decisive importance."
8. Einstein message for Ben Scheman dinner, Mar. 1952, AEA 28-931.

CHAPTER TWO: CHILDHOOD

1. Einstein to Sybille Blinoff, May 21, 1954, AEA 59-261; Ernst Straus, "Reminiscences," in Holton and Elkana, 419; Vallentin, 17; Maja Einstein, lviii.
2. See, for example, Thomas Sowell, *The Einstein Syndrome: Bright Children Who Talk Late* (New York: Basic Books, 2002).
3. Nobel laureate James Franck quoting Einstein in Seelig 1956b, 72.
4. Vallentin, 17; Einstein to psychologist Max Wertheimer, in Wertheimer, 214.
5. Einstein to Hans Muehsam, Mar. 4, 1953, AEA 60-604. Also: "I think we can dispense with this question of heritage," Einstein is quoted in Seelig 1956a, 11. See also Michelmore, 22.
6. Maja Einstein, xvi; Seelig 1956a, 10.
7. www.alemannia-judaica.de/synagoge_buchau.htm.
8. Einstein to Carl Seelig, Mar. 11, 1952, AEA 39-13; Highfield and Carter, 9.
9. Maja Einstein, xv; Highfield and Carter, 9; Pais 1982, 36.
10. Birth certificate, CPAE 1: 1; Fantova, Dec. 5, 1953.
11. Pais 1982, 36–37.
12. Maja Einstein, xviii. Maria was sometimes used as a stand-in for the name Miriam in Jewish families.
13. Frank 1947, 8.
14. Maja Einstein, xviii–xix; Fölsing, 12; Pais 1982, 37.
15. Some researchers view such a pattern as possibly being a mild manifestation of autism or Asperger's syndrome. Simon Baron-Cohen, the director of the Autism Research Center at Cambridge University, is among those who suggest that Einstein might have exhibited characteristics of autism. He writes that autism is associated with a "particularly intense drive to systemize and an unusually low drive to empathize." He also notes that this pattern "explains the 'islets of ability' that people with autism display in subjects like math or music or drawing—all skills that benefit from systemizing." See Simon Baron-Cohen, "The Male Condition," *New York Times*, Aug. 8, 2005; Simon Baron-Cohen, *The Essential Difference* (New York: Perseus, 2003), 167; Norm Ledgin, *Asperger's and Self-Esteem: Insight and Hope through Famous Role Models* (Arlington, TX: Future Horizons, 2002), chapter 7; Hazel Muir, "Einstein and Newton Showed Signs of Autism," *New Scientist*, Apr. 30, 2003; Thomas Marlin, "Albert Einstein and LD," *Journal of Learning Disabilities*, Mar. 1, 2000, 149. A Google search of Einstein + Asperger's results in 146,000 pages. I do not find such a long-distance diagnosis to be convincing. Even as a teenager, Einstein made close friends, had passionate relationships, enjoyed collegial discussions, communicated well verbally, and could empathize with friends and humanity in general.

16. Einstein 1949b, 9; Seelig 1956a, 11; Hoffmann 1972, 9; Pais 1982, 37; Vallentin, 21; Reiser, 25; Holton 1973, 359; author's interview with Shulamith Oppenheim, Apr. 22, 2005.

17. Overbye, 8; Shulamith Oppenheim, *Rescuing Albert's Compass* (New York: Crocodile, 2003).

18. Holton 1973, 358.

19. Fölsing, 26; Einstein to Philipp Frank, draft, 1940, CPAE 1, p. lxiii.

20. Maja Einstein, xxi; Bucky, 156; Einstein to Hans Albert Einstein, Jan. 8, 1917.

21. Hans Albert Einstein interview in Whitrow, 21; Bucky, 148.

22. Einstein to Paul Plaut, Oct. 23, 1928, AEA 28-65; Dukas and Hoffmann, 78; Moszkowski, 222. Einstein originally wrote that music and science "complement each other in the *release* they offer," but he later changed that to *Befriedigung*, or satisfaction, according to Barbara Wolff of Hebrew University.

23. Einstein to Otto Juliusburger, Sept. 29, 1942, AEA 38-238.

24. Clark, 25; Einstein 1949b, 3; Reiser, 28. (Anton Reiser was the pseudonym of Rudoph Kayser, who married Ilse Einstein, the daughter of Einstein's second wife, Elsa.)

25. Maja Einstein, xix, says he was 7; in fact he enrolled on Oct. 1, 1885, when he was 6.

26. According to the version later told by his stepson-in-law, the teacher then added that Jesus was nailed to the cross "by the Jews"; Reiser, 30. But Einstein's friend and physics colleague Philipp Frank makes a point of specifically noting that the teacher did not raise the role of the Jews; Frank 1947, 9.

27. Fölsing, 16; Einstein to unknown recipient, Apr. 3, 1920, CPAE 1: lx.

28. Reiser, 28–29; Maja Einstein, xxi; Seelig 1956a, 15; Pais 1982, 38; Fölsing, 20. Maja again has him only 8 when he enters the gymnasium, which he actually did in Oct. 1888, at age 9 and a half.

29. Brian 1996, 281. A Google search of *Einstein failed math*, performed in 2006, turned up close to 648,000 references.

30. Pauline Einstein to Fanny Einstein, Aug. 1, 1886; Fölsing, 18–20, citing Einstein to Sybille Blinoff, May 21, 1954, and Dr. H. Wieleitner in *Nueste Nachrichten*, Munich, Mar. 14, 1929.

31. Einstein to Sybille Blinoff, May 21, 1954, AEA 59-261; Maja Einstein, xx.

32. Frank 1947, 14; Reiser, 35; Einstein 1949b, 11.

33. Maja Einstein, xx; Bernstein 1996a, 24–27; Einstein interview with Henry Russo, *The Tower*, Princeton, Apr. 13, 1935.

34. Talmey, 164; Pais 1982, 38.

35. The first edition appeared in twelve volumes between 1853 and 1857. New editions, under a new title that is referred to in Maja's essay, appeared in the late 1860s. They were constantly updated. The version likely owned by Einstein had twenty-one volumes and was bound into four or five large books. The definitive study of this book's influence on Einstein is Frederick Gregory, "The Mysteries and Wonders of Science: Aaron Bernstein's *Naturwissenschaftliche Volksbücher* and the Adolescent Einstein," in Howard and Stachel 2000, 23–42. Maja Einstein, xxi; Einstein 1949b, 15; Seelig 1956a, 12.

36. Aaron Bernstein, *Naturwissenschaftliche Volksbücher*, 1870 ed., vols. 1, 8, 16, 19; Howard and Stachel 2000, 27–39.
37. Einstein 1949b, 5.
38. Talmey, 163. (Talmud wrote his small memoir after he had changed his name to Talmey in America.)
39. Einstein, "On the Method of Theoretical Physics," Herbert Spencer lecture, Oxford, June 10, 1933, in Einstein 1954, 270.
40. Einstein 1949b, 9, 11; Talmey, 163; Fölsing, 23 (he speculates that the "sacred" book may have been another text); Einstein 1954, 270.
41. Aaron Bernstein, vol. 12, cited by Frederick Gregory in Howard and Stachel 2000, 37; Einstein 1949b, 5.
42. Frank 1947, 15; Jammer, 15–29. "The meaning of a life of brilliant scientific activity drew on the remnants of his fervent first feelings of youthful religiosity," writes Gerald Holton in Holton 2003, 32.
43. Einstein 1949b, 5; Maja Einstein, xxi.
44. Einstein, "What I Believe," *Forum and Century* (1930): 194, reprinted as "The World As I See It," in Einstein 1954, 10. According to Philipp Frank, "He saw the parade as a movement of people compelled to be machines"; Frank 1947, 8.
45. Frank 1947, 11; Fölsing, 17; C. P. Snow, "Einstein," in *Variety of Men* (New York: Scribner's, 1966), 26.
46. Einstein to Jost Winteler, July 8, 1901.
47. Pais 1982, 17, 38; Hoffmann 1972, 24.
48. Maja Einstein, xx; Seelig 1956a, 15; Pais 1982, 38; Einstein draft to Philipp Frank, 1940, CPAE 1, p. lxiii.
49. Stefann Siemer, "The Electrical Factory of Jacob Einstein and Cie.," in Renn 2005b, 128–131; Pyenson, 40.
50. Overbye, 9–10; Einstein draft to Philipp Frank, 1940, CPAE 1, p. lxiii; Hoffmann, 1972, 25–26; Reiser, 40; Frank 1947, 16; Maja Einstein, xxi; Fölsing, 28–30.
51. Einstein to Marie Winteler, Apr. 21, 1896; Fölsing 34; *The Jewish Spectator*, Jan. 1969.
52. Frank 1947, 17; Maja Einstein, xxii; Hoffmann 1972, 27.
53. Einstein, "On the Investigation of the State of the Ether in a Magnetic Field," summer 1895, CPAE 1: 5.
54. Einstein to Caesar Koch, summer 1895.
55. Albin Herzog to Gustave Maier, Sept. 25, 1895, CPAE 1 (English), p. 7; Fölsing, 37; Seelig 1956a, 9.
56. This process of envisaging is what Kantian philosophers call *Anschauung*. See Miller 1984, 241–246.
57. Seelig 1956b, 56; Fölsing, 38.
58. Miller 2001, 47; Maja Einstein, xxii; Seelig 1956b, 9; Fölsing, 38; Holton, "On Trying to Understand Scientific Genius," in Holton 1973, 371.
59. Bucky, 26; Fölsing, 46. Einstein provides a fuller description in his "Autobiographical Notes," in Schilpp, 53.

60. Gustav Maier to Jost Winteler, Oct. 26, 1895, CPAE 1: 9; Fölsing, 39; High-field and Carter, 22–24.
61. Vallentin, 12; Hans Byland, *Neue Bündner Zeitung*, Feb. 7, 1928, cited in Seelig 1956a, 14; Fölsing, 39.
62. Pauline Einstein to the Winteler family, Dec. 30, 1895, CPAE 1: 15.
63. Einstein to Marie Winteler, Apr. 21, 1896.
64. Entrance report, Aarau school, CPAE 1: 8; Aarau school record, CPAE 1: 10; Hermann Einstein to Jost Winteler, Oct. 29, 1995, CPAE 1: 11, and Dec. 30, 1895, CPAE 1: 14.
65. Report on a Music Examination, Mar. 31, 1896, CPAE 1: 17; Seelig 1956a, 15; Overbye, 13.
66. Release from Würtemberg citizenship, Jan. 28, 1896, CPAE 1: 16.
67. Einstein to Julius Katzenstein, Dec. 27, 1931, cited in Fölsing, 41.
68. *Israelitisches Wochenblatt*, Sept. 24, 1920; Einstein, "Why Do They Hate the Jews?," *Collier's*, Nov. 26, 1938.
69. Einstein to Hans Muehsam, Apr. 30, 1954, AEA 38-434; Fölsing 42.
70. Examination results, Sept. 18–21, 1896, CPAE 1: 20–27.
71. Overbye, 15; Maja Einstein, xvii.
72. Einstein to Heinrich Zangger, Aug. 11, 1918.

CHAPTER THREE: THE ZURICH POLYTECHNIC

1. Cahan, 42; editor's note, CPAE 1 (German), p. 44.
2. Einstein 1949b, 15.
3. Record and Grade Transcript, Oct. 1896–Aug. 1900, CPAE 1: 28; Bucky, 24; Einstein to Arnold Sommerfeld, Oct. 29, 1912; Fölsing, 50.
4. Einstein to Mileva Marić, Feb. 1898; Cahan, 64.
5. Louis Kollros, "Albert Einstein en Suisse," *Helvetica Physica*, Supplement 4 (1956): 22, in AEA 5-123; Adolf Frisch, in Seelig 1956a, 29; Cahan, 67; Clark, 55.
6. Seelig 1956a, 30; Overbye, 43; Miller 2001, 52; Charles Seife, "The True and the Absurd," in Brockman, 63.
7. Record and Grade Transcript, CPAE 1: 28.
8. Seelig 1956a, 30; Bucky, 25 (a slightly different version); Fölsing, 57.
9. Seelig 1956a, 30.
10. Einstein to Julia Niggli, July 28, 1899.
11. Seelig 1956a, 28; Whitrow, 5.
12. Einstein 1949b, 15–17.
13. Einstein interview in Bucky, 27; Einstein to Elizabeth Grossmann, Sept. 20, 1936, AEA 11-481; Seelig 1956a, 34, 207; Fölsing, 53.
14. Holton 1973, 209–212. Einstein's stepson-in-law Rudolph Kayser and colleague Philipp Frank both say that Einstein read Föppl in his spare time while at the Polytechnic.
15. Clark, 59; Galison, 32–34. Galison's book on Poincaré and Einstein is a fascinating exposition on how they developed their concepts and how Poincaré's

observations were "an anticipatory note to Einstein's special theory of relativity, a brilliant move by an author lacking the intellectual courage to pursue it to its logical, revolutionary end" (Galison, 34). Also very useful is Miller 2001, 200–204.

16. Seelig 1956a, 37; Whitrow, 5; Bucky, 156.

17. Miller 2001, 186; Hoffmann, 1972, 252; interview with Lili Foldes, *The Etude*, Jan. 1947, in Calaprice, 150; Einstein to Emil Hilb questionnaire, 1939, AEA 86-22; Dukas and Hoffmann, 76.

18. Seelig 1956a, 36.

19. Fölsing, 51, 67; Reiser, 50; Seelig 1956a, 9.

20. Clark, 50. Diana Kormos Buchwald points out that a careful examination of the picture of him at the Aarau school shows holes in his jacket.

21. Einstein to Maja Einstein, 1898.

22. Einstein to Maja Einstein, after Feb. 1899.

23. Marie Winteler to Einstein, Nov. 4–25, 1896.

24. Marie Winteler to Einstein, Nov. 30, 1896.

25. Pauline Einstein to Marie Winteler, Dec. 13, 1896.

26. Einstein to Pauline Winteler, May 1897.

27. Marie Winteler to Einstein, Nov. 4–25, Nov. 30.

28. Novi Sad, the cultural center of the Serbian people, had long been a "free royal city," then part of a Serbian autonomous region of the Hapsburg Empire. By the time Marić was born, it was in the Hungarian part of Austria-Hungary. Approximately 40 percent of the citizens there spoke Serbian when she was growing up, 25 percent spoke Hungarian, and about 20 percent spoke German. It is now the second largest city, after Belgrade, in the Republic of Serbia.

29. Desanka Trbuhovic-Gjuric, 9–38; Dord Krstic, "Mileva Einstein-Marić," in Elizabeth Einstein, 85; Overbye, 28–33; Highfield and Carter, 33–38; Marriage certificate, CPAE 5: 4.

30. Dord Krstic, "Mileva Einstein-Marić," in Elizabeth Einstein, 88 (Krstic's piece is based partly on interviews with school friends); Barbara Wolff, an expert on Einstein's life at the Hebrew University archives, says, "I imagine that Einstein was the main reason Mileva fled Zurich."

31. Mileva Marić to Einstein, after Oct. 20, 1897.

32. Einstein to Mileva Marić, Feb. 16, 1898.

33. Einstein to Mileva Marić, after Apr. 16, 1898, after Nov. 28, 1898.

34. Recollection of Suzanne Markwalder, in Seelig 1956a, 34; Fölsing, 71.

35. Einstein to Mileva Marić, Mar. 13 or 20, 1899.

36. Einstein to Mileva Marić, Aug. 10, 1899, Mar. 1899, Sept. 13, 1900.

37. Einstein to Mileva Marić, Sept. 13, 1900, early Aug. 1899, Aug. 10, 1899.

38. Einstein to Mileva Marić, ca. Sept. 28, 1899.

39. Mileva Marić to Einstein, 1900.

40. Intermediate Diploma Examinations, Oct. 21, 1898, CPAE 1: 42.

41. Einstein to Mileva Marić, Sept. 10, 1899; Einstein 1922c (see bibliography for explanation about this Dec. 14, 1922, lecture in Kyoto, Japan).

42. Einstein, 1922c; Reiser, 52; Einstein to Mileva Marić, ca. Sept. 28, 1899; Renn and Schulmann, 85, footnotes 11: 3, 11: 4. Wilhelm Wien's paper was delivered in Sept. 1898 in Düsseldorf and published in the *Annalen der Physik* 65, no. 3 of that year.
43. Einstein to Mileva Marić, Oct. 10, 1899; Seelig 1956a, 30; Fölsing, 68; Overbye, 55; final diploma examinations, CPAE 1: 67. The essay marks as recorded in CPAE are multiplied by 4 to reflect their weight in the final results.
44. Final diploma examinations, CPAE 1: 67.
45. Einstein to Walter Leich, Apr. 24, 1950, AEA 60-253; Walter Leich memo describing Einstein, Mar. 6, 1957, AEA 60-257.
46. Einstein, 1949b, 17.
47. Einstein to Mileva Marić, Aug. 1, 1900.

CHAPTER FOUR: THE LOVERS

1. Einstein to Mileva Marić, ca. July 29, 1900.
2. Einstein to Mileva Marić, Aug. 6, 1900.
3. Einstein to Mileva Marić, Aug. 1, Sept. 13, Oct. 3, 1900.
4. Einstein to Mileva Marić, Aug. 30, 1900.
5. Einstein to Mileva Marić, Aug. 1, Aug. 6, ca. Aug. 14, Aug. 20, 1900.
6. Einstein to Mileva Marić, Aug. 6, 1900.
7. Einstein to Mileva Marić, ca. Aug. 9, Aug. 14?, Aug. 20, 1900.
8. Einstein to Mileva Marić, ca. Aug. 9, ca. Aug. 14, 1900. Both of these letters came from this visit to Zurich.
9. Einstein to Mileva Marić, Sept. 13, 1900.
10. Einstein to Mileva Marić, Sept. 19, 1900.
11. Einstein to Adolf Hurwitz, Sept. 26, Sept. 30, 1900.
12. Einstein to Mileva Marić, Oct. 3, 1900; Einstein to Mrs. Marcel Grossmann, 1936; Seelig 1956a, 208.
13. Einstein's municipal citizenship application, Zurich, Oct. 1900, CPAE 1: 82; Einstein to Helene Kaufler, Oct. 11, 1900; minutes of the naturalization commission of Zurich, Dec. 14, 1900, CPAE 1: 84.
14. Einstein to Mileva Marić, Sept. 13, 1900.
15. Einstein to Mileva Marić, Oct. 3, 1900.
16. Einstein, "Conclusions Drawn from the Phenomena of Capillarity," *Annalen der Physik*, CPAE 2: 1, received Dec. 13, 1900, published Mar. 1, 1901. "The paper is very difficult to understand, not least because of the large number of obvious misprints; from its lack of clarity we can only assume that it had not been independently refereed . . . Yet it was an extraordinarily advanced paper for a recent graduate who was receiving no independent scientific advice." John N. Murrell and Nicole Grobert, "The Centenary of Einstein's First Scientific Paper," *The Royal Society* (London), Jan. 22, 2002, www.journals.royalsoc.ac .uk/app/home/content.asp.
17. Dudley Herschbach, "Einstein as a Student," Mar. 2005, unpublished paper provided to the author.

18. Einstein to Mileva Marić, Apr. 15, Apr. 30, 1901; Mileva Marić to Helene Savić, Dec. 20, 1900.
19. Einstein to G. Wessler, Aug. 24, 1948, AEA 59-26.
20. Maja Einstein, sketch, 19; Reiser, 63; minutes of the Municipal Naturalization Commission of Zurich, Dec. 14, 1900, CPAE 1: 84; Report of the Schweitzerisches Informationsbureau, Jan. 30, 1901, CPAE 1: 88; Military Service Book, Mar. 13, 1901, CPAE 1: 91.
21. Mileva Marić to Helene Savić, Dec. 20, 1900; Einstein to Mileva Marić, Mar. 23, Mar. 27, 1901.
22. Einstein to Mileva Marić, Apr. 4, 1901.
23. Einstein to Heike Kamerlingh Onnes, Apr. 12, 1901; Einstein to Marcel Grossmann, Apr. 14, 1901; Fölsing, 78; Clark, 66; Miller 2001, 68.
24. Einstein to Wilhelm Ostwald, Mar. 19, Apr. 3, 1901.
25. Hermann Einstein to Wilhelm Ostwald, Apr. 13, 1901.
26. Einstein to Mileva Marić, Mar. 23, Mar. 27, 1901; Einstein to Marcel Grossmann, Apr. 14, 1901.
27. Einstein to Mileva Marić, Mar. 27, 1901; Mileva Marić to Helene Savić, Dec. 9, 1901.
28. Einstein to Mileva Marić, Apr. 4, 1901; Einstein to Michele Besso, June 23, 1918; Overbye, 25; Miller 2001, 78; Fölsing, 115.
29. Einstein to Mileva Marić, Mar. 27, Apr. 4, 1901.
30. Einstein to Marcel Grossmann, Apr. 14, 1901; Einstein to Mileva Marić, Apr. 15, 1901.
31. Einstein to Mileva Marić, Apr. 30, 1901. The official translation is "blue nightshirt," but the word that Einstein actually used, *Schlafrock*, translates more accurately as "dressing gown."
32. Mileva Marić to Einstein, May 2, 1901.
33. Mileva Marić to Helene Savić, second half of May, 1901.
34. Einstein to Mileva Marić, second half of May, 1901.
35. Einstein to Mileva Marić, tentatively dated in CPAE as May 28?, 1901. The actual date is probably a week or so later.
36. Overbye, 77–78.
37. Einstein to Mileva Marić, July 7, 1901.
38. Mileva Marić to Einstein, after July 7, 1901 (published in CPAE vol. 8 as 1: 116, because it was discovered after vol. 1 had been printed).
39. Mileva Marić to Einstein, ca. July 31, 1901; Highfield and Carter, 80.
40. Einstein to Jost Winteler, July 8, 1901; Einstein to Marcel Grossmann, Apr. 14, 1901. The comparison to the compass needle comes from Overbye, 65.
41. Renn 2005a, 109. Jürgen Renn is the director of the Max Planck Institute for the History of Science in Berlin and an editor of the *Collected Papers of Albert Einstein*. I am grateful to him for help with this topic.
42. Einstein to Mileva Marić, Apr. 15, 1901; Einstein to Marcel Grossmann, Apr. 15, 1901.
43. Renn 2005a, 124.
44. Einstein to Mileva Marić, Apr. 4, ca. June 4, 1901. The letters to and from

Drude no longer exist, so it is not known precisely what Einstein's objections were.

45. Einstein to Mileva Marić, ca. July 7, 1901; Einstein to Jost Winteler, July 8, 1901.

46. Renn 2005a, 118. Renn's source notes say, "I gratefully acknowledge the kindness of Mr. Felix de Marez Oyens, from Christie's, who pointed my attention to the missing page of the letter by Einstein to Mileva Marić, ca. 8 July 1901. As, unfortunately, no copy of the page is available to me, my interpretation had to be based on a raw transcription of the passage in question."

47. Einstein to Marcel Grossmann, Sept. 6, 1901.

48. Overbye, 82–84. This includes a good synopsis of the Boltzmann-Ostwald dispute.

49. Einstein, "On the Thermodynamic Theory of the Difference in Potentials between Metals and Fully Dissociated Solutions of Their Salts," Apr. 1902. Renn does not mention this paper in his analysis of Einstein's dispute with Drude, and instead focuses only on the June 1902 paper.

50. Einstein, "Kinetic Theory of Thermal Equilibrium and the Second Law of Thermodynamics," June 1902; Renn 2005a, 119; Jos Uffink, "Insuperable Difficulties: Einstein's Statistical Road to Molecular Physics," *Studies in the History and Philosophy of Modern Physics* 37 (2006): 38; Clayton Gearhart, "Einstein before 1905: The Early Papers on Statistical Mechanics," *American Journal of Physics* (May 1990): 468.

51. Mileva Marić to Helene Savić, ca. Nov. 23, 1901; Einstein to Mileva Marić, Nov. 28, 1901.

52. Einstein to Mileva Marić, Dec. 17 and 19, 1901.

53. Receipt for the return of Doctoral Fees, Feb. 1, 1902, CPAE 1: 132; Fölsing, 88–90; Reiser, 69; Overbye, 91. From Einstein to Mileva Marić, ca. Feb. 8, 1902: "I'm explaining to [Conrad] Habicht the paper I submitted to Kleiner. He's very enthusiastic about my good ideas and is pestering me to send Boltzmann the part of the paper which relates to his book. I'm going to do it."

54. Einstein to Marcel Grossmann, Sept. 6, 1901.

55. Einstein to Mileva Marić, Nov. 28, 1901.

56. Mileva Marić to Einstein, Nov. 13, 1901; Highfield and Carter, 82.

57. Einstein to Mileva Marić, Dec. 12, 1901; Fölsing, 107; Zackheim, 35; Highfield and Carter, 86.

58. Pauline Einstein to Pauline Winteler, Feb. 20, 1902.

59. Mileva Marić to Helene Savić, ca. Nov. 23, 1901.

60. Einstein to Mileva Marić, Dec. 11 and 19, 1901.

61. Einstein to Mileva Marić, Dec. 28, 1901.

62. Einstein to Mileva Marić, Feb. 4, 1902, Dec. 12, 1901.

63. Einstein to Mileva Marić, Feb. 4, 1902.

64. Mileva Marić to Einstein, Nov. 13, 1901. For some context, see Popović, which includes a collection of letters between Marić and Savić collected by Savić's grandson.

65. Einstein to Mileva Marić, Feb. 17, 1902.
66. Swiss Federal Council to Einstein, June 19, 1902.
67. See Peter Galison's treatment of the synchronization of time in Europe at that period, in Galison, 222–248. Also, see chapter 6 below for a fuller discussion of the role this might have played in Einstein's development of special relativity.
68. Einstein to Hans Wohlwend, autumn 1902; Fölsing, 102.
69. Einstein interview, Bucky, 28; Reiser, 66.
70. Einstein to Michele Besso, Dec. 12, 1919.
71. Einstein interview, Bucky, 28; Einstein 1956, 12. Both say essentially the same thing, with variations in wording and translation. Reiser, 64.
72. Alas, as a rule, all applications were destroyed after eighteen years, and even though Einstein was by then world-famous, his comments on inventions were disposed of during the 1920s; Fölsing, 104.
73. Galison, 243; Flückiger, 27.
74. Fölsing, 103; C. P. Snow, "Einstein," in Goldsmith et al., 7.
75. Einstein interview, Bucky, 28; Einstein 1956, 12. See Don Howard, "A kind of vessel in which the struggle for eternal truth is played out," AEA Cedex-H.
76. Solovine, 6.
77. Maurice Solovine, Dedication of the Olympia Academy, "A.D. 1903," CPAE 2: 3.
78. Solovine, 11–14.
79. Einstein to Maurice Solovine, Nov. 25, 1948; Seelig 1956a, 57; Einstein to Conrad Habicht and Maurice Solovine, Apr. 3, 1953; Hoffmann 1972, 243.
80. The editors of Einstein's papers, in the introduction to vol. 2, xxiv–xxv, describe the books and specific editions read by the Olympia Academy.
81. Einstein to Moritz Schlick, Dec. 14, 1915. In a 1944 essay about Bertrand Russell, Einstein wrote, "Hume's clear message seemed crushing: the sensory raw material, the only source of our knowledge, through habit may lead us to belief and expectation but not to the knowledge and still less to the understanding of lawful relations." Einstein 1954, 22. See also Einstein, 1949b, 13.
82. David Hume, *Treatise on Human Nature*, book 1, part 2; Norton 2005a.
83. There are varying interpretations of Kant's *Critique of Pure Reason* (1781). I have tried here to stick closely to Einstein's own view of Kant. Einstein, "Remarks on Bertrand Russell's Theory of Knowledge," (1944) in Schilpp; Einstein 1954, 22; Einstein, 1949b, 11–13; Einstein, "On the Methods of Theoretical Physics," the Herbert Spencer lecture, Oxford, June 10, 1933, in Einstein 1954, 270; Mara Beller, "Kant's Impact on Einstein's Thought," in Howard and Stachel 2000, 83–106. See also Einstein, "Physics and Reality" (1936) in Einstein 1950a, 62; Yehuda Elkana, "The Myth of Simplicity," in Holton and Elkana, 221.
84. Einstein 1949b, 21.
85. Einstein, Obituary for Ernst Mach, Mar. 14, 1916, CPAE 6: 26.
86. Philipp Frank, "Einstein, Mach and Logical Positivism," in Schilpp, 272;

Overbye, 25, 100–104; Gerald Holton, "Mach, Einstein and the Search for Reality," *Daedalus* (spring 1968): 636–673, reprinted in Holton 1973, 221; Clark, 61; Einstein to Carl Seelig, Apr. 8, 1952; Einstein, 1949b, 15; Norton 2005a.

87. Spinoza, *Ethics*, part I, proposition 29 and passim; Jammer 1999, 47; Holton 2003, 26–34; Matthew Stewart, *The Courtier and the Heretic* (New York: Norton, 2006).

88. Pais 1982, 47; Fölsing, 106; Hoffmann 1972, 39; Maja Einstein, xvii; Overbye, 15–17.

89. Marriage Certificate, CPAE 5: 6; Miller 2001, 64; Zackheim, 47.

90. Einstein to Michele Besso, Jan. 22, 1903; Mileva Marić to Helene Savić, Mar. 1903; Solovine, 13; Seelig 1956a, 46; Einstein to Carl Seelig, May 5, 1952; AEA 39-20.

91. Mileva Marić to Einstein, Aug. 27, 1903; Zackheim, 50.

92. Einstein to Mileva Marić, ca. Sept. 19, 1903; Zackheim; Popović; author's discussions and e-mails with Robert Schulmann.

93. Popović, 11; Zackheim, 276; author's discussions and e-mails with Robert Schulmann.

94. Michelmore, 42.

95. Einstein to Mileva Marić, ca. Sept. 19, 1903.

96. Mileva Marić to Helene Savić, June 14, 1904; Popović, 86; Whitrow, 19.

97. Overbye, 113, citing Desanka Trbuhovic-Gjuric, *Im Schatten Albert Einstein* (Bern: Verlag Paul Haupt, 1993), 94.

CHAPTER FIVE: THE MIRACLE YEAR

1. This quote is attributed in a variety of books and sources to an address Lord Kelvin gave to the British Association for the Advancement of Science in 1900. I have not found direct evidence for it, which is why I qualify it as "reportedly" said. It is not in the two-volume biography by Silvanus P. Thompson, *The Life of Lord Kelvin* (New York: Chelsea Publishing, 1976), originally published in 1910.

2. Pierre-Simon Laplace, *A Philosophical Essay on Probabilities* (1820; reprinted, New York: Dover, 1951). This famous statement of determinism comes in the preface of a work devoted to probability theory. The fuller line is that in ultimate reality we have determinism, but in practice we have probabilities. The achievement of full knowledge is not reachable, he says, so we need probabilities.

3. Einstein, Letter to the Royal Society on Newton's bicentennial, Mar. 1927.

4. Einstein 1949b, 19.

5. For the influence of Faraday's induction theories on Einstein, see Miller 1981, chapter 3.

6. Einstein and Infeld, 244; Overbye, 40; Bernstein 1996a, 49.

7. Einstein to Conrad Habicht, May 18 or 25, 1905.

8. Sent on Mar. 17, 1905, and published in *Annalen der Physik* 17 (1905). I want to thank Yale professor Douglas Stone for help with this section.

9. Max Born, obituary for Max Planck, Royal Society of London, 1948.

10. John Heilbron, *The Dilemmas of an Upright Man* (Berkeley: University of California Press, 1986). Lucid explanations of Einstein's quantum paper, from which this section is drawn, include Gribbin and Gribbin; Bernstein 1996a, 2006; Overbye, 118–121; Stachel 1998; Rigden; A. Douglas Stone, "Genius and Genius²: Planck, Einstein and the Birth of Quantum Theory," Aspen Center for Physics, unpublished lecture, July 20, 2005.

11. Planck's approach was probably a bit more complex and involved assuming a group of oscillators and positing a total energy that is an integer multiple of a quantum unit. Bernstein 2006, 157–161.

12. Max Planck, speech to the Berlin Physical Society, Dec. 14, 1900. See Lightman 2005, 3.

13. Einstein 1949b, 46. Miller 1984, 112; Miller 1999, 50; Rynasiewicz and Renn, 5.

14. Einstein, "On the General Molecular Theory of Heat," Mar. 27, 1904.

15. Einstein to Conrad Habicht, Apr. 15, 1904. Jeremy Bernstein discussed the connections between the 1904 and 1905 papers in an e-mail, July 29, 2005.

16. Einstein, "On a Heuristic Point of View Concerning the Production and Transformation of Light," Mar. 17, 1905.

17. "We are startled, wondering what happened to the waves of light of the 19th century theory and marveling at how Einstein could see the signature of atomic discreteness in the bland formulae of thermodynamics," says the science historian John D. Norton. "Einstein takes what looks like a dreary fragment of the thermodynamics of heat radiation, an empirically based expression for the entropy of a volume of high-frequency heat radiation. In a few deft inferences he converts this expression into a simple, probabilistic formula whose unavoidable interpretation is that the energy of radiation is spatially localized in finitely many, independent points." Norton 2006c, 73. See also Lightman 2005, 48.

18. Einstein's paper in 1906 noted clearly that Planck had not grasped the full implications of the quantum theory. Apparently, Besso encouraged Einstein not to make this criticism of Planck too explicit. As Besso wrote much later, "In helping you edit your publications on the quanta, I deprived you of a part of your glory, but, on the other hand, I made a friend for you in Planck." Michele Besso to Einstein, Jan. 17, 1928. See Rynasiewicz and Renn, 29; Bernstein 1991, 155.

19. Holton and Brush, 395.

20. Gilbert Lewis coined the name "photon" in 1926. Einstein in 1905 discovered a quantum of light. Only later, in 1916, did he discuss the quantum's momentum and its zero rest mass. Jeremy Bernstein has noted that one of the most interesting discoveries Einstein did *not* make in 1905 was the photon. Jeremy Bernstein, letter to the editor, *Physics Today*, May 2006.

21. Gribbin and Gribbin, 81.

22. Max Planck to Einstein, July 6, 1907.
23. Max Planck and three others to the Prussian Academy, June 12, 1913, CPAE 5: 445.
24. Max Planck, *Scientific Autobiography* (New York: Philosophical Library, 1949), 44; Max Born, "Einstein's Statistical Theories," in Schilpp, 163.
25. Quoted in Gerald Holton, "Millikan's Struggle with Theory," *Europhysics News* 31 (2000): 3.
26. Einstein to Michele Besso, Dec. 12, 1951, AEA 7-401.
27. Completed Apr. 30, 1905, submitted to the University of Zurich on July 20, 1905, submitted to *Annalen der Physik* in revised form on Aug. 19, 1905, and published by *Annalen der Physik* Jan. 1906. See Norton 2006c and www.pitt .edu/~jdnorton/Goodies/Einstein_stat_1905/.
28. Jos Uffink, "Insuperable Difficulties: Einstein's Statistical Road to Molecular Physics," *Studies in the History and Philosophy of Modern Physics* 37 (2006): 37, 60.
29. bulldog.u-net.com/avogadro/avoga.html.
30. Rigden, 48–52; Bernstein 1996a, 88; Gribbin and Gribbin, 49–54; Pais 1982, 88.
31. Hoffmann 1972, 55; Seelig 1956b, 72; Pais 1982, 88–89.
32. Brownian motion introduction, CPAE 2 (German), p. 206; Rigden, 63.
33. Einstein, "On the Motion of Small Particles Suspended in Liquids at Rest Required by the Molecular-Kinetic Theory of Heat," submitted to the *Annalen der Physik* on May 11, 1905.
34. Einstein 1949b, 47.
35. The root mean square average is asymptotic to $\sqrt{2n/\pi}$. Good analyses of the relationship of random walks to Einstein's Brownian motion are in Gribbin and Gribbin, 61; Bernstein 2006, 117. I am grateful to George Stranahan of the Aspen Center for Physics for his help on the mathematics behind this relationship.
36. Einstein, "On the Theory of Brownian Motion," 1906, CPAE 2: 32 (in which he notes Seidentopf's results); Gribbin and Gribbin, 63; Clark, 89; Max Born, "Einstein's Statistical Theories," in Schilpp, 166.

CHAPTER SIX: SPECIAL RELATIVITY

1. Contemporary historical research on Einstein's special theory begins with Gerald Holton's essay, "On the Origins of the Special Theory of Relativity" (1960), reprinted in Holton 1973, 165. Holton remains a guiding light in this field. Most of his earlier essays are incorporated in his books *Thematic Origins of Scientific Thought: Kepler to Einstein* (1973), *Einstein, History and Other Passions* (2000), and *The Scientific Imagination*, Cambridge, Mass.: Harvard University Press, 1998.

   Einstein's popular description is his 1916 book, *Relativity: The Special and the General Theory,* and his more technical description is his 1922 book, *The Meaning of Relativity.*

For good explanations of special relativity, see Miller 1981, 2001; Galison; Bernstein 2006; Calder; Feynman 1997; Hoffmann 1983; Kaku; Mermin; Penrose; Sartori; Taylor and Wheeler 1992; Wolfson.

This chapter draws on these books along with the articles by John Stachel; Arthur I. Miller; Robert Rynasiewicz; John D. Norton; John Earman, Clark Glymour, and Robert Rynasiewicz; and Michel Jannsen listed in the bibliography. See also Wertheimer 1959. Arthur I. Miller provides a careful and skeptical look at Max Wertheimer's attempt to reconstruct Einstein's development of special relativity as a way to explain Gestalt psychology; see Miller 1984, 189–195.

2. See Janssen 2004 for an overview of the arguments that Einstein's attempt to extend general relativity to arbitrary and rotating motion was not fully successful and perhaps less necessary than he thought.

3. Galileo Galilei, *Dialogue Concerning the Two Chief World Systems* (1632), translated by Stillman Drake, 186.

4. Miller 1999, 102.

5. Einstein, "Ether and the Theory of Relativity," address at the University of Leiden, May 5, 1920.

6. Ibid.; Einstein 1916, chapter 13.

7. Einstein, "Ether and the Theory of Relativity," address at the University of Leiden, May 5, 1920.

8. Einstein to Dr. H. L. Gordon, May 3, 1949, AEA 58-217.

9. See Alan Lightman's *Einstein's Dreams* for an imaginative and insightful fictional rumination on Einstein's discovery of special relativity. Lightman captures the flavor of the professional, personal, and scientific thoughts that might have been swirling in Einstein's mind.

10. Peter Galison, the Harvard science historian, is the most compelling proponent of the influence of Einstein's technological environment. Arthur I. Miller presents a milder version. Among those who feel that these influences are overstated are John Norton, Tilman Sauer, and Alberto Martinez. See Alberto Martinez, "Material History and Imaginary Clocks," *Physics in Perspective* 6 (2004): 224.

11. Einstein 1922c. I rely on a corrected translation of this 1922 lecture that gives a different view of what Einstein said; see bibliography for an explanation.

12. Einstein, 1949b, 49. For other versions, see Wertheimer, 214; Einstein 1956, 10.

13. Miller 1984, 123, has an appendix explaining how the 1895 thought experiment affected Einstein's thinking. See also Miller 1999, 30–31; Norton 2004, 2006b. In the latter paper, Norton notes, "[This] is untroubling to an ether theorist. Maxwell's equations *do* entail quite directly that the observer would find a frozen waveform; and the ether theorist does not expect frozen waveforms in our experience since we do not move at the velocity of light in the ether."

14. Einstein to Erika Oppenheimer, Sept. 13, 1932, AEA 25-192; Moszkowski, 4.

15. Gerald Holton was the first to emphasize Föppl's influence on Einstein, citing the memoir by his son-in-law Anton Reiser and the German edition of Philipp Frank's biography. Holton 1973, 210.

16. Einstein, "Fundamental Ideas and Methods of the Theory of Relativity" (1920), unpublished draft of an article for *Nature*, CPAE 7: 31. See also Holton 1973, 362–364; Holton 2003.

17. Einstein to Mileva Marić, Aug. 10, 1899.

18. Einstein to Mileva Marić, Sept. 10 and 28, 1899; Einstein 1922c.

19. Einstein to Robert Shankland, Dec. 19, 1952, says that he read Lorentz's book before 1905. In his 1922 Kyoto lecture (Einstein 1922c) he speaks of being a student in 1899 and says, "Just at that time I had a chance to read Lorentz's paper of 1895." Einstein to Michele Besso, Jan. 22?, 1903, says he is beginning "comprehensive, extensive studies in electron theory." Arthur I. Miller provides a good look at what Einstein had already learned. See Miller 1981, 85–86.

20. This section draws from Gerald Holton, "Einstein, Michelson, and the 'Crucial' Experiment," in Holton 1973, 261–286, and Pais 1982, 115–117. Both assess Einstein's varying statements. The historical approach has evolved over the years. For example, Einstein's longtime friend and fellow physicist Philipp Frank wrote in 1957, "Einstein started from the most prominent case in which the old laws of motion and light propagation had failed to yield to the observed facts: the Michelson experiment" (Frank 1957, 134). Gerald Holton, the Harvard historian of science, wrote in a letter to me about this topic (May 30, 2006): "Concerning the Michelson/Morley experiment, until three or four decades ago practically everyone wrote, particularly in textbooks, that there was a straight line between that experiment and Einstein's special relativity. All this changed of course when it became possible to take a careful look at Einstein's own documents on the matter . . . Even non-historians have long ago given up the idea that there was a crucial connection between that particular experiment and Einstein's work."

21. Einstein 1922c; Einstein toast to Albert Michelson, the Athenaeum, Caltech, Jan. 15, 1931, AEA 8-328; Einstein message to Albert Michelson centennial, Case Institute, Dec. 19, 1952, AEA 1-168.

22. Wertheimer, chapter 10; Miller 1984, 190.

23. Robert Shankland interviews and letters, Feb. 4, 1950, Oct. 24, 1952, Dec. 19, 1952. See also Einstein to F. G. Davenport, Feb. 9, 1954: "In my own development, Michelson's result has not had a considerable influence, I even do not remember if I knew of it at all when I wrote my first paper on the subject. The explanation is that I was, for general reasons, firmly convinced that there does not exist absolute motion."

24. Miller 1984, 118: "It was unnecessary for Einstein to review every extant ether-drift experiment, because in his view their results were ab initio [from the beginning] a foregone conclusion." This section draws on Miller's work and on suggestions he made to an earlier draft.

25. Einstein saw the null results of the ether-drift experiments as support for the

relativity principle, not (as is sometimes assumed) support for the postulate that light always moves at a constant velocity. John Stachel, "Einstein and Michelson: The Context of Discovery and Context of Justification," 1982, in Stachel 2002a.

26. Professor Robert Rynasiewicz of Johns Hopkins is among those who emphasize Einstein's reliance on inductive methods. Even though Einstein in his later career wrote often that he relied more on deduction than on induction, Rynasiewicz calls this "highly contentious." He argues instead, "My view of the annus mirabilis is that it is a triumph of what can be secured inductively in the way of fixed points from which to carry on despite the lack of a fundamental theory." Rynasiewicz e-mail to me, commenting on an earlier draft of this section, June 29, 2006.

27. Miller 1984, 117; Sonnert, 289.

28. Holton 1973, 167.

29. Einstein, "Induction and Deduction in Physics," *Berliner Tageblatt*, Dec. 25, 1919, CPAE 7: 28.

30. Einstein to T. McCormack, Dec. 9, 1952, AEA 36-549. McCormack was a Brown University undergraduate who had written Einstein a fan letter.

31. Einstein 1949b, 89.

32. The following analysis draws from Miller 1981 and from the work of John Stachel, John Norton, and Robert Rynasiewicz cited in the bibliography. Miller, Norton, and Rynasiewicz kindly read drafts of my work and suggested corrections.

33. Miller 1981, 311, describes a connection between Einstein's papers on light quanta and special relativity. In section 8 of his special relativity paper, Einstein discusses light pulses and declares, "It is remarkable that the energy and the frequency of a light complex vary with the state of motion of the observer in accordance with the same law."

34. Norton 2006a.

35. Einstein to Albert Rippenbein, Aug. 25, 1952, AEA 20-46. See also Einstein to Mario Viscardini, Apr. 28, 1922, AEA 25-301: "I rejected this hypothesis at that time, because it leads to tremendous theoretical difficulties (e.g., the explanation of shadow formation by a screen that moves relative to the light source)."

36. Mermin, 23. This was finally proven conclusively by Willem de Sitter's study of double stars that rotate around each other at great speeds, which was published in 1913. But even before then, scientists had noted that no evidence could be found for the theory that the velocity of light from moving stars, or any other source, varied.

37. Einstein to Paul Ehrenfest, Apr. 25, June 20, 1912. By taking this approach, Einstein was continuing to lay the foundation for a quandary about quantum theory that would bedevil him for the rest of his life. In his light quanta paper, he had praised the wave theory of light while at the same time proposing that light could also be regarded as particles. An emission theory of light could have fit nicely with that approach. But both facts and intuition made him abandon

that approach to relativity, just at the same moment he was finishing his light quanta paper. "To me, it is virtually inconceivable that he would have put forward two papers in the same year which depended upon hypothetical views of Nature that he felt were in contradiction with each other," says physicist Sir Roger Penrose. "Instead, he must have felt (correctly, as it turned out) that 'deep down' there was no real contradiction between the accuracy—indeed 'truth'—of Maxwell's wave theory and the alternative 'quantum' particle view that he put forward in the quantum paper. One is reminded of Isaac Newton's struggles with basically the same problem—some 300 years earlier—in which he proposed a curious hybrid of a wave and particle viewpoint in order to explain conflicting aspects of the behavior of light." Roger Penrose, foreword to *Einstein's Miraculous Year* (Princeton: Princeton University Press, 2005), xi. See also Miller 1981, 311.

38. Einstein, "On the Electrodynamics of Moving Bodies," June 30, 1905, CPAE 2: 23, second paragraph. Einstein originally used V for the constant velocity of light, but seven years later began using the term now in common use, c.

39. In section 2 of the paper, he defines the light postulate more carefully: "Every light ray moves in the 'rest' coordinate system with a fixed velocity V, independently of whether this ray of light is emitted by a body at rest or in motion." In other words, the postulate says that the speed of light is the same *no matter how fast the light source is moving*. Many writers, when defining the light postulate, confuse this with the stronger assertion that light always moves in any inertial frame at the same velocity no matter how fast the light source *or the observer* is moving toward or away from each other. That statement is also true, but it comes only by *combining* the relativity principle with the light postulate.

40. Einstein 1922c. In his popular 1916 book *Relativity: The Special and General Theory*, Einstein explains this in chapter 7, "The Apparent Incompatibility of the Law of Propagation of Light with the Principle of Relativity."

41. Einstein 1916, chapter 7.

42. Einstein 1922c; Reiser, 68.

43. Einstein 1916, chapter 9.

44. Einstein 1922c; Heisenberg 1958, 114.

45. Sir Isaac Newton, *Philosophiae Naturalis Principia Mathematica* (1689), books 1 and 2; Einstein, "The Methods of Theoretical Physics," Herbert Spencer lecture, Oxford, June 10, 1933, in Einstein 1954, 273.

46. Fölsing, 174–175.

47. Poincaré went on to reference himself, saying that he had discussed this idea in an article called "The Measurement of Time." Arthur I. Miller notes that Einstein's friend Maurice Solovine may have read this paper, in French, and discussed it with Einstein. Einstein would later cite it, and his analysis of the synchronizations of clocks reflects some of Poincaré's thinking. Miller 2001, 201–202.

48. Fölsing, 155: "He was observed gesticulating to friends and colleagues as he pointed to one of Bern's bell towers and then to one in the neighboring village of Muri." Galison, 253, picks up this tale. Both cite as their source Max Flück-

iger, *Einstein in Bern* (Bern: Paul Haupt, 1974), 95. In fact, Flückiger merely quotes a colleague saying that Einstein referred to these clocks as a hypothetical example. See Alberto Martinez, "Material History and Imaginary Clocks," *Physics in Perspective* 6 (2004): 229. Martinez does concede, however, that it is indeed interesting that there was a steeple clock in Muri not synchronized with the clocks in Bern and that Einstein referred to this in explaining the theory to friends.

49. Galison, 222, 248, 253; Dyson. Galison's thesis is based on his original research into the patent applications.

50. Norton 2006a, 3, 43: "Another oversimplification pays too much attention to the one part of Einstein's paper that especially fascinates us now: his ingenious use of light signals and clocks to mount his conceptual analysis of simultaneity. This approach gives far too much importance to notions that entered briefly only at the end of years of investigation . . . They are not necessary to special relativity or to the relativity of simultaneity." See also Alberto Martinez, "Material History and Imaginary Clocks," *Physics in Perspective* 6 (2004): 224–240; Alberto Martinez, "Railways and the Roots of Relativity," *Physics World*, Nov. 2003; Norton 2004. For a good assessment, which gives more credit to Galison's research and insights, see Dyson. Also see Miller 2001.

51. Einstein interview, Bucky, 28; Einstein 1956, 12.

52. Moszkowski, 227.

53. Overbye, 135.

54. Miller 1984, 109, 114. Miller 1981, chapter 3, explains the influence of Faraday's experiments with rotating magnets on Einstein's special theory.

55. Einstein, "On the Electrodynamics of Moving Bodies," *Annalen der Physik* 17 (Sept. 26, 1905). There are many available editions. For a web version, see www.fourmilab.ch/etexts/einstein/specrel/www/. Useful annotated versions include Stachel 1998; Stephen Hawking, ed., *Selections from the Principle of Relativity* (Philadelphia: Running Press, 2002); Richard Muller, ed., *Centennial Edition of* The Theory of Relativity (San Francisco: Arion Press, 2005).

56. Einstein, unused addendum to 1916 book *Relativity*, CPAE 6: 44a.

57. Einstein 1916.

58. Bernstein 2006, 71.

59. This example is lucidly described in Miller 1999, 82–83; Panek, 31–32.

60. James Hartle, lecture at the Aspen Center for Physics, June 29, 2005; British National Measurement Laboratory, report on time dilation experiments, spring 2005, www.npl.co.uk/publications/metromnia/issue18/.

61. Einstein to Maurice Solovine, undated, in Solovine, 33, 35.

62. Krauss, 35–47.

63. Seelig 1956a, 28. For a full mathematical description of the special theory, see Taylor and Wheeler 1992.

64. Pais, 1982, 151, citing Hermann Minkowski, "Space and Time," lecture at the University of Cologne, Sept. 21, 1908.

65. Clark, 159–60.

66. Thorne, 79. This is also explained well in Miller 2001, 200: "Neither Lorentz, Poincaré, nor any other physicist was willing to grant Lorentz's local time any physical reality . . . Only Einstein was willing to go beyond appearances." See also Miller 2001, 240: "Einstein inferred a meaning Poincaré did not. His thought experiment enabled him to *interpret* the mathematical formalism as a new theory of space and time, whereas for Poincaré it was a generalized version of Lorentz's electron theory." Miller has also explored this topic in "Scientific Creativity: A Comparative Study of Henri Poincaré and Albert Einstein," *Creativity Research Journal* 5 (1992): 385.

67. Arthur Miller e-mail to the author, Aug. 1, 2005.

68. Hoffmann 1972, 78. Prince Louis de Broglie, the quantum theorist who theorized that particles could behave as waves, said of Poincaré in 1954, "Yet Poincaré did not take the decisive step; he left to Einstein the glory of grasping all the consequences of the principle of relativity." See Schilpp, 112; Galison, 304.

69. Dyson.

70. Miller 1981, 162.

71. Holton 1973, 178; Pais 1982, 166; Galison, 304; Miller 1981. All four authors have done important work on Poincaré and the credit he deserves, from which some of this section is drawn. I am grateful to Prof. Miller for a copy of his paper "Why Did Poincaré Not Formulate Special Relativity in 1905?" and for helping to edit this section.

72. Miller 1984, 37–38; Henri Poincaré lecture, May 4, 1912, University of London, cited in Miller 1984, 37; Pais 1982, 21, 163–168. Pais writes: "In all his life, Poincaré never understood the basis of special relativity . . . It is apparent that Poincaré either never understood or else never accepted the Theory of Relativity." See also Galison, 242 and passim.

73. Einstein to Mileva Marić, Mar. 27, 1901.

74. Michelmore, 45.

75. Overbye, 139; Highfield and Carter, 114; Einstein and Mileva Marić to Conrad Habicht, July 20, 1905.

76. Overbye, 140; Trbuhovic-Gjuric, 92–93; Zackheim, 62.

77. The issue of whether Marić's name was in any way ever on a manuscript of the special theory is a knotted one, but it turns out that the single source for such reports, a late Russian physicist, never actually said precisely that, and there is no other evidence at all to support the contention. For an explanation, see John Stachel's appendix to the introduction of *Einstein's Miraculous Year*, centennial reissue edition (Princeton: Princeton University Press, 2005), lv.

78. "The Relative Importance of Einstein's Wife," *The Economist*, Feb. 24, 1990; Evan H. Walker, "Did Einstein Espouse His Spouse's Ideas?", *Physics Today*, Feb. 1989; Ellen Goodman, "Out from the Shadows of Great Men," *Boston Globe*, Mar. 15, 1990; *Einstein's Wife*, PBS, 2003, www.pbs.org/opb/einsteins wife/index.htm; Holton 2000, 191; Robert Schulmann and Gerald Holton, "Einstein's Wife," letter to the *New York Times Book Review*, Oct. 8, 1995; Highfield and Carter, 108–114; Svenka Savić, "The Road to Mileva Marić-Einstein," www.zenskestudie.edu.yu/wgsact/e-library/e-lib0027.html#_ftn1;

Christopher Bjerknes, *Albert Einstein: The Incorrigible Plagiarist*, home.com cast.net/~xtxinc/CIPD.htm; Alberto Martínez, "Arguing about Einstein's Wife," *Physics World*, Apr. 2004, physicsweb.org/articles/world/17/4/2/1; Alberto Martínez, "Handling Evidence in History: The Case of Einstein's Wife," *School Science Review*, Mar. 2005, 51–52; Zackheim, 20; Andrea Gabor, *Einstein's Wife: Work and Marriage in the Lives of Five Great Twentieth-Century Women* (New York: Viking, 1995); John Stachel, "Albert Einstein and Mileva Marić: A Collaboration That Failed to Develop," in H. Prycior et al., eds., *Creative Couples in Science* (New Brunswick, N.J.: Rutgers University Press, 1995), 207–219; Stachel 2002a, 25–37.

79. Michelmore, 45.
80. Holton 2000, 191.
81. Einstein to Conrad Habicht, June 30–Sept. 22, 1905 (almost certainly in early September, after returning from vacation and getting to work on the $E=mc^2$ paper).
82. Einstein, "Does the Inertia of a Body Depend on Its Energy Content?," *Annalen der Physik* 18 (1905), received Sept. 27, 1905, CPAE 2: 24.
83. For an insightful look at the background and ramifications of Einstein's equation, see Bodanis. Bodanis also has a useful website that includes further details: davidbodanis.com/books/emc2/notes/relativity/sigdev/index.html. The calculation about the mass of a raisin is in Wolfson, 156.

CHAPTER SEVEN: THE HAPPIEST THOUGHT

1. Maja Einstein, xxi.
2. Fölsing, 202; Max Planck, *Scientific Autobiography and Other Papers* (New York: Philosophical Library, 1949), 42.
3. More precisely, the definition that Richard Feynman uses in his *Lectures on Physics* (Boston: Addison-Wesley, 1989), 19-1 is, "Action in physics has a precise meaning. It is the time average of the kinetic energy of a particle minus the potential energy. The principle of least action then states that a particle will travel along the path that minimizes the difference between its kinetic and potential energies."
4. Fölsing, 203; Einstein to Maurice Solovine, Apr. 27, 1906; Einstein tribute to Planck, 1913, CPAE 2: 267.
5. Max Planck to Einstein, July 6, 1907; Hoffmann 1972, 83.
6. Max Laue to Einstein, June 2, 1906.
7. Hoffmann 1972, 84; Seelig 1956a, 78; Fölsing, 212.
8. Arnold Sommerfeld to Hendrik Lorentz, Dec. 26, 1907, in Diana Kormos Buchwald, "The First Solvay Conference," in *Einstein in Context* (Cambridge, England: Cambridge University Press, 1993), 64. Sommerfeld is referring to the German physicist Emil Cohn, an expert in electrodynamics.
9. Jakob Laub to Einstein, Mar. 1, 1908.
10. Swiss Patent Office to Einstein, Mar. 13, 1906.
11. Mileva Marić to Helene Savić, Dec. 1906.

12. Einstein, "A New Electrostatic Method for the Measurement of Small Quantities of Electricity," Feb. 13, 1908, CPAE 2: 48; Overbye, 156.

13. Einstein to Paul and/or Conrad Habicht, Aug. 16, Sept. 2, 1907, Mar. 17, June, July 4, Oct. 12, Oct. 22, 1908, Jan. 18, Apr. 15, Apr. 28, Sept. 3, Nov. 5, Dec. 17, 1909; Overbye, 156–158.

14. Einstein, "On the Inertia of Energy Required by the Relativity Principle," May 14, 1907, CPAE 2: 45; Einstein to Johannes Stark, Sept. 25, 1907.

15. Einstein to Bern Canton Education Department, June 17, 1907, CPAE 5: 46; Fölsing, 228.

16. Einstein 1922c.

17. Einstein, "Fundamental Ideas and Methods of Relativity Theory," 1920, unpublished draft of a paper for *Nature* magazine, CPAE 7: 31. The phrase he used was "glücklichste Gedanke meines Lebens."

18. "Einstein Expounds His New Theory," *New York Times*, Dec. 3, 1919.

19. Bernstein 1996a, 10, makes the point that Newton's thought experiments involving a falling apple and Einstein's involving an elevator "were liberating insights that revealed unexpected depths in commonplace experiences."

20. Einstein 1916, chapter 20.

21. Einstein, "The Fundaments of Theoretical Physics," *Science*, May 24, 1940, in Einstein 1954, 329. See also Sartori, 255.

22. Einstein first used the phrase in a paper he wrote for the *Annalen der Physik* in Feb. 1912, "The Speed of Light and the Statics of the Gravitational Field," CPAE 4: 3.

23. Janssen 2002.

24. The gravitational field would have to be static and homogeneous and the acceleration would have to be uniform and rectilinear.

25. Einstein, "On the Relativity Principle and the Conclusions Drawn from It," *Jahrbuch der Radioaktivität and Elektronik*, Dec. 4, 1907, CPAE 2: 47; Einstein to Willem Julius, Aug. 24, 1911.

26. Einstein to Marcel Grossmann, Jan. 3, 1908.

27. Einstein to the Zurich Council of Education, Jan. 20, 1908; Fölsing, 236.

28. Einstein to Paul Gruner, Feb. 11, 1908; Alfred Kleiner to Einstein, Feb. 8, 1908.

29. Flückiger, 117–121; Fölsing, 238; Maja Einstein, xxi.

30. Alfred Kleiner to Einstein, Feb. 8, 1908.

31. Friedrich Adler to Viktor Adler, June 19, 1908; Rudolph Ardelt, *Friedrich Adler* (Vienna: Österreichischer Bundesverlag, 1984), 165–194; Seelig 1956a, 95; Fölsing, 247; Overbye, 161.

32. Frank 1947, 75; Einstein to Michele Besso, Apr. 29, 1917.

33. Einstein to Jakob Laub, May 19, 1909; Reiser, 72.

34. Friedrich Adler to Viktor Adler, July 1, 1908; Einstein to Jakob Laub, July 30, 1908.

35. Einstein to Jakob Laub, May 19, 1909.

36. Alfred Kleiner, report to the faculty, Mar. 4, 1909; Seelig 1956a, 166; Pais 1982, 185; Fölsing, 249.

37. Alfred Kleiner, report to faculty, Mar. 4, 1909.

38. Einstein to Jakob Laub, May 19, 1909.

39. Einstein, verse in the album of Anna Schmid, Aug. 1899, CPAE 1: 49.

40. Einstein to Anna Meyer-Schmid, May 12, 1909.

41. Mileva Marić to Georg Meyer, May 23, 1909; Einstein to Georg Meyer, June 7, 1909; Einstein to Erika Schaerer-Meyer, July 27, 1951; Highfield and Carter, 125; Overbye, 164.

42. Mileva Marić to Helene Savić, late 1909, Sept. 3, 1909, in Popović, 26–27.

43. Seelig 1956a, 92; Dukas and Hoffmann, 5–7.

44. Einstein to Arnold Sommerfeld, Jan. 14, 1908. I am grateful to Douglas Stone of Yale, who helped me with Einstein's early work on the quanta.

45. Einstein lecture in Salzburg, "On the Development of Our Views Concerning the Nature and Constitution of Radiation," Sept. 21, 1909, CPAE 2: 60; Schilpp, 154; Armin Hermann, *The Genesis of the Quantum Theory* (Cambridge, Mass.: MIT Press, 1971), 66–69.

46. Einstein to Arnold Sommerfeld, July 1910. As Einstein's friend Banesh Hoffmann quipped in *The Strange Story of the Quantum* (New York: Dover, 1959), "They could but make the best of it, and went around with woebegone faces sadly complaining that on Mondays, Wednesdays, and Fridays they must look upon light as a wave; on Tuesdays, Thursdays and Saturdays, as a particle. On Sundays they simply prayed."

47. Discussion following Sept. 21, 1909, lecture in Salzburg, CPAE 2: 61.

48. Einstein to Jakob Laub, Nov. 4 and 11, 1910.

49. Einstein to Heinrich Zangger, May 20, 1912.

CHAPTER EIGHT: THE WANDERING PROFESSOR

1. The best and original work about Duhem's influence on Einstein is by Don Howard. See Howard 1990a, 2004.

2. Friedrich Adler to Viktor Adler, Oct. 28, 1909, in Fölsing, 258.

3. Seelig 1956a, 97.

4. Seelig 1956a, 113.

5. Seelig 1956a, 99–104; Brian 1996, 76.

6. Seelig 1956a, 102; Einstein to Arnold Sommerfeld, Jan. 19, 1909.

7. Overbye, 185; Miller 2001, 229–231.

8. Hans Albert Einstein interview, *Gazette and Daily* (York, Pa.), Sept. 20, 1948; Seelig 1956a, 104; Highfield and Carter, 129.

9. Einstein to Pauline Einstein, Apr. 28, 1910.

10. Student petition, University of Zurich, June 23, 1910, CPAE 5: 210.

11. Repeated in lecture by Max Planck, Columbia University, spring 1909; Pais 1982, 192; Fölsing, 271.

12. Einstein to Jakob Laub, Aug. 27, Oct. 11, 1910; Count Karl von Stürgkh to Einstein, Jan. 13, 1911; Frank 1947, 98–101; Clark, 172–176; Fölsing, 271–273; Pais 1982, 192.

13. Frank 1947, 104. Frank has the visit occuring in 1913, but in fact it occurred in Sept. 1910 when Einstein was in Vienna for his official interview about the Prague professorship. See notes in CPAE 5 (German version), p. 625.
14. Einstein to Hendrik Lorentz, Jan. 27, 1911.
15. Einstein to Jakob Laub, May 19, 1909.
16. Einstein to Hendrik Lorentz, Feb. 15, 1911.
17. Pais 1982, 8; Brian 1996, 78; Klein 1970a, 303. The Ehrenfest description is from a draft of his eulogy for Lorentz.
18. Einstein, "Address at the Grave of Lorentz" (1928), in Einstein 1954, 73; Einstein, "Message for Hundredth Anniversary of the Birth of Lorentz" (1953), in Einstein 1954, 73. See also Bucky, 114.
19. Mileva Marić to Helene Savić, Jan. 1911, in Popović, 30; Einstein to Heinrich Zangger, Apr. 7, 1911.
20. Frank 1947, 98.
21. Max Brod, *The Redemption of Tycho Brahe* (New York: Knopf, 1928); Seelig 1956a, 121; Clark, 179; Highfield and Carter, 138.
22. Einstein to Paul Ehrenfest, Jan. 26, Feb. 12, 1912.
23. Einstein, "Paul Ehrenfest: In Memoriam," written in 1934 for a Leiden almanac and reprinted in Einstein 1950a, 132.
24. Klein 1970a, 175–178; Seelig 1956a, 125; Fölsing, 294; Clark, 194; Brian 1996, 83; Highfield and Carter, 142.
25. Einstein to Paul Ehrenfest, Mar. 10, 1912; Einstein to Alfred Kleiner, Apr. 3, 1912; Einstein to Paul Ehrenfest, Apr. 25, 1912. Einstein to Heinrich Zangger, Mar. 17, 1912: "I would like to see him my successor here. But his fanatical atheism makes that impossible." Zangger's letter was part of material released in 2006 and is published as CPAE 5: 374a in a supplement to vol. 10.
26. Dirk van Delft, "Albert Einstein in Leiden," *Physics Today*, Apr. 2006, 57.
27. Einstein to Heinrich Zangger, Nov. 7, 1911.
28. An invitation from Ernest Solvay, June 9, 1911, CPAE 5: 269; Einstein to Michele Besso, Sept. 11, Oct. 21, 1911.
29. Einstein, "On the Present State of the Problem of Specific Heats," Nov. 3, 1911, CPAE 3: 26; the quote about "really exist in nature" appears on p. 421 of the English translation of vol. 3.
30. Discussion following Einstein lecture, Nov. 3, 1911, CPAE 3: 27.
31. Einstein to Heinrich Zangger, Nov. 7 and 15, 1911.
32. Einstein to Michele Besso, Dec. 26, 1911.
33. Bernstein 1996b, 125.
34. Einstein to Heinrich Zangger, Nov. 7, 1911.
35. Einstein to Marie Curie, Nov. 23, 1911. (This letter is included at the beginning of CPAE vol. 8, not vol. 5, where it would have fit chronologically had this letter been available when that volume was published.)
36. Mileva Marić to Einstein, Oct. 4, 1911.
37. Overbye, 201. Einstein's quote is from a letter to Carl Seelig, May 5, 1952.
38. Reiser, 126.

39. Highfield and Carter, 145.

40. Einstein to Elsa Einstein Löwenthal, Apr. 30, 1912; regarding her keeping the letters, CPAE 5: 389 (German edition), footnote 12.

41. Einstein to Elsa Einstein, Apr. 30, 1912; Einstein "scratch notebook," CPAE 3 (German edition), appendix A; CPAE 5: 389 (German edition), footnote 4.

42. Einstein to Elsa Einstein, May 7 and 12, 1912.

43. Einstein to Michele Besso, May 13, 1911; Einstein to Hans Tanner, Apr. 24, 1911; Einstein to Alfred and Clara Stern, Mar. 17, 1912.

44. Mileva Marić to Helene Savić, Dec. 1912, in Popović, 106.

45. Willem Julius to Einstein, Sept. 17, 1911; Einstein to Willem Julius, Sept. 22, 1911.

46. Heinrich Zangger to Ludwig Forrer, Oct. 9, 1911; CPAE 5: 291 (German edition), footnote 2; CPAE 5: 305 (German edition), footnote 2.

47. Einstein to Heinrich Zangger, Nov. 15, 1911.

48. Einstein to Willem Julius, Nov. 16, 1911.

49. Marie Curie, letter of recommendation, Nov. 17, 1911; Seelig 1956a, 134; Fölsing, 291; CPAE 5: 308 (German edition), footnote 3.

50. Henri Poincaré, letter of recommendation, Nov. 1911; Seelig 1956a, 135; Galison, 300; Fölsing, 291; CPAE 5: 308 (German edition), footnote 3.

51. Einstein to Alfred and Clara Stern, Feb. 2, 1912.

52. Articles appeared in Vienna's weekly paper *Montags-Revue* on July 29, 1912, and Prague's *Prager Tagblatt* on May 26 and Aug. 5, 1912. CPAE 5: 414 (German edition), footnotes 2, 3, 11; Einstein statement, Aug. 3, 1912.

53. Einstein to Ludwig Hopf, June 12, 1912.

54. Overbye, 234, 243; Highfield and Carter, 153; Seelig 1956a, 112.

55. In a letter from Einstein to Elsa Einstein, July 30, 1914, he recalls how she kidded him for including his new address in the May 7, 1912, letter in which he declared they must quit corresponding.

56. Einstein to Elsa Einstein, ca. Mar. 14, 1913.

57. Einstein to Elsa Einstein, Mar. 23, 1913.

58. Seelig 1956a, 244; Levenson, 2; CPAE 5: 451 (German edition), footnote 2; Clark, 213; Overbye, 248; Fölsing, 329. The editors of the collected papers use the white handkerchief, based on a letter by Nernst's daughter, while other accounts use the red rose, based on the account that Seelig was given.

59. Max Planck, Walther Nernst, Heinrich Rubens, and Emil Warburg to the Prussian Academy, June 12, 1913, CPAE 5: 445.

60. Seelig 1956a, 148.

61. Einstein to Jakob Laub, July 22, 1913.

62. Einstein to Paul Ehrenfest, late Nov. 1913.

63. Einstein to Hendrik Lorentz, Aug. 14, 1913.

64. Einstein to Heinrich Zangger, June 27, 1914, CPAE 8: 5a, released in 2006 and published as a supplement to CPAE vol. 10.

65. Einstein to Elsa Einstein, July 14, 19, before July 24, and Aug. 13, 1913.

66. Einstein to Elsa Einstein, after Aug. 11, 1913.

67. Einstein to Elsa Einstein, after Aug. 11 and Aug. 11, 1913.

68. Eve Curie, *Madame Curie* (New York: Doubleday, 1937), 284; Fölsing, 325; Highfield and Carter, 157.

69. The baptism took place at the St. Nicholas Church in Novi Sad on Sept. 21, 1913. Hans Albert Einstein to Dord Krstic, Nov. 5, 1970; Elizabeth Einstein, 97; Highfield and Carter, 159; Overbye, 255; Einstein to Heinrich Zangger, Sept. 20, 1913; Seelig 1956a, 113.

70. Einstein to Elsa Einstein, Oct. 10, 1913.

71. Einstein to Elsa Einstein, Oct. 16, 1913.

72. Einstein to Elsa Einstein, before Dec. 2, 1913.

73. Einstein to Elsa Einstein, after Dec. 21 and Aug. 11, 1913.

74. Einstein to Elsa Einstein, after Dec. 21, 1913.

75. Einstein to Elsa Einstein, after Feb. 11, 1914; Lisbeth Hurwitz diary, cited in Overbye, 265.

76. Marianoff, 1; Einstein to Mileva Marić, Apr. 2, 1914.

77. Einstein to Paul Ehrenfest, ca. Apr. 10, 1914; Paul Ehrenfest to Einstein, ca. Apr. 10, 1914; Highfield and Carter, 167.

78. Whitrow, 20.

79. Einstein to Heinrich Zangger, June 27, 1914, CPAE 8: 16a, made available in 2006 and printed in a supplement to vol. 10.

80. Einstein, Memorandum to Mileva Marić, ca. July 18, 1914, CPAE 8: 22. See also appendix, CPAE 8b (German edition), p. 1032, for a memo from Anna Besso-Winteler to Heinrich Zangger, Mar. 1918, about the Einstein breakup.

81. Einstein to Mileva Marić, ca. July 18 and July 18, 1914.

82. CPAE 8a: 26 (German edition), footnote 3; memo from Anna Besso-Winteler to Heinrich Zangger, Mar. 1918, CPAE 8b (German edition), p. 1032; Overbye, 268.

83. Einstein to Elsa Einstein, July 26, 1914.

84. Einstein to Elsa Einstein, after July 26, 1914.

85. Einstein to Elsa Einstein, July 30, 1914 (two letters); Michele Besso to Einstein, Jan. 17, 1928 (recalling the breakup); Pais 1982, 242; Fölsing, 338.

86. Einstein to Elsa Einstein, after Aug. 3, 1914.

87. Einstein to Mileva Marić, Sept. 15, 1914, contains the poisoning allegation. Many other letters in 1914 detail their struggle over money, furniture, and treatment of the children.

CHAPTER NINE: GENERAL RELATIVITY

1. Renn and Sauer 2006, 117.

2. The description of the equivalence principle follows the formulation that Einstein used in his yearbook article of 1907 and his comprehensive general relativity paper of 1916. Others have subsequently modified it slightly. See also Einstein, "Fundamental Ideas and Methods of Relativity Theory," 1920, unpublished draft of a paper for *Nature* magazine, CPAE 7: 31.

Some of this chapter draws from a dissertation by one of the editors of the

Einstein Papers Project: Jeroen van Dongen, "Einstein's Unification: General Relativity and the Quest for Mathematical Naturalness," 2002. He provided a copy to me along with guidance and editing for this chapter. This chapter also follows the research findings of other scholars studying Einstein's general relativity work. I am grateful to van Dongen and others who met with me and helped me on this chapter, including Tilman Sauer, Jürgen Renn, John D. Norton, and Michel Janssen. This chapter draws on their work and also that of John Stachel, all listed in the bibliography.

3. Einstein, "The Speed of Light and the Statics of the Gravitational Field," *Annalen der Physik* (Feb. 1912), CPAE 4: 3; Einstein 1922c; Janssen 2004, 9. In his 1907 and 1911 papers, Einstein refers to it as the "equivalence hypothesis," but in this 1912 paper, he raises it to the status of an *Aequivalenzprinzip*.

4. Einstein, "On the Influence of Gravitation on the Propagation of Light," *Annalen der Physik* (June 21): 1911, CPAE 3: 23.

5. Einstein to Erwin Freundlich, Sept. 1, 1911.

6. Stachel 1989b.

7. Record and grade transcript, CPAE 1: 25; Adolf Hurwitz to Hermann Bleuler, July 27, 1900, CPAE 1: 67; Einstein to Mileva Marić, Dec. 28, 1901.

8. Fölsing, 314; Pais 1982, 212.

9. Hartle, 13.

10. Einstein to Arnold Sommerfeld, Oct. 29, 1912.

11. Einstein, foreword to the Czech edition of his popular book *Relativity*, 1923; see utf.mff.cuni.cz/Relativity/Einstein.htm. In it Einstein writes, "The decisive idea of the analogy between the mathematical formulation of the theory and the Gaussian theory of surfaces came to me only in 1912 after my return to Zurich, without being aware at that time of the work of Riemann, Ricci, and Levi-Civita. This was first brought to my attention by my friend Grossmann." Einstein 1922c: "I realized that the foundations of geometry have physical significance. My dear friend the mathematician Grossmann was there when I returned from Prague to Zurich. From him I learned for the first time about Ricci and later about Riemann."

12. Sartori, 275.

13. Amir Aczel, "Riemann's Metric," in Aczel 1999, 91–101; Hoffmann 1983, 144–151.

14. I am grateful to Tilman Sauer and Craig Copi for help with this section.

15. Janssen 2002; Greene 2004, 72.

16. Calaprice, 9; Flückiger, 121.

17. The Zurich Notebook is in CPAE 4: 10. An online facsimile is available at echo.mpiwg-berlin.mpg.de/content/relativityrevolution/jnul. See also Janssen and Renn.

18. Norton 2000, 147. See also Renn and Sauer 2006, 151. I am grateful to Tilman Sauer for his editing of this section.

19. Einstein, Zurich Notebook, CPAE 4: 10 (German edition), p. 39 has the first notations of what became known as the Einstein tensor.

20. An explanation of this dilemma is in Renn and Sauer 1997, 42–43. The

mystery of why Einstein in early 1913 could not find the correct gravitational tensor—and the issue of his understanding of coordinate condition options—is addressed nicely in Renn 2005b, 11–14. He builds on, and suggests some revisions to, the conclusions of Norton 1984.

21. Norton, Janssen, and Sauer have all suggested that Einstein's bad experience in 1913 of abandoning a mathematical strategy for a physical one, and his subsequent belated success with a mathematical strategy, is reflected in the views he expressed in his 1933 Spencer lecture at Oxford and also his approach in the later decades of his life to finding a unified field theory.

22. Einstein, "Outline [*Entwurf*] of a Generalized Theory of Relativity and of a Theory of Gravitation" (with Marcel Grossmann), before May 28, 1913, CPAE 4: 13; Janssen 2004; Janssen and Renn.

23. Einstein to Elsa Einstein, Mar. 23, 1913.

24. Einstein-Besso manuscript, CPAE 4: 14; Janssen, 2002.

25. Einstein, "On the Foundations of the General Theory of Relativity," *Annalen der Physik* (Mar. 6, 1918), CPAE 7: 4. A vivid explanation of Newton's bucket and how it connects to relativity is in Greene 2004, 23–74. Einstein is largely responsible for inferring how Mach would regard an empty universe. See Norton 1995c; Julian Barbour, "General Relativity as a Perfectly Machian Theory," Carl Hoefer, "Einstein's Formulation of Mach's Principle," and Hubert Goenner, "Mach's Principle and Theories of Gravity," all in Barbour and Pfister.

26. Janssen 2002, 14; Janssen 2004, 17; Janssen 2006. Janssen has done important work analyzing the Einstein-Besso collaborations of 1913. Reproductions of the Einstein-Besso manuscript and other related documents, along with an essay by Janssen on their significance, is in a 288-page catalogue from Christie's, which auctioned the originals on Oct. 4, 2002. (The 50-page Einstein-Besso manuscript sold for $595, 000.) For an example of how Einstein dismissed Besso's suggestion that the Minkowski metric in rotating coordinates wasn't a valid solution to the *Entwurf* field equations—and how Einstein kept feeling that the *Entwurf* did indeed comply with Mach's principle—see Einstein to Michele Besso, ca. Mar. 10, 1914.

27. Einstein to Ernst Mach, June 25, 1913; Misner, Thorne, and Wheeler, 544.

28. Einstein to Hendrik Lorentz, Aug. 14, 1913. But two days later, he writes Lorentz again to say that he has resigned himself to the belief that covariance is impossible: "Only now, after this ugly dark spot seems to have been eliminated, does the theory give me pleasure." Einstein to Hendrik Lorentz, Aug. 16, 1913.

29. The hole argument basically said that a generally covariant gravitational theory would be indeterministic. Generally covariant field equations could not determine the metric field uniquely. A full specification of the metric field outside of some small region that was devoid of matter, known as "the hole," would not be able to fix the metric field within that region. See Stachel 1989b; Norton 2005b; Janssen 2004.

30. Einstein to Ludwig Hopf, Nov. 2, 1913. See also Einstein to Paul Ehrenfest, Nov. 7, 1913: "It can be proved that *generally covariant* equations that deter-

mine the field completely from the matter tensor cannot exist at all. Can there be anything more beautiful than this, that the necessary specialization follows from the conservation laws? Thus, the conservation laws determine the surfaces that, from among all the surfaces, are to be privileged as coordinate surfaces. We can designate these privileged surfaces as planes, since we are left with linear substitutions as the only ones that are justified." Einstein's clearest explanation of the hole argument is "On the Foundations of the Generalized Theory of Relativity and the Theory of Gravitation," Jan. 1914, CPAE 4: 25.

31. When Einstein appeared at the annual convocation of German-speaking scientists in Sept. 1913, the rival gravitation theorist Gustav Mie rose to launch a "lively" attack on him and subsequently published a violent polemic that displayed a vitriol far beyond anything explained by scientific disagreements. Einstein also engaged in a bitter debate with Max Abraham, whose own gravitational theory Einstein had attacked with great relish throughout 1912. Report on the Vienna conference, Sept. 23, 1913, CPAE 4: 17.

32. Einstein to Heinrich Zangger, ca. Jan. 20, 1914.

33. Einstein to Heinrich Zangger, Mar. 10, 1914. Jürgen Renn has pointed out that the 1913–1915 period of defending and refining the *Entwurf*, even though it did not save that theory, did help Einstein to better understand the difficulties that seemed to bedevil the tensors he had explored in the mathematical strategy. "Practically all of the technical problems Einstein had encountered in the Zurich notebook with candidates derived from the Riemann tensor were actually resolved in this period in the course of his examination of problems associated with the *Entwurf* theory." Renn 2005b, 16.

34. Einstein to Erwin Freundlich, Jan. 8, 1912, mid-Aug. 1913; Einstein to George Hale, Oct. 14, 1913; George Hale to Einstein, Nov. 8, 1913.

35. Clark, 207.

36. Einstein to Erwin Freundlich, Dec. 7, 1913.

37. Einstein to Erwin Freundlich, Jan. 20, 1914.

38. Fölsing, 356–357.

39. Einstein to Paul Ehrenfest, Aug. 19, 1914.

40. Ibid.

41. Einstein to Paolo Straneo, Jan. 7, 1915.

42. For a good description from which this is drawn, see Levenson, especially 60–65.

43. Elon, 277, 303–304.

44. Fölsing, 344.

45. Einstein to Hans Albert Einstein, Jan. 25, 1915.

46. Nathan and Norden, 4; Elon, 326. Also translated as the "Manifesto to the Civilized World."

47. Einstein to Georg Nicolai, Feb. 20, 1915. The full text is in CPAE 6: 8, and Nathan and Norden, 5. Clark, 228, makes the case that some of the writing was Einstein's. See also Wolf William Zuelzer, *The Nicolai Case* (Detroit: Wayne State University Press, 1982); Overbye, 273; Levenson, 63; Fölsing, 346–347; Elon, 328.

48. Nathan and Norden, 9; Overbye, 275–276; Fölsing, 349; Clark, 238.
49. Einstein to Romain Rolland, Sept. 15, 1915; CPAE 8a: 118 (German edition), footnote 2; Romain Rolland diary, cited in Nathan and Norden, 16; Fölsing, 366.
50. Einstein to Paul Hertz, before Oct. 8, 1915; Paul Hertz to Einstein, Oct. 8, 1915; Einstein to Paul Hertz, Oct. 9, 1915.
51. Einstein, "My Opinion on the War," Oct. 23–Nov. 11, 1915, CPAE 6: 20.
52. Einstein to Heinrich Zangger, after Dec. 27, 1914, CPAE 8: 41a, in supplement to vol. 10.
53. Hans Albert Einstein to Einstein, two postcards, before Apr. 4, 1915, part of the family correspondence trust that was under seal until 2006. CPAE 8: 69a, 8: 69b, in supplement to vol. 10.
54. Einstein to Hans Albert Einstein, ca. Apr. 4, 1915.
55. Einstein to Heinrich Zangger, July 16, 1915.
56. Einstein to Elsa Einstein, Sept. 11, 1915; Einstein to Heinrich Zangger, Oct. 15, 1915; Einstein to Hans Albert Einstein, Nov. 4, 1915. For Einstein's complaint that he was barely able to see his boys during the Sept. 1916 visit, see Einstein to Mileva Marić, Apr. 1, 1916: "I hope that this time you will not again withhold the boys almost entirely from me."
57. Einstein to Heinrich Zangger, Oct. 15, 1915; Michele Besso to Einstein, ca. Oct. 30, 1915.
58. Once again, I have drawn on the works of Jürgen Renn, Tilman Sauer, John Stachel, Michel Janssen, and John D. Norton.
59. Horst Kant, "Albert Einstein and the Kaiser Wilhelm Institute for Physics in Berlin," in Renn 2005d, 168–170.
60. Wolf-Dieter Mechler, "Einstein's Residences in Berlin," in Renn 2005d, 268.
61. Janssen 2004, 29.
62. Einstein to Heinrich Zangger, July 7, ca. July 24, 1915; Einstein to Arnold Sommerfeld, July 15, 1915.
63. Specifically, the issue was whether the *Entwurf* field equations were invariant under the non-autonomous transformation to rotating coordinates in the case of the Minkowski metric in its standard diagonal form. Janssen 2004, 29.
64. Michele Besso memo to Einstein, Aug. 28, 1913; Janssen 2002; Norton 2000, 149; Einstein to Erwin Freundlich, Sept. 30, 1915.
65. Einstein to Hendrik Lorentz, Oct. 12, 1915. Einstein describes his October 1915 breakthroughs in a subsequent letter to Lorentz and another one to Arnold Sommerfeld. Einstein to Hendrik Lorentz, Jan. 1, 1916: "Trying times awaited me last fall as the inaccuracy of the older gravitational field equations gradually dawned on me. I had already discovered earlier that Mercury's perihelion motion had come out too small. In addition, I found that the equations were not covariant for substitutions corresponding to a uniform rotation of the new reference system. Finally, I found that the consideration I made last year on the determination of Lagrange's H function for the gravitational field was thoroughly illusory, in that it could easily be modified such that no restricting conditions had to be attached to H, thus making it possible to choose

it completely freely. In this way I came to the conviction that introducing adapted systems was on the wrong track and that a more broad-reaching covariance, preferably a *general* covariance, must be required. Now general covariance has been achieved, whereby nothing is changed in the subsequent specialization of the frame of reference . . . I had considered the current equations in essence already three years ago together with Grossmann, who had brought my attention to the Riemann tensor." Einstein to Arnold Sommerfeld, Nov. 28, 1915: "In the last month I had one of the most stimulating and exhausting times of my life, and indeed also one of the most successful. For I realized that my existing gravitational field equations were untenable! The following indications led to this: 1) I proved that the gravitational field on a uniformly rotating system does not satisfy the field equations. 2) The motion of Mercury's perihelion came to 18″ rather than 45″ per century. 3) The covariance considerations in my paper of last year do not yield the Hamiltonian function *H*. When it is properly generalized, it permits an arbitrary *H*. From this it was demonstrated that covariance with respect to 'adapted' coordinate systems was a flop."

66. Norton 2000, 152.
67. There is a subtle divergence of opinion among the group of general relativity historians about the extent of his purported shift from the physical to the mathematical strategy in Oct.–Nov. 1915. John Norton has argued that Einstein's "new tactic was to reverse his decision of 1913" and go back to a mathematical strategy, emphasizing a tensor analysis that would produce general covariance (Norton 2000, 151). Likewise, Jeroen van Dongen says the shift in tactics was clear: "Einstein immediately got hold of the way out of the *Entwurf*'s quagmire: he returned to the mathematical requirement of general covariance that he had abandoned in the Zurich notebook" (van Dongen, 25). Both scholars produce quotes from Einstein's later years in which he claims that the big lesson he learned was to trust a mathematical strategy. On the other side, Jürgen Renn and Michel Janssen say that Norton and van Dongen (and the older Einstein in his hazy memory) make too much of this shift. Physical considerations still played a major role in finding the final theory in Nov. 1915. "In our reconstruction, however, Einstein found his way back to the generally-covariant field equations by making one important adjustment to the *Entwurf* theory, a theory born almost entirely out of physical considerations . . . That mathematical considerations pointed in the same direction undoubtedly inspired confidence that this was the right direction, but guiding him along this path were physical not mathematical considerations" (Janssen and Renn, 13; the quote I use in the text is on p. 10). Also, Janssen 2004, 35: "Whatever he believed, said, or wrote about it later on, Einstein only discovered the mathematical high road to the Einstein field equations after he had already found these equations at the end of a poorly paved road through physics."
68. Einstein to Arnold Sommerfeld, Nov. 28, 1915.
69. Einstein, "On the General Theory of Relativity," Nov. 4, 1915, CPAE 6: 21.

70. Einstein to Michele Besso, Nov. 17, 1915; Einstein to Arnold Sommerfeld, Nov. 28, 1915.

71. Einstein to Hans Albert Einstein, Nov. 4, 1915.

72. Einstein to David Hilbert, Nov. 7, 1915.

73. Overbye, 290.

74. Einstein, "On the General Theory of Relativity (Addendum)," Nov. 11, 1915, CPAE 6: 22; Renn and Sauer 2006, 276; Pais 1982, 252.

75. Einstein to David Hilbert, Nov. 12, 1915.

76. Einstein to Hans Albert Einstein, Nov. 15, 1915; Einstein to Mileva Marić, Nov. 15, 1915; Einstein to Heinrich Zangger, Nov. 15, 1915 (released in 2006 and printed in supplement to vol. 10).

77. Einstein to David Hilbert, Nov. 15, 1915.

78. Einstein, "Explanation of the Perihelion Motion of Mercury from the General Theory of Relativity," Nov. 18, 1915, CPAE 6: 24.

79. Pais 1982, 253; Einstein to Paul Ehrenfest, Jan. 17, 1916; Einstein to Arnold Sommerfeld, Dec. 9, 1915.

80. Einstein to David Hilbert, Nov. 18, 1915.

81. David Hilbert to Einstein, Nov. 19, 1915.

82. The equation has been expressed in many ways. The one I use follows the formulation Einstein used in his 1921 Princeton lectures. The entire left-hand side of the equation can be expressed more compactly as what is now known as the Einstein tensor: $G_{\mu\nu}$.

83. Overbye, 293; Aczel 1999, 117; archive.ncsa.uiuc.edu/Cyberia/NumRel/Ein steinEquations.html#intro. A variation of Wheeler's quote is on p. 5 of the book he coauthored with Charles Misner and Kip Thorne, *Gravitation*.

84. Greene 2004, 74.

85. Einstein, "The Foundations of the General Theory of Relativity," *Annalen der Physik* (Mar. 20, 1916), CPAE 6: 30.

86. Einstein to Heinrich Zangger, Nov. 26, 1915; Einstein to Michele Besso, Nov. 30, 1915.

87. Thorne, 119.

88. For an analysis of Hilbert's contribution, see Sauer 1999, 529–575; Sauer 2005, 577–590. Papers describing Hilbert's revisions include Corry, Renn, and Stachel; Sauer 2005. For a flavor of the controversy, see also John Earman and Clark Glymour, "Einstein and Hilbert: Two Months in the History of General Relativity," *Archive for History of Exact Sciences* (1978): 291; A. A. Logunov, M. A. Mestvirishvili, and V. A. Petrov, "How Were the Hilbert-Einstein Equations Discovered?," *Uspekhi Fizicheskikh Nauk* 174, no. 6 (June 2004): 663–678; Christopher Jon Bjerknes, *Albert Einstein: The Incorrigible Plagiarist*, available at home.comcast.net/~xtxinc/AEIPBook.htm; John Stachel, "Anti-Einstein Sentiment Surfaces Again," *Physics World*, Apr. 2003, physicsweb .org/articles/review/16/4/2/1; Christopher Jon Bjerknes, "The Author of *Albert Einstein: The Incorrigible Plagiarist* Responds to John Stachel's Personal Attack," home.comcast.net/~xtxinc/Response.htm; Friedwardt Winterberg, "On 'Belated Decision in the Hilbert-Einstein Priority Dispute,' " *Zeitschrift*

fuer Naturforschung A, (Oct. 2004): 715–719, www.physics.unr.edu/faculty/winterberg/Hilbert-Einstein.pdf; David Rowe, "Einstein Meets Hilbert: At the Crossroads of Physics and Mathematics," *Physics in Perspective* 3, no. 4 (Nov. 2001): 379.

89. Reid, 142. Although this comment is cited in other secondary sources as well, Tilman Sauer of the Einstein Papers Project, who is writing a book on Hilbert, says he has never found a primary source for it.
90. Einstein to David Hilbert, Dec. 20, 1915.
91. Einstein to Arnold Sommerfeld, Dec. 9, 1915; Einstein to Heinrich Zangger, Nov. 26, 1915.
92. It is a contentious question as to whether general relativity actually succeeds in making all forms of motion and all frames of reference equivalent. It can certainly be said that two observers in nonuniform relative motion can each legitimately view himself as "at rest" and the other as affected by a gravitational field. That does not necessarily mean (as Einstein sometimes seemed to believe and at other times not) that two observers in nonuniform relative motion are always physically equivalent, especially when it comes to rotation. See, for example, Norton 1995b, 223–245; Janssen 2004, 8–12; Don Howard, "Point Coincidences and Pointer Coincidences," in Goenner et al. 1999, 463; Robert Rynasiewicz, "Kretschmann's Analysis of Covariance and Relativity Principles," in Goenner et al. 1999, 431; Dennis Diek, "Another Look at General Covariance and the Equivalence of Reference Frames," *Studies in the History and Philosophy of Modern Physics* 37 (Mar. 2006): 174.
93. Fölsing, 374; Clark, 252.
94. Einstein to Michele Besso, Dec. 10, 1915.

CHAPTER TEN: DIVORCE

1. Michele Besso to Einstein, Nov. 29, 1915; Einstein to Michele Besso, Nov. 30, 1915; Neffe, 192.
2. Hans Albert Einstein to Einstein, before Nov. 30, 1915; Einstein to Hans Albert Einstein, Nov. 30, 1915.
3. Michele Besso to Einstein, Nov. 30, 1915. See also Einstein to Heinrich Zangger, Dec. 4, 1915: "The boy's soul is being systematically poisoned to make sure that he doesn't trust me."
4. Einstein to Mileva Marić, Dec. 1 and 10, 1915.
5. Einstein to Hans Albert Einstein, Dec. 23 and 25, 1915. Einstein wrote a similar postcard to Hans Albert on Dec. 18, 1915. Einstein to Hans Albert Einstein, Mar. 11, 1916.
6. Einstein to Heinrich Zangger, Nov. 26, 1915; Einstein to Michele Besso, Jan. 3, 1916.
7. Overbye, 300.
8. Einstein to Mileva Marić, Feb. 6, 1916.
9. Einstein to Mileva Marić, Mar. 12, Apr. 1, 1916; Neffe, 194.

10. Einstein to Mileva Marić, Apr. 1 and 8, 1916; Einstein to Michele Besso, Apr. 6, 1916; Michele Besso to Heinrich Zangger, Apr. 12, 1916, CPAE 8: 211 (German edition), footnote 2.

11. Einstein to Elsa Einstein, Apr. 12 and 15, 1916. See also Einstein to Elsa Einstein, Apr. 10, 1916, in the sealed family correspondence released in 2006, CPAE 8: 211a: "My relationship with him is becoming very warm."

12. Einstein to Elsa Einstein, Apr. 21, 1916. See also Einstein to Heinrich Zangger, July 11, 1916: "Following an exceedingly nice Easter excursion, the subsequent days in Zurich brought on a complete chilling in a way that is not quite explicable to me."

13. Einstein to Heinrich Zangger, July 11, 1916; Einstein to Michele Besso, July 14, 1916. See CPAE 8: 233 (German edition), footnote 4, for Zangger being the other person referred to in the letter.

14. Pauline Einstein to Elsa Einstein, Aug. 6, 1916, in Overbye, 301.

15. Einstein to Michele Besso, July 14, 1916; Michele Besso to Einstein, July 17, 1916; CPAE 8: 239 (German version), footnote 2.

16. Einstein to Michele Besso, July 21, 1916, two letters.

17. CPAE 8: 241 (German edition), footnotes 3, 4; Einstein to Heinrich Zangger, July 25, 1916; Heinrich Zangger to Michele Besso, July 31, 1916.

18. Einstein to Heinrich Zangger, Aug. 18, 1916; Einstein to Hans Albert Einstein, July 25, 1916. See also Einstein to Heinrich Zangger, Mar. 10, 1917.

19. Einstein to Michele Besso, Aug. 24, 1916; Einstein to Hans Albert Einstein, Sept. 26, 1916.

20. Hans Albert Einstein to Einstein, before Nov. 26, 1916.

21. Einstein to Michele Besso, Oct. 31, 1916.

22. Einstein to Helene Savić, Sept. 8, 1916.

23. Einstein, "The Foundation of the General Theory of Relativity," Mar. 20, 1916, CPAE 6: 30.

24. Einstein, *On the Special and the General Theory of Relativity*, Dec. 1916, CPAE 6: 42, and many popular editions; Michelmore, 63. For an Internet version of Einstein's book, see bartleby.com/173/ or www.gutenberg.org/etext/5001.

25. Einstein, "Principles of Research," 1918, in Einstein 1954, 224.

26. Einstein to Heinrich Zangger, Jan. 16, 1917; Clark, 241.

27. Clark, 248; Highfield and Carter, 183; Overbye, 327; Einstein to Paul Ehrenfest, Feb. 14, 1917; Einstein to Heinrich Zangger, Dec. 6, 1917.

28. Einstein to Michele Besso, Mar. 9, 1917; Einstein to Heinrich Zangger, Feb. 16 and Mar. 10, 1917.

29. Einstein to Paul Ehrenfest, May 25, 1917.

30. Einstein to Heinrich Zangger, June 12, 1917.

31. Einstein to Mileva Marić, Jan. 31, 1918.

32. Mileva Marić to Einstein, Feb. 9, 1918, from family trust correspondence, CPAE 8: 461a, in supplement to vol. 10.

33. Mileva Marić to Einstein, after Feb. 6, 1918. The Feb. 9 letter from the family trust correspondence, footnote 32 above, was unsealed in 2006. It clearly

comes before the one that was dated "after Feb. 6" by the Einstein papers editors.

34. Overbye, 338–339.
35. Mileva Marić to Einstein, Apr. 22, 1918.
36. Einstein to Mileva Marić, Apr. 15, 23, 26, 1918.
37. Maja Winteler-Einstein to Einstein, Mar. 6, 1918, family foundation correspondence, unsealed in 2006, CPAE 8: 475b, in supplement to vol. 10.
38. Einstein to Anna Besso, after Mar. 4, 1918.
39. Anna Besso to Einstein, after Mar. 4, 1918.
40. Mileva Marić to Einstein, before May 23, 1918; Einstein to Mileva Marić, June 4, 1918. See also Vero Besso (Anna and Michele's son) to Einstein, Mar. 28, 1918, family trust correspondence: "The postcard you sent to my mother was really not nice . . . Her words would not have offended you in any way if you had heard them yourself; you would just have laughed and would have toned down their sense a little."
41. Mileva Marić to Einstein, Mar. 17, 1918: "My state of health is now such that I can lie down quite well at home; although I can't get up, I can very well occupy myself quite a considerable amount with the children, and this makes me very happy and contributes much to my well-being." Einstein to Heinrich Zangger, May 8, 1918.
42. Einstein to Heinrich Zangger, May 8, 1918.
43. Einstein to Max Born, after June 29, 1918; Einstein to Michele Besso, July 29, 1918.
44. Einstein to Hans Albert Einstein, after June 4, 1918.
45. Einstein to Hans Albert Einstein, after June 19, 1918.
46. Hans Albert Einstein to Einstein, ca. July 17, 1918; Einstein to Eduard Einstein, ca. July 17, 1918.
47. Edgar Meyer to Einstein, Aug. 11, 1918; Einstein to Michele Besso, Aug. 20, 1918.
48. Einstein to Heinrich Zangger, Aug. 16, 1918; Einstein to Michele Besso, Sept. 6, 1918; Fölsing, 424.
49. Reiser, 140.
50. Nathan and Norden, 24. See also Rowe and Schulmann.
51. Born 2005, 145–147. My description relies on Born's recollection, which accompanies Einstein's references to the event in a letter to Born, Sept. 7, 1944. See also Bolles, 3–11; Seelig 1956a, 178; Fölsing, 423; Levenson, 198.
52. Einstein, "On the Need for a National Assembly," Nov. 13, 1918, CPAE 8: 14; Nathan and Norden, 25. Otto Nathan says that Einstein delivered these remarks to the student radicals at the university. There is no evidence of this, and Born does not mention it. The newspapers report it as a New Fatherland League speech later that day. See CPAE 8: 14 (German edition), footnote 2.
53. Einstein to Max Born, Sept. 7, 1944.
54. Einstein, Deposition in Divorce, Dec. 23, 1918, CPAE 8: 676.
55. Einstein to Mileva Marić and Hans Albert Einstein, Jan. 10, 1919; Einstein to Hedwig and Max Born, Jan. 15 and 19, 1919; Theodor Vetter to Einstein, Jan.

28, 1919. Vetter was the president of Zurich University, and he responded to Einstein's complaint about a guard being posted at the door of the lectures.

56. Divorce Decree, Feb. 14, 1919, CPAE 9: 6.

57. Overbye, 273–280.

58. Einstein to Georg Nicolai, ca. Jan. 22 and Feb. 28, 1917; Georg Nicolai to Einstein, Feb. 26, 1917.

59. Ilse Einstein to Georg Nicolai, May 22, 1918, CPAE 8: 545.

60. Einstein to Elsa Einstein, July 12 and 17, 1919.

61. Einstein to Elsa Einstein, July 28, 1919.

62. "Professor Einstein Here," *New York Times*, Apr. 3, 1921.

63. "Pronounced Sense of Humor," *New York Times*, Dec. 22, 1936.

64. Fölsing, 429; Highfield and Carter, 196.

65. Reiser, 127; Marianoff, 15, 174. Both of these authors married daughters of Elsa. Reiser's real name was Rudolph Kayser.

66. Elias Tobenkin, "How Einstein, Thinking in Terms of the Universe, Lives from Day to Day," *New York Evening Post*, Mar. 26, 1921.

67. Frank 1947, 219; Marianoff, 1; Fölsing, 428; Reiser, 193.

CHAPTER ELEVEN: EINSTEIN'S UNIVERSE

1. Overbye, 314; Einstein to Karl Schwarzschild, Jan. 9, 1916.

2. Einstein, "On a Stationary System with Spherical Symmetry Consisting of Many Gravitating Masses," *Annals of Mathematics*, 1939.

3. For a description of the history, math, and science of black holes, see Miller 2005; Thorne, 121–139.

4. Freeman Dyson in Robinson, 8–9.

5. Einstein to Karl Schwarzschild, Jan. 9, 1916.

6. CPAE vol. 8 brings together all of the correspondence between Einstein and de Sitter, with a good commentary on the dispute. Michel Janssen (uncredited author), "The Einstein–De Sitter–Weyl–Klein debate," CPAE 8a (German edition), p. 351.

7. Einstein to Willem de Sitter, Feb. 2, 1917.

8. Einstein to Paul Ehrenfest, Feb. 4, 1917.

9. Einstein, "Cosmological Considerations in the General Theory of Relativity," Feb. 8, 1917, CPAE 6: 43.

10. Einstein 1916, chapter 31.

11. Clark, 271.

12. For a delightful fictional tale along these lines (so to speak), see Edwin Abbott's *Flatland*, first published in 1880 and available in many paperback editions.

13. Edward W. Kold, "The Greatest Discovery Einstein Didn't Make," in Brockman, 205.

14. Lawrence Krauss and Michael Turner, "A Cosmic Conundrum," *Scientific American* (Sept. 2004): 71; Aczel 1999, 155; Overbye, 321. Einstein's famous blunder quote is from Gamow, 1970, 44.

15. Overbye, 327.

16. Einstein 1916, chapter 22.

17. There is a wonderful reprint now available in paperback of Eddington's classic book first published in 1920: Arthur Eddington, *Space, Time and Gravitation: An Outline of the General Relativity Theory* (Cambridge, England: Cambridge Science Classics, 1995). Page 141 describes the Principe expedition. See also an award-winning article: Matthew Stanley, "An Expedition to Heal the Wounds of War: 1919 Eclipse and Eddington as Quaker Adventurer," *Isis* 94 (2003): 57–89. A comprehensive account of all the tests is in Crelinsten.

18. Douglas, 40; Aczel 1999, 121–137; Clark, 285–287; Fölsing, 436–437; Overbye, 354–359.

19. Douglas, 40.

20. Einstein to Pauline Einstein, Sept. 5, 1919; Einstein to Paul Ehrenfest, Sept. 12, 1919.

21. Einstein to Pauline Einstein, Sept. 27, 1919; Bolles, 53.

22. Ilse Rosenthal-Schneider, *Reality and Scientific Truth: Discussions with Einstein, von Laue, and Planck* (Detroit: Wayne State University Press, 1980), 74. She reports mistakenly that the telegram was from Eddington when it was from Lorentz. Einstein's remark is famous, and is translated in many ways. The German sentence, as recorded by Rosenthal-Schneider, is "Da könnt' mir halt der Liebe Gott leid tun, die Theorie stimmt doch."

23. Max Planck to Einstein, Oct. 4, 1919; Einstein to Max Planck, Oct. 23, 1919.

24. Zurich Physics Colloquium to Einstein, Oct. 11, 1919.

25. Einstein to Zurich Physics Colloquium, Oct. 16, 1919.

26. Alfred North Whitehead, *Science and the Modern World* (1925; New York: Free Press, 1997), 13. See also pp. 29 and 113.

27. *The Times* of London, Nov. 7, 1919; Pais 1982, 307; Fölsing, 443; Clark, 289.

28. *The Times* of London, Nov. 7, 1919.

29. Einstein 1949b, 31. Purchase of violin is in Einstein to Paul Ehrenfest, Dec. 10, 1919.

30. Douglas, 41; Subrahmanyan Chandrasekhar, *Truth and Beauty: Aesthetics and Motivations in Science* (Chicago: University of Chicago Press, 1987), 117. (David Hilbert certainly would have been a third, though there were, of course, many others.) Chandrasekhar, who later worked with Eddington, told Jeremy Bernstein he heard this directly from Eddington; Bernstein 1973, 192.

CHAPTER TWELVE: FAME

1. Clark, 309. For a good overview, see David Rowe, "Einstein's Rise to Fame," Perimeter Institute, Oct. 15, 2005, www.mediasite.com.

2. "Fabric of the Universe," *The Times* of London, editorial, Nov. 7, 1919.

3. *New York Times*, Nov. 9, 1919.

4. Brian 1996, 100, from Meyer Berger, *The Story of the New York Times* (New York: Simon & Schuster, 1951), 251–252.

5. *New York Times*, Nov. 9, 1919.

6. The *New York Times* deserves praise, of course, for taking the theory seriously.

7. "Einstein Expounds His New Theory," *New York Times*, Dec. 3, 1919.

8. Einstein to Heinrich Zangger, Dec. 15, 1919.

9. Einstein to Marcel Grossmann, Sept. 12, 1920. Einstein went on to make the point to Grossmann that the issue, amid rising nationalism and anti-Semitism, had become politicized: "Their conviction is determined by what political party they belong to."

10. Leopold Infeld, "To Albert Einstein on His 75th Birthday," in Goldsmith et al., 24.

11. *New York Times*, Dec. 4 and 21, 1919.

12. *The Times* of London, Nov. 28, 1919.

13. Paul Ehrenfest to Einstein, Nov. 24, 1919; Maja Einstein to Einstein, Dec. 10, 1919.

14. Einstein to Max Born, Dec. 8, 1919; Einstein to Ludwig Hopf, Feb. 2, 1920.

15. C. P. Snow, "On Einstein," in *The Variety of Men* (New York: Scribner's, 1966), 108.

16. Freeman J. Dyson, "Wise Man," *New York Review of Books*, Oct. 20, 2005.

17. Clark, 296.

18. Born 2005, 41.

19. Hedwig Born to Einstein, Oct. 7, 1920.

20. Max Born to Einstein, Oct. 13, 1920.

21. Max Born to Einstein, Oct. 28, 1920.

22. Einstein to Max Born, Oct. 26, 1920. Einstein wrote to Maurice Solovine, when the book actually appeared a few months later, that Moszkowski was "abominable" and "wretched" and that "he committed a forgery" by using some of Einstein's letters in an unauthorized way to imply that Einstein had written an introduction to the book. Einstein to Maurice Solovine, Mar. 8 and 19, 1921. He was also dismayed when he heard that Hans Albert had bought it, and said, "I was unable to prevent its publication, and it has caused me a lot of grief"; Einstein to Hans Albert Einstein, June 18, 1921. See also Highfield and Carter, 199.

23. Brian 1996, 114–116; Moszkowski, 22–58.

24. Born 2005, 41.

25. Frank 1947, 171–174.

26. Michelmore, 95; Fölsing, 485.

27. Einstein to Heinrich Zangger, Dec. 24, 1919.

28. Einstein, "My First Impressions of the U.S.A.," *Nieuwe Rotterdamsche Courant*, July 4, 1921, CPAE 7, appendix D; Einstein 1954, 3–7.

29. Einstein, "Einstein on His Theory," *The Times* of London, Nov. 28, 1919.

30. Einstein to Hedwig and Max Born, Jan. 27, 1920; Einstein to Arthur Eddington, Feb. 2, 1920. Einstein graciously told an embarrassed Eddington, "The tragicomical outcome of the medal affair [is] insignificant compared to the self-sacrificing and fruitful labors you and your friends devoted to the theory of relativity and its verification."

31. Frida Bucky, quoted in Brian 1996, 230.

32. Einstein, "The World as I See It" (1930), in Einstein 1954, 8. A different translation is in Einstein 1949a, 3.

33. This appraisal appears with slight variations in Infeld, 118; Infeld, "To Albert Einstein on His 75th Birthday," in Goldsmith et al., 25; and in the *Bulletin of the World Federation of Scientific Workers*, July 1954.

34. Editorial note by Max Born in Born 2005, 127.

35. Abraham Pais, "Einstein and the Quantum Theory," *Reviews of Modern Physics* (Oct. 1979). See also Pais, "Einstein, Newton and Success," in French, 35; Pais 1982, 39.

36. Einstein, "Why Socialism?," *Monthly Review*, May 1949, reprinted in Einstein 1954, 151.

37. Erik Erikson, "Psychoanalytic Reflections on Einstein's Centenary," in Holton and Elkana, 151.

38. This idea is from Barbara Wolff of the Einstein archives at Hebrew University.

39. Levenson, 149.

40. Einstein to Paul Ehrenfest, Jan. 17, 1922; Fölsing, 482.

41. Einstein to Eduard Einstein, June 25, 1923, Einstein family correspondence trust, unpublished, letter in possession of Bob Cohn, who provided me a copy. Cohn is a collector of Einstein material. The letters in his possession have been translated by Dr. Janifer Stackhouse. I am grateful for their help.

42. Michelmore, 79.

43. Einstein to Mileva Marić, May 12, 1924, AEA 75-629.

44. Einstein to Michele Besso, Jan. 5, 1924, AEA 7-346; Einstein to Hans Albert Einstein, Mar. 7, 1924.

45. Einstein to Heinrich Zangger, Mar. 1920; Fölsing, 474; Highfield and Carter, 192; Clark, 243.

46. Paul Johnson, *Modern Times* (New York: HarperCollins, 1991), 1–3. This section is adapted from an essay I wrote when Einstein was chosen as *Time's* Person of the Century: "Who Mattered and Why," *Time*, Dec. 31, 1999. For a critique of this idea, which I also draw on in this section, see David Greenberg, "It Didn't Start with Einstein," *Slate*, Feb. 3, 2000, www.slate.com/id/74164/. Miller 2001 is also an important resource.

47. Charles Poor, professor of celestial mechanics, Columbia University, in the *New York Times*, Nov. 16, 1919.

48. *New York Times*, Dec. 7, 1919.

49. Isaiah Berlin, "Einstein and Israel," in Holton and Elkana, 282. See also, from his stepson-in-law Reiser, 158: "The word relativity was confused in lay circles and, today, is still confused with the word relativism. Einstein's work and personality, however, are far removed from the ambiguity and the concept of relativism, both in the theory of knowledge and in ethics . . . Ethical relativism, which denies all the generally obligatory moral norms, totally contradicts the high social idea which Einstein stands for and always follows."

50. Haldane, 123. For a contemporary book treating, in more sophisticated depth, many of the same topics, and sharing a title, see Ryckman 2005.

51. Frank 1947, 189–190; Clark, 339–340.

52. Gerald Holton, "Einstein's Influence on the Culture of Our Time," in Holton 2000, 127, and also Holton and Elkana, xi.

53. Miller 2001, especially 237–241.

54. Damour 34; Marcel Proust to Armand de Guiche, Dec. 1921.

55. Philip Courtenay, "Einstein and Art," in Goldsmith et al., 145; Richard Davenport-Hines, *Proust at the Majestic* (New York: Bloomsbury, 2006).

CHAPTER THIRTEEN: THE WANDERING ZIONIST

1. *The Times* of London, Nov. 28, 1919.

2. Kurt Blumenfeld, "Einstein and Zionism," in Seelig 1956b, 74; Kurt Blumenfeld, *Erlebte Judenfrage* (Stuttgart: Verlags-Anstalt, 1962), 127–128.

3. Einstein to Paul Epstein, Oct. 5, 1919.

4. Einstein to German Citizens of the Jewish Faith, Apr. 5, 1920, CPAE 7: 37.

5. Einstein, "Anti-Semitism: Defense through Knowledge," after Apr. 3, 1920, CPAE 7: 35.

6. Einstein, "Assimilation and Anti-Semitism," Apr. 3, 1920, CPAE 7: 34. See also Einstein, "Immigration from the East," Dec. 30, 1919, an article in *Berliner Tageblatt*, CPAE 7:29.

7. Einstein, "Anti-Semitism: Defense through Knowledge," after Apr. 3, 1920, CPAE 7: 35; Hubert Goenner, "The Anti-Einstein Campaign in Germany in 1920," in Beller et al., 107.

8. Elon, 277.

9. Hubert Goenner, "The Anti-Einstein Campaign in Germany in 1920," in Beller et al., 121.

10. *New York Times*, Aug. 29, 1920.

11. Frank 1947, 161; Clark, 318; Fölsing, 462; Brian 1996, 111.

12. "Einstein to Leave Berlin," *New York Times*, Aug. 29, 1920; the story, datelined Berlin, begins, "Local newspapers state that Professor Albert Einstein will leave the German capital on account of the many unfair attacks made against his relativity theory and himself."

13. Einstein, "My Response," Aug. 27, 1920, CPAE 7: 45.

14. See, in particular, Philipp Lenard to Einstein, June 5, 1909.

15. Einstein, "My Response," Aug. 27, 1920, CPAE 7: 45.

16. Seelig 1956a, 173.

17. Hedwig Born to Einstein, Sept. 8, 1920.

18. Paul Ehrenfest to Einstein, Sept. 2, 1920.

19. Einstein to Max and Hedwig Born, Sept. 9, 1920.

20. Einstein to Paul Ehrenfest, before Sept. 9, 1920.

21. Arnold Sommerfeld to Einstein, Sept. 11, 1920.

22. Jerome, 206–208, 256–257.

23. Born 2005, 35; Einstein to Max Born, Oct. 26, 1920.

24. Clark, 326–327; Fölsing, 467; Bolles, 73.

25. Fölsing, 523; Adolf Hitler, *Völkischer Beobachter*, Jan. 3, 1921.

26. *Dearborn* (Mich.) *Independent*, Apr. 30, 1921, on display at the "Chief Engi-

neer of the Universe" exhibit, Kronprinzenpalais, Berlin, May–Sept. 2005. A headline at the bottom of the page reads, "Jew Admits Bolshevism!"

27. Einstein to Paul Ehrenfest, Nov. 26, 1920, Feb. 12, 1921, AEA 9-545; Fölsing, 484. The Einstein letters after 1920 have not yet been published in the CPAE series, and I identify these unpublished letters by the Albert Einstein Archives (AEA) call numbers.

28. Clark, 465–466.

29. Einstein to Maurice Solovine, Mar. 8, 1921, AEA 9-555.

30. Einstein statement to Abba Eban, Nov. 18, 1952, AEA 28-943.

31. Fritz Haber to Einstein, Mar. 9, 1921, AEA 12-329.

32. Einstein to Fritz Haber, Mar. 9, 1921, AEA 12-331.

33. Seelig 1956a, 81; Fölsing, 500; Clark, 468.

34. *New York Times*, Apr. 3, 1921.

35. Illy, 29.

36. *Philadelphia Public Ledger*, Apr. 3, 1921.

37. These quotes and descriptions are taken from the Apr. 3, 1921, stories in the *New York Times, New York Call, Philadelphia Public Ledger*, and *New York American*.

38. Weizmann, 232.

39. "Einstein Sees End of Time and Space," *New York Times*, Apr. 4, 1921.

40. "City's Welcome for Dr. Einstein," *New York Evening Post*, Apr. 5, 1921.

41. Talmey, 174.

42. *New York Times*, Apr. 11 and 16, 1921.

43. The memorial, at the corner of Constitution Avenue and Twenty-second Street N.W. near the Mall, is a hidden treasure of Washington. (See picture on p. 605.) The sculptor was Robert Berks, who also did the bust of John Kennedy at the Kennedy Center nearby, and the landscape architect was James Van Sweden. On the tablet that Einstein holds are three equations, describing the photoelectric effect, general relativity, and of course $E=mc^2$. On the marble steps where the statue reclines are three quotes, including: "As long as I have any choice in the matter, I shall live only in a country where civil liberty, tolerance, and equality of all citizens before the law prevail." See www.nasonline.org.

44. *Washington Post*, Apr. 7, 1921; *New York Times*, Apr. 26 and 27, 1921; Frank 1947, 184. An account of the Academy dinner by Caltech astronomer Harlow Shapley is at the Einstein papers in Pasadena.

45. Charles MacArthur, "Einstein Baffled in Chicago: Seeks Pants in Only Three Dimensions, Faces Relativity of Trousers," *Chicago Herald and Examiner*, May 3, 1921.

46. *Chicago Daily Tribune*, May 3, 1921.

47. Memorandum of Agreement, Einstein and Princeton University Press, May 9, 1921. The deal was an exclusive one; no other venue in the United States was permitted to publish any of his lectures. The four lectures appeared as *The Meaning of Relativity*. It is now in its fifth edition.

48. *Philadelphia Evening Bulletin*, May 14, 1921.

49. Einstein to Oswald Veblen, Apr. 30, 1930, AEA 23-152. Pais 1982, 114, gives

a history of this phrase, which is recounted in a memo prepared for the Einstein archives by Einstein's secretary Helen Dukas. The fireplace is in room 202, the faculty lounge of what is now called Jones Hall at Princeton and was earlier known as Fine Hall, until that name moved to a newer math building.

50. Seelig 1956a, 183; Frank 1947, 285; Clark, 743.
51. *New York Times*, July 31, 1921.
52. Einstein to Felix Frankfurter, May 28, 1921, AEA 36-210.
53. See Ben Halpern, *A Clash of Heroes: Brandeis, Weizmann and American Zionism* (New York: Oxford University Press, 1987).
54. *Boston Herald*, May 19, 1921.
55. *New York Times*, May 18, 1921; Frank 1947, 185; Brian 1996, 129; Illy, 25–32.
56. *Hartford* (Conn.) *Daily Times*, May 23, 1921. Also, *Hartford Daily Courant*, May 23, 1921.
57. *Cleveland Press*, May 26, 1921.
58. Illy, 185.
59. Fölsing, 51.
60. Einstein, "How I Became a Zionist," interview in *Jüdische Rundschau*, June 21, 1921, conducted on May 30, CPAE 7: 57.
61. Einstein to Mileva Marić, Aug. 28, 1921, Einstein family trust correspondence, letter in possession of Bob Cohn. On this trip, in deference to Elsa's feelings, he decided at the last moment not to stay at Marić's apartment.
62. Einstein to Walther Rathenau, Mar. 8, 1917; Walther Rathenau to Einstein, May 10, 1917.
63. Reiser, 146, describes the Weizmann-Rathenau-Einstein discussions. See also Fölsing, 519; Elon, 364.
64. Weizmann, 288; Elon, 268.
65. Frank 1947, 192.
66. Reiser, 145.
67. Milena Wazeck, "Einstein on the Murder List," in Renn 2005d, 222; Einstein to Max Planck, July 6, 1922, AEA 19-300.
68. Einstein to Maurice Solovine, July 16, 1922, AEA 21-180.
69. Einstein to Marie Curie, July 4, 1922, AEA 34-773; Marie Curie to Einstein, July 7, 1922, AEA 34-775.

70. Fölsing, 521.

71. Nathan and Norden, 54.

72. Hermann Struck to Pierre Comert, July 12, 1922; Nathan and Norden, 59. (Einstein sent word to League press official Comert through their mutual friend, the painter Struck.)

73. Nathan and Norden, 70.

74. Einstein, "Travel Diary: Japan-Palestine-Spain," AEA 29-129. All quotes in this section from Einstein's diary are from this document.

75. Joan Bieder, "Einstein in Singapore," 2000, www.onthepage.org/outsiders/einstein_in_singapore.htm.

76. Fölsing, 527; Clark, 368; Brian 1996, 143; Frank 1947, 199.

77. Einstein to Hans Albert and Eduard Einstein, Dec. 12, 1922, AEA 75-620.

78. Frank 1947, 200.

79. Einstein, "Travel Diary: Japan-Palestine-Spain," AEA 29-129.

80. Clark, 477–480; Frank 1947, 200–201; Brian 1966, 145; Fölsing, 528–532.

CHAPTER FOURTEEN: NOBEL LAUREATE

1. Svante Arrhenius to Einstein, Sept. 1, 1922, AEA 6-353; Einstein to Svante Arrhenius, Sept. 20, 1922, AEA 6-354.

2. Pais 1982, 506–507; Elzinga, 82–84.

3. R. M. Friedman 2005, 129. See also Friedman's book, *The Politics of Excellence: Behind the Nobel Prize in Science* (New York: Henry Holt, 2001), especially chapter 7, "Einstein Must Never Get a Nobel Prize!"; Elzinga; Pais 1982, 502.

4. Pais 1982, 508; Hendrik Lorentz and Dutch colleagues to the Swedish Academy, Jan. 24, 1920; Niels Bohr to the Swedish Academy, Jan. 30, 1920; Elzinga, 134.

5. Brian 1996, 143, citing research and interviews by the writer Irving Wallace for his novel *The Prize*.

6. Elzinga, 144.

7. R. M. Friedman, 130. See also Pais 1982, 508.

8. Arthur Eddington to the Swedish Academy, Jan. 1, 1921.

9. Pais 1982, 509; R. M. Friedman, 131; Elzinga, 151.

10. Marcel Brillouin to the Swedish Academy, Jan. 1922; Arnold Sommerfeld to the Swedish Academy, Jan. 11, 1922.

11. Christopher Aurivillius to Einstein, Nov. 10, 1922. In another translation and version, the actual Nobel citation sent to Einstein includes the phrase "independent of the value that (after eventual confirmation) may be credited to the relativity and gravitation theory."

12. Elzinga, 182.

13. Svante Arrhenius, Nobel Prize presentation speech, Dec. 10, 1922, nobelprize.org/physics/laureates/1921/press.html.

14. Einstein, "Fundamental Ideas and Problems of the Theory of Relativity," Nobel lecture, July 11, 1923.

15. Einstein to Hans Albert and Eduard Einstein, Dec. 22, 1922, AEA 75-620.

The full story of the Nobel money was complex and over the years caused considerable disputes, as became clear in letters between Einstein and Marić released in 2006. According to the divorce agreement, the Nobel money was to go to a Swiss bank account. Marić was supposed to have use of the interest, but she could spend the capital only with Einstein's consent. In 1923, after consultation with a financial adviser, Einstein decided to place only part of the money in Switzerland and have the rest invested in an American account. That scared Marić and caused frictions that were calmed by friends. With Einstein's consent she bought a Zurich apartment house in 1924 using the Swiss money and a big loan. The rents covered the loan payments, as well as the maintenance of the house and a part of the family's livelihood. Two years later, again with Einstein's consent, Marić bought two more houses using another 40,000 Swiss francs from the Nobel money and an additional loan. The two new houses turned out to be bad investments and had to be sold to avoid endangering ownership of the first house, where Marić lived with Eduard. In the meantime, the Great Depression in America reduced the value of the account and investments made there. Einstein continued to pay considerable sums to Marić and Eduard, but Marić's fears for her financial security were understandable. At the end of the 1930s, Einstein created a holding company to buy from Marić the remaining apartment house, where she still lived, and to take over her debts in order to save the house from being repossessed by the bank. Marić could continue to live in the same apartment and receive the excess rental proceeds. In addition, Einstein sent a monthly contribution for Eduard's support. This arrangement lasted until the late 1940s, when Mileva was no longer able to care for the house and the income from the rents no longer covered the expenses. With Einstein's consent Marić sold the house but not the right to her apartment. The money from that sale was eventually found under Marić's mattress. Some critics have accused Einstein of allowing Marić to die impoverished. Although Marić at times certainly felt impoverished, Einstein did try to protect her and Eduard from financial worries, not only by paying what he was obliged to pay, but also by subsidizing their living expenses. I am grateful to Barbara Wolff of the Hebrew University Einstein archives for help researching this topic. See also Alexis Schwarzenbach, *Das verschmähte Genie: Albert Einstein und die Schweiz* (Berlin: DVA, 2003).

16. Einstein to Heinrich Zangger, Dec. 6, 1917.
17. "*All* the really great discoveries in *theoretical* physics—with a few exceptions that stand out because of their oddity—have been made by men *under thirty*." Bernstein 1973, 89, emphasis in the original. Einstein finished his work on general relativity when he was 36, but his initial step, what he called his "happiest thought" about the equivalence of gravity and acceleration, came when he was 28. Max Planck was 42 when, in Dec. 1900, he gave his lecture on the quantum.
18. Einstein to Heinrich Zangger, Aug. 11, 1918; Clive Thompson, "Do Scientists Age Badly?," *Boston Globe*, Aug. 17, 2003. John von Neumann, a founder of modern computer science, once claimed that the intellectual

powers of mathematicians peaked at the age of 26. One study of a random group of scientists showed that 80 percent did their best work before their early forties.

19. Einstein to Maurice Solovine, Apr. 27, 1906.
20. Aphorism for a friend, Sept. 1, 1930, AEA 36-598.
21. Einstein to Hendrik Lorentz, June 17, 1916; Miller 1984, 55–56.
22. Einstein, "Ether and the Theory of Relativity," speech at University of Leiden, May 5, 1920, CPAE 7: 38.
23. Einstein to Karl Schwarzschild, Jan. 9, 1916.
24. Einstein, "Ether and the Theory of Relativity," speech at University of Leiden, May 5, 1920, CPAE 7: 38.
25. Greene 2004, 74.
26. Janssen 2004, 22. Einstein made this clearer in his 1921 Princeton lectures, but also continued to say, "It appears probable that Mach was on the right road in his thought that inertia depends on a mutual action of matter." Einstein 1922a, chapter 4.
27. Einstein, "Ether and the Theory of Relativity," speech at University of Leiden, May 5, 1920, CPAE 7: 38.
28. Einstein, "On the Present State of the Problem of Specific Heats," Nov. 3, 1911, CPAE 3: 26; the quote about "really exist in nature" appears on p. 421 of the English translation of vol. 3.
29. Robinson, 84–85.
30. Holton and Brush, 435.
31. Lightman 2005, 151.
32. Clark 202; George de Hevesy to Ernest Rutherford, Oct. 14, 1913; Einstein 1949b, 47.
33. Einstein, "Emission and Absorption of Radiation in Quantum Theory," July 17, 1916, CPAE 6: 34; Einstein, "On the Quantum Theory of Radiation," after Aug. 24, 1916, CPAE 6: 38, and also in *Physikalische Zeitschrift* 18 (1917). See Overbye, 304–306; Rigden, 141; Pais 1982, 404–412; Fölsing, 391; Clark, 265; Daniel Kleppner, "Rereading Einstein on Radiation," *Physics Today* (Feb. 2005): 30. In addition, in 1917 Einstein wrote a paper on the quantization of energy in mechanical theories called "On the Quantum Theorem of Sommerfeld and Epstein." It shows the problems that the classical quantum theory encountered when applied to mechanical systems we would now call chaotic. It was cited by earlier pioneers of quantum mechanics, but has since been largely forgotten. A good description of it and its importance in the development of quantum mechanics is Douglas Stone, "Einstein's Unknown Insight and the Problem of Quantizing Chaos," *Physics Today* (Aug. 2005).
34. Einstein to Michele Besso, Aug. 11, 1916.
35. I am grateful to Professor Douglas Stone of Yale for help with the wording of this.
36. Einstein to Michele Besso, Aug. 24, 1916.
37. Einstein, "On the Quantum Theory of Radiation," after Aug. 24, 1916, CPAE 6: 38.

38. Einstein to Max Born, Jan. 27, 1920.

39. Einstein to Max Born, Apr. 29, 1924, AEA 8-176.

40. Niels Bohr, "Discussion with Einstein," in Schilpp, 205–206; Clark, 202.

41. Einstein to Niels Bohr, May 2, 1920; Einstein to Paul Ehrenfest, May 4, 1920.

42. Niels Bohr to Einstein, Nov. 11, 1922, AEA 8-73.

43. Fölsing, 441.

44. John Wheeler, "Memoir," in French, 21; C. P. Snow, "Albert Einstein," in French, 3.

45. Bohr's quip is often quoted. One source I can find for it, in a less pithy fashion, is from Bohr's own descriptions of being with Einstein at the 1927 Solvay Conference: "Einstein mockingly asked us whether we could really believe that the providential authorities took recourse to dice-playing ('. . . ob der liebe Gott würfelt'), to which I replied by pointing at the great caution, already called for by ancient thinkers, in ascribing attributes to Providence in everyday language." Niels Bohr, "Discussion with Einstein," in Schilpp, 211. Werner Heisenberg, who was at these discussions, also recounts the quip: "To which Bohr could only answer: 'But still, it cannot be for us to tell God how he is to run the world.' " Heisenberg 1989, 117.

46. Holton and Brush, 447; Pais 1982, 436.

47. Pais 1982, 438. Wolfgang Pauli recalled, "In a discussion at the physics meeting in Innsbruck in the autumn of 1924, Einstein proposed to search for interference and diffraction phenomena with molecular beams." Pauli, 91.

48. Einstein, "Quantum Theory of Single-Atom Gases," part 1, 1924, part 2, 1925. This quote occurs in part 2, section 7. The manuscript of this paper was found in Leiden in 2005.

49. I am grateful to Professor Douglas Stone of Yale for helping to craft this section and explaining the fundamental importance of what Einstein did. A theoretical condensed matter physicist, he is writing a book on Einstein's contributions to quantum mechanics and how far-reaching they really were, despite Einstein's later rejection of the theory. According to Stone, "99% of the credit for this fundamental discovery called Bose-Einstein condensation is really owed to Einstein. Bose did not even realize that he had counted in a different way." Regarding the Nobel Prize for achieving Bose-Einstein condensation, see www.nobelprize.org/physics/laureates/2001/public.html.

50. Bernstein 1973, 217; Martin J. Klein, "Einstein and the Wave-Particle Duality," *Natural Philosopher* (1963): 26.

51. Max Born, "Einstein's Statistical Theories," in Schilpp, 174.

52. Einstein to Erwin Schrödinger, Feb. 28, 1925, AEA 22-2.

53. Don Howard, "Spacetime and Separability," 1996, AEA Cedex H; Howard 1985; Howard 1990b, 61–64; Howard 1997. The 1997 essay identifies the philosophy of Arthur Schopenhauer as an influence on Einstein's theories of spatial separability.

54. Bernstein 1996a, 138.

55. More precisely, it is the square of the wave function that is proportional to the probability. Holton and Brush, 452.

56. Einstein to Hedwig Born, Mar. 7, 1926, AEA 8-266; Einstein to Max Born, Dec. 4, 1926, AEA 8-180.

57. aip.org/history/heisenberg/p07.htm; Born 2005, 85.

58. Max Born to Einstein, July 15, 1925, AEA 8-177; Einstein to Hedwig Born, Mar. 7, 1926, AEA 8-178; Einstein to Paul Ehrenfest, Sept. 25, 1925, AEA 10-116.

59. Werner Heisenberg to Einstein, June 10, 1927, AEA 12-174.

60. Heisenberg 1971, 63; Gerald Holton, "Werner Heisenberg and Albert Einstein," *Physics Today* (2000), www.aip.org/pt/vol-53/iss-7/p38.html.

61. Frank 1947, 216.

62. Aage Petersen, "The Philosophy of Niels Bohr," *Bulletin of the Atomic Scientists* (Sept. 1963): 12.

63. Dugald Murdoch, *Niels Bohr's Philosophy of Physics* (Cambridge, England: Cambridge University Press, 1987), 47, citing the Niels Bohr Archives: Scientific Correspondence, 11: 2.

64. Einstein, "To the Royal Society on Newton's Bicentennial," Mar. 1927.

65. Einstein to Michele Besso, Apr. 29, 1917; Michele Besso to Einstein, May 5, 1917; Einstein to Michele Besso, May 13, 1917. For a good analysis, see Gerald Holton, "Mach, Einstein, and the Search for Reality," in Holton 1973, 240.

66. "Belief in an external world independent of the perceiving subject is the basis of all natural science." Einstein, "Maxwell's Influence on the Evolution of the Idea of Physical Reality," 1931, in Einstein 1954, 266.

67. Einstein to Max Born, Jan. 27, 1920.

68. Einstein's introduction to Rudolf Kayser, *Spinoza* (New York: Philosophical Library, 1946). Kayser was married to Einstein's stepdaughter and wrote a semi-authorized memoir of Einstein.

69. Fölsing, 703–704; Einstein to Fritz Reiche, Aug. 15, 1942, AEA 20-19.

70. Einstein to Max Born, Dec. 4, 1926, AEA 8-180.

CHAPTER FIFTEEN: UNIFIED FIELD THEORIES

1. Einstein, "Ideas and Problems of the Theory of Relativity," Nobel lecture, July 11, 1923. Available at nobelprize.org/nobel_prizes. This section draws from these papers on Einstein's unified field quest: van Dongen 2002, courtesy of the author; Tilman Sauer, "Dimensions of Einstein's Unified Field Theory Program," forthcoming in the *Cambridge Companion to Einstein*, courtesy of the author; Norton 2000; Goenner 2004.

2. Einstein, "The Principles of Research," a toast in honor of Max Planck, Apr. 26, 1918, CPAE 7: 7.

3. Einstein to Hermann Weyl, Apr. 6, 1918.

4. Einstein to Hermann Weyl, Apr. 8, 1918. In a letter to Heinrich Zangger, May 8, 1918, Einstein called Weyl's theory "ingenious" but "physically incorrect." It did, however, later become one of the recognized precursors of Yang-Mills gauge theory.

5. My description of the work of Kaluza and Klein relies on Krauss, 94–104,

which is an engaging book on the role extra dimensions have played in explaining the universe.

6. Einstein to Theodor Kaluza, Apr. 21, 1919.

7. Einstein to Niels Bohr, Jan. 10, 1923, AEA 8-74.

8. Einstein to Hermann Weyl, May 26, 1923, AEA 24-83.

9. Einstein, "On the General Theory of Relativity," Prussian Academy, Feb. 15, 1923.

10. *New York Times*, Mar. 27, 1923.

11. Pais 1982, 466; Einstein, "On the General Theory of Relativity," the Prussian Academy, Feb. 15, 1923.

12. Einstein, "Unified Field Theory of Gravity and Electricity," July 25, 1925; Hoffmann 1972, 225.

13. Steven Weinberg, "Einstein's Mistakes," *Physics Today* (Nov. 2005).

14. Einstein, "On the Unified Theory," Jan. 30, 1929.

15. Einstein to Michele Besso, Jan. 5, 1929, AEA 7-102.

16. *New York Times*, Nov. 4, 1928; Vallentin, 160.

17. Clark, 494; *London Daily Chronicle*, Jan. 26, 1929.

18. "Einstein's Field Theory," *Time*, Feb. 18, 1929. Einstein also appeared on *Time*'s cover on Apr. 4, 1938, July 1, 1946, and posthumously Feb. 19, 1979, and Dec. 31, 1999. Elsa appeared on the cover Dec. 22, 1930.

19. Fölsing, 605; Clark, 496; Brian 1996, 174.

20. *New York Times*, Feb. 4, 1929.

21. Einstein to Maja Winteler-Einstein, Oct. 22, 1929, AEA 29-409.

22. Wolfgang Pauli to Einstein, Dec. 19, 1929, AEA 19-163.

23. *New York Times*, Jan. 23, Oct. 26, 1931; Einstein to Wolfgang Pauli, Jan. 22, 1932, AEA 19-169.

24. Goenner 2004; Elie Cartan, "Absolute Parallelism and the Unified Theory," *Review Metaphysic Morale* (1931).

25. For a two-minute home movie of the conference shot by Irving Langmuir, the 1932 Nobel Prize winner in chemistry, see www.maxborn.net/index.php?page=filmnews.

26. Einstein to Hendrik Lorentz, Sept. 13, 1927, AEA 16-613.

27. Pauli, 121.

28. John Archibald Wheeler and Wojciech Zurek, *Quantum Theory and Measurement* (Princeton: Princeton University Press, 1983), 7.

29. Fölsing, 589; Pais 1982, 445, from Proceedings of the Fifth Solvay Conference.

30. Heisenberg 1989, 116.

31. Niels Bohr, "Discussion with Einstein," in Schilpp, 211–219, offers a detailed and loving description of the Solvay and other discussions; Otto Stern recollections, in Pais 1982, 445; Fölsing, 589.

32. "Reports and Discussions," in *Solvay Conference of 1927* (Paris: Gauthier-Villars, 1928), 102. See also Travis Norsen, "Einstein's Boxes," *American Journal of Physics*, vol. 73, Feb. 2005, pp. 164-176.

33. Louis de Broglie, "My Meeting with Einstein," in French, 15.

34. Einstein, "Speech to Professor Planck," Max Planck award ceremony, June 28, 1929.

35. Léon Rosenfeld, "Niels Bohr in the Thirties," in Rozental 1967, 132.

36. Niels Bohr, "Discussion with Einstein," in Schilpp, 225–229; Pais 1982, 447–448. I am grateful to Murray Gell-Mann and David Derbes for the phrasing of this section.

37. Einstein, "Maxwell's Influence on the Evolution of the Idea of Physical Reality," 1931, in Einstein 1954, 266.

38. Einstein, "Reply to Criticisms" (1949), in Schilpp, 669.

39. A fuller discussion of Einstein's realism is in chapter 20 of this book. For contrasting views on this issue, see Gerald Holton, "Mach, Einstein, and the Search for Reality," in Holton 1973, 219, 245 (he argues that there is a very clear change in Einstein's philosophy: "For a scientist to change his philosophical beliefs so fundamentally is rare"); Fine, 123 (he argues that "Einstein underwent a philosophical conversion, turning away from his positivist youth and becoming deeply committed to realism"); Howard 2004 (which argues, "Einstein was never an ardent 'Machian' positivist, and he was never a scientific realist"). This section also draws on van Dongen 2002 (he argues, "Broadly speaking, one can say that Einstein moved from Mach's empiricism, earlier in his career, to a strong realist position later on"). See also Anton Zeilinger, "Einstein and Absolute Reality," in Brockman, 121–131.

40. Einstein, "On the Method of Theoretical Physics," the Herbert Spencer lecture, Oxford, June 10, 1933, in Einstein 1954, 270.

41. Einstein 1949b, 89.

42. Einstein, "Principles of Theoretical Physics," inaugural address to the Prussian Academy, 1914, in Einstein 1954, 221.

43. Einstein to Hermann Weyl, May 26, 1923, AEA 24-83.

44. John Barrow, "Einstein as Icon," *Nature*, Jan. 20, 2005, 219. See also Norton 2000.

45. Einstein, "On the Method of Theoretical Physics," the Herbert Spencer lecture, Oxford, June 10, 1933, in Einstein 1954, 274.

46. Steven Weinberg, "Einstein's Mistakes," *Physics Today* (Nov. 2005): "Since Einstein's time, we have learned to distrust this sort of aesthetic criterion. Our experience in elementary-particle physics has taught us that any term in the field equations of physics that is allowed by fundamental principles is likely to be there in the equations."

47. Einstein, "Latest Developments of the Theory of Relativity," May 23, 1931, the third of three Rhodes Lectures at Oxford, this one coming on the day he was awarded his honorary doctorate there. Reprinted in the *Oxford University Gazette*, June 3, 1931.

48. Einstein, "On the Method of Theoretical Physics," Oxford, June 10, 1933, in Einstein 1954, 270.

49. Marcia Bartusiak, "Beyond the Big Bang," *National Geographic* (May 2005). Elsa's quip is widely reported but never fully sourced. See Clark, 526.

50. Associated Press, Dec. 30, 1930.

51. Einstein to Michele Besso, Mar. 1, 1931, AEA 7-125.
52. Greene 2004, 279: "That would certainly have ranked among the greatest discoveries—it may have been *the* greatest discovery—of all time." See also Edward W. Kolb, "The Greatest Discovery Einstein Didn't Make," in Brockman, 201.
53. Einstein, "On the Cosmological Problem of the General Theory of Relativity," Prussian Academy, 1931; "Einstein Drops Idea of 'Closed' Universe," *New York Times*, Feb. 5, 1931.
54. Einstein 1916, appendix IV (first appears in the 1931 edition).
55. Gamow 1970, 149.
56. Steven Weinberg, "The Cosmological Constant Problem," in *Morris Loeb Lectures in Physics* (Cambridge, Mass.: Harvard University Press 1988); Steven Weinberg, "Einstein's Mistakes," *Physics Today* (Nov. 2005); Aczel 1999, 167; Krauss 117; Greene 2004, 275–278; Dennis Overbye, "A Famous Einstein 'Fudge' Returns to Haunt Cosmology," *New York Times*, May 26, 1998; Jeremy Bernstein, "Einstein's Blunder," in Bernstein 2001, 86–89.
57. Lawrence Krauss of Case Western Reserve and Michael Turner of the University of Chicago have argued that an explanation of the universe requires use of a cosmological term that is different from the one Einstein added into his field equations and then discarded. Their version arises from quantum mechanics, not general relativity, and is based on the premise that even "empty" space does not necessarily possess zero energy. See Krauss and Turner, "A Cosmic Conundrum," *Scientific American* (Sept. 2004).
58. "Einstein's Cosmological Constant Predicts Dark Energy," *Universe Today*, Nov. 22, 2005. This particular headline was based on a research project known as the Supernova Legacy Survey (SNLS). According to a press release from Caltech, SNLS "aims to discover and examine 700 distant supernovae to map out the history of the expansion of the universe. The survey confirms earlier discoveries that the expansion of the universe proceeded more slowly in the past and is speeding up today. However, the crucial step forward is the discovery that Einstein's 1917 explanation of a constant energy term for empty space fits the new supernova data very well."

CHAPTER SIXTEEN: TURNING FIFTY

1. Vallentin, 163.
2. *New York Times*, Mar. 15, 1929.
3. Reiser, 205.
4. Reiser, 207; Frank 1947, 223; Fölsing, 611.
5. www.einstein-website.de/z_biography/caputh-e.html; Jan Otakar Fischer, "Einstein's Haven," *International Herald Tribune*, June 30, 2005; Fölsing, 612; Einstein to Maja Einstein, Oct. 22, 1929; Erika Britzke, "Einstein in Caputh," in Renn 2005d, 272.
6. Vallentin, 168.
7. Reiser, 221.

8. Einstein to Betty Neumann, Nov. 5 and 13, 1923. These letters are part of a set given to Hebrew University and are not catalogued in the Einstein archives.

9. Einstein to Betty Neumann, Jan. 11, 1924; Pais 1982, 320.

10. Einstein to Elsa Einstein, Aug. 14, 1924, part of sealed correspondence released in 2006; Einstein to Betty Neumann, Aug. 24, 1924. I am grateful to Ze'ev Rosenkranz of the Einstein archives in Jerusalem and Caltech for helping me find and translate these letters.

11. Einstein to Ethel Michanowski, May 16 and 24, 1931, in private collection.

12. Einstein to Elsa Einstein and Einstein to Margot Einstein, May 1931, part of sealed correspondence released in 2006. I am grateful for the help of Ze'ev Rosenkranz of the Einstein Papers Project for providing context and translation.

13. Einstein to Margot Einstein, May 1931, sealed correspondence released in 2006.

14. This is a sentiment that lasted through his life. Einstein to Eugenia Anderman, June 2, 1953, AEA 59-097: "You must be aware that most men (and many women) are by nature not monogamous. This nature is asserted more forcefully when tradition stands in the way."

15. Fölsing, 617; Highfield and Carter, 208; Marianoff, 186. (Note: Fölsing spells her name Lenbach, which is not correct according to the Einstein archive copies.)

16. Elsa Einstein to Hermann Struck, 1929.

17. George Dyson, "Helen Dukas: Einstein's Compass," in Brockman, 85–94 (George Dyson was the son of Freeman Dyson, a physicist at the Institute for Advanced Study in Princeton, and Dukas worked as his babysitter after Einstein died). See also Abraham Pais, "Eulogy for Helen Dukas," 1982, American Institute of Physics Library, College Park, Md.

18. Einstein to Maurice Solovine, Mar. 4, 1930, AEA 21-202.

19. Einstein to Mileva Marić, Feb. 23, 1927, AEA 75-742.

20. Ibid.

21. Einstein to Hans Albert Einstein, Feb. 2, 1927, AEA 75-738, and Feb. 23, 1927, AEA 75-739.

22. Highfield and Carter, 227.

23. Einstein to Eduard Einstein, Dec. 23, 1927, AEA 75-748.

24. Einstein to Eduard Einstein, July 10, 1929, AEA 75-782.

25. Eduard Einstein to Einstein, May 1, Dec. 10, 1926. Both are in sealed correspondence folders that were released in 2006 and not catalogued in the archives.

26. Eduard Einstein to Einstein, Dec. 24, 1935. Also in the sealed correspondence folders released in 2006 and not catalogued in the archives.

27. Sigmund Freud to Sandor Ferenczi, Jan. 2, 1927. For an analysis of the interwoven influence of Freud and Einstein, see Panek 2004.

28. Viereck, 374; Sayen, 134. See also Bucky, 113: "I have many doubts about some of his theories. I think Freud placed too much emphasis on dream theo-

ries. After all, a junk closet does not bring everything forth . . . On the other hand, Freud was very interesting to read and he was also very witty. I certainly do not mean to be overly critical."

29. Einstein to Eduard Einstein, 1936 or 1937, AEA 75-939.
30. Einstein to Eduard Einstein, Feb. 5, 1930, not catalogued; Highfield and Carter, 229, 234. See translation in epigraph source note on p. 565.
31. Einstein to Eduard Einstein, Dec. 23, 1927, AEA 75-748.
32. Einstein to Mileva Marić, Aug. 14, 1925, AEA 75-693.
33. Marianoff, 12. He apparently mistakes the year of his own wedding, as he refers to the fall of 1929 when it was in fact just before Einstein's second visit to the United States in late 1930. Barbara Wolff of the Einstein archives at Hebrew University says she believes this anecdote to be embellished.
34. Elsa Einstein to Antonina Vallentin, undated, in Vallentin, 196.
35. Einstein, Trip Diary to the U.S.A., Nov. 30, 1930, AEA 29-134.
36. "Einstein Works at Sea," *New York Times*, Dec. 5, 1930.
37. "Einstein Puzzled by Our Invitations," *New York Times*, Nov. 23, 1930.
38. "Einstein Consents to Face Reporters," *New York Times*, Dec. 10, 1930.
39. Einstein, Trip Diary, Dec. 11, 1930, AEA 29-134.
40. "Einstein on Arrival Braves Limelight for Only 15 Minutes," *New York Times*, Dec. 12, 1930.
41. "He Is Worth It," *Time*, Dec. 2, 1930.
42. Brian 1996, 204; "Einstein Receives Keys to the City," *New York Times*, Dec. 14, 1930.
43. "Einstein Saw His Statue in Church Here," *New York Times*, Dec. 28, 1930.
44. George Sylvester Viereck, profile of John D. Rockefeller, *Liberty*, Jan. 9, 1932; Nathan and Norden, 157. Einstein also mentions his visit to Rockefeller in a letter to Max Born, May 30, 1933, AEA 8-192.
45. Einstein, New History Society speech, Dec. 14, 1930, in Nathan and Norden, 117; "Einstein Advocates Resistance to War," *New York Times*, Dec. 15, 1930, p. 1; Fölsing, 635.
46. "Einstein Considers Seeking a New Home," Associated Press, Dec. 16, 1930.
47. Einstein, Trip Diary, Dec. 15–31, 1931, AEA 29-134; "Einstein Welcomed by Leaders of Panama," *New York Times*, Dec. 24, 1930; "Einstein Heard on Radio," *New York Times*, Dec. 26, 1930.
48. Brian 1996, 206.
49. Hedwig Born to Einstein, Feb. 22, 1931, AEA 8-190.
50. Amos Fried to Robert Millikan, Mar. 4, 1932; Robert Millikan to Amos Fried, Mar. 8, 1932; cited in Clark, 551.
51. Brian 1996, 216.
52. Seelig 1956a, 194. At the movie, Einstein "stared bewildered, utterly absorbed, like a child at a Christmas pantomime," according to a vivid report by Cissy Patterson, an ambitious young journalist who had also described him sunbathing nude. She would later own the *Washington Herald*. Brian 1996, 214, citing *Washington Herald*, Feb. 10, 1931.
53. Einstein address, Feb. 16, 1931, in Nathan and Norden, 122.

54. "At Grand Canyon Today," *New York Times*, Feb. 28, 1931; Einstein at Hopi House, www.hanksville.org/sand/Einstein.html.

55. "Einstein in Chicago Talks for Pacifism," *New York Times*, Mar. 4, 1931; Nathan and Norden, 123.

56. Fölsing, 641; Einstein talk to War Resisters' League, Mar. 1, 1931, in Nathan and Norden, 123.

57. Nathan and Norden, 124.

58. Marianoff, 184.

59. Einstein to Mrs. Chandler and the Youth Peace Federation, Apr. 5, 1931; Nathan and Norden, 124; Fölsing, 642. For an image of the note, see www .alberteinstein.info/db/ViewImage.do?DocumentID=21007&Page=1.

60. Einstein interview with George Sylvester Viereck, Jan. 1931, in Nathan and Norden, 125.

61. Einstein to Women's International League, Jan. 4, 1928, AEA 48-818.

62. Einstein to London chapter of War Resisters' International, Nov. 25, 1928; Einstein to the League for the Organization of Progress, Dec. 26, 1928.

63. Einstein statement, Feb. 23, 1929, in Nathan and Norden, 95.

64. Manifesto of the Joint Peace Council, Oct. 12, 1930; Nathan and Norden, 113.

65. Einstein, "The 1932 Disarmament Conference," *The Nation*, Sept. 23, 1931; Einstein 1954, 95; Einstein, "The Road to Peace," *New York Times*, Nov. 22, 1931.

66. Nathan and Norden, 168; "Einstein Assails Arms Conference," *New York Times*, May 24, 1931.

67. Einstein to Kurt Hiller, Aug. 21, 1931, AEA 46-693; Nathan and Norden, 143.

68. Jerome, 144. See in particular chapter 11, "How Red?"

69. Einstein, "The Road to Peace," *New York Times*, Nov. 22, 1931; Einstein 1954, 95.

70. Thomas Bucky interview with Denis Brian, in Brian 1996, 229.

71. Einstein to Henri Barbusse, June 1, 1932, AEA 34-543; Nathan and Norden, 175–179.

72. Einstein to Isaac Don Levine, after Jan. 1, 1925, AEA 28-29.00 (for image of handwritten document, see www.alberteinstein.info/db/ViewImage.do? DocumentID=21154&Page=1; Roger Baldwin and Isaac Don Levine, *Letters from Russian Prisons* (New York: Charles Boni, 1925); Robert Cottrell, *Roger Nash Baldwin and the American Civil Liberties Union* (New York: Columbia, 2001), 180.

73. Einstein to Isaac Don Levine, Mar. 15, 1932, AEA 50-922.

74. Einstein, "The World As I See It," originally published in 1930, reprinted in Einstein 1954, 8.

75. "Ask Pardon for Eight Negroes," *New York Times*, Mar. 27, 1932; "Einstein Hails Negro Race," *New York Times*, Jan. 19, 1932, citing an Einstein piece in the forthcoming *Crisis* magazine of Feb. 1932.

76. Brian 1996, 219.

77. Einstein to Chaim Weizmann, Nov. 25, 1929, AEA 33-411.

78. Einstein, "Letter to an Arab," Mar. 15, 1930; Einstein 1954, 172; Clark, 483; Fölsing, 623.

79. Einstein to Sigmund Freud, July 30, 1932, www.cis.vt.edu/modernworld/d/Einstein.html.

80. Sigmund Freud to Einstein, Sept. 1932, www.cis.vt.edu/modernworld/d/Einstein.html.

CHAPTER SEVENTEEN: EINSTEIN'S GOD

1. Charles Kessler, ed., *The Diaries of Count Harry Kessler* (New York: Grove Press, 2002), 322 (entry for June 14, 1927); Jammer 1999, 40. Jammer 1999 provides a thorough look at the biographical, philosophical, and scientific aspects of Einstein's religious thought.

2. Einstein, "Ueber den Gegenwertigen Stand der Feld-Theorie," 1929, AEA 4-38.

3. Neil Johnson, *George Sylvester Viereck: Poet and Propagandist* (Iowa City: University of Iowa Press, 1968); George S. Viereck, *My Flesh and Blood: A Lyric Autobiography with Indiscreet Annotations* (New York: Liveright, 1931).

4. Viereck, 372–378; Viereck first published the interview as "What Life Means to Einstein," *Saturday Evening Post*, Oct. 26, 1929. I have generally followed the translation and paraphrasing in Brian 2005, 185–186 and in Calaprice. See also Jammer 1999, 22.

5. Einstein, "What I Believe," originally written in 1930 and recorded for the German League for Human Rights. It was published as "The World As I See It" in *Forum and Century*, 1930; in *Living Philosophies* (New York: Simon & Schuster, 1931); in Einstein 1949a, 1–5; in Einstein 1954, 8–11. The versions are all translated somewhat differently and have slight revisions. For an audio version, see www.yu.edu/libraries/digital_library/einstein/credo.html.

6. Einstein to M. Schayer, Aug. 5, 1927, AEA 48-380; Dukas and Hoffmann, 66.

7. Einstein to Phyllis Wright, Jan. 24, 1936, AEA 52-337.

8. "Passover," *Time*, May 13, 1929.

9. Einstein to Herbert S. Goldstein, Apr. 25, 1929, AEA 33-272; "Einstein Believes in Spinoza's God," *New York Times*, Apr. 25, 1929; Gerald Holton, "Einstein's Third Paradise," *Daedalus* (fall 2002): 26–34. Goldstein was the rabbi of the Institutional Synagogue in Harlem and the longtime president of the Union of Orthodox Jewish Congregations of America.

10. Rabbi Jacob Katz of the Montefiore Congregation, quoted in *Time*, May 13, 1929.

11. Calaprice, 214; Einstein to Hubertus zu Löwenstein, ca. 1941, in Löwenstein's book, *Towards the Further Shore* (London: Victor Gollancz, 1968), 156.

12. Einstein to Joseph Lewis, Apr. 18, 1953, AEA 60-279.

13. Einstein to unknown recipient, Aug. 7, 1941, AEA 54-927.

14. Guy Raner Jr. to Einstein, June 10, 1948, AEA 57-287; Einstein to Guy Raner

Jr., July 2, 1945, AEA 57-288; Einstein to Guy Raner Jr., Sept. 28, 1949, AEA 57-289.

15. Einstein, "Religion and Science," *New York Times*, Nov. 9, 1930, reprinted in Einstein 1954, 36–40. See also Powell.

16. Einstein, speech to the Symposium on Science, Philosophy and Religion, Sept. 10, 1941, reprinted in Einstein 1954, 41; "Sees No Personal God," Associated Press, Sept. 11, 1941. A yellowed clipping of this story was given to me by Orville Wright, who was a young naval officer at the time and had kept it for sixty years; it had been passed around his ship and had notations from various sailors saying such things as, "Tell me, what do you think of this?"

17. "In the mind there is no absolute or free will, but the mind is determined by this or that volition, by a cause, which is also determined by another cause, and this again by another, and so on *ad infinitum.*" Baruch Spinoza, *Ethics*, part 2, proposition 48.

18. Einstein, statement to the Spinoza Society of America, Sept. 22, 1932.

19. Sometimes translated as "A man can do what he wants, but not want what he wants." I cannot find this quote in Schopenhauer's writings. The sentiment, nevertheless, comports with Schopenhauer's philosophy. He said, for example, "A man's life, in all its events great and small, is as necessarily predetermined as are the movements of a clock." Schopenhauer, "On Ethics," in *Parerga and Paralipomena: Short Philosophical Essays* (New York: Oxford University Press, 2001), 2:227.

20. Einstein, "The World As I See It," in Einstein 1949a and Einstein 1954.

21. Viereck, 375.

22. Max Born to Einstein, Oct. 10, 1944, in Born 2005, 150.

23. Hedwig Born to Einstein, Oct. 9, 1944, in Born 2005, 149.

24. Viereck, 377.

25. Einstein to the Rev. Cornelius Greenway, Nov. 20, 1950, AEA 28-894.

26. Sayen, 165.

CHAPTER EIGHTEEN: THE REFUGEE

1. Einstein trip diary, Dec. 6, 1931, AEA 29-136.

2. Einstein trip diary, Dec. 10, 1931, AEA 29-141.

3. Flexner, 381–382; Batterson, 87–89.

4. Abraham Flexner to Robert Millikan, July 30, 1932, AEA 38-007; Abraham Flexner to Louis Bamberger, Feb. 13, 1932, in Batterson, 88.

5. Einstein trip diary, Feb. 1, 1932, AEA 29-141; Elsa Einstein to Rosika Schwimmer, Feb. 3, 1932; Nathan and Norden, 163.

6. Einstein to Paul Ehrenfest, Apr. 3, 1932, AEA 10-227.

7. Clark, 542, citing Sir Roy Harrod.

8. Flexner, 383.

9. Einstein to Abraham Flexner, July 30, 1932; Batterson, 149; Brian 1996, 232.

10. Elsa Einstein to Robert Millikan, June 22, 1932, AEA 38-002.

11. Robert Millikan to Abraham Flexner, July 25, 1932, AEA 38-006; Abraham Flexner to Robert Millikan, July 30, 1932, AEA 38-007; Batterson, 114.

12. "Einstein Will Head School Here," *New York Times*, Oct. 11, 1932, p. 1.

13. Frank 1947, 226.

14. Woman Patriot Corporation memo to the U.S. State Department, Nov. 22, 1932, contained in Einstein's FBI file, section 1, available at foia.fbi.gov /foiaindex/einstein.htm. This episode is nicely detailed in Jerome, 6–11.

15. Reprinted in Einstein 1954, 7. Einstein's relationship with Louis Lochner of United Press is detailed in Marianoff, 137.

16. *New York Times*, Dec. 4, 1932.

17. "Einstein's Ultimatum Brings a Quick Visa," "Consul Investigated Charge," and "Women Made Complaint," all in *New York Times*, Dec. 6, 1932; Sayen, 6; Jerome, 10.

18. This was uncovered by Richard Alan Schwartz of Florida International University, who did the original research into Einstein's FBI files. The versions he received were redacted by 25 percent. Fred Jerome was able to get fuller versions under the Freedom of Information Act, which he used in his book. Schwartz's articles on the topic include "The F.B.I. and Dr. Einstein," *The Nation*, Sept. 3, 1983, 168–173, and "Dr. Einstein and the War Department," *Isis* (June 1989): 281–284. See also Dennis Overbye, "New Details Emerge from the Einstein Files," *New York Times*, May 7, 2002.

19. "Einstein Resumes Packing," *New York Times*, Dec. 7, 1932; "Einstein Embarks, Jests about Quiz" and "Stimson Regrets Incident," *New York Times*, Dec. 11, 1932.

20. Einstein (from Caputh) to Maurice Solovine, Nov. 20, 1932, AEA 21-218; Frank 1947, 226; Pais 1982, 318, 450. Both Frank and Pais recount Einstein's prophetic words to Elsa about Caputh, and each likely heard the anecdote directly from them. Pais, among others, says they carried thirty pieces of luggage. Elsa, in her call to reporters after the U.S. consulate interrogation, said she had packed six trunks, but she may not have been finished packing, or may have been referring only to trunks, or may have understated the number so as not to inflame German authorities (or Pais may have been wrong). Barbara Wolff of the Einstein archives in Jerusalem thinks the tale that she packed thirty trunks is a fabrication, as is the tale that Einstein told her to "take a very good look at it" when they left Caputh (private correspondence with the author).

21. "Einstein Will Urge Amity with Germany," *New York Times*, Jan. 8, 1933.

22. Nathan and Norden, 208; Clark, 552.

23. "Einstein's Address on World Situation" (text of speech) and "Einstein Traces Slump to Machine," *New York Times*, Jan. 24, 1933.

24. Fölsing, 659.

25. Einstein to Margarete Lebach, Feb. 27, 1933, AEA 50-834.

26. Evelyn Seeley, interview with Einstein, *New York World-Telegram*, Mar. 11, 1933; Brian 1996, 243.

27. Marianoff, 142–144.

28. Michelmore, 180. Michelmore got much of his material from Hans Albert Einstein, though this quote may have been exaggerated.

29. Einstein, Statement against the Hitler regime, Mar. 22, 1933, AEA 28-235.

30. Einstein to the Prussian Academy, Mar. 28, 1933, AEA 36–55.

31. Max Planck to Einstein, Mar. 31, 1933.

32. Max Planck to Heinrich von Ficker, Mar. 31, 1933, cited in Fölsing, 663.

33. Prussian Academy declaration, Apr. 1, 1933. The exchanges are reprinted in Einstein 1954, 205–209.

34. Einstein to Prussian Academy, Apr. 5, 1933.

35. Frank 1947, 232.

36. Prussian Academy to Einstein, Apr. 7 and 13, 1933; Einstein to Prussian Academy, Apr. 12, 1933.

37. Max Planck to Einstein, Mar. 31, 1933, AEA 19-389; Einstein to Max Planck, Apr. 6, 1933, AEA 19-392.

38. Einstein to Max Born, May 30, 1933, AEA 8-192; Max Born to Einstein, June 2, 1933, AEA 8-193.

39. Einstein to Fritz Haber, May 19, 1933, AEA 12-378. For a good profile of the Einstein-Haber relationship and this final episode, see Stern, 156–160. Also very useful is John Cornwall, *Hitler's Scientists* (New York: Viking, 2003), 137–139.

40. Fritz Haber to Einstein, Aug. 1, 1933, AEA 385; Einstein to Fritz Haber, Aug. 8, 1933, AEA 12-388.

41. Einstein to Willem de Sitter, Apr. 5, 1933, AEA 20-575; Frank 1947, 232; Clark, 573.

42. Vallentin, 231.

43. Frank 1947, 240–242.

44. Einstein to Maurice Solovine, Apr. 23, 1933, AEA 21-223.

45. Einstein to Paul Langevin, May 5, 1933, AEA 15-394.

46. "Einstein Will Go to Madrid," *New York Times*, Apr. 11, 1933; Abraham Flexner to Einstein, Apr. 13, 1933, AEA 38-23; Pais 1982, 493.

47. Abraham Flexner to Einstein, Apr. 26 and 28, 1933, AEA 38-25, 38-26.

48. "Einstein Lists Contracts; Princeton, Paris, Madrid, Oxford Lectures Are Only Engagements," *New York Times*, Aug. 5, 1933; Einstein to Frederick Lindemann, May 1, 1933, AEA 16-372.

49. Hannoch Gutfreund, "Albert Einstein and Hebrew University," in Renn 2005d, 318.

50. Einstein to Fritz Haber, Aug. 9, 1933, AEA 37-109; Einstein to Max Born, May 30, 1933, AEA 8-192.

51. *Jewish Chronicle*, Apr. 8, 1933; Chaim Weizmann to Einstein, Apr. 3, 1933, AEA 33-425; Einstein to Paul Ehrenfest, June 14, 1933, AEA 10-255.

52. Einstein to Herbert Samuel, Apr. 15, 1933, AEA 21-17; Einstein to Chaim Weizmann, June 9, 1933, AEA 33-435.

53. "Weizmann Scores Einstein's Stand," *New York Times*, June 30, 1933.

54. "Albert Einstein Definitely Takes Post at Hebrew University," Jewish Tele-

graphic Agency, July 3, 1933; Abraham Flexner to Elsa Einstein, July 19, 1933, AEA 33-033; "Einstein Accepts Chair: Dr. Weizmann Announces He Has Made Peace with Hebrew University in Jerusalem," *New York Times*, July 4, 1933.

55. Einstein to the Rev. Johannes B. Th. Hugenholtz, July 1, 1933, AEA 50-320.
56. Nathan and Norden, 225.
57. The queen's name has been spelled Elizabeth in many books, but as carved on her statue and national monument in Brussels, and in most official sources, it is Elisabeth.
58. Einstein to Elsa Einstein, Nov. 1, 1930, uncatalogued new material provided to author.
59. Einstein to King Albert I of Belgium, Nov. 14, 1933, in Nathan and Norden, 230.
60. Einstein to Alfred Nahon, July 20, 1933, AEA 51-227.
61. *New York Times*, Sept. 10, 1933.
62. Einstein to E. Lagot, Aug. 28, 1933, AEA 50-477.
63. Einstein to Lord Ponsonby, Aug. 28, 1933, AEA 51-400.
64. Einstein to A. V. Frick, Sept. 9, 1933, AEA 36-567.
65. Einstein to G. C. Heringa, Sept. 11, 1933, AEA 50-199.
66. Einstein to P. Bernstein, Apr. 5, 1934, AEA 49-276.
67. Romain Rolland, Sept. 1933 diary entry, in Nathan and Norden, 232.
68. Michele Besso to Einstein, Sept. 18, 1932, AEA 7-130; Einstein to Michele Besso, Oct. 21, 1932, AEA 7-370.
69. Einstein to Frederick Lindemann, May 9, 1933, AEA 16-377.
70. Einstein to Elsa Einstein, July 21, 1933, AEA 143-250.
71. Locker-Lampson speech, House of Commons, July 26, 1933; "Einstein a Briton Soon: Home Secretary's Certificate Preferred to Palestine Citizenship," *New York Times*, July 29, 1933; Marianoff, 159.
72. *New York World Telegram*, Sept. 19, 1933, in Nathan and Norden, 234.
73. "Dr. Einstein Denies Communist Leanings," *New York Times*, Sept. 16, 1933; "Professor Einstein's Political Views," *Times* of London, Sept. 16, 1933, in Brian 1996, 251.
74. Einstein, Appreciation of Paul Ehrenfest, written in 1934 for a Leiden almanac and reprinted in Einstein 1950a, 236.
75. Clark, 600–605; Marianoff, 160–163; Jacob Epstein, *Let There Be Sculpture* (London: Michael Joseph, 1940), 78.
76. Dukas and Hoffmann, 56.
77. Einstein, "Civilization and Science," Royal Albert Hall, Oct. 3, 1933; *Times* of London, Oct. 4, 1933; Calaprice, 198; Clark, 610–611. Clark's version is more faithful to the way the speech was given than the written version, which had two references to Germany that Einstein, diplomatically, decided to omit.

CHAPTER NINETEEN: AMERICA

1. Abraham Flexner telegram to Einstein, Oct. 1933, AEA 38-049; Abraham Flexner to Einstein, Oct. 13, 1933, AEA 38-050.
2. "Einstein Arrives; Pleads for Quiet / Whisked from Liner by Tug at Quarantine," *New York Times*, Oct. 18, 1933.
3. "Einstein Views Quarters," *New York Times*, Oct. 18, 1933; Rev. John Lampe interview, in Clark, 614; "Einstein to Princeton," *Time*, Oct. 30, 1933.
4. Brian 1996, 251.
5. "Einstein Has Musicale," *New York Times*, Nov. 10, 1933. The sketches that Einstein made for Seidel are now in the Judah Magnes Museum, endowed by the president of Hebrew University with whom Einstein fought.
6. Bucky, 150.
7. Thomas Torrance, "Einstein and God," Center for Theological Inquiry, Princeton, ctinquiry.org/publications/reflections_volume_1/torrance.htm. Torrance says a friend related the tale to him.
8. Eleanor Drorbaugh interview with Jamie Sayen, in Sayen, 64, 74.
9. Sayen, 69; Bucky, 111; Fölsing, 732.
10. "Had Pronounced Sense of Humor," *New York Times*, Dec. 22, 1936.
11. Brian 1996, 265.
12. Abraham Flexner to Einstein, Oct. 13, 1933, in Regis, 34.
13. "Einstein, the Immortal, Shows Human Side," (Newark) *Sunday Ledger*, Nov. 12, 1933.
14. Abraham Flexner to Elsa Einstein, Nov. 14, 1933, AEA 38-055.
15. Abraham Flexner to Elsa Einstein, Nov. 15, 1933, AEA 38-059. Flexner also wrote to Herbert Maass, an Institute trustee, on Nov. 14, 1933: "I am beginning to weary a little of this daily necessity of 'sitting down' on Einstein and his wife. They do not know America. They are the merest children, and they are extremely difficult to advise and control. You have no idea the barrage of publicity I have intercepted." Batterson, 152.
16. Abraham Flexner to Einstein, Nov. 15, 1933, AEA 38-061.
17. "Fiddling for Friends," *Time*, Jan. 29, 1934; "Einstein in Debut as Violinist Here," *New York Times*, Jan. 18, 1934.
18. Stephen Wise to Judge Julian Mack, Oct. 20, 1933.
19. Col. Marvin MacIntyre report to the White House Social Bureau, Dec. 7, 1933, AEA 33-131; Abraham Flexner to Franklin Roosevelt, Nov. 3, 1933; Einstein to Eleanor Roosevelt, Nov. 21, 1933, AEA 33-129; Eleanor Roosevelt to Einstein, Dec. 4, 1933, AEA 33-130; Elsa Einstein to Eleanor Roosevelt, Jan. 16, 1934, AEA 33-132; Einstein to Queen Elisabeth of Belgium, Jan. 25, 1934, AEA 33-134; "Einstein Chats about Sea," *New York Times*, Jan. 26, 1934.
20. Einstein to Board of Trustees of the IAS, Dec. 1–31, 1933.
21. Johanna Fantova, Journal of conversations with Einstein, Jan. 23, 1954, in Calaprice, 354.

22. Einstein to Max Born, Mar. 22, 1934; Erwin Schrödinger to Frederick Lindemann, Mar. 29, 1934, Jan. 22, 1935.

23. Einstein to Queen Elisabeth of Belgium, Nov. 20, 1933, AEA 32-369. The line is usually translated as "puny demigods on stilts." The word Einstein uses, *stelzbeinig*, means stiff-legged, *as if* the legs were wooden stilts. It has nothing to do with height. Instead, it evokes the gait of a peacock.

24. Einstein, "The Negro Question," *Pageant*, Jan. 1946. In this essay, he was juxtaposing the generally democratic social tendency of Americans to the way they treated blacks. That became more of an issue for him than it was back in 1934, as will be noted later in this book.

25. Bucky, 45; "Einstein Farewell," *Time*, Mar. 14, 1932.

26. Vallentin, 235. See also Elsa Einstein to Hertha Einstein (wife of music historian Alfred Einstein, a distant cousin), Feb. 24, 1934, AEA 37-693: "The place is charming, altogether different from the rest of America . . . Here everything is tinged with Englishness—downright Oxford style."

27. "Einstein Cancels Trip Abroad," *New York Times*, Apr. 2, 1934.

28. Marianoff, 178. Other sources report that Ilse's ashes, or at least some of them, were brought to a cemetery in Holland, to a place chosen by the widower Rudi Kayser.

29. This entire story is from an interview given by the Blackwoods' son James to Denis Brian on Sept. 7, 1994, and is detailed in Brian 1996, 259–263.

30. Ibid. See also James Blackwood, "Einstein in the Rear-View Mirror," *Princeton History*, Nov. 1997.

31. "Einstein Inventor of Camera Device," *New York Times*, Nov. 27, 1936.

32. Bucky, 5. Bucky's book is written, in part, as a running conversation, though there are sections that actually draw from other Einstein interviews and writings.

33. Bucky, 16–21.

34. *New York Times*, Aug. 4, 1935; Brian 1996, 265, 280.

35. Vallentin, 237.

36. Brian 1996, 268.

37. Fölsing, 687; Brian 1996, 279.

38. Calaprice, 251.

39. Bucky, 25.

40. Clark, 622.

41. Pais 1982, 454.

42. Jon Blackwell, "The Genius Next Door," *The Trentonian*, www.capitalcentury.com/1933.html; Seelig 1956a, 193; Sayen, 78; Brian 1996, 330.

43. Einstein to Barbara Lee Wilson, Jan. 7, 1943, AEA 42-606; Dukas and Hoffmann, 8; "Einstein Solves Problem That Baffled Boys," *New York Times*, June 11, 1937.

44. "Einstein Gives Advice to a High School Boy," *New York Times*, Apr. 14, 1935; Sayen, 76.

45. Elsa Einstein to Leon Watters, Dec. 10, 1935, AEA 52-210.

46. Vallentin, 238.
47. Bucky, 13.
48. Einstein to Hans Albert Einstein, Jan. 4, 1937, AEA 75-926.
49. Hoffmann 1972, 231.
50. Einstein, "Lens-like Action of a Star by Deviation of Light in the Gravitational Field," *Science* (Dec. 1936); Einstein with Nathan Rosen, "On Gravitational Waves," *Journal of the Franklin Institute* (Jan. 1937). The gravitational wave paper was originally submitted to *Physical Review*. Editors there sent it to a referee, who noted flaws. Einstein was outraged, withdrew the paper, and had it published instead by the Franklin Institute. He then realized he was wrong after all (after the anonymous referee indirectly let him know), and he and Rosen juggled many modifications, just as Elsa was dying. Daniel Kinneflick uncovered the details of this saga and provides a fascinating acount in "Einstein versus the Physical Review," *Physics Today* (Sept. 2005).
51. Einstein to Max Born, Feb. 1937, in Born 2005, 128.
52. Einstein, "The Causes of the Formation of Meanders in the Courses of Rivers and of the So-Called Baer's Law," Jan. 7, 1926.
53. "Dr. Einstein Welcomes Son to America," *New York Times*, Oct. 13, 1937.
54. Bucky, 107.
55. Einstein to Mileva Marić, Dec. 21, 1937, AEA 75-938.
56. Einstein to Frieda Einstein, Apr. 11, 1937, AEA 75-929.
57. Robert Ettema and Cornelia F. Mutel, "Hans Albert Einstein in South Carolina," *Water Resources and Environmental History*, June 27, 2004; "Einstein's Son Asks Citizenship," *New York Times*, Dec. 22, 1938. He applied for citizenship on Dec. 21, 1938, at the U.S. District Court in Greenville, S.C. Some biographies have him living in Greensboro, N.C., at the time, but that is incorrect.
58. Einstein to Hans Albert and Frieda Einstein, Jan. 1939; James Shannon, "Einstein in Greenville," *The Beat* (Greenville, S.C.), Nov. 17, 2001.
59. Highfield and Carter, 242.
60. "Hitler Is 'Greatest' in Princeton Poll: Freshmen Put Einstein Second and Chamberlain Third," *New York Times*, Nov. 28, 1939. The story reports that this was for the second year in a row.
61. *Collier's*, Nov. 26, 1938; Einstein 1954, 191.
62. Sayen, 344; "Einstein Fiddles," *Time*, Feb. 3, 1941. *Time* reported of a little concert in Princeton for the American Friends Service Committee: "Einstein proved that he could play a slow melody with feeling, turn a trill with elegance, jigsaw on occasion. The audience applauded warmly. Fiddler Einstein smiled his broad and gentle smile, glanced at his watch in fourth-dimensional worriment, played his encore, peered at the watch again, retired."
63. Jerome, 77.
64. Einstein to Isaac Don Levine, Dec. 10, 1934, AEA 50-928; Isaac Don Levine, *Eyewitness to History* (New York: Hawthorne, 1973), 171.

65. Sidney Hook to Einstein, Feb. 22, 1937, AEA 34-731; Einstein to Sidney Hook, Feb. 23, 1937, AEA 34-735.

66. Sidney Hook, "My Running Debate with Einstein," *Commentary*, July 1982, 39.

CHAPTER TWENTY: QUANTUM ENTANGLEMENT

1. Hoffmann 1972, 190; Rigden, 144; Léon Rosenfeld, "Niels Bohr in the Thirties," in Rozental 1967, 127; N. P. Landsman, "When Champions Meet: Rethinking the Bohr–Einstein Debate," *Studies in the History and Science of Modern Physics* 37 (Mar. 2006): 212.

2. Einstein 1949b, 85.

3. Ibid.

4. Einstein to Max Born, Mar. 3, 1947, in Born 2005, 155 (not in AEA).

5. Einstein to Erwin Schrödinger, June 19, 1935, AEA 22-47.

6. *New York Times*, May 4 and 7, 1935; David Mermin, "My Life with Einstein," *Physics Today* (Jan. 2005).

7. Albert Einstein, Boris Podolsky, and Nathan Rosen, "Can Quantum-Mechanical Description of Physical Reality Be Regarded as Complete?," *Physical Review*, May 15, 1935 (received Mar. 25, 1935); www.drchinese .com/David/EPR.pdf.

8. Another formulation of the experiment would be for one observer to measure the position of a particle while at the "same moment" another observer measures the momentum of its twin. Then they compare notes and, supposedly, know the position and momentum of both particles. See Charles Seife, "The True and the Absurd," in Brockman, 71.

9. Aczel 2002, 117.

10. Whitaker, 229; Aczel 2002, 118.

11. Niels Bohr, "Can Quantum-Mechanical Description of Physical Reality Be Regarded as Complete?," *Physical Review*, Oct. 15, 1935 (received July 13, 1935).

12. Greene 2004, 102. Note that Arthur Fine says that the synopsis of EPR used by Bohr "is closer to a caricature of the EPR paper than it is to a serious reconstruction." Fine says that Bohr and other interpreters of Einstein feature a "criterion of reality" that Einstein in his own later writings on EPR does not feature, even though the EPR paper as written by Podolsky does talk about determining "an element of reality." Brian Greene's book is among those that do emphasize the "criterion of reality" element. See Arthur Fine, "The Einstein-Podolsky-Rosen Argument in Quantum Theory," *Stanford Encyclopedia of Philosophy*, plato.stanford.edu/entries/qt-epr/, and also: Fine 1996, chapter 3; Mara Beller and Arthur Fine, "Bohr's Response to EPR," in Jann Faye and Henry Folse, eds., *Niels Bohr and Contemporary Philosophy* (Dordrecht: Kluwer Academic, 1994), 1–31.

13. Arthur Fine has shown that Einstein's own critique of quantum mechanics

was not fully captured in the way that Podolsky wrote in the EPR paper, and especially in the way that Bohr and the "victors" described it. Don Howard has built on Fine's work and emphasized the issues of "separability" and "locality." See Howard 1990b.

14. Einstein to Erwin Schrödinger, May 31, 1928, AEA 22-22; Fine, 18.
15. Erwin Schrödinger to Einstein, June 7, 1935, AEA 22-45, and July 13, 1935, AEA 22-48.
16. Einstein to Erwin Schrödinger, June 19, 1935, AEA 22-47.
17. Erwin Schrödinger, "The Present Situation in Quantum Mechanics," third installment, Dec. 13, 1935, www.tu-harburg.de/rzt/rzt/it/QM/cat.html.
18. More specifically, Schrödinger's equation shows the rate of change over time of the mathematical formulation of the probabilities for the outcome of possible measurements made on a particle or system.
19. Einstein to Erwin Schrödinger, June 19, 1935, AEA 22-47.
20. I am grateful to Craig Copi and Douglas Stone for helping to compose this section.
21. Einstein to Erwin Schrödinger, Aug. 8, 1935, AEA 22-49; Arthur Fine, "The Einstein-Podolsky-Rosen Argument in Quantum Theory," *Stanford Encyclopedia of Philosophy*, plato.stanford.edu/entries/qt-epr/. Note that Arthur Fine uncovered some of the Einstein-Schrödinger correspondence. Fine, chapter 3.
22. Erwin Schrödinger to Einstein, Aug. 19, 1935, AEA 22-51.
23. Erwin Schrödinger, "The Present Situation in Quantum Mechanics," Nov. 29, 1935, www.tu-harburg.de/rzt/rzt/it/QM/cat.html.
24. Einstein to Erwin Schrödinger, Sept. 4, 1935, AEA 22-53. Schrödinger's paper had not been published, but Schrödinger included its argument in his Aug. 19, 1935, letter to Einstein.
25. en.wikipedia.org/wiki/Schrodinger's_cat.
26. Einstein to Erwin Schrödinger, Dec. 22, 1950, AEA 22-174.
27. David Bohm and Basil Huey, "Einstein and Non-locality in the Quantum Theory," in Goldsmith et al., 47.
28. John Stewart Bell, "On the Einstein-Podolsky-Rosen Paradox," *Physic* 1, no. 1 (1964).
29. Bernstein 1991, 20.
30. For an explanation of how Bohm and Bell set up their analysis, see Greene 2004, 99–115; Bernstein 1991, 76.
31. Bernstein 1991, 76, 84.
32. *New York Times*, Dec. 27, 2005.
33. *New Scientist*, Jan. 11, 2006.
34. Greene 2004, 117.
35. In the decoherent-histories formulation of quantum mechanics, the coarse graining is such that the histories don't interfere with one another: if A and B are mutually exclusive histories, then the probability of A or B is the sum of the probabilities of A and of B as it should be. These "decoherent" histories form a tree-like structure, with each of the alternatives at one time branching out into alternatives at the next time, and so forth. In this theory, there is much less em-

phasis on measurement than in the Copenhagen version. Consider a piece of mica in which there are radioactive impurities emitting alpha particles. Each emitted alpha particle leaves a track in the mica. The track is real, and it makes little difference whether a physicist or other human being or a chinchilla or a cockroach comes along to look at it. What is important is that the track is correlated with the direction of emission of the alpha particle and *could be used* to measure the emission. Before the emission takes place, all directions are equally probable and contribute to a branching of histories. I am grateful to Murray Gell-Mann for his help with this section. See also Gell-Mann, 135–177; Murray Gell-Mann and James Hartle, "Quantum Mechanics in the Light of Quantum Cosmology," in W. H. Zurek, ed., *Complexity, Entropy and the Physics of Information* (Reading, Mass.: Addison-Wesley, 1990), 425–459, and "Equivalent Sets of Histories and Multiple Quasiclassical Realms," May 1996, www.arxiv.org/abs/gr-qc/9404013. This view is derived from the many-worlds interpretation pioneered in 1957 by Hugh Everett.

36. The literature on Einstein and realism is fascinating. This section relies on the works of Don Howard, Gerald Holton, Arthur I. Miller, and Jeroen van Dongen cited in the bibliography.

    Don Howard has argued that Einstein was never a true Machian nor a true realist, and that his philosophy of science did not change much over the years. "On my view, Einstein was never an ardent 'Machian' positivist, and he was never a scientific realist, at least not in the sense acquired by the term 'scientific realist' in later twentieth-century philosophical discourse. Einstein expected scientific theories to have the proper empirical credentials, but he was no positivist; and he expected scientific theories to give an account of physical reality, but he was no scientific realist. Moreover, in both respects his views remained more or less the same from the beginning to the end of his career." Howard 2004.

    Gerald Holton, on the other side, argues that Einstein underwent "a pilgrimage from a philosophy of science in which sensationalism and empiricism were at the center, to one in which the basis was a rational realism . . . For a scientist to change his philosophical beliefs so fundamentally is rare" (Holton 1973, 219, 245). See also Anton Zeilinger, "Einstein and Absolute Reality," in Brockman, 123: "Instead of accepting only concepts that can be verified by observation, Einstein insisted on the existence of a reality prior to and independent of observation."

    Arthur Fine's *The Shaky Game* explores all sides of the issue. He develops for himself what he calls a "natural ontological attitude" that is neither realist nor antirealist, but instead "mediates between the two." Of Einstein he says, "I think there is no backing away from the fact that Einstein's so-called realism has a deeply empiricist core that makes it a 'realism' more nominal than real." Fine, 130, 108.

37. Einstein to Jerome Rothstein, May 22, 1950, AEA 22-54.
38. Einstein to Donald Mackay, Apr. 26, 1948, AEA 17-9.
39. Einstein 1949b, 11.

40. Gerald Holton, "Mach, Einstein and the Search for Reality," in Holton 1973, 245. Arthur I. Miller disagrees with some of Holton's interpretation. He stresses that Einstein's point was that for something to be real it should be measurable *in principle*, even if not actually measurable in real life, and he was content using thought experiments to "measure" something. Miller 1981, 186.

41. Einstein 1949b, 81.

42. Einstein to Max Born, comments on a paper, Mar. 18, 1948, in Born 2005, 161.

43. Einstein, "The Fundamentals of Theoretical Physics," *Science*, May 24, 1940; Einstein 1954, 334.

44. For example, Arthur Fine argues, "Causality and observer-independence were *primary* features of Einstein's realism, whereas a space/time representation was an important but *secondary* feature." Fine, 103.

45. Einstein, "Physics, Philosophy and Scientific Progress," *Journal of the International College of Surgeons* 14 (1950), AEA 1-163; Fine, 98.

46. Einstein, "Physics and Reality," *Journal of the Franklin Institute* (Mar. 1936), in Einstein 1954, 292. Gerald Holton says that this is more properly translated: "The eternally incomprehensible thing about the world is its comprehensibility"; see Holton, "What Precisely Is Thinking?," in French, 161.

47. Einstein to Maurice Solovine, Mar. 30, 1952, in Solovine, 131 (not in AEA).

48. Einstein to Maurice Solovine, Jan. 1, 1951, in Solovine, 119.

49. Einstein to Max Born, Sept. 7, 1944, in Born 2005, 146, and AEA 8-207.

50. Born 2005, 69. He put Einstein in the category of "conservative individuals who were unable to free their minds from the prevailing philosophical prejudices."

51. Einstein to Maurice Solovine, Apr. 10, 1938, in Solovine, 85.

52. Einstein and Infeld, 296.

53. Ibid., 241.

54. Born 2005, 118, 122.

55. Brian 1996, 289.

56. Hoffmann 1972, 231.

57. Regis, 35.

58. Leopold Infeld, *Quest* (New York: Chelsea, 1980), 309.

59. Brian 1996, 303.

60. Infeld, introduction to the 1960 edition of Einstein and Infeld; Infeld, 112–114.

61. Pais 1982, 23.

62. Vladimir Pavlovich Vizgin, *Unified Field Theories in the First Third of the 20th Century* (Basel: Birkhäuser, 1994), 218. Matthew 19:6, King James Version: "What therefore God hath joined together, let not man put asunder."

63. Einstein to Max von Laue, Mar. 23, 1934, AEA 16-101.

64. From Whitrow, xii: "Einstein agreed that the chance of success was very small but the attempt must be made. He himself had established his name; his position was assured, so he could afford to take the risk of failure. A young man

with his way to make in the world could not afford to take a risk by which he might lose a great career, so Einstein felt that in this matter he had a duty."

65. Hoffmann 1972, 227.
66. Arthur I. Miller, "A Thing of Beauty," *New Scientist*, Feb. 4, 2006.
67. Einstein to Maurice Solovine, June 27, 1938. See also Einstein to Maurice Solovine, Dec. 23, 1938, AEA 21-236: "I have come across a wonderful subject which I am studying enthusiastically with two young colleagues. It offers the possibility of destroying the statistical basis of physics, which I have always found intolerable. This extension of the general theory of relativity is of very great logical simplicity."
68. William Laurence, "Einstein in Vast New Theory Links Atoms and Stars in Unified System," *New York Times*, July 5, 1935; William Laurence, "Einstein Sees Key to Universe Near," *New York Times*, Mar. 14, 1939.
69. Hoffmann 1972, 227; Bernstein 1991, 157.
70. William Laurence, "Einstein Baffled by Cosmos Riddle," *New York Times*, May 16, 1940.
71. Fölsing, 704.
72. *Pittsburgh Post-Gazette*, Dec. 29, 1934.
73. William Laurence, "Einstein Sees Key to Universe Near," *New York Times*, Mar. 14, 1939.

CHAPTER TWENTY-ONE: THE BOMB

1. FBI interview with Einstein regarding Leó Szilárd, Nov. 1, 1940, obtained by Gene Dannen under the Freedom of Information Act, www.dannen.com/ein stein.html. It is ironic that the FBI had such an extensive and friendly interview with Einstein to check out Szilárd's worthiness for a security clearance, because Einstein had been denied such a clearance himself. See also Gene Dannen, "The Einstein-Szilárd Refrigerators," *Scientific American* (Jan. 1997).
2. Recollections of Chuck Rothman, son of David Rothman, www.sff.net/peo ple/rothman/einstein.htm.
3. Weart and Szilard 1978, 83–96; Brian 1996, 316.
4. An authoritative narrative is in Rhodes, 304–308.
5. See Kati Marton, *The Great Escape: Nine Hungarians Who Fled Hitler and Changed the World* (New York: Simon & Schuster, 2006).
6. Leó Szilárd to Einstein, July 19, 1933, AEA 76-532.
7. Some popular accounts suggest that Einstein merely signed a letter that Szilárd wrote and brought with him. Along these lines, Teller told the writer Ronald W. Clark in 1969 that Einstein had signed, with "very little comment," a letter that Szilárd and Teller had brought that day. See Clark, 673. This is contradicted, however, by Szilárd's own detailed description of that day and the notes of the conversation made by Teller that day. The notes and new draft letter in German as dictated by Einstein are in the Teller archives and reprinted in Nathan and Norden, 293. It is true that the letter dictated by Ein-

stein was based on a draft Szilárd brought that day, but that was a translation of the one Einstein had dictated two weeks earlier. Some accounts, including occasional comments made later by Einstein himself, try to minimize his role and say he simply signed a letter that someone else wrote. In fact, even though Szilárd prompted and propelled the discussions, Einstein was fully involved in writing the letter that he alone signed.

8. Einstein to Franklin Roosevelt, Aug. 2, 1939. The longer version is in the Franklin Roosevelt archives in Hyde Park, New York (with a copy in AEA 33-143), the shorter one in the Szilárd archives at the University of California, San Diego.

9. Clark, 676; Einstein to Leó Szilárd, Aug. 2, 1939, AEA 39-465; Leó Szilárd to Einstein, Aug. 9, 1939, AEA 39-467; Leó Szilárd to Charles Lindbergh, Aug. 14, 1939, Szilárd papers, University of California, San Diego, box 12, folder 5.

10. Charles Lindbergh, "America and European Wars," speech, Sept. 15, 1939, www.charleslindbergh.com/pdf/9_15_39.pdf.

11. Leó Szilárd to Einstein, Sept. 27, 1933, AEA 39-471. Lindbergh later did not recall getting any letters from Szilárd.

12. Leó Szilárd to Einstein, Oct. 3, 1939, AEA 39-473.

13. Moore, 268. The Napoleon tale is clearly one that Sachs or someone garbled, as Robert Fulton did in fact work on building ships for Napoleon, including a failed submarine; see Kirkpatrick Sale, *The Fire of His Genius* (New York: Free Press, 2001), 68–73.

14. Sachs told this tale to a U.S. Senate special committee on atomic energy hearing, Nov. 27, 1945. It is recounted in most histories of the atom bomb, including Rhodes, 313–314.

15. Franklin Roosevelt to Einstein, Oct. 19, 1939, AEA 33-192.

16. Einstein to Alexander Sachs, Mar. 7, 1940, AEA 39-475.

17. Einstein to Lyman Briggs, Apr. 25, 1940, AEA 39-484.

18. Sherman Miles to J. Edgar Hoover, July 30, 1940, in the FBI files on Einstein, foia.fbi.gov/einstein/einstein1a.pdf. A good analysis and context for these files is Jerome.

19. J. Edgar Hoover to Sherman Miles, Aug. 15, 1940.

20. Einstein to Henri Barbusse, June 1, 1932, AEA 34-543. The FBI refers to this conference with a different translation of its name, the World Congress against War.

21. Jerome, 28, 295 n. 6. The Miles note is on the copy in the National Archives but not the FBI files.

22. Jerome, 40–42.

23. Einstein, "This Is My America," unpublished, summer 1944, AEA 72-758.

24. "Einstein to Take Test," *New York Times*, June 20, 1940; "Einstein Predicts Armed League," *New York Times*, June 23, 1940.

25. "Einstein Is Sworn as Citizen of U.S.," *New York Times*, Oct. 2, 1940.

26. Einstein, "This Is My America," unpublished, summer 1944, AEA 72-758.

27. Frank Aydelotte to Vannevar Bush, Dec. 19, 1941; Clark, 684.

28. Vannevar Bush to Frank Aydelotte, Dec. 30, 1941.

29. Pais 1982,12; George Gamow, "Reminiscence," in French, 29; Fölsing, 715.
30. Sayen, 150; Pais 1982, 147. The manuscripts were purchased by the Kansas City Life Insurance Co. and were subsequently donated to the Library of Congress.
31. Einstein to Niels Bohr, Dec. 12, 1944, AEA 8-95.
32. Clark, 698.
33. Einstein to Otto Stern, Dec. 26, 1944, AEA 22-240; Clark, 699–700.
34. Einstein to Franklin Roosevelt, Mar. 25, 1945, AEA 33-109.
35. Sayen, 151.
36. *Time*, July 1, 1946. The portrait was by the longtime cover artist for the magazine, Ernest Hamlin Baker.
37. *Newsweek*, Mar. 10, 1947.
38. Linus Pauling report of conversation, Nov. 16, 1954, in Calaprice, 185.

CHAPTER TWENTY-TWO: ONE-WORLDER

1. Brian 1996, 345; Helen Dukas to Alice Kahler, Aug. 8, 1945: "One of the young reporters who was a guest at the Sulzbergers from the *New York Times* came over late at night . . . Arthur Sulzberger also called constantly for a statement. But no dice." Arthur Ochs Sulzberger Sr. told me that his father, Arthur Hays Sulzberger, and uncle David summered at Saranac Lake and knew Einstein.
2. United Press interview, Sept. 14, 1945, reprinted in *New York Times*, Sept. 15, 1945.
3. Einstein to J. Robert Oppenheimer (care of a post office box in Santa Fe near Los Alamos), Sept. 29, 1945, AEA 57-294; J. Robert Oppenheimer to Einstein, Oct. 10, 1945, AEA 57-296.
4. When he realized that Oppenheimer had not written the statement he considered too timid, Einstein wrote to the scientists in Oak Ridge, Tennessee, who actually had. In the letter, he explained his thoughts about what powers a world government should and should not have. "There would be no immediate need for member nations to subordinate their own tariff and immigration legislation to the authority of world government," he said. "In fact, I believe the sole function of world government should be to have a monopoly over military power." Einstein to John Balderston and other Oak Ridge scientists, Dec. 3, 1945, AEA 56-493.
5. It is reprinted in Nathan and Norden, 347, and Einstein 1954, 118. See also Einstein, "The Way Out," in *One World or None*, Federation of Atomic Scientists, 1946, www.fas.org/oneworld/index.html. The book is an important look at the ideas of scientists at the time—including Einstein, Oppenheimer, Szilárd, Wigner, and Bohr—on how to use world federalism to control nuclear arms.
6. Einstein realized there was no lasting "secret" of the bomb to protect. As he said later, "America has temporary superiority in armament, but it is certain that we have no lasting secret. What nature tells one group of men, she will tell

in time to any other group." Einstein, "The Real Problem Is in the Hearts of Men," *New York Times Magazine*, June 23, 1946.

7. Einstein, remarks at the Nobel Prize dinner, Hotel Astor, Dec. 10, 1945, in Einstein 1954, 115.

8. Einstein, ECAS fund-raising telegram, May 23, 1946. Material relating to this is in folder 40-11 of the Einstein archives. The history and archives of the ECAS can be found through www.aip.org/history/ead/chicago_ecas/20010108_content.html#top.

9. Einstein, ECAS letter, Jan. 22, 1947, AEA 40-606; Sayen, 213.

10. *Newsweek*, Mar. 10, 1947.

11. Richard Present to Einstein, Jan. 30, 1946, AEA 57-147.

12. Einstein to Dr. J. J. Nickson, May 23, 1946, AEA 57-150; Einstein to Louis B. Mayer, June 24, 1946, AEA 57-152.

13. Louis B. Mayer to Einstein, July 18, 1946, AEA 57-153; James McGuinness to Louis B. Mayer, July 16, 1946, AEA 57-154.

14. Sam Marx to Einstein, July 1, 1946, AEA 57-155; Einstein to Sam Marx, July 8, 1946, AEA 57-156; Sam Marx to Einstein, July 16, 1946, AEA 57-158.

15. Einstein to Sam Marx, July 19, 1946, AEA 57-162; Leó Szilárd telegram to Einstein, and Einstein note on reverse, July 27, 1946, AEA 57-163, 57-164.

16. Bosley Crowther, "Atomic Bomb Film Starts," *New York Times*, Feb. 21, 1947.

17. William Golden to George Marshall, June 9, 1947, Foreign Relations of the U.S.; Sayen, 196.

18. Halsman's quote from Einstein, recounted by Halsman's widow, is in *Time*'s Person of the Century issue, Dec. 31, 1999, which has the portrait he took (shown on p. 487) as the cover.

19. Einstein comment on the animated antiwar film, *Where Will You Hide?*, May 1948, AEA 28-817.

20. Einstein interview with Alfred Werner, *Liberal Judaism*, Apr.–May 1949.

21. Norman Cousins, "As 1960 Sees Us," *Saturday Review*, Aug. 5, 1950; Einstein to Norman Cousins, Aug. 2, 1950, AEA 49-453. (A weekly magazine is actually published one week earlier than it is dated.)

22. Einstein talk (via radio) to the Jewish Council for Russian War Relief, Oct. 25, 1942, AEA 28-571. See also, among many examples, Einstein unsent message regarding the May-Johnson Bill, Jan. 1946; in Nathan and Norden, 342; broadcast interview, July 17, 1947, in Nathan and Norden, 418.

23. "Rankin Denies Einstein A-Bomb Role," United Press, Feb. 14, 1950.

24. Einstein to Sidney Hook, Apr. 3, 1948, AEA 58-300; Sidney Hook, "My Running Debate with Einstein," *Commentary* (July 1982).

25. Einstein to Sidney Hook, May 16, 1950, AEA 59-1018.

26. "Dr. Einstein's Mistaken Notions," in *New Times* (Moscow), Nov. 1947, in Nathan and Norden, 443, and Einstein 1954, 134.

27. Einstein, Reply to the Russian Scientists, *Bulletin of Atomic Scientists* (the publication of the Emergency Committee that he chaired), Feb. 1948, in Einstein 1954, 135; "Einstein Hits Soviet Scientists for Opposing World Government," *New York Times*, Jan. 30, 1948.

28. Einstein, "Atomic War or Peace," part 2, *Atlantic Monthly*, Nov. 1947.
29. Einstein to Henry Usborne, Jan. 9, 1948, AEA 58-922.
30. Einstein to James Allen, Dec. 22, 1949, AEA 57-620.
31. Otto Nathan contributed to this phenomenon with the 1960 book of excerpts he coedited from Einstein's political writings, *Einstein on Peace*. Nathan, as the coexecutor with Helen Dukas of Einstein's literary estate, had a lot of influence over what was published early on. He was a committed socialist and pacifist. His collection is valuable, but in searching through the full Einstein archives, it becomes noticeable that he tended to leave out some material in which Einstein was critical of Russia or of radical pacifism. David E. Rowe and Robert Schulmann, in their own anthology of Einstein's political writings published in 2007, *Einstein's Political World*, provide a counterbalance. They stress that Einstein "was not tempted to give up free enterprise in favor of a rigidly planned economy, least of all at the price of basic freedoms," and they also emphasize the realistic and practical nature of Einstein's evolution away from pure pacifism.
32. Einstein to Arthur Squires and Cuthbert Daniel, Dec. 15, 1947, AEA 58-89.
33. Einstein to Roy Kepler, Aug. 8, 1948, AEA 58-969.
34. Einstein to John Dudzik, Mar. 8, 1948, AEA 58-108. See also Einstein to A. Amery, June 12, 1950, AEA 59-95: "However much I may believe in the necessity of socialism, it will not solve the problem of international security."
35. "Poles Issue Message by Einstein: He Reveals Quite Different Text," *New York Times*, Aug. 29, 1948; Einstein to Julian Huxley, Sept. 14, 1948, AEA 58-700; Nathan and Norden, 493.
36. Einstein to A. J. Muste, Jan. 30, 1950, AEA 60- 636.
37. *Today with Mrs. Roosevelt,* NBC, Jan. 12, 1950, www.cine-holocaust.de/cgi-bin/gdq?efw00fbw002802.gd; *New York Post*, Feb. 13, 1950.
38. D. M. Ladd to J. Edgar Hoover, Feb. 15, 1950, and V. P. Keay to H. B. Fletcher, Feb. 13, 1950, both in Einstein's FBI files, box 1a, foia.fbi.gov/foiaindex/einstein.htm. Fred Jerome's book *The Einstein File* offers an analysis. Jerome says that when making Einstein the Person of the Century, *Time* refrained from noting that he was a socialist: "As if the executives at *Time* decided to go so far but no farther, their article makes no mention of Einstein's socialist convictions." As the person who was the magazine's managing editor then, I can attest that the omission may indeed have been a lapse on our part, but it was not the result of a policy decision.
39. Gen. John Weckerling to J. Edgar Hoover, July 31, 1950, Einstein FBI files, box 2a.
40. See foia.fbi.gov/foiaindex/einstein.htm. Herb Romerstein and Eric Breindel in *The Venona Secrets* (New York: Regnery, 2000), an attack on Soviet espionage based on the "Venona" secret cables sent by Russian agents in the United States, have a section called "Duping Albert Einstein" (p. 398). It says that he was regularly willing to be listed as the "honorary chairman" of a variety of groups that were fronts for pro-Soviet agendas, but the authors say there is no evidence that he ever went to communist meetings or did anything other than

lend his name to various worthy-sounding organizations, with names like "Workers International Relief," that occasionally were part of the "front apparatus" of international Comintern leaders.

41. Marjorie Bishop, "Our Neighbors on Eighth Street," and Maria Turbow Lampard, introduction, in Sergei Konenkov, *The Uncommon Vision* (New Brunswick, N.J.: Rutgers University Press, 2000), 52–54, 192–195.

42. Pavel Sudoplatov, *Special Tasks*, updated ed. (Boston: Back Bay, 1995), appendix 8, p. 493; Jerome, 260, 283; Sotheby's catalogue, June 26, 1988; Robin Pogrebin, "Love Letters by Einstein at Auction," *New York Times*, June 1, 1998. The role of Konenkova has been confirmed by other sources.

43. Einstein to Margarita Konenkova, Nov. 27, 1945, June 1, 1946, uncatalogued.

44. Einstein, "Why Socialism?," *Monthly Review*, May 1949, reprinted in Einstein 1954, 151.

45. *Princeton Herald*, Sept. 25, 1942, in Sayen, 219.

46. Einstein, "The Negro Question," *Pageant*, Jan. 1946, in Einstein 1950a, 132.

47. Jerome, 71; Jerome and Taylor, 88–91; "Einstein Is Honored by Lincoln University," *New York Times*, May 4, 1946.

48. Einstein, "To the Heroes of the Warsaw Ghetto," 1944, in Einstein 1950a, 265.

49. Einstein to James Franck, Dec. 6, 1945, AEA 11-60; Einstein to James Franck, Dec. 30, 1945, AEA 11-64.

50. Einstein to Verlag Vieweg, Mar. 25, 1947, AEA 42-172; Einstein to Otto Hahn, Jan. 28, 1949, AEA 12-72.

51. Brian 1996, 340; Milton Wexler to Einstein, Sept. 17, 1944, AEA 55-48; Roberto Einstein (cousin) to Einstein, Nov. 27, 1944, AEA 55-49.

52. Einstein to Clara Jacobson, May 7, 1945, AEA 56-900.

53. Sayen, 219.

CHAPTER TWENTY-THREE: LANDMARK

1. Seelig 1956b, 71.
2. Pais 1982, 473.
3. See Bird and Sherwin.
4. J. Robert Oppenheimer to Frank Oppenheimer, Jan. 11, 1935, in Alice Smith and Charles Weiner, eds., *Robert Oppenheimer: Letters and Recollections* (Cambridge, Mass.: Harvard University Press, 1980), 190.
5. Sayen, 225; J. Robert Oppenheimer, "On Albert Einstein," *New York Review of Books*, Mar. 17, 1966.
6. Jim Holt, "Time Bandits," *New Yorker*, Feb. 28, 2005; Yourgrau 1999, 2005; Goldstein. Yourgrau 2005, 3, discusses the connections of incompleteness, relativity, and uncertainty to the zeitgeist. Holt's piece explains the insights they shared.
7. Goldstein, 232 n. 8, says that, alas, various research efforts have failed to discover the precise flaw Gödel thought he had discovered.
8. Kurt Gödel, "Relativity and Idealistic Philosophy," in Schilpp, 558.
9. Yourgrau 2005, 116.

10. Einstein, "Reply to Criticisms," in Schilpp, 687–688.
11. Einstein to Han Muehsam, June 15, 1942, AEA 38-337.
12. Hoffmann 1972, 240.
13. Einstein 1949b, 33.
14. Einstein and Wolfgang Pauli, "Non-Existence of Regular Solutions of Relativistic Field Equations," 1943.
15. Einstein and Valentine Bargmann, "Bivector Fields," 1944. He is sometimes referred to as Valentin, but in America he signed his name Valentine.
16. Einstein to Erwin Schrödinger, Jan. 22, 1946, AEA 22-93.
17. Erwin Schrödinger to Einstein, Feb. 19, 1946, AEA 22-94; Einstein to Erwin Schrödinger, Apr. 7, 1946, AEA 22-103; Einstein to Erwin Schrödinger, May 20, 1946, AEA 22-106; Einstein, "Generalized Theory of Gravitation," 1948, with subsequent addenda.
18. Einstein, *The Meaning of Relativity*, 1950 ed., appendix 2, revised again for the 1954 ed.; William Laurence, "New Theory Gives a Master Key to the Universe," *New York Times*, Dec. 27, 1949; William Laurence, "Einstein Publishes His Master Theory: Long-Awaited Chapter to Relativity Volume Is Product of 30 Years of Labor; Revised at Last Minute," *New York Times*, Feb. 15, 1950.
19. Einstein to Maurice Solovine, Nov. 25, 1948, AEA 21-256; Einstein to Maurice Solovine, Mar. 28, 1949, AEA 21-260; Einstein to Maurice Solovine, Feb. 12, 1951, AEA 21-277.
20. Tilman Sauer, "Dimensions of Einstein's Unified Field Theory Program," courtesy of the author; Hoffmann 1972, 239; I am grateful for the help of Sauer, who is doing research in Einstein's late work on field theories.
21. Whitrow, xii.
22. Niels Bohr, "Discussion with Einstein," in Schilpp, 199.
23. Abraham Pais, in Rozental 1967, 225; Clark, 742.
24. John Wheeler, "Memoir," in French, 21; John Wheeler, "Mentor and Sounding Board," in Brockman, 31; Einstein quoted in Johanna Fantova journal, Nov. 11, 1953. In letters to Besso in 1952, Einstein defended his stubbornness. He insisted that a complete description of nature would describe reality, or a "deterministic real state," rather than merely describe observations. "The orthodox quantum theoreticians generally refuse to admit the notion of a real state (based on positivist considerations). One thus ends up with a situation that resembles that of the good Bishop Berkeley." Einstein to Michele Besso, Sept. 10, 1952, AEA 7-412. A month later he noted that quantum theory declared that "laws don't apply to things, but only to what observation informs us about things . . . Now, I can't accept that." Einstein to Michele Besso, Oct. 8, 1952, AEA 7-414.
25. Einstein to Mileva Marić, Dec. 22, 1946, AEA 75-845.
26. Fölsing, 731; Highfield and Carter, 253; Brian 1996, 371; Einstein to Karl Zürcher, July 29, 1947.
27. Einstein to Hans Albert Einstein, Jan. 21, 1948, AEA 75-959.
28. Einstein to Carl Seelig, Jan. 4, 1954, AEA 39-59; Fölsing, 731.
29. Sayen, 221; Pais 1982, 475.

30. *Sarasota Tribune*, Mar. 2, 1949, AEA 30-1097; Bucky, 131. Jeremy Bernstein writes, "Anyone who spent five minutes with Miss Dukas would understand what a lunatic accusation this is." Bernstein 2001, 109.

31. Hans Albert Einstein interview, in Whitrow, 22.

32. "Trouble is brewing between Maja and Paul. They ought to divorce as well. Paul is supposedly having an affair and the marriage is quite in pieces. One shouldn't wait too long (as I did) . . . No mixed marriages are any good (Anna says: oh!)." Einstein to Michele Besso, Dec. 12, 1919. The half-joking reference to Anna was about Anna Winteler Besso, who was Michele Besso's wife and Paul Winteler's sister. The Wintelers were not Jewish; Besso and the Einsteins were.

33. Highfield and Carter, 248.

34. Einstein to Solovine, Nov. 25, 1948, AEA 21-256; Sayen, 134.

35. Einstein to Lina Kocherthaler, July 27, 1951, AEA 38-303; Sayen, 231.

36. "Einstein Repudiates Biography Written by His Ex-Son-in-Law," *New York Times*, Aug. 5, 1944; Frieda Bucky, "You Have to Ask Forgiveness," *Jewish Quarterly* (winter 1967–68), AEA 37-513.

37. "Einstein Extolled by 300 Scientists," *New York Times*, Mar. 20, 1949; Sayen, 227; Fölsing, 735.

38. Einstein to Queen Mother Elisabeth of Belgium, Jan. 6, 1951, AEA 32-400; Sayen, 139.

39. Einstein to Max Born, Apr. 12, 1949, AEA 8-223.

40. "3,000 Hear Einstein at Seder Service," *New York Times*, Apr. 18, 1938; Einstein, "Our Debt to Zionism," in Einstein 1954, 190.

41. "Einstein Condemns Rule in Palestine," *New York Times*, Jan. 12, 1946; Sayen, 235–237; Stephen Wise to Einstein, Jan. 14, 1946, AEA 35-258; Einstein to Stephen Wise, Jan. 14, 1946, AEA 35-260.

42. "Einstein Statement Assails Begin Party," *New York Times*, Dec. 3, 1948; "Einstein Is Assailed by Menachim Begin," *New York Times*, Dec. 7, 1948.

43. Einstein to Hans Muehsam, Jan. 22, 1947, AEA 38-360, and Sept. 24, 1948, AEA 38-379.

44. Einstein to Lina Kocherthaler, May 4, 1948, AEA 38-302.

45. Dukas interview, in Sayen, 245; Abba Eban to Einstein, Nov. 17, 1952, AEA 41-84; Einstein to Abba Eban, Nov. 18, 1952, AEA 28-943.

46. Einstein's travails with Hebrew University are recounted in Parzen 1974. For his relationship with Brandeis, see Abram Sacher, *Brandeis University* (Waltham, Mass.: Brandeis University Press, 1995), 22. The one place with which he had a great relationship was Yeshiva University. He was made the honorary chair of the fund-raising drive to build the College of Medicine there in 1952, and the following year allowed the medical college to be named after him. I am grateful to Edward Burns for providing information. See www.yu.edu/libraries/digital_library/einstein/panel10.html.

47. Einstein to *Maariv* newspaper editor Azriel Carlebach, Nov. 21, 1952, AEA 41-93; Sayen, 247; Nathan and Norden, 574; Einstein to Joseph Scharl, Nov. 24, 1952, AEA 41-107.

48. Yitzhak Navon, "On Einstein and the Presidency of Israel," in Holton and Elkana, 295.

CHAPTER TWENTY-FOUR: RED SCARE

1. Einstein to Queen Mother Elisabeth of Belgium, Jan. 6, 1951, AEA 32-400.
2. Einstein to Leopold Infeld, Oct. 28, 1952, AEA 14-173; Einstein to Russian students in Berlin, Apr. 1, 1952, AEA 59-218.
3. Einstein to T. E. Naiton, Oct. 9, 1952, AEA 60-664.
4. Einstein to Judge Irving Kaufman, Dec. 23, 1952, AEA 41-547.
5. Newark FBI Field Office to J. Edgar Hoover, Apr. 22, 1953, in Einstein FBI files, box 7.
6. Einstein to Harry Truman, with fifteen lines of equations on the other side, Jan. 11, 1953, AEA 41-551.
7. *New York Times*, Jan. 13, 1953.
8. Marian Rawles to Einstein, Jan. 14, 1953, AEA 41-629; Charles Williams to Einstein, Jan. 17, 1953, AEA 41-651; Homer Greene to Einstein, Jan. 15, 1953, AEA 41-588; Joseph Heidt to Einstein, Jan. 13, 1953, AEA 41-589.
9. Einstein to William Douglas, June 23, 1953, AEA 41-576; William Douglas to Einstein, June 30, 1953, AEA 41-577.
10. Generosa Pope Jr. to Einstein, Jan. 15, 1953, AEA 41-625; Daniel James to Einstein, Jan. 14, 1953, AEA 41-614.
11. Einstein to Daniel James, Jan. 15, 1953, AEA 60-696; *New York Times*, Jan. 22, 1953.
12. Einstein, Acceptance of the Lord & Taylor Award, May 4, 1953, AEA 28-979. In a letter to Dick Kluger, then a student editor of *The Daily Princetonian*, he wrote: "As long as a person has not violated the 'social contract' nobody has the right to inquire about his or her convictions. If this principal is not followed free intellectual development is not possible." Einstein to Dick Kluger, Sept. 17, 1953, in Kluger's possession.
13. Einstein to William Frauenglass, May 16, 1953, AEA 41-112; "Refuse to Testify Einstein Advises," *New York Times*, June 12, 1953; *Time*, June 22, 1953.
14. All of these editorials ran on June 13, 1953, except the Chicago editorial, which ran on June 15.
15. Sam Epkin to Einstein, June 15, 1953, AEA 41-409; Victor Lasky to Einstein, June 1953, AEA 41-441; George Stringfellow to Einstein, June 15, 1953, AEA 41-470.
16. *New York Times*, June 14, 1953.
17. Bertrand Russell to *New York Times*, June 26, 1953; Einstein to Bertrand Russell, June 28, 1953, AEA 33-195.
18. Abraham Flexner to Einstein, June 12, 1953, AEA 41-174; Shepherd Baum to Einstein, June 17, 1953, AEA 41-202.
19. Richard Frauenglass to Einstein, June 20, 1953, AEA 41-181.
20. Sarah Shadowitz, "Albert Shadowitz," *Globe and Mail* (Toronto), May 26, 2004. The author is the subject's daughter.

21. Sayen, 273–276; Permanent Subcommittee on Investigations, Committee on Government Operations, "Testimony of Albert Shadowitz," Dec. 14, 1953, and "Report on the Proceedings against Albert Shadowitz for Contempt of the Senate," July 16, 1954; Albert Shadowitz to Einstein, Dec. 14, 1953, AEA 41-659; Einstein to Albert Shadowitz, Dec. 15, 1953, AEA 41-660. Shadowitz was cleared in July 1955, two years after his testimony, after the fall of McCarthy.

22. Jerome and Taylor, 120–121.

23. Bird and Sherwin, 133, 495.

24. Ibid., 495.

25. James Reston, "Dr. Oppenheimer Suspended by A.E.C. in Security Review," *New York Times*, Apr. 13, 1954. On Sunday, Apr. 11, Joseph and Stewart Alsop, in their *New York Herald Tribune* column, had speculated that "leading physicists" were now a target of security investigations, but they did not mention Oppenheimer by name.

26. Pais 1982, 11; Bird and Sherwin, 502–504.

27. Johanna Fantova's journal, June 3, 16, 17, 1954, in Calaprice, 359.

28. Einstein to Herbert Lehman, May 19, 1954, AEA 6-236.

29. Johanna Fantova's journal, June 17, 1954, in Calaprice, 359.

30. Einstein to Norman Thomas, Mar. 10, 1954, AEA 61-549; Einstein to W. Stern, Jan. 14, 1954, AEA 61-470. See also Einstein to Felix Arnold, Mar. 19, 1954, AEA 59-118: "The current investigations are an incomparably greater danger to our society than those few communists in the country could ever be."

31. Johanna Fantova journal, Mar. 4, 1954, in Calaprice, 356; Einstein to Queen Mother Elisabeth of Belgium, Mar. 28, 1954, AEA 32-410.

32. Theodore White, "U.S. Science," *The Reporter*, Nov. 11, 1954. White went on to write *The Making of the President* series of books.

CHAPTER TWENTY-FIVE: THE END

1. Johanna Fantova journal, Mar. 19, 1954, in Calaprice, 356.

2. Einstein eulogy for Rudolf Ladenberg, Apr. 1, 1952, AEA 5-160.

3. Einstein to Jakob Ehrat, May 12, 1952, AEA 59-554; Einstein to Ernesta Marangoni, Oct. 1, 1952, AEA 60-406; Einstein to Queen Mother Elisabeth of Belgium, Jan. 12, 1953, AEA 32-405.

4. Einstein interview with Lili Foldes, *The Etude*, Jan. 1947; Calaprice, 150. Information about his repeated playing of this record was given to me by someone who knew Einstein in his later years.

5. Einstein to Hans Muehsam, Mar. 30, 1954, AEA 38-434.

6. Einstein to Conrad Habicht and Maurice Solovine, Apr. 3, 1953, AEA 21-294; Einstein to Maurice Solovine, Feb. 27, 1955, AEA 21-306.

7. Sayen, 294.

8. Einstein to Hans Albert Einstein, May 1, 1954, AEA 75-918.

9. Einstein to Hans Albert Einstein, unfinished letter, Dec. 28, 1954, courtesy of Bob Cohn, purchased at Christie's sale, Einstein Family Correspondence.

10. Gertrude Samuels, "Einstein, at 75, Is Still a Rebel," *New York Times Magazine*, Mar. 14, 1954.

11. Johanna Fantova journal, 1954, in Calaprice, 354–363.

12. Wolfgang Pauli to Max Born, Mar. 3, 1954, in Born 2005, 213.

13. Einstein to Michele Besso, Aug. 10, 1954, AEA 7-420.

14. Einstein to Louis de Broglie, Feb. 8, 1954, AEA 8-311.

15. Einstein 1916, final appendix to the 1954 ed., 178.

16. Bertrand Russell to Einstein, Feb. 11, 1955, AEA 33-199; Einstein to Bertrand Russell, Feb. 16, 1955, AEA 33-200.

17. Einstein to Niels Bohr, Mar. 2, 1955, AEA 33-204.

18. Bertrand Russell, "Manifesto by Scientists for Abolition of War," sent to Einstein on Apr. 5, 1955, AEA 33-209, and issued publicly July 9, 1955.

19. Einstein to Farmingdale Elementary School, Mar. 26, 1955, AEA 59-632; Alice Calaprice, ed., *Dear Professor Einstein* (New York: Prometheus, 2002), 219.

20. Einstein to Vero and Bice Besso, Mar. 21, 1955, AEA 7-245.

21. Eric Rogers, "The Equivalence Principle Demonstrated," in French, 131; I. Bernard Cohen, "An Interview with Einstein," *Scientific American* (July 1955).

22. Whitrow, 90; Einstein to Bertrand Russell, Apr. 11, 1955, AEA 33-212.

23. Einstein to Zvi Lurie, Jan. 5, 1955, AEA 60-388; Abba Eban, *An Autobiography* (New York: Random House, 1977), 191; Nathan and Norden, 640.

24. Helen Dukas, "Einstein's Last Days," AEA 39-71; Calaprice, 369; Pais 1982, 477.

25. Helen Dukas, "Einstein's Last Days," AEA 39-71; Helen Dukas to Abraham Pais, Apr. 30, 1955, in Pais 1982, 477.

26. Michelmore, 261.

27. Nathan and Norden, 640.

28. Einstein, final calculations, AEA 3-12. The final page can be viewed at www.alberteinstein.info/db/ViewImage.do?DocumentID=34430&Page=12.

EPILOGUE: EINSTEIN'S BRAIN AND EINSTEIN'S MIND

1. Michelmore, 262. Einstein's will, which was witnessed by the logician Kurt Gödel, among others, gave Helen Dukas $20,000, most of his personal belongings and books, and the income from his royalties until she died, which she did in 1982. Hans Albert received only $10,000; he died while a visiting lecturer in Woods Hole, Mass., in 1973, survived by a son and daughter. Einstein's other son, Eduard, received $15,000 to assure his continued care at the Zurich asylum, where he died in 1965. His stepdaughter Margot got $20,000 and the Mercer Street house, which was actually already in her name, and she died there in 1986. Dukas and Otto Nathan were made literary executors, and they guarded his reputation and papers so zealously that biographers and the editors of his collected papers would for years be stymied when they attempted to print anything verging on the merely personal.

2. "Einstein the Revolutionist," *New York Times*, Apr. 19, 1955; *Time*, May 2,

1955. The lead story in the extra edition of *The Daily Princetonian* was written by R. W. "Johnny" Apple, a future *Times* correspondent.

3. The weird tale has produced two fascinating books: Carolyn Abraham's *Possessing Genius*, a comprehensive account of the odyssey of Einstein's brain, and Michael Paterniti's *Driving Mr. Albert*, a delightful narrative of a ride across America with Einstein's brain in the trunk of a rented Buick. There have also been some memorable articles, including Steven Levy's "My Search for Einstein's Brain," *New Jersey Monthly*, August 1978; Gina Maranto's "The Bizarre Fate of Einstein's Brain," *Discover*, May 1985; Scott McCartney, "The Hidden Secrets of Einstein's Brain Are Still a Mystery," *Wall Street Journal*, May 5, 1994. In addition, Einstein's ophthalmologist Henry Abrams happened to wander into the autopsy room, and he ended up taking with him his former patient's eyeballs, which he subsequently kept in a New Jersey safe deposit box.

4. Abraham, 22. Abraham interviewed the grown girl in 2000.

5. "Son Asked Study of Einstein's Brain," *New York Times*, Apr. 20, 1955; Abraham, 75. Harvey had indicated that he was going to send the brain to Montefiore Medical Center in New York to oversee the studies. But as doctors there waited in anticipation, he changed his mind and decided to keep it to himself. The dispute made headlines. "Doctors Row over Brain of Dr. Einstein," reported the *Chicago Daily Tribune*. Abraham, 83, citing *Chicago Daily Tribune*, Apr. 20, 1955.

6. Levy 1978. See also www.echonyc.com/~steven/einstein.html.

7. See Abraham, 214–230, for an account of this issue.

8. Bill Toland, "Doctor Kept Einstein's Brain in Jar 43 Years: Seven Years Ago, He Got 'Tired of the Responsibility,'" *Pittsburgh Post-Gazette*, Apr. 17, 2005.

9. Marian Diamond, "On the Brain of a Scientist," *Experimental Neurology* 88 (1985); www.newhorizons.org/neuro/diamond_einstein.htm.

10. Sandra Witelson et al., "The Exceptional Brain of Albert Einstein," *Lancet*, June 19, 1999; Lawrence K. Altman, "Key to Intellect May Lie in Folds of Einstein's Brain," *New York Times*, June 18, 1999; www.fhs.mcmaster.ca/psychiatryneuroscience/faculty/witelson; Steven Pinker, "His Brain Measured Up," *New York Times*, June 24, 1999.

11. Einstein to Carl Seelig, Mar. 11, 1952, AEA 39-013. See also Bucky, 29: "I am not more gifted than anybody else. I am just more curious than the average person, and I will not give up on a problem until I have found the proper solution."

12. Seelig 1956a, 70.

13. Born 1978, 202.

14. Einstein to William Miller, quoted in *Life* magazine, May 2, 1955, in Calaprice, 261.

15. Hans Tanner, quoted in Seelig 1956a, 103.

16. André Maurois, *Illusions* (New York: Columbia University Press, 1968), 35, courtesy of Eric Motley. Perse was the pseudonym of Marie René Auguste Alexis Léger, who won the Nobel Prize for literature in 1960.

17. Newton's *Principia*, book 3; Einstein, "On the Method of Theoretical

Physics," the Herbert Spencer lecture, Oxford, June 10, 1933, in Einstein 1954, 274.

18. Clark, 649.

19. Lee Smolin, "Einstein's Lonely Path," *Discover* (Sept. 2004).

20. Einstein's foreword to Galileo Galilei, *Dialogue Concerning the Two Chief World Systems* (Berkeley: University of California Press, 2001), xv.

21. Einstein, "Freedom and Science," in Ruth Anshen, ed., *Freedom, Its Meaning* (New York: Harcourt, Brace, 1940), 92, reprinted in part in Einstein 1954, 31.

22. Einstein to Phyllis Wright, Jan. 24, 1936, AEA 52-337.

23. Einstein to Herbert S. Goldstein, Apr. 25, 1929, AEA 33-272. For a discussion of Maimonides and divine providence in Jewish thought, see Marvin Fox, *Interpreting Maimonides* (Chicago: University of Chicago Press, 1990), 229–250.

24. Banesh Hoffmann, in Harry Woolf, ed., *Some Strangeness in the Proportion* (Saddle River, N.J.: Addison-Wesley, 1980), 476.

INDEX

Page numbers in *italics* refer to illustrations.

ABOUT THE AUTHOR

Walter Isaacson is the CEO of the Aspen Institute. He has been chairman and CEO of CNN and managing editor of *Time* magazine. He is the author of *Benjamin Franklin: An American Life* and *Kissinger: A Biography*, and he is the coauthor with Evan Thomas of *The Wise Men: Six Friends and the World They Made*. He lives with his wife and daughter in Washington, D.C.

ILLUSTRATION CREDITS

Numbers in roman type refer to illustrations in the insert; numbers in *italics* refer to book pages.

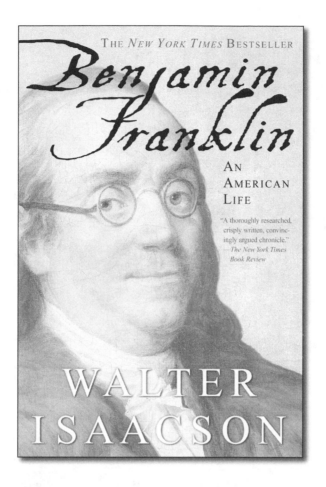

WALTER ISAACSON

Drawing on extensive interviews with Henry Kissinger and other sources, this first full-length biography is filled with surprising revelations and explores the relationship between this complex man's personality and the foreign policy he pursued.

"**Endlessly fascinating** . . . A brilliant and disturbing study of power." —*The New York Times*

A captivating blend of personal biography and public drama, *The Wise Men* introduces the original best and brightest leaders whose outsized personalities and actions brought order to the chaos that followed the end of World War II.

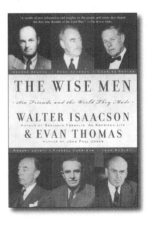

"**A wealth of new information and insights** on the people and events that shaped the first four decades of the Cold War."

—*The Boston Globe*

SIMON & SCHUSTER PAPERBACKS
A CBS COMPANY